NEUROMETHODS ☐ 10

Analysis of Psychiatric Drugs

NEUROMETHODS

Program Editors: Alan A. Boulton and Glen B. Baker

NEUROMETHODS

Program Editors: Alan A. Boulton and Glen B. Baker

Analysis of Psychiatric Drugs

Edited by

Alan A. Boulton

University of Saskatchewan, Saskatoon, Canada

Glen B. Baker

University of Alberta, Edmonton, Canada

and

Ronald T. Coutts

University of Alberta, Edmonton, Canada

Humana Press • Clifton, New Jersey

Library of Congress Cataloging in Publication Data

Analysis of psychiatric drugs / edited by Alan A. Boulton, Glen B.
    Baker, and Ronald T. Coutts.
        p. cm.—(Neuromethods ; 10)
    Includes bibliographies and index.
    ISBN 0-89603-121-7
    1. Antidepressants—Analysis. 2. Tranquilizing drugs—Analysis.
I. Boulton, A. A. (Alan A.) II. Baker, Glen B., 1947–
III. Coutts, Ronald Thomson. IV.   Series.
    [DNLM: 1. Antidepressive Agents—analysis. 2. Tranquilizing
Agents, Major—analysis. 3. Tranquilizing Agents, Minor—analysis.
W1 NE337G v. 10 / QV 77.9 A532]
RM332.A53 1988
615'.78—dc19
DNLM/DLC
for Library of Congress   88-133
                                                                       C

Printed in the United States of America

Preface to the Series

When the President of Humana Press first suggested that a series on methods in the neurosciences might be useful, one of us (AAB) was quite skeptical; only after discussions with GBB and some searching both of memory and library shelves did it seem that perhaps the publisher was right. Although some excellent methods books have recently appeared, notably in neuroanatomy, it is a fact that there is a dearth in this particular field, a fact attested to by the alacrity and enthusiasm with which most of the contributors to this series accepted our invitations and suggested additional topics and areas. After a somewhat hesitant start, essentially in the neurochemistry section, the series has grown and will encompass neurochemistry, neuropsychiatry, neurology, neuropathology, neurogenetics, neuroethology, molecular neurobiology, animal models of nervous disease, and no doubt many more "neuros." Although we have tried to include adequate methodological detail and in many cases detailed protocols, we have also tried to include wherever possible a short introductory review of the methods and/or related substances, comparisons with other methods, and the relationship of the substances being analyzed to neurological and psychiatric disorders. Recognizing our own limitations, we have invited a guest editor to join with us on most volumes in order to ensure complete coverage of the field and to add their specialized knowledge and competencies. We anticipate that this series will fill a gap; we can only hope that it will be filled appropriately and with the right amount of expertise with respect to each method, substance or group of substances, and area treated.

Alan A. Boulton
Glen B. Baker

Preface

In this volume, we hope to cover the major techniques that are presently being used to analyze the actions of drugs used in psychiatry. The contributors of the chapters are active researchers who have considerable practical experience with the techniques they are describing, and the emphasis in the chapters is on three types of psychiatric drugs, namely antidepressants, antipsychotics, and anxiolytics.

The first chapter, by Curry and Yu, discusses protein binding of psychotropic drugs, with special reference to equilibrium analysis as the method of assessment and to lipophilicity correlations. Since many of the drugs used in treating psychiatric disorders are bound extensively to protein, this aspect is of great importance with regard to their therapeutic actions and toxicity. Basic mathematical models, techniques for the study of protein binding, molecular aspects of protein binding, binding in relation to lipophilicity, indirect approaches to measurement of the fraction of drug bound, the function of protein binding, and tissue binding are among the topics discussed. Chapter 2, by Norman and Burrows, deals with the principles of isotope derivative assays and their applications to antidepressants and antipsychotics; extensive protocols are provided. The third chapter, by Cooper, deals with analysis by gas-liquid chromatography. Sample collection and storage, extraction procedures, column selection, use of internal standards, types of detectors, techniques of application, and applications to specific drugs are among the topics discussed. The chapter on gas chromatography–mass spectrometry by Reed deals with the instrumentation involved (interfaces, ion sources, mass analyzers, and selected ion monitoring), sample preparation, and quantitation. This is followed by a comprehensive section on the application of gas chromatography–mass spectrometry to the analysis of specific antidepressants, antipsychotics, benzodiazepines, and their metabolites. Chapter 5, by LeGatt, deals with another methodology that has had greatly increased application to psychotropic drugs, namely high-performance liquid chromatography. After an introductory section on theory, the author then discusses, in order, antidepressants, neuroleptics (antipsychotics), and benzodiazepines with regard to the following aspects: preanalytical concerns, extraction, and analysis.

In the chapter on principles of radioreceptor assays, Tune provides an introductory section on the theory of radioreceptor assays and then follows with a section on clinical applications, with a focus on anticholinergics and neuroleptics. The chapter ends with a section on possible clinical applications of the technique to antidepressants and benzodiazepines. Chapter 7, by Schmidt and Ebert, deals with the application of immunoassay techniques to psychopharmacology. A comprehensive discussion of general immunoassay principles is followed by sections on radioimmunoassays, enzyme immunoassays, luminescence immunoassays, electrochemical immunoassays, and fluorescent immunoassays. Tables summarizing specific applications of the assays are also provided. In chapter 8, Jones and Singer describe drug screening procedures for benzodiazepines, antidepressants, and neuroleptics as they are utilized in a toxicology laboratory setting. In the introduction, the authors discuss factors to be considered in a drug screening program; this is followed by sections on applications of immunoassays and related methods, radioreceptor assays, thin layer chromatography, high-performance liquid chromatography, gas chromatography, and mass spectrometry.

In chapter 9, Baker and Greenshaw discuss in vitro and ex vivo neurochemical screening procedures for psychiatric drugs. Sections on the following are provided: effects of drugs on uptake of radiolabeled biogenic amines into nerve ending preparations, inhibition of monoamine oxidase, neurotransmitter receptor binding, adenylate cyclase, and possible future directions. Chapter 10, by Greenshaw, Nguyen, and Sanger, deals with animal models for assessing anxiolytic, neuroleptic, and antidepressant drug action. A discussion of the criteria for such models is followed by sections on the three types of psychiatric drugs mentioned above and the tests that have been used for their evaluation.

Chapter 11, by Schoemaker and Langer, provides a discussion of high-affinity binding of antidepressants to platelets and brain tissue. Comprehensive sections on high affinity antidepressant binding sites and the serotonergic, noradrenergic, dopaminergic, and adrenergic transporters are presented. Information is provided about the regional and cellular localization of the binding sites, as well as their pharmacological profiles. An overview of clinical studies that have been carried out on the imipramine recognition binding site has also been provided.

The last chapter, by Borea, Hamor, and Martin, deals with structure–activity relationships at the benzodiazepine receptor. The emphasis is on the application of X-ray crystallography to such investigations.

It is hoped that this volume, which has descriptions of analytical techniques by researchers who use them on a day-to-day basis and are very familiar with their advantages and disadvantages, will be of great practical value to neuroscientists. In addition to being a rather comprehensive compilation of practical aspects of such analyses, the volume is rather unique in that behavioral and pharmacological techniques have been considered in the same volume as procedures that are more strictly chemical. In addition, the utilization of some of the procedures as screening methods for the study of potential new drugs and for toxicological studies on drugs is considered. In accord with the purpose of the other volumes in the *Neuromethods* series, we hope this one will prove to be useful both to experienced researchers and to those planning to enter a new aspect of the neurosciences, in this case analysis of drugs used in treating psychiatric disorders.

Glen B. Baker
Ronald T. Coutts

Contents

PROTEIN BINDING OF PSYCHOTROPIC DRUGS:
EQUILIBRIUM DIALYSIS AND LIPOPHILICITY
CORRELATIONS
Stephen H. Curry, Oliver Y-P Hu, and Robin Whelpton

ISOTOPE DERIVATIVE ASSAYS FOR PSYCHOTROPIC DRUGS
Trevor R. Norman and Graham D. Burrows

GAS-LIQUID CHROMATOGRAPHY OF ANTIDEPRESSANT, ANTIPSYCHOTIC, AND BENZODIAZEPINE DRUGS IN PLASMA AND TISSUES
Thomas B. Cooper

GAS CHROMATOGRAPHIC–MASS SPECTROMETRIC ANALYSIS OF ANTIDEPRESSANTS, NEUROLEPTICS, AND BENZODIAZEPINES
Kenton L. Reed

HIGH-PERFORMANCE LIQUID CHROMATOGRAPHIC
ANALYSIS OF ANTIDEPRESSANTS, NEUROLEPTICS,
AND BENZODIAZEPINES
Donald F. LeGatt

PRINCIPLES OF RADIORECEPTOR ASSAYS
Larry Tune

APPLICATION OF IMMUNOASSAY TECHNIQUES IN
PSYCHOPHARMACOLOGY
Dennis E. Schmidt and Michael H. Ebert

DRUG SCREENING FOR BENZODIAZEPINES,
ANTIDEPRESSANTS, AND NEUROLEPTICS
Graham R. Jones and Peter P. Singer

IN VITRO AND EX VIVO NEUROCHEMICAL SCREENING
PROCEDURES FOR ANTIDEPRESSANTS, NEUROLEPTICS,
AND BENZODIAZEPINES
Glen B. Baker and Andrew J. Greenshaw

ANIMAL MODELS FOR ASSESSING ANXIOLYTIC, NEUROLEPTIC, AND ANTIDEPRESSANT DRUG ACTION
Andrew J. Greenshaw, Tuong Van Nguyen, and David J. Sanger

HIGH-AFFINITY BINDING OF ANTIDEPRESSANTS TO PLATELETS AND BRAIN TISSUE

Hans Schoemaker and Salomon Z. Langer

STRUCTURE-ACTIVITY RELATIONSHIPS AT THE
BENZODIAZEPINE RECEPTOR
Pier A. Borea, Thomas A. Hamor, and Ian L. Martin

Contributors

GLEN B. BAKER • *Neurochemical Research Unit, Department of Psychiatry, University of Alberta, Edmonton, Alberta, Canada*

ALAN A. BOULTON • *Neuropsychiatric Research Unit, University of Saskatchewan, Saskatoon, Saskatchewan, Canada*

PIER A. BOREA • *Institute of Pharmacology, University of Ferrara, Ferrara, Italy*

GRAHAM D. BURROWS • *Department of Psychiatry, University of Melbourne, Austin Hospital, Heidelberg, Victoria, Australia*

THOMAS B. COOPER • *Nathan S. Klein Institute, Orangeburg, New York and New York State Psychiatric Institute, New York, New York*

RONALD T. COUTTS • *Faculty of Pharmacy and Pharmaceutical Sciences, University of Alberta, Edmonton, Alberta, Canada*

STEPHEN H. CURRY • *Division of Clinical Pharmacokinetics, College of Pharmacy, University of Florida, Gainesville, Florida*

MICHAEL H. EBERT • *Departments of Pharmacology and Psychiatry, Vanderbilt University School of Medicine, Nashville, Tennessee*

ANDREW J. GREENSHAW • *Department of Psychiatry, University of Alberta, Edmonton, Alberta, Canada*

THOMAS A. HAMOR • *Department of Chemistry, University of Birmingham, Birmingham, U.K.*

OLIVER Y-P HU • *School of Pharmacy, National Defense Medical Center, Taipei, Taiwan*

GRAHAM R. JONES • *Office of the Chief Medical Examiner, Edmonton, Alberta, Canada*

SALOMON Z. LANGER • *Department of Biology, LERS Synthelabo, Paris, France*

DONALD F. LeGATT • *Department of Laboratory Medicine, University of Alberta Hospitals, Edmonton, Alberta, Canada*

IAN L. MARTIN • *MRC Molecular Neurobiology Unit, MRC Centre, Cambridge, U.K.*

TREVOR R. NORMAN • *Department of Psychiatry, University of Melbourne, Austin Hospital, Heidelberg, Victoria, Australia*

KENTON L. REED • *Department of Psychiatry, Psychopharmacology Research Laboratories, University of Toronto, Toronto, Ontario, Canada*

DAVID J. SANGER • *Department of Neuropharmacology, LERS Synthelabo, Paris, France*

DENNIS E. SCHMIDT • *Departments of Pharmacology and Psychiatry, Vanderbilt University School of Medicine, Nashville, Tennessee*

HANS SCHOEMAKER • *Department of Biology, LERS Synthelabo, Paris, France*

PETER P. SINGER • *Office of the Chief Medical Examiner, Edmonton, Alberta, Canada*

LARRY TUNE • *Department of Psychiatry and Behavioral Sciences, The Johns Hopkins University School of Medicine, Baltimore, Maryland*

TUONG VAN NGUYEN • *Neuropsychiatric Research Unit, University of Saskatchewan, Saskatoon, Saskatchewan, Canada*

ROBIN WHELPTON • *Department of Pharmacology and Therapeutics, The London Hospital Medical College, London, U.K.*

Protein Binding of Psychotropic Drugs

Equilibrium Dialysis and Lipophilicity Correlations

Stephen H. Curry, Oliver Y-P Hu, and Robin Whelpton

1. Introduction

Binding of drugs to plasma protein has been known for many years to be significant in both drug effects and pharmacokinetics (Goldstein, 1949; Gillette, 1973; Curry, 1980; Meyer and Guttman, 1968; Vallner, 1974). However, in spite of rigorous study, binding remains contentious and poorly understood. This review is concerned with binding phenomena concerning psychotropic drugs, methods of study, and pharmacological significance. Tissue binding is also considered. Special attention is given to the most popular assessment technique—equilibrium dialysis. This technique has recently undergone rigorous reevaluation.

There are four fundamental questions—what is binding, how does it occur, why does it occur, and in what way does it matter? To answer two of those questions, binding is, generally, reversible adsorption of drug molecules on the surface of macromolecules. There is some covalent interaction between drugs and proteins, but this is commonly associated with toxicity, so that drugs subject to covalent binding tend to disappear from the therapeutic armamentarium. This review is confined to reversible binding phenomena. Binding occurs because many drugs have relatively low solubility in plasma, such that intravenous doses would precipitate in plasma water if they failed to become at least partially bound to transporting molecules. A molecule of low water solubility, pharmacologically active in vitro, but without protein binding, could fail to reach its site of action in vivo, and would then not show activity in clinical trials.

2. Basic Mathematical Models

The binding of small molecules by macromolecules is assumed to obey the law of mass action (Meyer and Guttman, 1968):

$$\text{drug} + \text{protein} \rightleftharpoons \text{drug–protein complex}$$

If D_t, D_b, and D_f are the molar concentrations of total, bound, and free drug, respectively, and P_t is the total molar concentration of protein, then:

$$D_f + (P_t - D_b) \rightleftharpoons D_b \tag{1}$$

Note that the free protein concentration is represented by the difference between P_t and D_b, since D_b is equal to the molar bound protein concentration. From this, the equilibrium constant K is given by:

$$K = D_b/D_f(P_t - D_b) \tag{2}$$
$$= k_1/k_{-1} \tag{3}$$

where k_1 and k_{-1} are the rate constants for the forward (complex formation) and reverse reactions, respectively. Note that the total concentration in plasma is given by:

$$D_t = D_b + D_f \tag{4}$$

and the fraction bound (β), is given by:

$$\beta = D_f/D_t \tag{5}$$
$$= D_f/(D_b + D_f) \tag{6}$$

These equations can be rearranged in various ways, Thus:

$$KD_fP_t - KD_fD_b = D_b \tag{7}$$
$$KD_fP_t = D_b + KD_fD_b \tag{8}$$
$$= D_b(1 + KD_f) \tag{9}$$

$$\frac{KD_fP_t}{1 + KD_f} = D_b \tag{10}$$

$$\frac{D_b}{P_t} = \frac{KD_f}{1 + KD_f} \tag{11}$$

$$= r \tag{12}$$

where r is the ratio of moles of drug bound to total moles of protein in the system. If there are n equivalent binding sites per mole of protein:

$$r = \frac{nKD_f}{1 + KD_f} \tag{13}$$

This equation can be rearranged to:

$$\frac{1}{r} = \frac{1}{n} + \frac{1}{nKD_f} \tag{14}$$

which is the basis for the double reciprocal plot of Klotz, in which $1/r$ is plotted against $1/D_f$, and from which estimates of K and n can be made from the slope and intercept.

Another rearrangement yields:

$$r/D_f = nK - rK \tag{15}$$

from which a plot of r/D_f vs. r has a slope of $-K$ and intercepts permitting calculation of K and n (Scatchard plot). Curvature of such plots is usually indicative of the existence of more than one class of sites.

A third rearrangement gives:

$$\frac{D_b}{D_f} = nKP_t - KD_b \tag{16}$$

from which a plot of D_b/D_f against D_b is independent of protein concentration and allows for estimation of K, from the slope (plot of Sandberg and Rosenthal). This plot will also be curved if more than one class of binding site on a single protein or more than one protein is involved. A small modification of this is:

$$\frac{C_b}{C_f} = nKP_t - \frac{KC_b}{1000 \times mw} \tag{17}$$

in which C_b and C_f are the concentrations of bound and unbound drug, respectively, in µg/mL, and mw is the molecular weight of the drug.

A plot of C_b/C_f against C_b is a straight line with a slope of $-K/(1000 \times mw)$, so that K can then be determined directly from raw data. If P_t is known, n can then be calculated. It is noteworthy that changing the protein concentration should yield a series of parallel lines of the same slope, with their intercepts proportional to P_t.

For clinical purposes, it is usually desirable to determine the "fraction bound" (β) or percent bound ($\beta \times 100$). This is obtained from the fact that fraction not bound (unbound, or "free," α) is given by:

$$\alpha = D_f/(D_b + D_f) \tag{18}$$

and

$$1 - \alpha = \beta \tag{19}$$

It can be shown that:

$$\beta = \frac{1}{1 + D_f/nP_t + 1/nKP_t} \tag{20}$$

However, within the limits of clinical work, D_f is usually much less than P_t, and n is unity or greater, so that D_f/nP_t is close to zero. Thus:

$$\beta = \frac{1}{1 + 1/nKP_t} \tag{21}$$

and

$$\alpha = \frac{1}{nKP_t + 1} \tag{22}$$

Usually: $nKP_t \gg 1$
so:

$$\alpha = 1/nKP_t \tag{23}$$

and:

$$\beta = 1 - 1/nKP_t \tag{24}$$

This shows that, in the clinical concentration range, β is increased proportionately by increases in n, K, and P_t, and that β is not affected by the concentration of the ligand. Clinical experience has repeatedly shown that the percent bound is constant throughout the clinical range in any particular individual, with large changes resulting from interpatient difference, intrapatient variations in P_t in particular.

3. Techniques for the Study of Protein Binding

A broad variety of techniques has been applied to the study of protein binding (Curry and Whelpton, 1983). These techniques can be classified as: (1) those that partially or completely separate bound and free drug, such as equilibrium dialysis, gel filtration ultrafiltration, and ultracentrifugation; (2) spectroscopic techniques that examine the molecular properties of the drug–protein complex, such as difference UV spectroscopy, fluorescence spectroscopy, circular dichroism, and nuclear magnetic resonance spectroscopy, and (3) microcalorimetry, which measures the heat of reaction of the drug–protein complex formation (Curry and Whelpton, 1983; Vallner, 1974).

Each of the techniques has its advantages and disadvantages. Equilibrium dialysis and ultrafiltration both require separation of the drug and protein molecules by means of a semipermeable membrane. Unfortunately, many drugs, especially lipophilic centrally acting drugs, bind to the membranes used. This in itself need not be a problem if the solutions on both sides of the membrane are analyzed for drug and protein content after separation is achieved. However, there is always the suspicion that the drug bound to the membrane alters the properties of the membrane, and it is important to realize that binding is studied at a drug concentration different from that originally in the plasma. Additionally, equilibrium dialysis involves bathing of the protein with an electrolyte solution different from that of plasma water, and binding has been shown to be sensitive to ionic changes. Also, ultrafiltration involves a dilute solution of a drug passing through a membrane of large surface area into an air space, with consequent risks of experimental artifacts from drug decomposition or evaporation. Ultrafiltration also leads to elevated protein concentrations behind the membrane and centrifugal force on the pores of the membrane.

Ultracentrifugation disturbs the protein concentration, which might change the binding characteristics of the protein, whereas gel filtration separates protein and drug in an artificial environment, using nonequilibrium dynamics. Spectroscopic techniques certainly measure the optical changes occurring when drugs and proteins interact. Microcalorimetry measures heats of reaction when drugs and proteins interact. However, physical methods in no sense measure the free fraction in vivo. In fact, no technique yet available can be presumed to measure the percent binding of the drug that prevailed in vivo before the plasma was collected. No "absolute" figure for this value can be presumed to be correct for any drug. However, differential binding data, such as free fraction in a series of individuals with samples handled in the same way, are entirely valid.

Equilibrium dialysis has been used in approximately half of all binding research published for psychotropic drugs. Equilibrium dialysis is the reference technique for almost all clinical work, so this approach will receive special attention in this review.

The principle of equilibrium dialysis is the use of a cellophane membrane (e.g., Visking tubing) that allows for the passage of small molecules (e.g., drugs of mw 300), but not macromolecules (e.g., albumin of mw 69,000). At equilibrium, one side of the membrane contains drug and protein and the other side (buffer side) contains only drug. Analysis of the drug concentration on the protein side gives the concentration of total drug, C_t, and analysis of the buffer side gives the concentration of nonbound drug, C_f, from which $C_b = C_t - C_f$. It has to be presumed that the concentration in the buffer equals the unbound concentration on the protein side at equilibrium.

A simple dialysis experiment consists of placing drug and protein in a piece of Visking tubing. The tubing is first knotted at one end to make a sac. After filling the sac, the open end is closed with a second knot. The sac is dialyzed in a test tube containing a suitable fluid, e.g., isotonic saline solution. Dialysis cells operate on the same principle. When the system has reached equilibrium, the drug concentrations inside and outside the dialysis sac are determined:

$$\text{percentage of drug bound} = \frac{\text{concentration inside} - \text{concentration outside}}{\text{concentration inside}} \times 100 \qquad (25)$$

Although, in theory, it is immaterial whether the protein solution is inside or outside the dialysis sac, in practice it is more convenient to place protein solutions inside and start the experiment with the drug solution outside. In this way samples of drug solution can be diluted directly in the experimental test tubes, and the outsides of the dialysis sacs can be cleaned of protein before they are added to the drug solutions. It is essential in any research that a preliminary experiment be conducted to ascertain that this approach is valid.

There is increasing interest in rigorous treatment of protein binding data obtained by means of equilibrium dialysis techniques. In particular, it recently became apparent that a water flux occurs during equilibrium dialysis (Abel et al., 1979; Boudinot and Jusko, 1984; Bowers et al., 1984; Curry and Hu, 1984; Huang, 1983; Hu and Curry, 1984; Lima et al., 1983; Lockwood and Wagner, 1983; Mueller and Potter, 1981; Nyberg et al., 1978; Tozer et al., 1983). This flux, from the protein-free side of the membrane to the plasma side, results from osmotic pressure differences, and has been termed "volume shift." When applicable, short dialysis times minimize this problem. Also, the use of isotonic buffer solutions reduces the degree of volume shift. However, no approach eradicates the problem.

By definition, the fraction bound (P_b) can be expressed as:

$$\beta = A_b/A \tag{26}$$

where A_b is the amount of drug bound in plasma, and A is the total amount of drug in plasma after dialysis. When a volume shift occurs, the amount of drug bound in the plasma, A_b, is given by:

$$A_b = A_i - D_f\,V_p - D_f V' \tag{27}$$

where A_i is the amount of drug in the plasma compartment after dialysis with volume shift, V_p is the plasma volume before dialysis, and V' is the extent of volume shift. A can be expressed as:

$$A = A_i - D_f\,V' \tag{28}$$

However,

$$A_i = A_0 - D_f\,(V_b - V') \tag{29}$$

where A_0 is the amount of drug in the plasma before dialysis, and V_b is the buffer volume before dialysis. Substituting:

$$\beta = \frac{A_0 - D_f(V_b - V') - D_f V_p - D_f V'}{A_0 - D_f(V_b - V') - D_f V'} \tag{30}$$

Rearrangement gives:

$$\beta = \frac{A_0 - D_f\,(V_p + V_b)}{A_0 - D_f V_b} \tag{31}$$

However, the fraction unbound (α) is given by $(1-\beta)$. Substituting:

$$\alpha = \frac{D_f V_p}{A_0 - D_f V_b} \tag{32}$$

This equation provides a means for evaluating fraction bound using final measured concentrations, initial defined volumes, and amounts, in any circumstances of volume shift, including zero. This equation was presented by Hu and Curry (1984), whereas Tozer (1983) presented a similar equation, without derivation, principally as a basis for more complex equations designed to evaluate the extent of volume shift.

There is a second methodological problem (Hu and Curry, 1986). Losses of drug from the system can occur during dialysis by drug decomposition, drug binding to apparatus, and drug binding to membranes. In this circumstance, the drug lost (A_{loss}) is defined by:

$$A_{\text{loss}} = A_0 - A_{\text{e}} \tag{33}$$

where A_{e} (the amount of drug in "plasma" and buffer at equilibrium) is determined experimentally by removal of the membrane, removal of the two equilibrium fluids quantitatively from the apparatus, combination of the two fluids, and assay of the mean fluid concentration. A_{e} is obtained by multiplication of the mean concentration by $(V_{\text{p}} + V_{\text{b}})$. A modified version of the earlier equation then gives α for any circumstances of volume shift and/or loss, including zero for both:

$$\alpha = \frac{D_{\text{f}}V_{\text{p}}}{A_0 - A_{\text{loss}} - D_{\text{f}}V_{\text{b}}} \tag{34}$$

The standard approach to calculation of the fraction bound involves measurement of the concentration of the drug in the buffer and the plasma at equilibrium, and calculation of the ratio of the concentrations. Although this gives a result numerically correct for any particular equilibrium system studied, its result bears little resemblance to the true value of fraction bound in the plasma before dialysis. Apart from the fact that the experiment causes dilution of the overall drug concentration in the system, compared with that originally in the plasma, with well-known implications for calculation of fraction bound when binding is saturable, a volume shift will inevitably cause a reduction in the protein concentration within the dialysis cell, with a resultant change in the concentration of the drug in the "plasma."

Huang (1983) proposed a mathematical method to correct for volume shift, but this method requires direct measurement of the equilibrium volumes. Lockwood and Wagner (1983) proposed measurement of the protein concentration on the plasma side of the membrane and calculation of the volume shift from the reduction in protein concentration, but this involves an additional analysis with consequent risks of error. Tozer et al. (1983) presented an exhaustive mathematical evaluation of the volume shift problem. Tozer's approach used three principles: (1) conservation of amount of drug and volume of solvent; (2) conservation of amount bound when a volume shift occurs; and (3) no binding to the dialysis cell or membrane occurs. This approach also requires the assumption that the unbound concentration remains the same with and without volume shift. This assumption, an analog of the second principle, is readily validated (Hu and Curry, 1984). It can then be easily shown that a volume shift of 0.1 to 1.2 mL, from a buffer volume of 10 mL into a plasma volume of 1 mL, can lead to an error in calculated fraction unbound of any number between approximately 1 and 100%, depending on the true fraction bound (considered over the range of 0.1 to 0.99).

We consider that there is little need to evaluate the magnitude of the volume shift when the objective is calculation of fraction bound. Furthermore, one of the three principles, i.e., that no losses result from binding to membranes and cells (or by decomposition), is suspect. In contrast, the equation presented earlier (34) provides a simple means of evaluation of fraction bound without direct or indirect measurement of volume shift, and with allowance for any losses. It should be recognized that much protein binding data in past literature is suspect as the result of failure to recognize the problem of volume shift.

4. Binding Data for Neuroleptics

Some early data for binding of chlorpromazine to bovine serum albumin is given in Table 1 (Curry, 1970a). Solutions of the drug were prepared in albumin at various concentrations over a wide range, including that of therapeutic concentrations in humans. A general trend toward a greater free fraction at higher chlorpromazne concentrations is evident. Values of n and K were calculated in accord with the various equations derived earlier. The computed constants are shown in Table 2. There is close agreement between the data derived from the Scatchard, Sandberg/Rosenthal, and modified Sandberg/Rosenthal equations. The double reciprocal plot, however, gave dramatically different values for n and K when all the data were used, in spite of a remarkably high correlation coefficient. When the terminal point was removed in the double reciprocal plot calculation, the data were in agreement with those from the other calculations. The discrepancy arose from the fact that these different plots give different degrees of weighting to particular experimental points. The double reciprocal plot gives undue importance to data points collected at low concentrations, and these data points are those most subject to analytical error.

The equation for β has been solved for a range of concentrations of D_f (C_f) (and by implication D_t and C_t), using the values of n

Table 1
Binding of Chlorpromazine to 1% Bovine Serum Albumin[a]

C_t	C_f	C_b
0.134	0.029	0.105
0.701	0.173	0.528
1.399	0.359	1.04
3.67	0.953	2.72
7.19	1.96	5.23
10.6	2.94	7.66
14.8	4.34	10.46
28.6	9.50	19.10

[a]Binding was determined by equilibrium dialysis. Both equilibrium solutions were assayed. Data are concentrations (μg/mL) (where C indicates for concentration, and t, f, and b indicate total, free, and bound.

(1) and K ($2.3 \times 10^{-2}/M$) obtained earlier, and assuming P_t to be 4 g/100 mL, which, for a mw of 68,000 (albumin), is $5.8 \times 10^{-4}/M$. The values obtained are given in Table 3. It is of great interest that β was found to be in the general area of 0.93, in the range 1–132 ng/mL (agreeing closely with the experimentally determined 0.94 for whole plasma), but of even greater interest was the fact that, on the basis of measured n and K values, it was shown that β varied by less than 0.5% for a range of D_f and D_t over two orders of magnitude covering a large proportion of the clinical range.

As mentioned in an earlier section, it can now be stated as a general principle that relatively large changes in D_t, within the clinically important range, can lead to virtually undetectable changes in β (these changes occur when the data are close to the maximum value for β, which can be anything from 0 to 99.99, with any integral value of n). It is therefore not surprising that research commonly fails to demonstrate a change in β when investigations are restricted to a particular range.

At one time it was thought that chlorpromazine binds only to albumin (Curry, 1981). This conclusion was based partly on the existence of a straight-line Scatchard plot. This is now known to be not so, since the free fraction has been shown to correlate, albeit weakly, with α-acid glycoprotein concentrations in patients with various clinical conditions (such as renal failure, arthritis, Crohns disease, and cirrhosis), which elevate this protein.

Table 2
Values of n and K for Chlorpromazine Bound to Bovine Serum Albumin
Computed According to a Number of Equations, with the Correlation
Coefficient for the Variables Considered in Each Case

Type of equation	Functions plotted	n	K, molar^{-1}	Correlation coefficient
Reciprocal (i)	$\dfrac{1}{r}$ vs. $\dfrac{1}{D_f}$	0.18	1.4×10^5	0.9993
(ii)[a]	$\dfrac{1}{r}$ vs. $\dfrac{1}{D_f}$	0.91	2.3×10^4	0.9999
Scatchard	$\dfrac{r}{D_f}$ vs. r	0.95	2.3×10^4	-0.8701
Sandberg/Rosenthal	$\dfrac{D_b}{D_f}$ vs. D_b	1.04	2.4×10^4	-0.8941
Modified Sandberg/ Rosenthal	$\dfrac{C_b}{C_f}$ vs. C_b	1.04	2.4×10^4	-0.8941

[a]Recomputed without the lowest concentration data point.

Table 3
Calculated Values for β at a Range of Values of C_t
(and C_f) for Chlorpromazine in Humans

C_t, ng/mL	β	C_f, ng/mL
1.38	0.9276	0.1
13.76	0.9273	1
131.93	0.9242	10
947.87	0.8945	100
3093.10	0.6767	1000

5. Molecular Aspects of Protein Binding

The characteristics of albumin have been extensively studied
(Chignell, 1981). Its molecular weight is approximately 66,000. It is
said to be a "prolate ellipsoid" at pH 5–8, meaning that it is a
three-dimensional body elongated along a line joining its poles,

with all surfaces ellipses or circles. The molecules of both human serum albumin (HSA) and bovine serum albumin (BSA) contain residues of glutamic acid, aspartic acid, and lysine, leading to over 200 positive and negative charges on the molecule. This confers hydrophilic character on the molecule. At neutral pH the net charge is –18. There is an isoionic point at pH 5.2 at which all basic groups are protonated and all carboxyl groups are negatively charged.

Human serum albumin has 18 tyrosine residues and one tryptophan residue. As a result, there is a uv light absorption maximum at 280 nm. When activation is at 280 nm, there is fluorescence at 343 nm, mainly caused by the tryptophan residue.

Albumin contains 35 cysteine residues, of which 34 are involved in disulfide bridges. The free SH group also reacts with cysteine, and there is some evidence of it moving over the surface of the albumin molecule.

The albumin molecule is apparently folded into three small double loops and five large loops, making possible its subdivision into three very similar repeat "domains," each consisting of a large loop, a short connecting segment, a small double loop, a long connecting segment, a second large loop, and a connecting segment to the next repeat unit. Because of its high helical content, and the fact that there is a proline residue at the end of each loop, it is thought that albumin is folded into a closed structure shaped like a pentahedral cylinder with five "helical rods" located on the corners. The helical rods are approximately 32 Å long; they are joined by a disulfide bond at one end, and at the other end as the result of junction with a polypeptide backbone. The significance of this is that it provides for a hydrophobic interior to the prolate ellipsoid body, permitting hydrophobic bonding.

Binding of lipid-soluble drugs seems to involve both hydrophobic and hydrophilic bonding, and is presumbably by means of a mixture of Van der Waals', ionic, hydrogen, and dipole bonds. Perhaps the hydrophobic nuclei of the molecules penetrate the hydrophobic interior of the prolate ellipsoid, whereas the hydrophilic substituents are attracted to the ionic groups. If this were so, there might be pH influences on protein binding. In fact, it has been shown that binding of chlorpromazine to plasma protein is a pH-dependent phenomenon. Unfortunately, pH-dependent data is difficult to interpret in isolation, since although pH changes alter the ionization of the ligands, they also change the degree of ionization of the acidic and basic groups in the protein, and also cause various conformational changes, such as the N-to-F and

N-to-B transitions that occur in albumin in the pH ranges of 2–4 and 7–9, respectively. The N-to-B transition is favored by the presence of calcium ions, which have been shown to affect determinations of K and n in vitro. They are, furthermore, hardly relevant to highly buffered physiological conditions.

6. Binding of Phenothiazines and Benzodiazepines in Relation to Lipophilicity

There is considerable evidence that the binding of psychotropic drugs, within the context of any one group, is quite nonspecific (Whelpton, 1978; Curry and Whelpton, 1979). Analogies with the Ferguson principle (Albert, 1981) have been invoked. There is ample precedent for this in barbiturate literature (Goodman and Gilman, 1965; Goldbaum and Smith, 1954) (Fig. 1). For example, there is a strong correlation between binding of medazepam, a precursor of diazepam, and four other related drugs, and the logarithm of the octanol partition coefficients and the pK_a values of the drugs (Figs. 2 and 3). There is also a near-perfect correlation between binding of promazine, chlorpromazine, promethazine, triflupromazine, and trimeprazine to both human serum albumin and bovine serum albumin and the octanol/pH 7.4 buffer logarithm of the partition coefficient (Gabay and Huang, 1974; Krieglstein et al., 1972) (Fig. 4). Interestingly, the potency of these same drugs in inhibiting dopamine receptors of the D-2 type also correlates with lipid solubility and therefore with plasma protein binding. Of even greater interest is the fact that the 7-hydroxylated metabolite of chlorpromazine has pharmacological potency, not in excess of that of chlorpromazine, but in excess of that predicted from its lipid solubility. In contrast, the plasma protein binding of 7-hydroxy-chlorpromazine is considerably below that of the parent drug, and, incidentally, below that of one of the virtually inactive metabolites, chlorpromazine N-oxide (Fig. 5). Presumably the hydroxyl group inhibits entry of the molecule into the hydrophobic interior of the albumin molecule. Thus lipid solubility and protein binding are poor predictors of potency. This is not really surprising. In fact, with reduced binding leading to higher free or active 7-hydroxychlorpromazine concentrations, and higher receptor potency than anticipated by lipid solubility considerations, it appears that a given concentration of 7-hydroxychlorpromazine in whole plasma might have far greater clinical significance than the same concentration of the parent drug in plasma. However, such

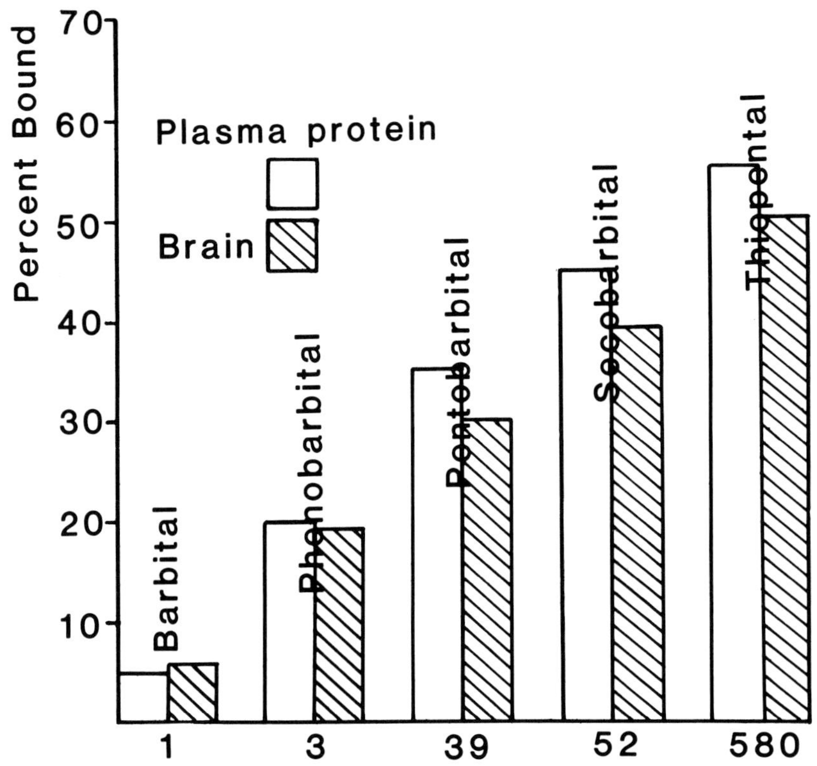

Fig. 1. Relation between binding to plasma protein (fraction bound in 0.001*M* drug solutions in M/15 phosphate buffer at pH 7.4 in 1% bovine serum albumin), binding to brain protein (fraction bound in rabbit brain homogenate) and partition coefficient (methylene chloride/aqueous phase) of nonionized form.

arguments must take into account tissue penetration, particularly blood–brain barrier transfer, which is relatively poorly understood.

7. Tricyclic Antidepressants

Most of the protein binding work with tricyclic antidepressants has been concerned with the issue of whether therapeutic

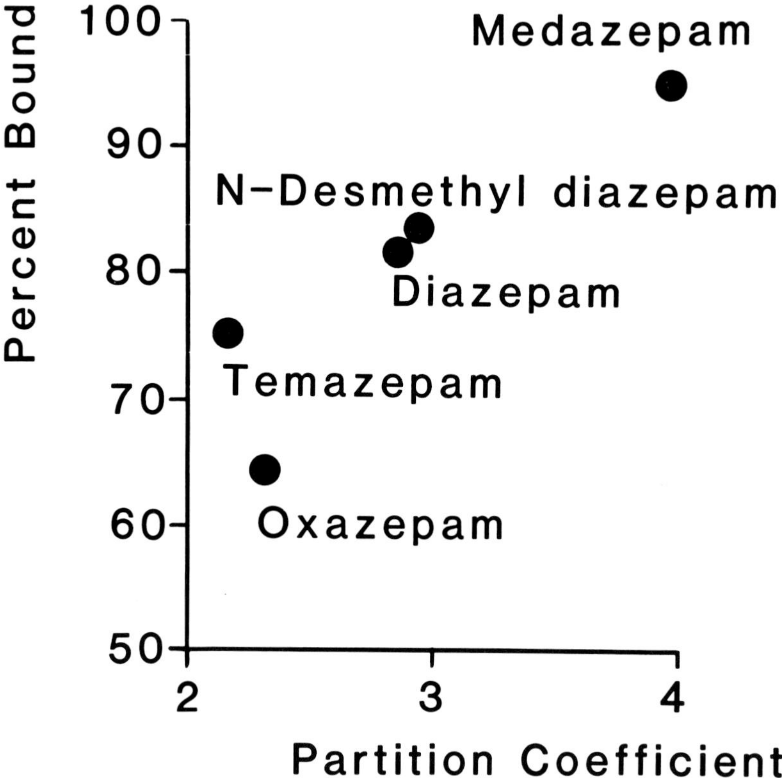

Fig. 2. Relation between protein binding (% bound) and partition coefficient for five benzodiazepine drugs.

monitoring should involve measurement of free concentrations (Bickel, 1975; Borga et al., 1969; Bertilsson et al., 1979; Alexanderson and Borga, 1972; Danon and Chen, 1979; Glassman et al., 1973; Piafsky and Borga, 1977; Piafsky et al., 1978; Potter et al., 1979; Borga et al., 1968). However, some fundamental observations have been made. For example, in what may have been the first paper in the field concerned with tricyclic antidepressants, it was shown that there are interspecies differences in binding of desipramine (mean percent of free figures were: cat, 2.2; dog, 2.9; rat, 8.0; human, 9.8; rabbit, 15.3). In a definitive paper in 1969, Borga et al. showed: (1) the dependence of desipramine and nortriptyline binding on the plasma protein concentration; (2) mean desipramine binding in humans of 90.5%; (3) temperature dependence (4°C, 92%; 20°C, 90.5%; 30°C, 89.6%); (4) weak dependence of desipramine binding on desipramine concentration; (5) the ex-

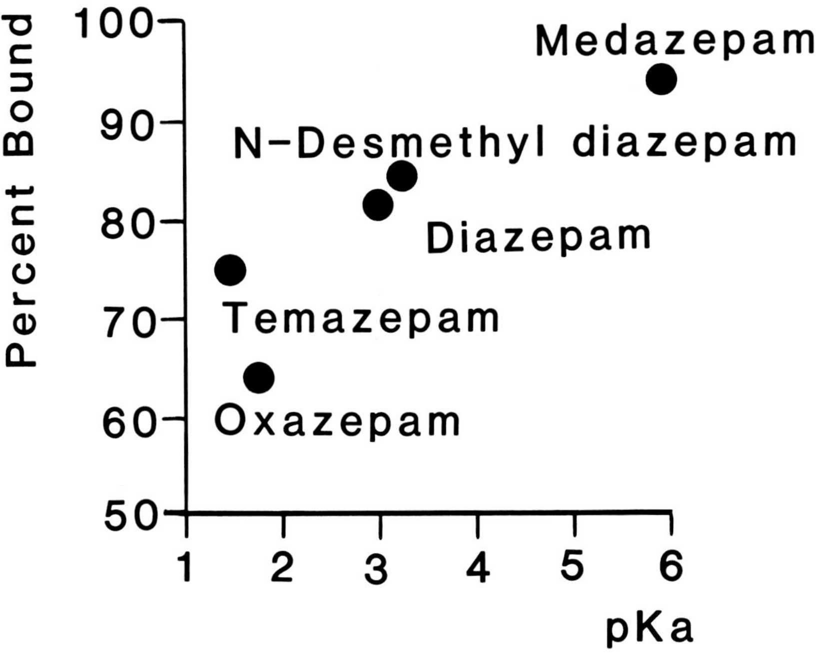

Fig. 3. Relation between protein binding (% bound) and pK_a for five benzodiazepine drugs.

istence of at least two populations of binding sites; (6) a higher percent bound for the two tertiary amine drugs amitriptyline and imipramine compared with their demethylated metabolites; (7) that nortriptyline concentrations in CSF and plasma water are similar; and (8) that a wide range of drugs had virtually no interactive binding effect. The existence of more than one population of binding sites in whole plasma was shown by a modified Scatchard plot (Fig. 6).

In a later study, Piafsky and Borga (1977) showed a two-fold variation in free fraction, and a correlation between free fraction and α-acid glycoprotein, but not albumin concentration. They also showed lack of a sex difference. Lipoprotein binding was studied with imipramine. Binding to lipoproteins occurred, and both specific binding to lipoproteins and overall binding were higher in hyperlipoproteinemic patients, correlating with plasma cholesterol and triglyceride levels. Thus tricyclic antidepressants bind to albumin, α-acid glycoprotein, and lipoprotein, providing three populations of binding sites, and the basis of the curved Scatchard plot.

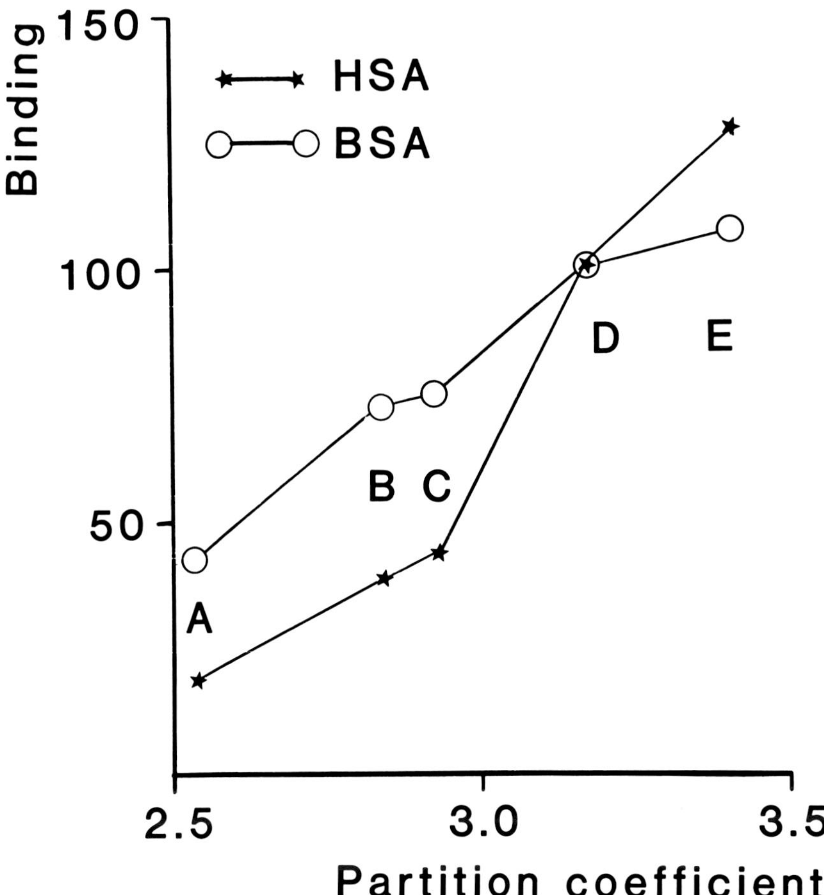

Fig. 4. Relation between binding of five phenothiazine drugs to bovine serum albumin (BSA) and human serum albumin (HSA) in relation to partition coefficient (log P, octanol/pH 7.4 buffer). (A) promazine; (B) promethazine; (C) trimeprazine; (D) chlorpromazine; (E) trifluoperazine.

One study has involved twins and nortriptyline. Again, there was a two-fold variation in free fraction, with significant relationships between binding in members of pairs of both identical and fraternal twins. It appeared that both genetic and environmental factors played a part in binding.

Glassman et al. (1970, 1973) showed much variation in binding of imipramine among individuals, with the percent free fraction varying from 5.4 to 23%. This was viewed as clinically important.

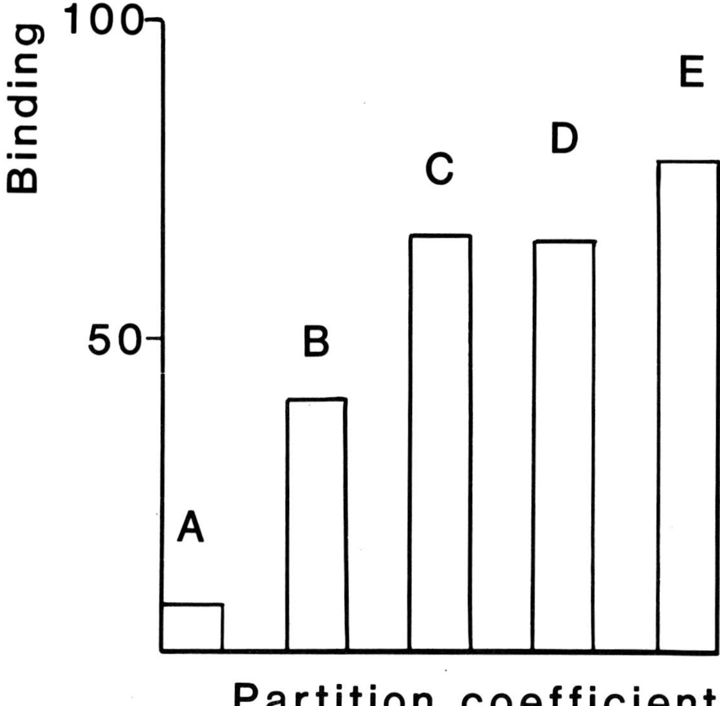

Fig. 5. Relation between binding (% bound to 1% bovine serum albumin) and partition coefficient for chlorpromazine and four of its metabolites. (A) chlorpromazine 5-oxide; (B) chlorpromazine *N*-oxide; (C) dedimethylchlorpromazine; (D) demonomethylchlorpromazine; (E) chlorpromazine. Partition coefficients ranking in ascending order, left to right.

In contrast, Potter et al. (1979) expressed the view that variations are "not so great," are unimportant, and when observed are experimental artifacts. There are indeed great problems in adequately assessing protein binding of tricyclic antidepressants (Bertilsson et al., 1979), since in both equilibrium dialysis and ultrafiltration there is binding to the membranes used. However, Potter et al. (1979) expressed the extraordinary view that plasma protein binding does not limit the entry of imipramine into the CSF (it clearly does limit the CSF/plasma distribution ratio, if not the rate of CNS entry) and the CSF concentrations are controlled entirely by metabolism, while noting that CSF concentrations were proportional to plasma concentrations. Potter et al. (1979) went on to

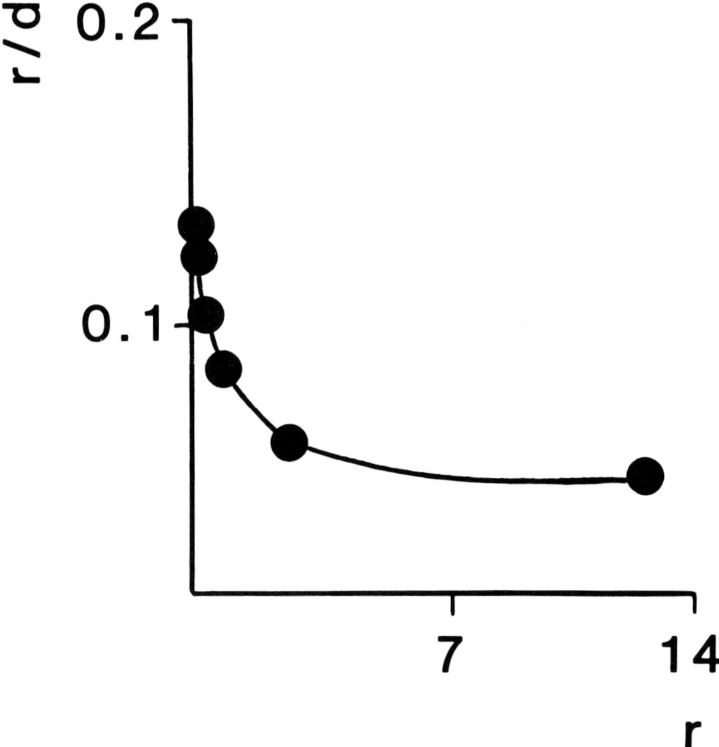

Fig. 6. Scatchard plot for binding of desmethylimipramine (DMI) to mixed plasma protein (redrawn, with permission, from Borga et al., 1969). (Note: r = moles of DMI bound per gram of protein; D = molar concentration of unbound DMI).

report a 24.0–37.1 range in free percentage, a near two-fold range, which most investigators would consider to be a great and important variation. Generally speaking, the case for monitoring free concentrations of tricyclic antidepressants is strong, and is supported, not refuted, by this work.

8. Indirect Approaches

There has been a considerable amount of interest in indirect approaches to measurement of the fraction bound. These have mostly involved assessment of drug concentration in saliva. The principle is that plasma concentrations of drugs are presumed to be in equilibrium with those in saliva. Since saliva is virtually protein-free, the salivary concentration might be expected to equal the concentration in plasma water. This could then provide a measure of fraction bound from:

$$\beta = \text{drug concentration in saliva/drug concentration in plasma} \quad (35)$$

Additionally, since drug effects are generally supposed to relate more closely to concentrations in plasma water than to those in plasma, therapeutic monitoring might be based on salivary concentrations. It has been stated by some investigators that saliva is relatively easy to obtain, lacking the trauma of venepuncture, and requiring less technical expertise. The saliva approach has been promoted particularly for phenytoin, phenobarbital, primidone, and lithium monitoring.

Although it may be true that saliva is easier to study than blood (although saliva is psychologically less acceptable), the same is certainly not true of the second indirect approach to the study of protein binding of psychotropic drugs involving measurement of levels in CSF. CSF levels are obviously of intrinsic interest. Whether they are a good approach to measurement of β is another matter. In principle, CSF levels equal those in plasma water, and this is more likely to be so than is the analogous argument for saliva, since CSF is more adequately buffered to pH 7.4 than is saliva—according to the pH-partition hypothesis of transfer of drugs through membranes, these concentration ratios will be highly pH-dependent for weak electrolytes, and most psychotropic drugs are weak electrolytes.

In fact, there is a risk of a fallacious circular argument occurring with salivary and CSF approaches to protein binding. Scientists are at risk from wanting theories validated by other theories. For example, we can either assume that the distributions are fully understood and use them to study protein binding, or make assumptions in regard to protein binding to validate our theories in regard to membranes. In fact, neither are fully understood, and it is not surprising that less than 100% credence has been placed on data obtained by these indirect approaches.

Diazepam concentrations in saliva have received particular attention (Hallstrom et al., 1979). Quite early in the work with single doses, an inconstant saliva-to-plasma concentration was observed. Twenty-four hours after the diazepam doses, the salivary concentrations were approximately 2% of plasma concentrations. However, diazepam is usually over 99% bound, although in another study diazepam concentrations were actually below those in plasma water. In patients, a significant correlation was observed between salivary diazepam and plasma diazepam ($r = 0.89$) and between salivary nordiazepam and plasma nordiazepam ($r = 0.81$). Thus, with these relationships accounting for 65–85% of the variance in their systems, it was possible to say that salivary monitoring might prove to be a good rough-and-ready tool for epidemiological studies. However, support for the salivary approach to measurement of β was lacking.

Of particular interest was the fact that, although the diazepam saliva-to-plasma water concentration was not greatly dissimilar from the plasma-to-plasma water concentration ratio, this was not the case for nordiazepam, for which the ratio unbound/total in plasma was 1.40 and the ratio saliva/plasma was 5.78. This difference remains unexplained. It may indicate that nordiazepam, being less lipophilic, is less likely to equilibrate rapidly between plasma water and saliva than is the case with diazepam. The fact that saliva and blood run counter-current in the salivary gland could be a factor in this. Whatever the explanation, the salivary approach to protein binding emerges as weak. In contrast, for diazepam, the correlation coefficient between CSF levels and unbound concentrations in plasma was 0.99. Thus CSF measurement might well be an excellent approach to measuring for diazepam. It is, however, hardly convenient.

9. What Does Protein Binding Do?

The fundamental role of reversible binding to plasma protein is as a highly efficient transport mechanism for essential nutrients. Thus, long before useful drugs were found to be protein-bound, substances such as cortisol, iron, and others were effectively and rapidly transferred from site to site in the body, the dynamic nature of their binding making possible rapid pick-up and deposit at different sites. Also, the high equilibrium constants and soluble nature of the complexes involved rendered relatively insoluble materials transferrable. Drugs are apparently handled similarly. Thus drugs of low aqueous solubility are efficiently and rapidly transferred from sites of absorption to sites of action and sites of elimination.

A somewhat negative attitude toward protein binding exists. This has resulted from two historical observations: (1) that of

Brodie (1967), who highlighted the fact that many drugs that are highly protein bound are slowly eliminated from the body, which led to the popular idea that protein binding is a restraining influence on drug metabolism and excretion; and (2) that of Veldstra (1956), who classified protein binding as a "site of loss," highlighting the fact that, at the time it is bound, a drug molecule is unavailable for exertion of an effect. Both of these observations deserve scrutiny.

The influence of protein binding on drug elimination has to be considered in relation to models of hepatic and renal drug elimination. Two models of hepatic drug elimination have been widely discussed in recent years (Rowland et al., 1973; Pang and Rowland, 1977; Bass et al., 1976; Bass and Brackan, 1978; Bass and Winkler, 1980; Winkler et al., 1973). In one of these models, the venous equilibration ("well-stirred") model, the liver is envisaged as behaving as if it were a single homogeneous compartment; drugs are presumed to distribute between sites of metabolism and blood as if there were just one single equilibrium distribution during each passage of a particular volume of blood through the liver. In the other model, the sinusoidal perfusion ("parallel tube") model, the liver is envisaged as consisting of a series of parallel cylinders, side by side, much like the sections of a radiator. A considerable volume of evidence has been presented in favor of both of these models (Ahmad et al., 1983; Bass and Winkler, 1980; Wilkinson and Shand, 1975). However, both models have their weaknesses, especially since neither model fully accounts for all observations connected with the influence of protein binding on drug elimination. Indeed, in the view of certain authorities, they are essentially the same (Wagner et al., 1984; Wagner 1985).

It appears that an analogy with theories of chromatography might be more appropriate (Curry, 1985). Chromatography columns are visualized as consisting of series of "theoretical plates," one above another, each one responsible for the distribution of molecules between stationary phase and mobile phase in accord with some physical principle. The greater the number of plates, the more efficient is the separation of a compound from its solvent, or separation of a pair of solutes. The fundamental feature of such a model is the requirement that a compound pass serially through a continuum of plates. Applying this idea to drug elimination, the compound is a drug dissolved in blood passing through a series of plates in the liver, with each plate facilitating distribution of drug between a group of drug-metabolizing sites and the blood. With such a model it is easy to envisage a situation in which a greater number of plates would lead to a higher level of hepatic activity, assessed as a higher extraction ratio or higher hepatic clearance. In relation to such a model, the single well-stirred pool model is a single theoretical plate. The parallel tube model is a group of single plates side by side.

The extraction ratio (E), using the single well-stirred pool assumption, is defined as:

$$E = \frac{Cl_i}{Q_H + Cl_i} \tag{36}$$

where Cl_i is the intrinsic hepatic clearance of the drug and Q_H is hepatic blood flow. To permit an influence of reversible plasma protein binding of drugs, this equation is commonly revised to:

$$E = \frac{\alpha Cl_i'}{Q_H + \alpha Cl_i'} \tag{37}$$

where α is the unbound fraction (free fraction in plasma and Cl_i' is the intrinsic hepatic clearance of the drug in the absence of protein binding restrictions. Since α has a maximum value of unity, $Cl_i' > Cl_i$ in almost all cases. Equation 37 clearly implies a protein-binding effect on clearance. However, since hepatic clearance (Cl_H) is given by:

$$Cl_H = Q_H E \tag{38}$$

there will not always be a protein-binding effect. For high extraction ratio drugs, it can be shown that $Cl_H = Q_H$, whereas for low extraction ratio drugs, $Cl_H = \alpha\, Cl_i'$ so that only for drugs with certain properties is a protein-binding effect predicted, even with the single well-stirred pool model. Only if Cl_i is relatively low will the liver fail to remove bound drug from protein-binding sites and transfer the molecules to drug-metabolizing enzyme surfaces in the single passage through the well-stirred pool.

Although all high E drugs with relatively high values of α (e.g., lidocaine, meperidine, morphine) provide data supporting these theories, there are examples of highly protein bound drugs that are rapidly metabolized (e.g., propranolol). The multiple plate hypothesis would allow this to happen, since, with sufficient plates, the series of equilibria would permit serial removal of molecules from binding sites when the volume of blood in question passed through the multiple plate liver. Mathematically, this would require a further modification of Eq. 36, as follows:

$$E = \frac{\alpha\, n\, Cl_i''}{Q_H + \alpha\, n\, Cl_i''} \tag{39}$$

in which Cl_i'' is the intrinsic clearance of the drug in each theoretical plate, and n is the number of theoretical plates. Since n has a minimum value of unity, $Cl_i'' < Cl_i'$, as $n\, Cl_i'' = Cl_i'$, or $\alpha n Cl_i'' = Cl_i$. The extraction ratio is now seen to be a function of the activity of the liver, liver blood flow, fraction unbound, and the number of theoretical plates, such that a large value of n could offset a small value of α.

The question arises as to what value of n would overcome a low value of α. Equation 39 has been rearranged and solved for n, given $E = 0.9$, for α varied from 0.01 to 1, and for Cl_i'' varied from 1 to 10^6 mL/min. Liver blood flow was taken to be 1500 mL/min. The results of this calculation show that relatively high values of n are needed to offset relatively low values of α, with the higher values of n needed when Cl_i'' is relatively low. They also show that, if Cl_i'' is relatively high, there is a point at which just one, or even less than one, theoretical plate is sufficient even for drugs with quite moderate values of α. It is with drugs with high intrinsic clearance values and moderate free fractions (e.g., lidocaine, meperidine, morphine) that the single well-stirred pool model has been seen to be most applicable. Similarly, it is with drugs with low intrinsic clearance values and low free fractions (e.g., warfarin, tolbutamide) that protein binding effects have been most clearly seen.

A similar approach is possible with renal clearance (Cl_R), which is considered to be the sum of two clearances, glomerular filtration clearance (Cl_G) and tubular clearance (Cl_T) (Gibaldi and Perrier, 1982; Levy, 1980). Thus:

$$Cl_R = Cl_G + Cl_T \tag{40}$$

In this, Cl_T can be negative. It has been suggested that the rate of filtration depends on the volume of blood filtered and on the unbound fraction in blood. Since the volume of fluid filtered approximately equals the creatinine clearance (Cl_{cr}):

$$Cl_G = \alpha\, Cl_{cr} \tag{41}$$

Similarly, Cl_T will depend on the ability of the drug to transfer in each direction across the membranes of the renal tubule, and possibly on α. An expression for Cl_T can then be developed:

$$Cl_{\rm T} = \frac{Q_{\rm k}\,\alpha Cl_{\rm T}{}'}{Q_{\rm k} + \alpha Cl_{\rm T}{}'} \qquad (42)$$

in which $Q_{\rm k}$ is renal blood flow, and $Cl_{\rm T}{}'$ is an intrinsic renal tubular clearance value for unbound drug. Combining equations:

$$Cl_{\rm R} = \alpha\left[Cl_{\rm cr} + \frac{Q_{\rm k}\,Cl_{\rm T}{}'}{Q_{\rm k} + \alpha Cl_{\rm T}{}'}\right] \qquad (43)$$

If $Q_{\rm k}$ greatly exceeds $\alpha Cl_{\rm T}{}'$, this equation reduces to:

$$Cl_{\rm R} = \alpha(Cl_{\rm cr} + Cl_{\rm T}{}') \qquad (44)$$

and if $\alpha Cl_{\rm T}$ greatly exceeds $Q_{\rm k}$, the equation reduces to:

$$Cl_{\rm R} = \alpha\left[Cl_{\rm cr} + \frac{Q_{\rm k}}{\alpha}\right] \qquad (45)$$

The protein binding assessment thus remains in the definition for $Cl_{\rm R}$ in all conditions.

The last five equations assume two single well-stirred pool equilibrations, one at the glomerulus and the other across the membrane of the renal tubule. However, blood flows *over* these sites, *rather than equilibrating once with them.* The idea of a static partition in each case is simplistic. Also, there are examples of highly bound drugs with high values of $Cl_{\rm R}$. As with the liver, a theoretical plate hypothesis seems appropriate. With enough theoretical plates in the glomerulus and/or renal tubule, α could become insignificant. This would be represented as follows:

$$Cl_{\rm G} = \alpha Cl_{\rm G}{}'$$

where $Cl_{\rm G}{}'$ is an intrinsic clearance value for the glomerulus.

Thus:

$$Cl_{\rm G} = \alpha n'\,Cl_{\rm G}{}'' \qquad (46)$$

where n' is the number of theoretical plates in the glomerulus and $Cl_{\rm G}{}''$ is the intrinsic clearance value for each plate. Note that if n' is high, α becomes insignificant.

Similarly, the equation for $Cl_{\rm T}$ is revised to:

$$Cl_{\rm T} = \frac{Q_{\rm k}\,\alpha\,n''\,Cl_{\rm T}{}''}{Q_{\rm k} + \alpha\,n''Cl_{\rm T}{}''} \qquad (47)$$

where n'' is the number of theoretical plates in the glomerulus, and $Cl_{\rm T}{}''$ is the intrinsic tubular renal clearance for unbound drug in a single theroretical plate. The full equation for renal clearance now becomes:

$$Cl_R = \alpha \left[n' \, Cl_G{}'' + \frac{Q_k n'' Cl_T{}''}{Q_k + \alpha \, n'' Cl_T{}''} \right] \qquad (48)$$

According to this, in all cases a relatively high value of n'' and/or n will lead to a reduction in the significance of α. This variation on renal clearance theory provides a basis for understanding why drugs with high equilibrium constants for protein binding can have high renal clearances.

In the case of chlorpromazine, it has been shown that, at 0.1 μg/mL, the free fraction is 0.106, 0.059, and 0.043 in rats, dogs, and humans, respectively. The apparent volume of distribution ($V_{d_{ss}}$), as a multiple of body weight, is correlated with this data (rat, 29.1; dog, 18.6; human, 11.2), in keeping with the idea that protein binding controls equilibrium distribution of drugs between plasma and tissues (Curry, 1970b, 1972). In contrast, two measures of elimination were examined, k_{10}, the output rate constant applicable to the central compartment, and β, the terminal phase plasma level decay rate constant. The k_{10} values did not correlate with binding. The β values did, with higher values of β (rat, 0.125; dog, 0.064; human, 0.023/h) in the species with the highest free fraction *and* highest $V_{d_{ss}}$.

As discussed earlier, chlorpromazine binding was thought at one time to be exclusively to albumin. This resulted partly from the fact that the fraction bound was the same in whole plasma and in a 4% albumin solution. However, there is more recent evidence that binding to the membranes of red cells to α_1-acid glycoprotein and to lipoproteins occurs. There is competition between the binding sites, so that albumin alone can indeed replicate plasma binding, but in whole blood all three types of site are involved. Something similar occurs with imipramine.

10. Tissue Binding

Compared with binding to plasma protein, tissue binding remains an unexplored field (Curry, 1970b). That is not to say that absolutely no work has been done. However, our understanding of tissue binding is weak.

The following apparently occur: (1) adsorption of lipophilic drugs to surfaces of membranes, such as the outer surface of red cells and linings of blood vessels; (2) penetration of membranes to give significant drug concentrations inside cells; (3) adsorption of drugs to constituents of cells; (4) solution of drugs in fatty areas. It should be recognized that some of the intracellular adsorption is to specific sites within cells; and some is to nonspecific sites. Adsorption to specific sites is thought to involve receptors for pharmacological action.

It is reasonable to presume that binding phenomena will show the kinetic characteristics shown by adsorption to plasma protein. However, it can be presumed that intracellular binding sites comprise mixed populations, with large numbers of individual classes of sites, each with different K and n values. Cell penetration, and especially penetration of lipid deposits, tends to show partition coefficient-type kinetics. Since nonspecific, partition-type distribution between tissues and plasma seems to dominate the tissue-to-plasma concentration ratio in any particular case, this ratio tends to appear to be a constant characteristic of the drug and tissue combination in question. This leads to evaluation of the mean tissue-to-plasma concentration by means of the "apparent volume of distribution (V_d), given by the formula:

$$V_d = \text{mean tissue concentration/plasma concentration} \quad (49)$$

which is viewed by most pharmacokineticists as a constant for the individual drug and patient, even though it is much affected by binding to plasma protein (Curry, 1970b).

11. Conclusion

Protein binding emerges as a ubiquitious problem. It is easy to observe that it occurs, but it is difficult to evaluate it with confidence that the evaluation is relevant in vivo. It is easy to hypothesize that it is significant, but it is difficult to prove that significance in any particular case. It is easy to propose simple models and theories of its mechanisms, but the exact molecular events involved remain mysterious. It is easy to invoke protein binding as the answer to all pharmacological problems, but this merely provides a hiding place when other explanations have failed us. Protein binding, especially of psychotropic drug, remains a relatively unexplored jungle for the pharmacologist.

References

Abel J. G., Sellers E. M., Naranjo C. A., Shaw J., Kader L., and Romach M. K. (1979) Inter- and intrasubject variation in diazepam free fraction. *Clin. Pharmacol. Ther.* **26** 247–255.

Ahmad A. B., Bennett P. N., and Rowland M. (1983) Models of hepatic clearance: Discrimination between the "well stirred" and parallel-tube models. *J. Pharm. Pharmacol.* **35** 219–224.

Albert A. (1981) *Selective Toxicity* (6th Ed.) Chapman & Hall, London.

Alexanderson B. and Borga O. (1972) Interindividual differences in plasma protein binding of nortriptyline in man—a twin study. *Eur. J. Clin. Pharmacol.* **4** 196–200.

Bass L. and Brackan A. J. (1978) Hepatic elimination of flowing substrates: The distributed model. *J. Theor. Biol.* **72,** 161–184.

Bass L. and Winkler K. (1980) A method of determining intrinsic hepatic clearance from the first-pass effect. *Clin. Exp. Pharmacol. Physiol.* **7,** 339–343.

Bass L., Keiding S., Winkler K., and Tygstrup H. (1976) Enzymatic elimination of substrates flowing through the intact liver. *J. Theor. Biol.* **61,** 393–409.

Bertilsson L., Braithwaite R., Tybring G., Garle M., and Borga O. (1979) Techniques for plasma protein binding of demethylchlorimipramine. *Clin. Pharmacol. Ther.* **26,** 265–271.

Bickel M. H. (1975) Binding of chlorpromazine and imipramine to red cells, albumin, lipoproteins and other blood components. *J. Pharm. Pharmacol.* **27,** 733–738.

Boudinot F. D. and Jusko W. J. (1984) Fluid shifts and other factors affecting plasma protein binding of prednisolone by equilibrium dialysis *J. Pharm. Sci.* **73,** 774–780.

Borga O., Azarnoff D. L., Forshell G. P., and Sjoqvist F. (1969) Plasma protein binding of tricyclic antidepressants in man. *Biochem. Pharmacol.* **18,** 2135–2143.

Borga O., Azarnoff D. L., and Sjoqvist F. (1968) Species differences in the plasma protein binding of desipramine. *J. Pharm. Pharmacol.* **20,** 571.

Bowers W. F., Fulton S., and Thompson J. (1984) Ultrafiltration vs. equilibrium dialysis for determination of free fraction. *Clin. Pharmacokinet.* **9** (suppl. 1), 48–60.

Brodie B. B. (1967) Physicochemical and biochemical aspects of pharmacology. *J. Am. Med. Assoc.* **202,** 600–609.

Chignell C. F. (1981) Molecular Features of Plasma Protein Binding, in *Clinical Pharmacology in Psychiatry* (E. Usdin, ed.) Elsevier, New York.

Curry S. H. (1970a) Plasma protein binding of chlorpromazine. *J. Pharm. Pharmacol.* **22,** 193–197.

Curry S. H. (1970b) Theroretical changes in drug distribution resulting from changes in binding to plasma proteins and to tissues. *J. Pharm. Pharmacol.* **22**, 753–757.

Curry S. H. (1972) Relation between binding to plasma protein, apparent volume of distribution, and rate constants of disposition and elimination for chlorpromazine in three species. *J. Pharm. Pharmacol.* **24**, 818–819.

Curry S. H. (1980) *Drug Disposition and Pharmacokinetics* (3rd Ed.) Blackwell, Oxford.

Curry S. H. (1981) Binding of Psychotropic Drugs to Plasma Protein and its Influence on Drug Distribution, in *Clinical Pharmacology in Psychiatry* (E. Usdin, ed.) Elsevier, New York.

Curry S. H. (1985) A multiple plate hypothesis of drug elimination. Abstracts: III World Conference on Clinical Pharmacology and Therapeutics, Stockholm (July 27–August 1, 1986). (Suppl.) *Acte Pharmacol. Toxicol.*

Curry S. H. and Hu O. Y-P (1984) Evaluation of equilibrium dialysis volume shifts: A comment. *J. Pharmacokinet. Biopharmacol.* **12**, 463–465.

Curry S. H. and Whelpton R. (1979) Pharmacokinetics of closely related benzodiazepines. *Br. J. Clin. Pharmacol.* **8**, 155–215

Curry S. H. and Whelpton R. (1983) *Manual of Laboratory Pharmacokinetics* Wiley, Cichester.

Danon A. and Chen Z. (1979) Binding of imipramine to plasma proteins: Effect of hyperlipoproteinemia. *Clin. Pharmacol. Ther.* **25**, 316–321.

Gabay S. and Huang P. C. (1984) The Binding Behavior of Phenothiazines and Structurally Related Compounds to Albumin from Several Species, in *Phenothiazines and Structurally Related Drugs* (I. S. Forrest, C. J. Carr, and E. Usdin, ed.) Raven, New York.

Gibaldi M. and Perrier D. (1982) *Pharmacokinetics* (3rd Ed.) Marcel Dekker, New York.

Gillette J. R. (1973) Overview of Protein Binding, in *Drug-Protein Binding* (A. H. Anton and H. M. Solomon, ed.) *Ann. NY Acad. Sci.* **226**, 1–362.

Glassman A. H., Hurwic M. J., Kanzler M., Shostak M., and Perel J. M. (1970) Imipramine steady-state studies and plasma binding, in *Phenothiazines and Structurally Related Drugs* (I. S. Forrest, C. J. Carr, E. Usdin, eds.) Raven, New York.

Glassman A. H., Hurwic J. J., and Perel J. M. (1973) Plasma binding of imipramine and clinical outcome. *Am. J. Psychiat.* **130**, 1367–1369.

Goldbaum L. R. and Smith P. K. (1954) The interaction of barbiturates with serum albumin and its possible relation to their disposition and pharmacological actions. *J. Pharmacol. Exp. Ther.* **111**, 197–209.

Goldstein A. (1949) The interactions of drugs and plasma proteins. *Pharmacol. Rev.* **1,** 102–165.

Goodman L. S. and Gilman A. (1965) *The Pharmacological Basis of Therapeutics* (3rd Ed.) Macmillan, New York.

Hallstrom C., Lader M. H., and Curry S. H. (1979) Diazepam and *N*-desmethyldiazepam concentrations in saliva, plasma and CSF. *Br. J. Clin. Pharmacol* **9,** 333–339.

Hu O. Y-P and Curry S. H. (1984), A Simple Approach for Estimation of the Unbound Fraction of Drugs in Equilibrium Dialysis with Consideration of Volume Shift, in *Proceedings of 10th Asian Congress of Pharmaceutical Sciences,* Taipei, Taiwan, R. O. C. 178–183.

Hu O. Y-P and Curry S. H. (1986) Calculation of fraction bound in equilibrium dialysis with special reference to drug losses by decomposition and adsorption. *Biopharmaceut. Drug Dispos.* **7,** 211–214.

Huang J. D. (1983) Errors in estimating the unbound fraction of drugs due to the volume shift in equilibrium dialysis. *J. Pharm. Sci.* **72,** 1368–1369.

Krieglstein J., Meiller W., and Staab J. (1972) Hydrophobic and ionic interactions of phenothiazine derivatives with BSA. *Biochem. Pharmacol.* **21,** 985–997.

Levy G. (1980) Effect of plasma protein binding on renal clearance of drugs. *J. Pharm. Sci.* **69,** 482–483.

Lima J. J., Mackichan J. J., Libertin N., and Sabino J. (1983) Influence of volume shifts on drug binding during equilibrium dialysis. *J. Pharmacokin. Biopharm.* **11,** 483–498.

Lockwood G. F. and Wagner J. G. (1983) Plasma volume changes as the result of equilibrium dialysis. *J. Pharm. Pharmacol.* **35,** 387–388.

Meyer M. C. and Guttman D. E. (1968) The binding of drugs by plasma proteins. *J. Pharmacol. Sci.* **57,** 895–918.

Mueller V. W. and Potter J. M. (1981). Binding of cortisol to human albumin and serum: The effect of protein concentration. *Biochem. Pharmacol.* **30,** 727–733.

Nyberg G., Axelsson R., and Martensson E. (1978) Binding of thioridazine and thioridazine metabolites to serum proteins in psychiatric patients. *Eur. J. Clin. Pharmacol.* **14,** 341–350.

Pang S. and Rowland M. (1977) Hepatic clearance of drugs. Theoretical considerations of a "well-stirred" model and a "parallel-tube" model. Influence of hepatic blood flow, plasma and blood cell binding, and the hepatocellular enzymatic activity in hepatic drug clearance. *J. Pharmacokin. Biopharm.* **5,** 625–653.

Piafsky K. M. and Borga O. (1977) Plasma protein binding of basic drugs. II. Importance of α_1-acid glycoprotein for interindividual variation. *Clin. Pharmacol. Ther.* **22,** 545–549.

Piafsky K. M., Borga O., Odar-Cederlof I., Johannsson C., and Sjoqvist F. (1978) Increased plasma protein binding of propranolol and chlorpromazine mediated by disease-induced elevations of plasma α_1-acid glycoprotein. *N. Eng. J. Med.* **299,** 1435–1439.

Potter W. Z., Muscettola G., and Goodwin F. K. (1979) Binding of imipramine to plasma protein and to brain tissue: Relationship to CSF tricyclic levels in man. *Psychopharmacology* **63,** 187–192.

Rowland M., Benet L. Z., and Graham G. G. (1973) Clearance concepts in pharmacokinetics. *J. Pharmacokin. Biopharm.* **1,** 123–136.

Tozer T. N., Gambertoglio J. G., Furst D. E., Avery D. S., and Holford N. H. G. (1983) Volume shift and protein binding estimates using equilibrium dialysis: Application to prednisolone binding in humans. *J. Pharm. Sci.* **72,** 1441–1446.

Vallner J. (1974) Binding of drugs by albumin and plasma protein. *J. Pharm. Sci.* **66,** 447–465.

Veldstra H. (1956) Synergism and potentiation. *Pharmacol. Rev.* **8,** 339–387.

Wagner J. G. (1985) Commentary: Relationships among the venous equilibration ("well-stirred") model, the sinusoidal perfusion ("parallel-tube") model, and a specific two-compartment open model. *Drug Metab. Dispos.* **13,** 119–120.

Wagner J. G., Szpuner G. J., and Ferry J. J. (1984) Commentary: Exact mathematical equivalence of the venous equilibration ("well-stirred") model, the sinusoidal perfusion ("parallel-tube") model, and a specific two-compartment open model. *Drug Metab. Dispos.* **12,** 385–388.

Whelpton R. (1978) Lipophilicity as a factor in the biochemical pharmacology of tranquillizing drugs. PhD thesis, University of London, London, UK.

Wilkinson G. R. and Shand D. G. (1975) A physiological approach to hepatic drug clearance. *Clin. Pharmacol. Ther.* **18,** 377–390.

Winkler K., Keiding S., and Tygstrup H. (1973) Clearance as a Quantitative Measure of Liver Function, in *The Liver, Quantitative Aspects of Structure and Function* (G. Paumgartner and R. Preisig, eds.) Kaiger, Basel.

Isotope Derivative Assays for Psychotropic Drugs

Trevor R. Norman and Graham D. Burrows

1. Introduction

Measurement of psychotropic drugs has become an integral part of the evaluation of new agents and, for older drugs, is often a routine part of clinical practise. Much of the interest in the measurement of psychotropic drugs can be traced to the notion that plasma concentrations may in some way reflect concentrations at vital central receptor sites and so be related to clinical efficacy (Brodie, 1967). Furthermore, an understanding of the pharmacokinetics of drugs is necessary for rational prescribing practices, with an obvious relationship to drug measurement. In some clinical situations, e.g., in cases of suspected noncompliance or drug toxicity, measurement of the plasma drug concentration is useful for management.

The implementation of clinically useful routine drug monitoring services is based on two assumptions. First, a simple, reliable, reproducible method (or methods) for drug analysis is available, and second, a relationship between plasma concentration and clinical effect is demonstrable. For the psychotropic drugs, with the notable exception of lithium, it has not been equivocally established that a simple plasma concentration–effect relationship exists. Antidepressants have been most often studied in methodologically diverse investigations with different relationships reported and no clear consensus emerging on the value of plasma concentration monitoring (Burrows and Norman, 1981; Norman and Burrows, 1980a; 1983; Risch et al., 1979a, b; Norman et al., 1982; Preskorn and Mac, 1984). Few studies have been carried out for the antipsychotic and antianxiety agents, but such studies that have also fail to establish that routine drug monitoring is clinically useful (Norman and Burrows, 1980b; 1984a,b; Newgreen et al., 1985; Davis et al., 1985; Rivera-Calimlim and Hershey, 1984). Because of the contradictory evidence, plasma concentration–effect relationships remains an active research topic. The lack of agreement between studies has been attributed to many factors, with heterogeneity of patient populations and drug assay methodology being the most critical areas of difference. Clearly, diagnostic

33

issues will remain a problem in the absence of specific biochemical tests, whereas the issue of assay methodology is somewhat more easily resolved.

Several reviews of methods for the measurement of psychotropic drugs have appeared (Cimbura, 1972; Tompsett, 1974; Clifford and Franklin-Smyth, 1974; Hailey, 1974; Riess, 1974a; Scoggins and Maguire, 1977; Gupta and Molnar, 1979; Riess et al., 1979; Scoggins et al., 1980; Norman and Maguire, 1985). The present paper is restricted to isotope derivative assays, which provided the first useful means of psychotropic drug analysis, although they have been superseded by other techniques, e.g., gas liquid chromatography (GC) and high performance liquid chromatography (HPLC), described elsewhere in this volume. The structures of the drugs mentioned in this paper are shown in Fig. 1.

2. Principles of Isotope Derivative Analysis

2.1. Single-Isotope Derivative Assay

Two types of isotope derivative assays can be utilized. One is the single-isotope derivative assay, in which nonradioactively labeled drugs are extracted from the biological matrix into a nonpolar organic solvent. The solution is treated with a radiolabeled compound of known specific activity. Most often [³H]-acetic anhydride is used to determine secondary amines, and [¹⁴C]-methyl iodide is used for the formation of quaternary amines from tertiary amines (*see* Fig. 2). Drug concentrations can be determined from standard curves constructed from plasma samples carried through the entire procedure.

Some alternative approaches to the determination of tertiary amines, based on either biochemical or chemical principles, have been explored. The formation of a secondary amine from a tertiary amine by reaction with trichloroethylchloroformate and hydrolysis was used for the determination of clomipramine (Carnis et al., 1976). Another chemical approach, which has not been utilized, is the reaction of a tertiary amine with cyanogen bromide (CNBr), and hydrolysis of the product:

$$R_3N + CNBr \rightarrow R_2N{\cdot}CN + RBr$$

$$R_2NCN \xrightarrow{\text{H}_2\text{O}} R_2N{\cdot}CO_2H \longrightarrow R_2NH + CO_2$$

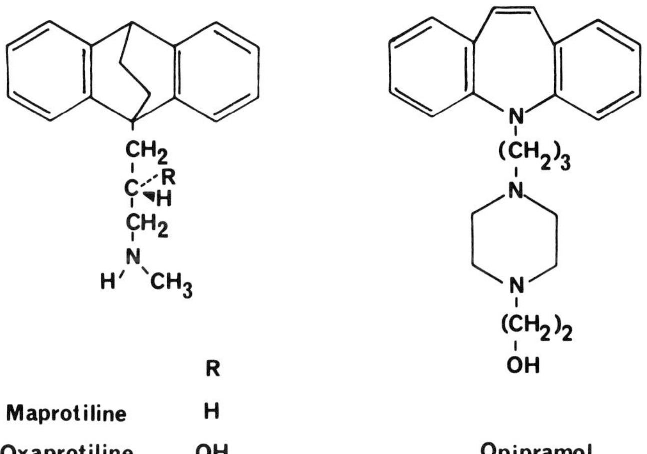

	R	X			R	X
Amitriptyline	CH₃	CH₂	Imipramine		CH₃	H
Nortriptyline	H	CH₂	Desipramine		H	H
Doxepin	CH₃	O	Clomipramine		CH₃	Cl
Desmethyldoxepin	H	O	Desmethylclomipramine		H	Cl

	R
Maprotiline	H
Oxaprotiline	OH

Opipramol

Fig. 1. Structures of drugs mentioned in this chapter.

It should be possible to utilize this reaction scheme as an alternative to reaction with trichloroethylchloroformate. The one difficulty with the CNBr approach is that heterocyclic tertiary amines also undergo the reaction; its use should therefore be limited to drugs, such as amitriptyline, that do not contain this grouping.

An alternative biochemical approach converts the tertiary amine to a secondary amine by incubation with an enzyme prepara-

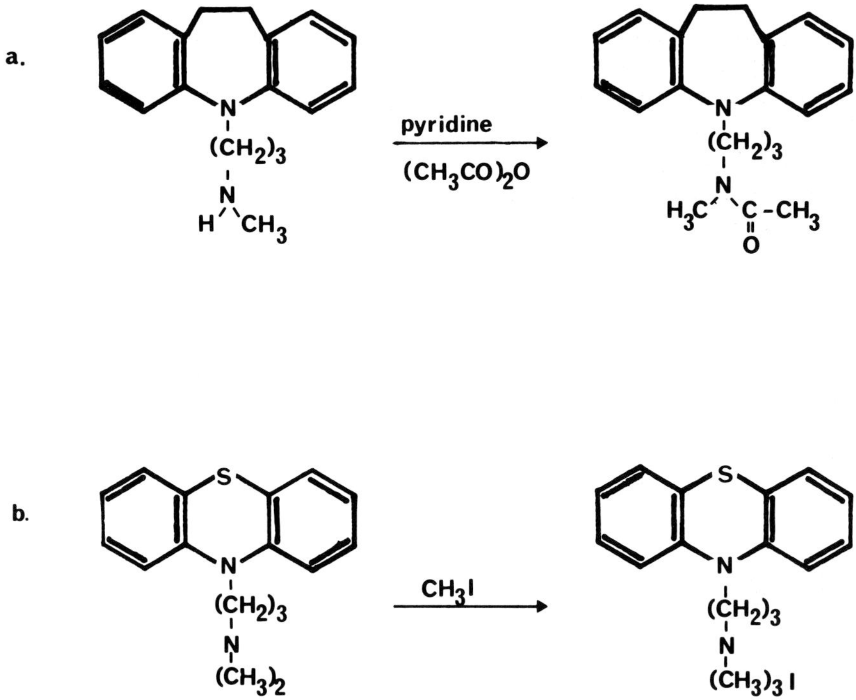

Fig. 2. Some derivatization reactions utilized for single-isotope derivative analysis. (a) Acetylation of desipramine with [^{3}H]-acetic anhydride, and (b) quaternary amine formation from chlorpromazine with [^{14}C]-methyl iodide.

tion. Both liver and brain homogenates have been used for the study of antidepressant metabolism and could be adapted (Dingell et al., 1964; Bickel and Baggiolini, 1966; Crammer and Rolfe, 1970). Studies in our laboratory indicate that this is unsatisfactory because of the poor yields of the secondary amine.

2.2. Double-Isotope Derivative Assay

In the double-Isotope procedure a known amount of the radioactive analog of the drug to be measured is added to the biological samples. The labeled and unlabeled drugs are then extracted from the sample with a suitable organic solvent. Derivatization of the extracted compounds with a suitable labeled reagent (containing a label different from that of the added drug) is carried out. After separation of the products from excess derivatizing agent, an excess of unlabeled carrier is added to the sample and the products are separated by chromatography. The activities of both isotopes are analyzed. Recovery of the method is determined from the activity of the isotope added initially. Although some authors advocate the construction of standard curves to calculate the mass of the unknown samples (Riess, 1974a; Riess et al., 1979), this is unnecessary with the double-isotope technique (Coghlan and Scoggins, 1967).

3. Application of Isotope Derivative Analysis

Isotope derivative analysis has been widely applied in biochemical analyses, particularly steroid hormone analysis, for which it was first developed for aldosterone (Kliman and Peterson, 1960). The application to pharmaceutical analysis has not been as extensive, and for the psychotropic drugs it has been widely applied only to the antidepressants.

3.1. Antidepressants

Historically the use of single-isotope derivative analysis represented the first practical method for the measurement of nanogram quantities of antidepressants present in human plasma samples. Hammer and Brodie (1967) used derivatization with [^{3}H]-acetic anhydride to determine desipramine plasma concentrations. The method is sensitive (5 μg/L for a 2-mL sample), specific, readily adaptable to any secondary or primary amine antidepressant, and has adequate precision for the measurement of clinical samples. The method was subsequently applied to the measurement of brain (Riva et al., 1975) steady-state plasma concentrations and pharmacokinetic studies (Hammer and Sjoqvist, 1967; Hammer et al., 1967; Sjoqvist et al., 1968; Hammer et al., 1969). The method was also applied to the measurement of nortriptyline plasma concentrations and was used extensively in the pharmacokinetic evaluations of this drug (Alexanderson, 1972a,b; 1973; Alexanderson et al., 1969; 1973). Studies of the relationship between plasma nortriptyline concentration and clinical effect made use of the single-isotope derivative assay (Asberg et al., 1970; 1971; Burrows et al., 1972). Assays for maprotiline (Norman et al., 1979) and Lu 5-003 (Fredricson-Overo, 1972), using derivatization with [^{3}H]-acetic anhydride, have also been developed. The R(–) and S(+)

steroisomers of oxaprotiline were separated after derivatization with N-trifluoroacetyl-S(–)-prolyl chloride (TPC) and high performance liquid chromatography (Dieterle and Faigle, 1982). This separation technique provided a convenient method of determining both isomers after the administration of racemic [^{14}C]-oxaprotiline. Using [^{3}H]-TPC, either a single- or double-isotope assay for human plasma samples could readily be developed. Although the single-isotope method potentially has wide applicability, it has not, except for the examples noted here, been utilized for as many antidepressants as possible.

By comparison with other methods used for plasma antidepressant measurements, the single-isotope method is obviously less time-efficient than a direct assay such as radioimmunoassay. Compared with the three-step extraction procedures normally employed in gas or high pressure liquid chromatographic methods, isotope derivative methods are about as time-consuming. A scheme that was used in our laboratories for desipramine, nortriptyline, or maprotiline plasma determinations is outlined in Fig. 3. Details of the sensitivity and precision of the assay are presented in Table 1 for each of the three drugs. The specificity of the assay may present a problem, since on administration to humans the secondary amine tricyclics undergo extensive first-pass metabolism to produce both primary amine and hydroxylated metabolites (Gram, 1974; Rubinstein et al., 1983). The inclusion of a thin-layer chromatography (TLC) step and judicious choice of developing solvent can resolve the hydroxylated and nonhydroxylated derivatives (e.g., in 1 : 1 benzene:acetone on 20 cm × 20 cm × 0.25 mm silica gel GF 254 plates; R_F values for hydroxynortriptyline acetate and nortriptyline acetate were 0.32 and 0.37, respectively). This TLC step also resolves the compounds of interest and endogenous acetylated components of normal plasma that may interfere with the assay (Zuleski et al., 1977). Acetylation of primary amines can be blocked by formation of a Schiff base with salicylaldehyde, which secondary and tertiary amines do not form. This has been demonstrated in assays of Lu 5-003 (Fredricson-Overo, 1972) and nortriptyline (Gram and Fredricson-Overo, 1972; Maguire et al., 1976).

As noted above, the isotope derivative assay for desipramine and nortriptyline has been extensively applied in clinical situations. It is not appropriate to review all such studies here, but a few illustrative examples are presented from work in our laboratories.

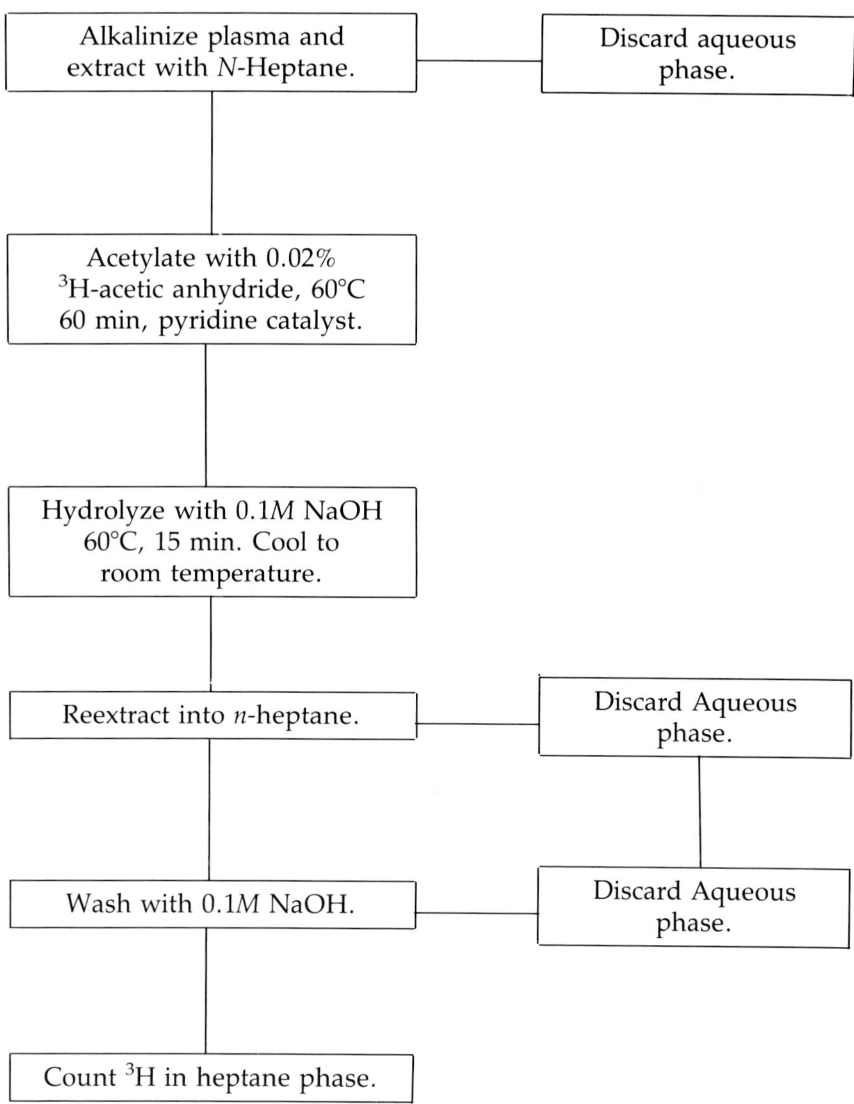

Fig. 3. An outline of the single-isotope derivative technique for the measurement of nortriptyline, desipramine, or maprotiline (*see* the Protocols section for a more detailed description of the experimental methods).

Table 1

Single-Isotope Derivative Assays for Antidepressants[a]

Drug	Derivatizing agent	Volume of plasma, mL	Sensitivity, μg/L	Precision, CV%	Authors
DMI	^{3}H-(AcO)$_2$	2	5	4	Hammer and Brodie (1967)
NT	^{3}H-(AcO)$_2$	2	5	10	UD
MAP	^{3}H-(AcO)$_2$	2	10	11	UD
DMI	^{3}H-(AcO)$_2$	2	2	10	UD
Lu 5-003	^{3}H-(AcO)$_2$	2	NS	NS	Fredricson-Overo (1972)
IMI	^{14}C-CH$_3$-I	3	15–20	NS	Harris et al. (1970)
DMI	^{3}H-(AcO)$_2$				
DOX	^{14}C-CH$_3$-I	3	15–20	6.5	Kline and Cooper (1974)
DDOX	^{3}H-(AcO)$_2$				
OXAP R(+)	TPC	1	2	3.2	Dieterle and Faigle (1982)
OXAP S(−)					

[a]Abbreviations: CV, coefficient of variation; DMI, desipramine; DDOX, desmethyldoxepin; DOX, doxepin; IMI, imipramine; MAP, maprotiline; NS, not stated; NT, nortriptyline; OXAP, oxaprotiline; (AcO)$_2$, acetic anhydride; TPC, N-trifluoroacetyl-S(−)-prolylchloride; UD, unpublished data from authors' laboratory.

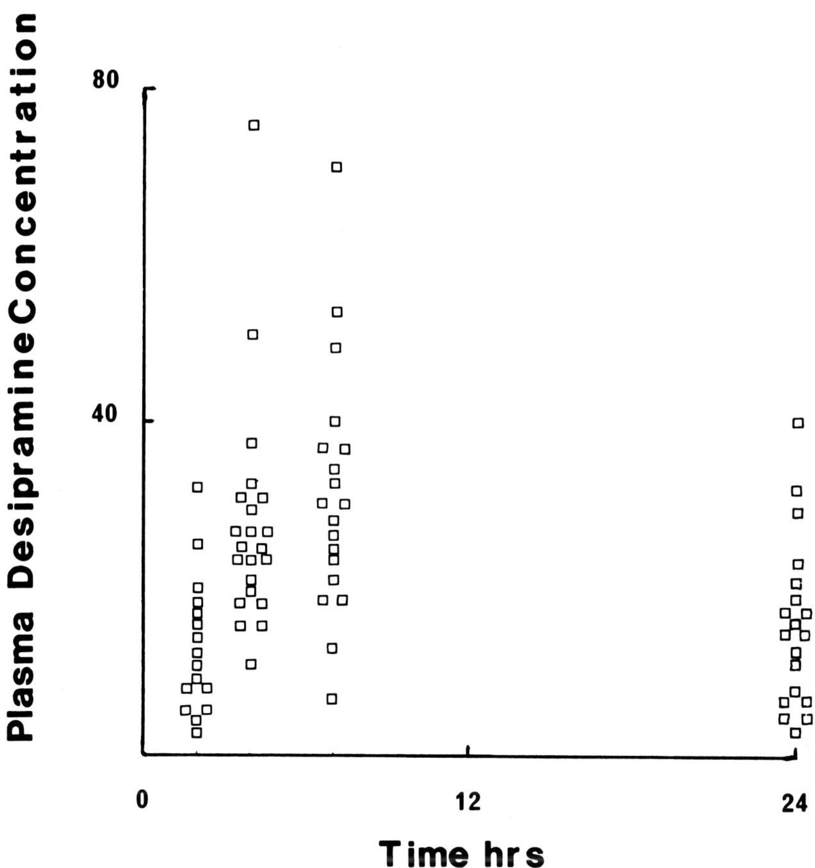

Fig. 4. Plasma desipramine concentrations (μg/L) measured by the single-isotope derivative technique following a single oral dose of 100 mg.

Single-dose studies were performed in both patient and normal healthy volunteers. Subjects received 100 mg of either desipramine or nortriptyline 2 h after a light breakfast. Plasma samples were drawn before drug administration and at 2, 4, 7, 12, and 24 h after the dose. A marked interindividual variability in plasma concentrations was observed (*see* Figs. 4 and 5).

Studies of the plasma concentrations achieved after repeated oral administration have also been performed using the single-isotope method. Nortriptyline concentrations have been measured most often, and Fig. 6 shows the plasma concentrations achieved following 2 wk of administration of 150 mg/d. Some individual

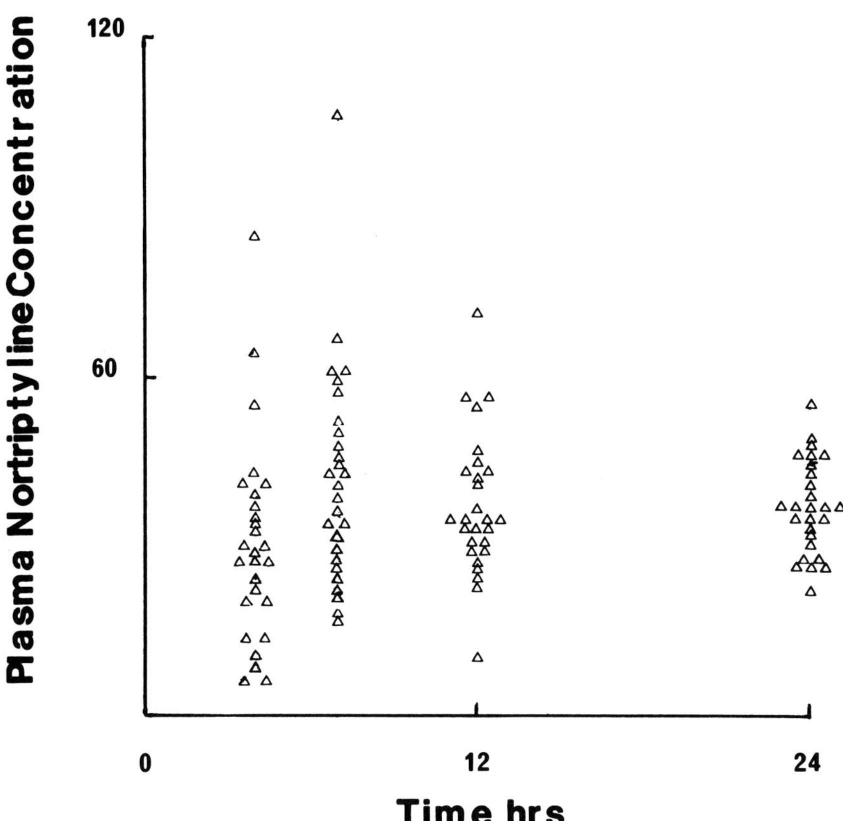

Fig. 5. Plasma nortriptyline concentration (μg/L) measured by the single-isotope derivative technique following a single oral dose of 100 mg.

plasma concentration data are shown for patients receiving desipramine (Fig. 7) and maprotiline (Fig. 8).

The single-isotope assay has been used for tertiary as well as secondary amine antidepressants, but requires an additional derivatization step. Formation of a quaternary ammonium salt from the tertiary amine and [^{14}C]-methyliodide was first used to determine chlorpromazine (see below) and, later, imipramine (Harris et al., 1970). Following extraction into an organic solvent, acetylation and methylation are performed sequentially and the labeled derivatives recovered (*see* Fig. 9). The method is reliable and reproducible, and the sensitivity (15–20 μg/L) is sufficient for application to routine clinical samples. The application of this technique to doxepin and desmethyldoxepin plasma determina-

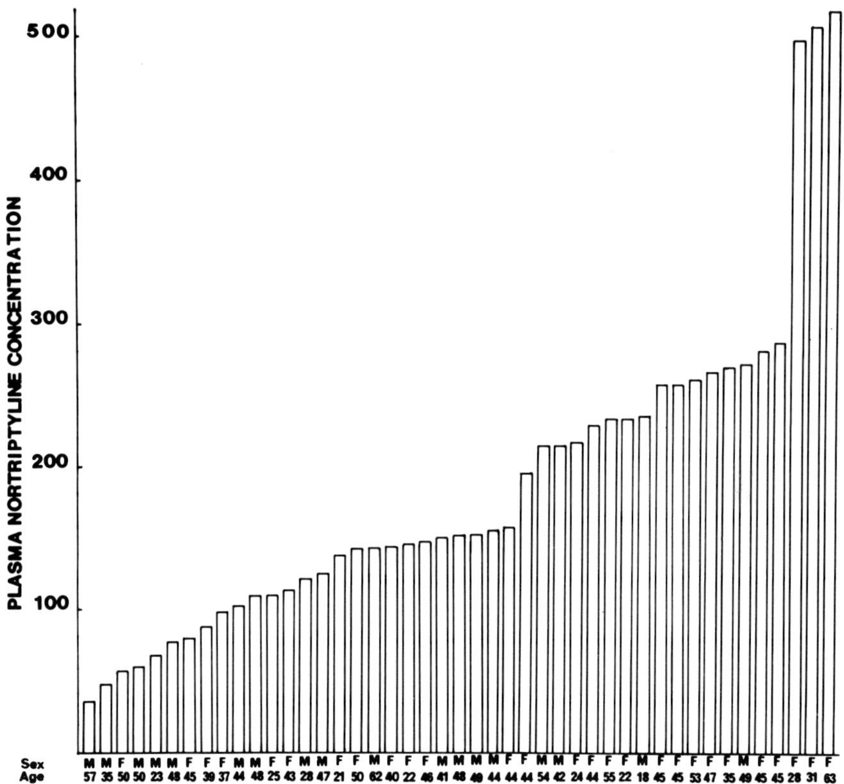

Fig. 6. Interindividual variability in steady-state plasma nortriptyline concentrations (μg/L) after 2 wk treatment with 150 mg/day in 45 patients. Drug concentrations were measured by the single-isotope derivative assay (*see* Table 1 for assay characteristics and the Protocols section for experimental details).

tion has been described (Kline and Cooper, 1974), but not reported, for other antidepressants. As with the Hammer-Brodie technique, the sensitivity of the assay can be considerably improved by using derivatizing agents with higher specific activity. Further, the specificity of the assay can also be improved by the incorporation of a TLC step. Hydroxy or didesmethylated metabolites will be derivatized and included as part of appropriate ^{14}C or ^{3}H counts. These derivatives can be removed by one- or two-dimensional TLC.

A further application of single-isotope methodology has been in the measurement of plasma nortriptyline concentrations by radiochemical gas chromatography (GC) (Loh et al., 1977). Plasma

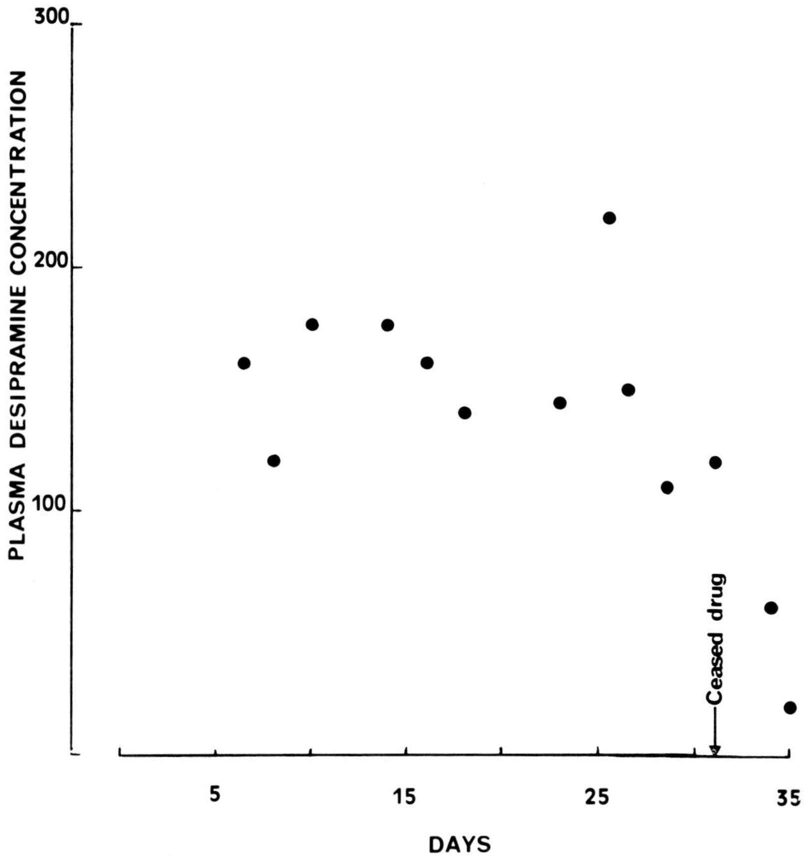

Fig. 7. Plasma desipramine concentration (μg/L) in a patient receiving 200 mg/d of imipramine. Desipramine was determined by single-isotope derivative analysis with [^{3}H]-acetic anhydride.

samples are extracted into hexane and the combined hexane extracts evaporated to dryness. The residue is redissolved in benzene, acetylated with [^{3}H]-acetic anhydride, and evaporated to dryness. The residue is treated with sodium hydroxide solution, and extracted twice with heptane. The heptane extracts are concentrated, and aliquots are injected into a gas chromatograph equipped with a gas proportional counter. The method is sensitive to 5 μg/L of nortriptyline. This represents a more sensitive assay than early GC methods based on flame ionization detection, but is less sensitive than nitrogen-specific GC methods (*see* Scoggins et al., 1980).

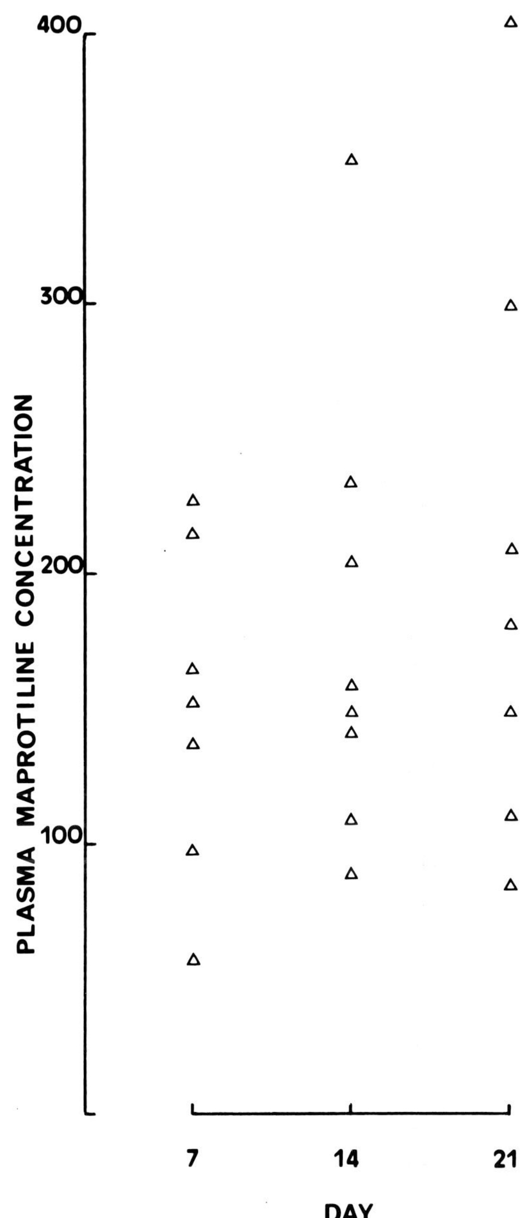

Fig. 8. Plasma maprotiline concentrations (μg/L) in patients receiving 150 mg/d of the drug. Maprotiline was determined by single-isotope derivative analysis with [^{3}H]-acetic anhydride.

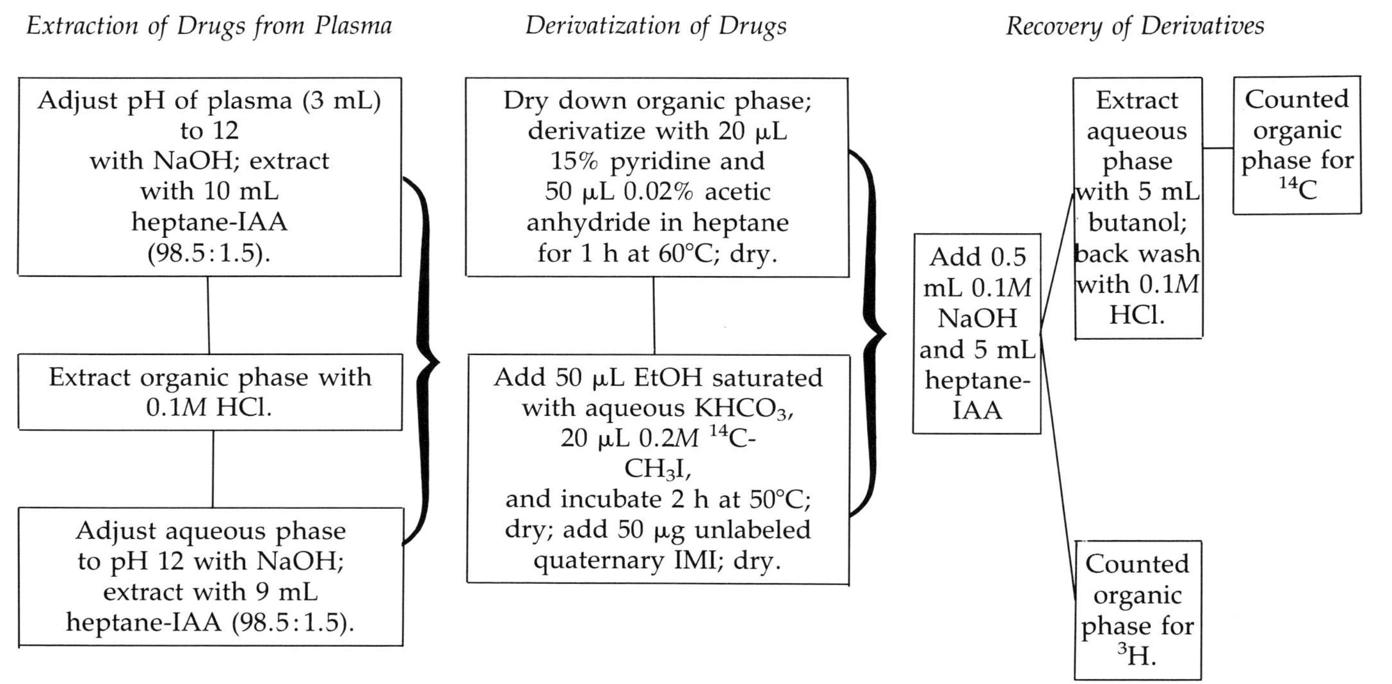

Fig. 9. Single-isotope derivative assay for the simultaneous determination of imipramine and desipramine. Details of experimental conditions are taken from Harris et al. (1970). Abbreviation: IAA = isoamyl alcohol.

Table 2
Double-Isotope Derivative Assays for Antidepressants[a]

Drug	Volume of plasma, mL	Sensitivity, μg/L	Precision, CV%	Authors
NT	1	5	7	Maguire et al. (1976)
MAP	1	5	10	Norman et al. (1979)
MAP	3	2	8	Riess (1974b)
OPIP	3	2	2–10	Riess (1977)
CMI	2	15 (CMI)	5–10 (CMI)	Carnis et al. (1976)
DCM1		2 (DCM1)	4–7 (DCM1)	

[a]Abbreviations: CMI, clomipramine; DCMI, desmethylclomipramine; OPIP, opipramol; see also footnote to Table 1.

The double-isotope procedure for antidepressant drugs was first described for maprotiline analysis (Riess, 1974b). It was later applied to the analysis of nortriptyline (Maguire et al., 1976), clomipramine (Carnis et al., 1976; Jones and Luscombe, 1976), and opipramol (Riess, 1977). The technique is sensitive, precise, and as rapid as most gas chromatographic methods. Specificity of the assay is ensured by the inclusion of a TLC step to provide separation of the derivatized metabolites. A comparison of methods is presented in Table 2. As with the single-isotope assay, the technique is most readily applied to secondary amines, although use could be made of the formation of the quaternary ammonium salt with methyl iodide and tertiary amines. Carnis and coworkers (1976) determined the tertiary amine clomipramine by conversion to the secondary amine desmethylclomipramine with trichloro-ethyl-chloroformate. The intermediate carbamate formed was hydrolyzed with sodium hydroxide and acetylated with [^{3}H]-acetic anhydride. This method could quite readily be adapted to other antidepressants provided suitable radiolabeled compounds were available. An outline of the method for the determination of plasma concentrations of both clomipramine and its desmethyl metabolite is presented in Fig. 10.

In our laboratories, using a scheme similar to that of Fig. 10, both nortriptyline and maprotiline have been measured in single-

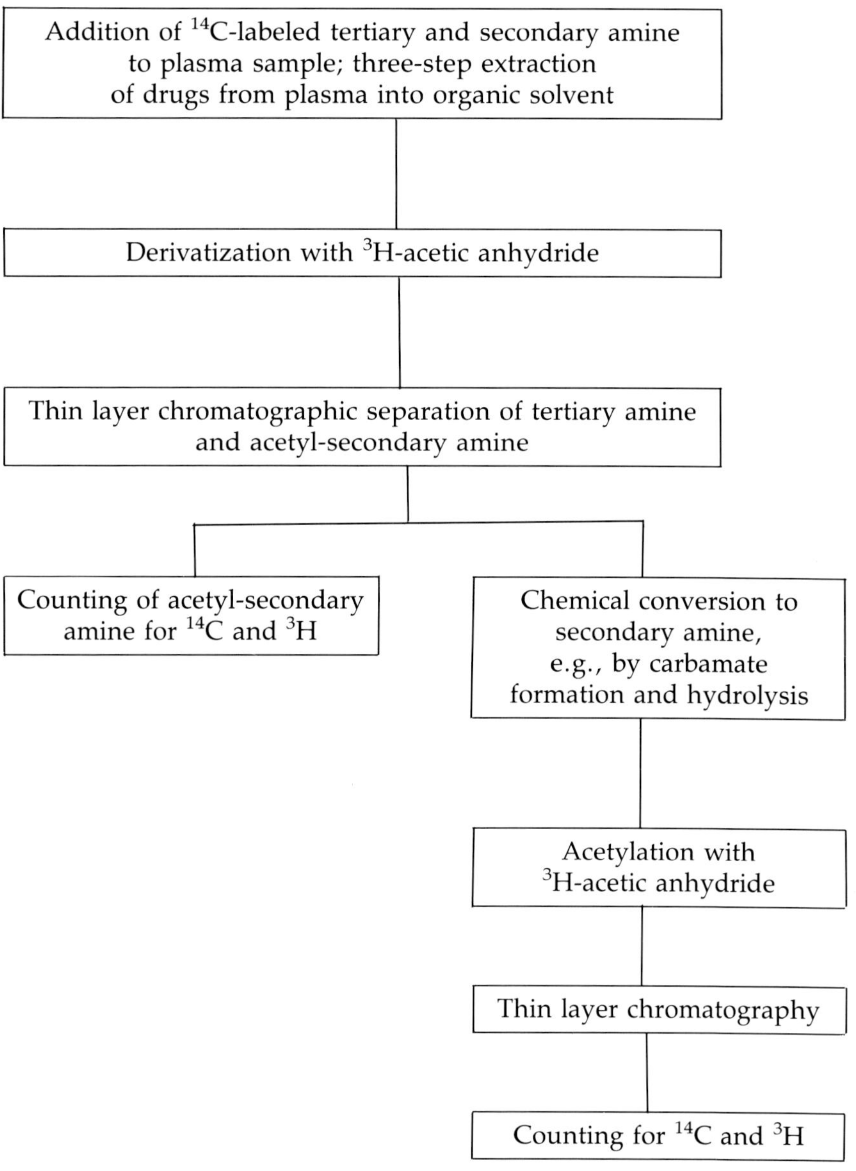

Fig. 10. Double-isotope derivative assay for the simultaneous determination of tertiary and secondary amine antidepressants by chemical conversion of the tertiary amine to a secondary amine. For an application to the determination of clomipramine and desmethylclomipramine, *see* Carnis et al., 1976.

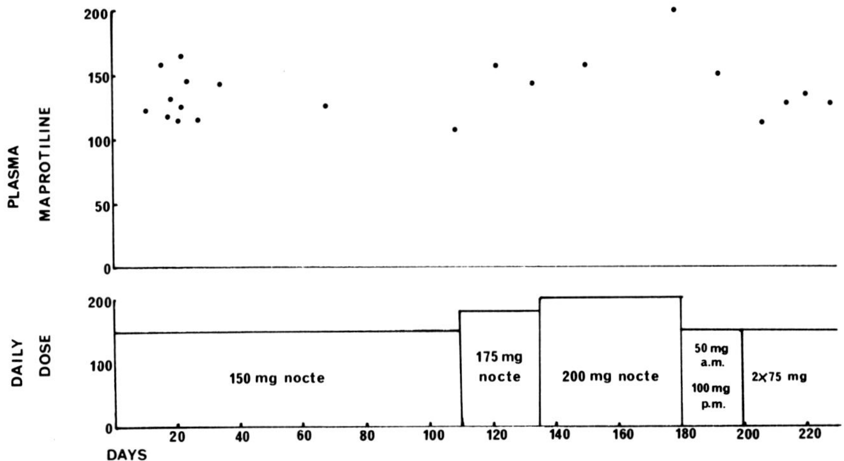

Fig. 11. Day-to-day variability of maprotiline plasma concentrations in a 60-year-old male patient receiving long-term treatment, with various drug regimens. Maprotiline concentrations were determined by double isotope derivative dilution analysis.

and multiple-dose studies and for plasma concentration and clinical effect relationships. Unlike the methods of Riess (1974) or Carnis et al. (1976), the specific activity of the derivatizing agent is used to determine mass, thereby utilizing to full advantage the principles of the isotope dilution technique. Application of the method to a study of day-to-day variability of plasma drug concentrations in a patient receiving varying doses of maprotiline is presented in Fig. 11. For nortriptyline, the utility of the method is demonstrated by an examination of the effects of added barbiturates on steady-state plasma concentrations presented in Fig. 12. Barbiturates are well known for their induction of hepatic drug-metabolizing enzymes and clearly have an effect on nortriptyline concentrations.

3.2. Antipsychotics

Isotope assays have not been widely applied to the measurement of antipsychotic drug plasma concentrations. Efron and associates in a series of papers described the radioassay of chlorpromazine and some of its metabolites (Efron et al., 1968; 1970; 1971). Chlorpromazine, desmethylchlorpromazine, and didesmethylchlorpromazine are extracted from alkanized plasma into heptane, back-extracted into acid, and finally extracted into heptane-isoamyl alcohol. The corresponding sulfoxides are ex-

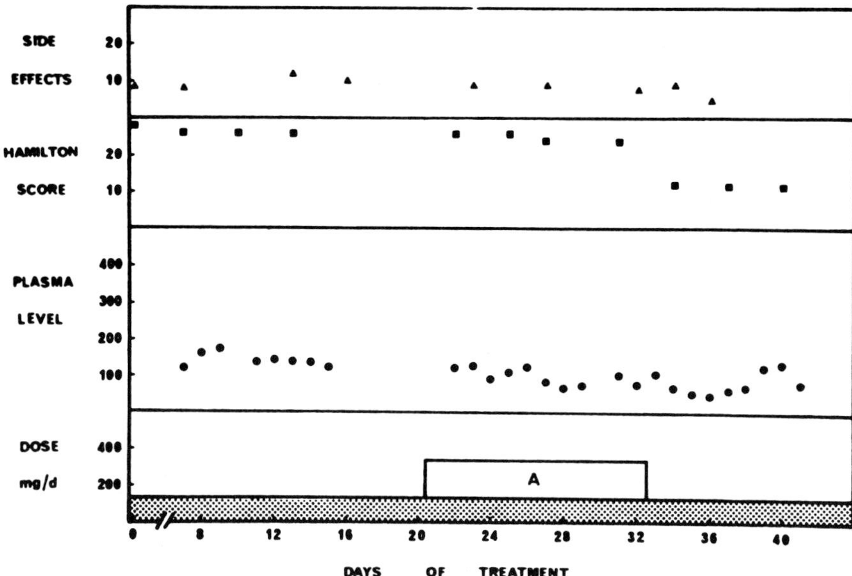

Fig. 12. Barbiturates and plasma nortriptyline concentration in a 41-year-old female patient receiving 150 mg/d of nortriptyline. Amylobaritone (A, 200 mg nocte) was administered for 14 d. Mean plasma nortriptyline concentration before bartiturate was 145 ± 7 µg/L; after bartiturate, 93 ± 6 µg/L ($P < 0.001$; paired t-test).

tracted from the residual plasma with ethyl acetate, back-extracted into acid, and finally extracted from acid into ethyl acetate. Both sets of extracts are acetylated with [3H]-acetic anhydride and quaternized with [14C]-methyl iodide. The primary and secondary amines react with acetic hydride, and the tertiary amines with methyl iodide. The derivatives in each tube are separated by extraction into butanol or heptane-isoamyl alcohol. The respective fractions are counted in a liquid scintillation counter and unknown samples can be analyzed from a standard curve. When the method as described was applied to the measurement of human plasma samples, the sensitivity for chlorpromazine was 10–15 µg/L. A better sensitivity can be achieved with higher specific activity radiotracers. Specificity may be improved with the inclusion of a two-dimensional TLC step for separation of the derivatives, rather than differential solvent extraction. Recovery may also be improved by this modification.

This method can be applied more widely to other antipsychotic drugs, but does not appear to have been pursued. The complex pattern of phenothiazine metabolism increases the difficulties of assay development for parent drug and active metabolites. The double-isotope derivative technique has not been applied to antipsychotic drug measurement, but could readily be developed to do so. The necessary radioisotopically labeled compounds are commercially available for most of the commonly used antipsychotics.

3.3. Other Psychotropic Drugs

Isotope derivative assays have not been used for the other major classes of psychotropic drugs, antianxiety agents, and monoamine oxidase inhibitors. Among the antianxiety agents, the benzodiazepines are preeminent, and several members of this class of drug could readily be quantitated by isotope derivative analysis, e.g., chlordiazepoxide, desmethyldiazepam, nitrazepam, oxazepam, flurazepam. Gas chromatography has been widely applied in the analysis of benzodiazepines and will most probably remain the first choice of analytical technique (Norman and Burrows, 1981).

Interest in the determination of plasma concentrations of the monoamine oxidase inhibitors is surpassed by the interest in determining their "in vivo" effects on platelet monoamine oxidase activity. Consequently, there are relatively few published methods for their determination and none using isotope derivative assays. The hydrazine-type inhibitors such as phenelzine or isocarboxazid or the primary amine tranylcypromine could clearly be derivatized with appropriate radiolabeled agents and quantitated.

4. Comparison with Other Methodologies

A particular concern in the development of a new assay method is its performance compared to an established assay. Interlaboratory studies for the determination of antiepileptic (Richens, 1975; Pippenger et al., 1976) or chlorpromazine (Turner et al., 1976) plasma concentrations have given cause for concern. In the antiepileptic study, which involved about 200 laboratories, wide variations from the mean value determined by the reference centers were observed (Pippenger et al., 1976). For chlorpromazine, consistent results across centers were observed, but results for the metabolites were considerably less consistent (Turner et al., 1976).

Both inter- and intralaboratory comparisons for the isotope derivative assays have been carried out. In an intralaboratory comparison, the single- and double-isotope methods for the determination of nortriptyline were compared by measuring 19 patients' samples by both methods. For the single-isotope assay, a mean concentration of 158 ± 83 (SD) µg/L was found, whereas the double-isotope method gave 181 ± 96 (SD) µg/L; these values were not significantly different (Maguire et al., 1976). Similarly, for 28 plasma samples containing nortriptyline, a close agreement between the double-isotope assay and radioimmunoassay was observed (Maguire et al., 1978). In an interlaboratory comparison, 24 nortriptyline plasma samples were assayed by GC (Burch et al., 1979) and the double-isotope derivative method. There was a good correlation between the two methods ($r = 0.994$), and a paired t-test showed no significant difference between the values determined by both techniques (Burrows et al., 1978). Intralaboratory comparisons of double-isotope and gas chromatographic determinations of plasma maprotiline concentrations have been carried out (Geiger et al., 1975; Fig. 13). In both studies good agreement between techniques was observed. Kline and Cooper (1974) showed a satisfactory correlation between a single-isotope and gas liquid chromatographic assay for desmethyldoxepin in their laboratory. Intralaboratory comparisons are less satisfactory than interlaboratory comparison of assays. Nevertheless they do give a measure of confidence in the new method. In this context, isotope derivative assays, when skilfully applied, give results comparable to other standard analytical techniques.

5. Protocols

5.1. Experimental

5.1.1. Preparation of [^{3}H]-Acetic Anhydride

Ampules with 1Ci [^{3}H]-acetic anhydride (500 mCi/mmol) were reduced in specific activity (to approximately 330 mCi/mmol) with anhydrous acetic anhydride and then diluted to a concentration of 0.5% with anhydrous benzene. Portions of 1 mL were distilled six times on a vacuum manifold at 200 mm Hg with the aid of liquid air. The distilled material was stored in tightly stoppered tubes at –20°C. Prior to use, it was redistilled and diluted to 0.02% with *n*-hexane.

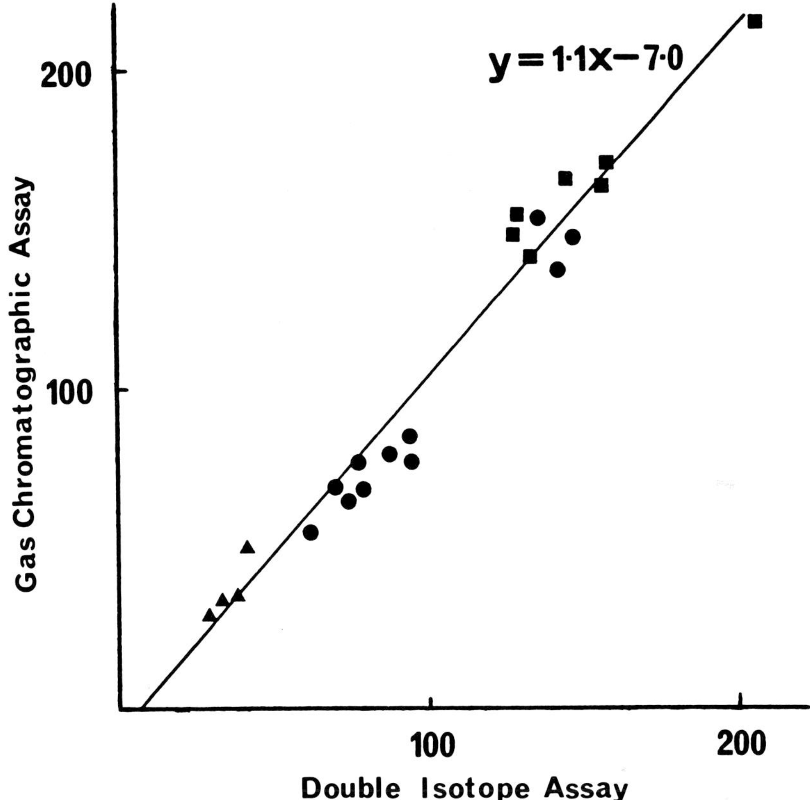

Fig. 13. A comparison of plasma maprotiline concentrations determined by both double-isotope derivative analysis and gas liquid chromatography with a nitrogen selective detector. Samples were taken from three patients receiving variable doses of maprotiline over a 2-mo interval.

The specific activity of [^{3}H]-acetic anhydride was determined by formation of cortisol-^{3}H-acetate (Coghlan and Scoggins, 1967). This derivative was purified by paper chromatography in cyclohexane:benzene:methanol:water (100:100:100:25) and a second chromatography in cyclohexane:dioxane:methanol:water (100:100:50:25) on ethanol-washed paper. The specific activity was derived by counting aliquots of the cortisol-^{3}H-acetate in a liquid scintillation spectrometer and determining the concentration from UV absorption at 254 m (ϵ = 16200) by Porter-Silber chromogen formation compared with a recrystallized cortisol-acetate standard.

5.1.2. Thin-Layer Chromatography (TLC)

One-dimensional TLC was carried out on aluminum sheets (20 × 20 cm) precoated with 0.25 mm silica gel (Silica-gel GF254/5554, Merck). Samples were applied with 3 × 0.2 mL washes of 40% methanol in methylene chloride. All systems were developed ascending at room temperature. Radioactive compounds were located by scanning (Packard Radiochromatogram Scanner, model 7201) or by comparison with nonradioactive reference compounds run as side-lane markers. The appropriate area of silica gel was removed under vacuum into drawn-out glass pipets plugged with glass wool. Samples were eluted by the passage of 10 mL of 40% methanol in methylene chloride through the pipets. This solvent has been found to give 95% recovery when used for application and elution of samples.

5.1.3. Description of the Single Isotope Derivative Assay for Nortriptyline or Desipramine

5.1.3.1. EXTRACTION.   Duplicate samples of 2.0 mL of plasma were transferred to the 7-mL test tubes. To each sample was added 0.4 mL of $1M$ NaOH and 3 mL of n-heptane. The tubes were shaken gently for 10 min on a Matburn wheel. After settling, any emulsion that formed was broken by freezing the tube in liquid nitrogen, followed by thawing. Using a Pasteur pipet, 2 mL of the heptane phase was transferred to a clean 7-mL tube.

5.1.3.2. ACETYLATION.   To each sample, 100 μL of pyridine and 50 μL of 0.02% [^{3}H]-acetic anhydride reagent were added to the 2 mL of heptane extract. The tubes were stoppered and heated for 60 min in a waterbath at 60°C. Stoppers were then removed and the solvent evaporated under an airstream at 60°C.

5.1.3.3. HYDROLYSIS.   To each sample 2 mL of $0.1M$ NaOH was added. The tubes were stoppered and kept to 60°C in the waterbath for 15 min. The tubes were allowed to cool to room temperature, 2 mL of n-heptane were added, and the tubes were shaken on a Matburn wheel for 10 min to extract the acetylated nortriptyline/ desipramine into the organic phase. The aqueous phase was aspirated and discarded, and a further 2 mL of $0.1M$ NaOH were added to the h-heptane extract. The tubes were shaken for 10 min and allowed to settle. Aliquots (0.2 mL) of the heptane phase were transferred to glass counting vials using an Eppendorf pipet.

5.1.3.4. COUNTING OF LABELED DERIVATIVES.   After addition of 10 mL of phosphor to each vial, all samples were counted in a Packard Tricarb Liquid Scintillation Spectrometer (Model 314Ex) at a ^{3}H efficiency of 30% with a heptane/phosphor background of 40 cpm. Vials were counted sequentially for 10 min each, and this counting was then repeated. Standard vials counted with each run of samples included a heptane background vial, a cortisol-^{3}H-acetate vial to determine the specific activity of the [^{3}H]-acetic anhydride, and a tritium efficiency vial.

5.1.4. Description of the Double-Isotope Derivative Dilution Assay for Nortriptyline or Maprotiline

5.1.4.1. PURIFICATION OF ^{14}C-NORTRIPTYLINE/MAPROTILINE. [^{14}C]-Maprotiline (9.5 mCi/mmol) or [^{14}C]-nortriptyline (5.4 mCi/mmol) was dissolved in ethanol to a concentration of 10 μg/mL and stored in the dark at 4°C. Portions of this stock solution were purified by TLC in the system benzene:dioxane:ethanol:diethylamine (50:40:5:5) for 1.5 h. There was a single peak of radioactivity upon scanning that had the same R_F as a nonradioactive nortriptyline/maprotiline standard. The eluted material was stored in ethanolic solution (concentration of approximately 100 ng/mL) in the dark at 4°C. [^{14}C]-nortriptyline/maprotiline solutions were prepared every 10–14 d.

5.1.4.2. EXTRACTION. After addition of [^{14}C]-nortriptyline/maprotiline (approximately 10 ng) as internal recovery indicator, 1 mL plasma samples were made alkaline (pH 10) by addition of 0.4 mL 1M NaOH. Treated samples were extracted with 3 mL n-heptane by rotation on a Matburn wheel for 15 min. Emulsions were broken by freezing the samples in liquid nitrogen and allowing them to thaw. Aliquots of 2 mL of the heptane phase were transferred to clean tubes (silanized glass tubes, 7 mL), were used for all steps of the assay.

5.1.4.3. DERIVATIVE FORMATION. Added to each sample were 100 μL of pyridine and 50 μL of 0.02% [^{3}H]-acetic anhydride solution. The tubes were stoppered and the contents well mixed. The tubes were placed in a waterbath set at 60°C for 60 min.

5.1.4.4. HYDROLYSIS. The solvent was evaporated under an airstream and 2 mL of 0.1M NaOH were added to each sample. The tubes were stoppered, returned to the water bath for 15 min, then cooled to room temperature.

5.1.4.5. EXTRACTION OF THE DERIVATIVE. Samples of 2 mL of n-heptane were added to each tube and the derivative extracted by rotation on a Matburn wheel for 10 min. After phase separation, the aqueous phase was aspirated off. Samples of 2 mL of 0.1M NaOH were added to the n-heptane extracts, and the samples were returned to the Matburn wheel for 10 min. After phase-separation, 1-mL aliquots of the heptane phase were transferred to clean tubes for TLC.

5.1.4.6. PURIFICATION OF THE DERIVATIVE.   After addition of 10 μg of nortriptyline/maprotiline-acetate to the heptane aliquot as UV marker, the solvent was evaporated under an airstream. The derivative was then purified by TLC using benzene:acetone (1:1) (Hammer and Brodie, 1967) as developing solvent. The samples were eluted into glass vials, the solvent was evaporated under air, and 10 mL of phosphor solution were added.

5.1.4.7. DETERMINATION OF RADIOACTIVITY.   Samples were counted for 60 min in a Packard Tricarb Liquid Scintillation Spectrometer (Model 3330). In a high-energy channel, ^{14}C was counted with an efficiency of 55% and a phosphor background of 9 cpm. In a low-energy channel, ^{3}H was counted with an efficienty of 35% and a background of 24 cpm. Standard vials were counted with each assay: phosphor background vial, cortisol-^{3}H-acetate vial for determination of specific activity of [^{3}H]-acetic anhydride, and pure ^{3}H and ^{14}C standards for determination of leakage between channels, ^{3}H and ^{14}C-efficiency vials, and a ^{14}C-nortriptyline/maprotiline recovery vial.

5.2. Calculation of Results for the Double-Isotope Assay

Equations for estimation of the mass of double-labeled samples have been described previously (Coghlan and Scoggins, 1967; Riess, 1974a). The following counts are needed for the calculations:

H   = Background cpm in ^{3}H channel

C   = Background cpm in ^{14}C channel

H1 = cpm of ^{3}H-standard in ^{3}H-channel

H2 = cpm of ^{3}H-standard in ^{14}C-channel

C1 = cpm of ^{14}C standard in ^{3}H-channel

C2 = cpm of ^{14}C standard in ^{14}C-channel

N1 = cpm for extracted, derivatized samples in ^{3}H channel

N2 = cpm for extracted, derivatized samples in ^{14}C channel

where H1 to N2 are corrected for background cpm obtained, as at H and C, respectively.

It can be shown that:

$$h = \frac{b\mathrm{N1} - \mathrm{N2}}{b - a} \tag{1}$$

and

$$c = \frac{b(\text{N2} - a\text{N1})}{b - a} \tag{2}$$

where

$$h = \text{actual } ^3\text{H cpm}$$
$$c = \text{actual } ^{14}\text{C cpm}$$

$$b = \frac{\text{H2}}{\text{H1}} \text{ for a pure } ^3\text{H compound}$$

$$a = \frac{\text{C2}}{\text{C1}} \text{ for a pure } ^{14}\text{C compound}$$

Mass in each sample can be calculated as:

$$M = \text{dilution factor} \times \text{mass indicator added} \tag{3}$$

where

$$\text{dilution factor} = \frac{\text{specific activity (cpm/ng) indicator}}{\text{specific activity (cpm/ng) sample}} \tag{4}$$

Then the actual mass per sample

$$= \text{total mass } (M) - \text{mass indicator added} \tag{5}$$

and the mass of the indicator added

$$= \left[\frac{C}{S}\right] \times \left[\frac{h}{c}\right]^{\text{I}} \tag{6}$$

where

$$C = \text{cpm } ^{14}\text{C added,}$$
$$S = \text{specific activity of acetic anhydride, and}$$

$$\left[\frac{h}{c}\right]^{\text{I}} = \text{ratio of } ^3\text{H to } ^{14}\text{C cpm in indicator sample}$$

This contribution of the ^{14}C-maprotiline or nortriptyline added as internal indicator is determined from a larger aliquot than is actually added to each sample to reduce the error associated with its determination.

$$\left[\frac{h}{c}\right]^S = \text{ratio of } {}^3\text{H to } {}^{14}\text{C in actual samples (this of course includes the indicator sample).}$$

From Eqs. (3) and (4):

$$\text{Total mass per sample} = \left[\frac{c}{s}\right] \times \left[\frac{h}{c}\right]^S \tag{7}$$

and from Eq. (5)

$$\text{Actual mass per sample} = \left[\frac{c}{s}\right] \times \left[\frac{h}{c}\right]^S - \left[\frac{c}{s}\right] \times \left[\frac{h}{c}\right]^I \tag{8}$$

$$= \left[\frac{c}{s}\right]\left(\left[\frac{b}{c}\right]^S - \left[\frac{h}{c}\right]^I\right) \tag{9}$$

In experimental practice the blank may be significant and also contains the same mass of ^{14}C-maprotiline or nortriptyline and Eq. (10) can be used instead of Eq. (9)

$$\text{Mass per sample} = \left[\frac{c}{s}\right]\left(\left[\frac{b^S}{c}\right] - \left[\frac{h}{c}\right]^B\right) \tag{10}$$

The following results were obtained in a determination of maprotiline:

$$
\begin{aligned}
\text{Efficiency for } {}^3\text{H} &= 35.0\% \\
\text{Efficiency for } {}^{14}\text{C} &= 55.5\% \\
\text{Background for } {}^3\text{H} &= 24 \text{ cpm} \\
\text{Background for } {}^{14}\text{C} &= 9 \text{ cpm} \\
b &= 0.28 \\
a &= 2.08 \times 10^{-5}
\end{aligned}
$$

Specific activity of [^{3}H]-acetic anhydride is:

$$
\begin{aligned}
&\text{Cortisol acetate concentration} = 9.77 \text{ µg/mL} \\
&\text{and } {}^3\text{H-cpm for } 0.1h\text{mL aliquot} = 306{,}128, \\
&\text{then specific activity} = 457.5 \text{ cpm/ng}
\end{aligned}
$$

Amount of ^{14}C-maprotiline added to each sample is 223 cpm = 5.9 ng. For the blank samples, 58 cpm and 786 cpm were obtained in the ^{14}C and ^{3}H channels, respectively.

$$\left[\frac{h}{c}\right]^B = 14.96$$

$$C = 223$$
$$S = 458$$

For a typical sample 69 and 10,948 cpm were obtained in the ^{14}C and ^{3}H channels, respectively.

The mass in the sample is obtained from Eq. (10):

$$= \frac{223}{458}\left[\frac{10706}{60} - \frac{587}{49}\right]$$

$$= 81 \text{ ng}$$

6. Summary and Conclusion

Isotope derivative assays have not been widely applied to the analysis of psychotropic drugs, despite the fact that they are sensitive, reliable, specific, and no more tedious to perform than gas chromatographic assays. For most applications in clinical psychopharmacology, e.g., pharmacokinetics or steady-state plasma concentration measurements, isotope derivative assays are well suited. Although not technically demanding, the assays require access to a liquid scintillation spectrometer and a well-equipped laboratory capable of handling radioisotopes. Lack of appropriate facilities may explain the apparent reluctance to use these techniques. Although requiring somewhat more sophisticated technical skills, chromatographic assays will probably be the first choice of analytical technique for psychotropic drugs. Chromatographic assays are available for virtually all the drugs, generally well evaluated, and much more widely disseminated than isotope derivative assays.

Acknowledgments

The studies reported from our laboratories were performed with the cooperation of the staff of the Howard Florey Institute of Experimental Physiology and Medicine, University of Melbourne, and our colleagues and students. In particular we wish to acknowledge the contributions of Dr. B. A. Scoggins and Dr. J. P. Coghlan (HFI), Dr. K. P. Maguire, Mr. J. M. E. Wurm, Ms. B. Maddison, and Ms. J. Barr. Professor B. Davies has provided encouragement and support for these studies over many years. The studies were supported, in part, by grants from the National Health and Medical Research Council of Australia.

References

Alexanderson B. (1972a) Pharmacokinetics of nortriptyline in man after single and multiple oral doses. The predictability of steady state plasma concentrations from single-dose plasma-level data. *Eur. J. Clin. Pharmacol.* **4**, 82–91.

Alexanderson B. (1972b) Pharmacokinetics of desmethylimipramine and nortriptyline in man after single and multiple oral doses. A crossover study. *Eur. J. Clin. Pharmacol.* **5**, 1–10.

Alexanderson B. (1973) Prediction of steady-state plasma levels of nortriptyline from single oral dose kinetics: A study in twins. *Eur. J. Clin. Pharmacol.* **6**, 44–53.

Alexanderson B., Evans D. A. P., and Sjoqvist F. (1969) Steady-state plasma levels of nortriptyline in twins. Influence of genetic factors and drug therapy. *Br. Med. J.* **iv**, 764–768.

Alexanderson B., Borga O., and Alvan G. (1973) The availability of orally administered nortriptyline. *Eur. J. Clin. Pharmacol.* **5**, 181–185.

Asberg M., Cronholm B., Sjoqvist F., and Tuck D. (1970) Correlation of subjective side effects with plasma concentrations of nortriptyline. *Br. Med. J.* **iv**, 18–21.

Asberg M., Cronholm B., Sjoqvist F., and Tuck D. (1971) Relationship between plasma level and therapeutic effect of nortriptyline. *Br. Med. J.* **iii**, 331–334.

Bickel M. H. and Baggiolini M. (1966) The metabolism of imipramine and its metabolites by rat liver microsomes. *Biochem. Pharmacol.* **15**, 1155–1169.

Brodie B. B. (1967) Physicochemical and biochemical aspects of pharmacology. *J. Am. Med. Assoc.* **202**, 600–609.

Burch J. E., Raddats M. A., and Thompson S. G. (1979) Reliable routine method for the determination of plasma amitriptyline and nortriptyline by gas chromatography. *J. Chromatogr.* **162**, 351–366

Burrows G. D. and Norman T. R. (1981) Tricyclic Antidepressants: Plasma Levels and Clinical Response, in *Psychotropic Drugs, Plasma Concentration and Clinical Response* (Burrows G. D. and Norman T. R. eds.), Marcel Dekker, New York.

Burrows G. D., Davies B. M., and Scoggins B. A. (1972) Plasma concentration of nortriptyline and clinical response in depressive illness. *Lancet* **ii,** 619–623

Burrows G. D., Davies B., Maguire K. P., Norman T. R., Scoggins B. A., Hullin R. P., and Burch J. (1978) Interlaboratory comparison of nortriptyline levels. *Biochem. Med.* **20,** 125–127.

Carnis G., Godbillon J., and Metayer J. P. (1976) Determination of clomipramine and desmethylclomipramine in plasma or urine by the double radioisotope derivative technique. *Clin. Chem.* **22,** 817–823.

Cimbura G. (1972) Review of methods of analysis for phenothiazine drugs. *J. Chromatogr Sci.* **10,** 287–293.

Clifford J. M. and Franklin-Smyth W. (1974) The determination of some 1,4-benzodiazepines and their metabolites in body fluids: A review. *Analyst* **99,** 241–272.

Coghlan J. P. and Scoggins B. A. (1967) Measurement of aldosterone in peripheral blood of man and sheep. *J. Clin. Endocrin.* **27,** 1470–1486.

Crammer J. L. and Rolfe B. (1970) Metabolism of ^{14}C-imipramine. III. Conversions by rat tissues. *Psychopharmacology* **18,** 26–37.

Davis J. M., Javaid J. I., Janicak P. G., and Mostert M. A. (1985) Antipsychotics: Plasma Levels and Clinical Response, in *Drugs in Psychiatry* vol. 3. *Antipsychotics* (Burrows G. D., Norman T. R., and Davies B. M., eds.), Elsevier Science, Amsterdam.

Dieterle W. and Faigle J. W. (1982) Multiple inverse isotope dilution assay for the stereospecific quantitative determination of R(–) and S(+) oxaprotiline in biological fluids. *J. Chromatogr.* **242,** 289–297.

Dingell J. V., Sulser F., and Gillette J. R. (1964) Species difference in the metabolism of impipramine and desmethylimipramine. *J. Pharmacol. Exp. Ther.* **143,** 14–22.

Efron D. H., Gaudette L. E., and Harris S. R. (1968) A new method for the measurement of minute amounts of chlorpromazine and some of its metabolites in plasma. *Agressologie* **9,** 103–107.

Efron D. H., Manian A. A., and Harris S. R. (1970) The simultaneous measurement of chlorpromazine, chlorpromazine sulphoxide and their demethylated analogs in plasma by radioactive derivative formation. *Psychopharmacology Bull.* **6,** 73.

Efron D. H., Harris S. R., Manian A. A., and Gaudette L. E. (1971) Radioassay of chlorpromazine and its metabolites in plasma. *Psychopharmacology* **19,** 207–223.

Fredricson-Overo K. (1972) Increased specificity of the ^{3}H-acetic anhydride coupling method for plasma analysis of drugs containing secondary amino groups. *Acta. Pharmacol. Toxicol.* **31,** 433–440.

Geiger U. P., Rajagopalan T. G., and Riess W. (1975) Quantitative assay of maprotiline in biological fluids by gas liquid chromatography. *J. Chromatogr.* **114,** 167–173.

Gram L. F. (1974) Metabolism of tricyclic antidepressants. A review. *Dan. Med. Bull.* **21**, 218–231.

Gram L. F. and Fredricson-Overo K. (1972) Drug interaction: Inhibitory effect of neuroleptics on metabolism of tricyclic antidepressants in man. *Br. Med. J.* **1**, 463–465.

Gupta R. and Molnar G. (1979) Measurement of therapeutic concentrations of tricyclic antidepressants in serum. *Drug. Metab. Rev.* **9**, 79–97.

Hailey D. M. (1974) Chromatography of the 1,4-benzodiazepines. *J. Chromatogr.* **98**, 527–568.

Hammer W. H. and Brodie B. B. (1967) Application of isotope-derivative technique to assay of secondary amines: Estimation of desipramine by acetylation with ^{3}H-acetic anhydride. *J. Pharmacol. Exp. Ther.* **157**, 503–508.

Hammer W. H. and Sjoqvist F. (1967) Plasma levels of monomethylated tricyclic antidepressants during treatment with imipramine-like compounds. *Life Sci.* **6**, 1895–1903.

Hammer W. H., Idestrom G. M., and Sjoqvist F. (1967) Chemical Control of Antidepressant Drug Therapy, in *Antidepressant Drugs* (Grattini S. and Dukes M. N. G., eds.), Excerpta Medica, Amsterdam.

Hammer W. H., Martens S., and Sjoqvist F. (1969) A comparative study of the metabolism of desmethylimipramine, nortriptyline and oxyphenylbutazone in man. *Clin. Pharmacol. Ther.* **10**, 44–49.

Harris S. R., Gaudette L. E., Efron D. H., and Marian A. A. (1970) A method for the measurement of plasma imipramine and desmethylimipramine concentrations. *Life Sci.* **9**, 781–788.

Jones R. B. and Luscombe D. K. (1976) Single dose studies with clomipramine in normal subjects. *Postgrad. Med J.* (suppl 3) **52**, 62–67.

Kliman B. and Peterson R. E. (1960) Double isotope derivative assay of aldosterone in biological extracts. *J. Biol. Chem.* **235**, 1639–1648.

Kline N. S. and Cooper T. B. (1974) Methods for the Evaluation and Interpretation of Drug Plasma Levels, in *Classification and Prediction of Outcome of Depression* (Angst J., ed.), Symposia Medica Hoechst 8. Schattauer, Stuttgart.

Loh A., Zuleski F. R., and Di Carlo F. J. (1977) Radiochemical GLC assay for nortriptyline in human plasma. *J. Pharm. Sci.* **66**, 1056–1057.

Maguire K. P., Burrows G. D., Coghlan J. P., and Scoggins B. A. (1976) Rapid radio-isotopic procedure for the determination of nortriptyline in plasma. *Clin. Chem.* **22**, 761–764.

Maguire K. P., Burrows G. D., Norman T. R., and Scoggins B. A. (1978) A radioimmunoassay for nortriptyline (and other tricyclic antidepressants) in plasma. *Clin. Chem.* **24**, 549–554.

Newgreen D., Norman T. R., and Burrows G. D. (1985) Antipsychotic Drugs, Pharmacokinetics, Plasma Concentrations and Clinical

Effects, in *Handbook of Studies on Schizophrenia* (Burrows G. D., Norman T. R., and Rubinstein G., eds.), Elsevier, Amsterdam, in press.

Norman T. R., Burrows G. D., Maguire K. P., and Scoggins B. A. (1979) Clinical applications of isotope derivative assays for maprotiline. *Clin Exp. Pharmacol. Physiol.* **6,** 211–212.

Norman T. R. and Burrows G. D. (1980a) Plasma Levels of Psychotropic Drugs and Clinical Response, in *Advances in Human Psychopharmacology* vol. 1 (Burrows G. D. and Werry J. eds.), JAI Press, Connecticut.

Norman T. R. and Burrows G. D. (1980b) Plasma Levels of Benzodiazepine Antianxiety Drugs and Clinical Response, in *Handbook of Studies on Anxiety* (Burrows G. D. and Davies B., eds.), Elsevier, Amsterdam.

Norman T. R. and Burrows G. D. (1981) Methods for the Measurement of Psychotropic Drugs. Antipsychotics and Antianxiety Agents, in *Psychotropic Drugs: Plasma Concentration and Clinical Response* (Burrows G. D. and Norman T. R., eds.), Marcel Dekker, New York.

Norman T. R. and Burrows G. D. (1983) Plasma Concentrations of Antidepressant Drugs and Clinical Response, in *Drugs and Psychiatry* vol. 1 *Antidepresants.* (Burrows G. D., Norman T. R., and Davies B. M., eds.), Elsevier, Amsterdam.

Norman T. R. and Burrows G. D. (1984a) Benzodiazepine Plasma Concentrations and Anxiolytic Response, in *Drugs in Psychiatry* vol. 2 *Antianxiety Agents* (Burrows G. D., Norman T. R., and Davies B. M., eds.), Elsevier, Amsterdam.

Norman T. R. and Burrows G. D. (1984b) Plasma concentrations of benzodiazepines—a review of clinical findings and implications. *Prog. Neuropsychopharmacol. Biol. Psychiat.* **8,** 115–126.

Norman T. R. and Maguire K. P. (1985) Analysis of tricyclic antidepressant drugs in plasma and serum by chromatographic techniques. *J. Chromatogr.* **340,** 173–197.

Norman T. R., Maguire K. P., Scoggins B. A., and Burrows G. D. (1982) Monitoring and interpretation of antidepressant plasma concentration. *Aust. N. Z. J. Psychiatry* **16,** 74–78.

Pippenger C. E., Kiffin-Penry J., White P. G., Daly D. D., and Buddington R. (1976) Interlaboratory variability in determination of plasma antiepileptic drug concentration. *Arch Neurol.* **33,** 351–356.

Preskorn S. H. and Mac D. S. (1984) The implication of concentration/response studies of tricyclic antidepressants for psychiatric research and practice. *Psychiat. Devel.* **3,** 201–222.

Richens A. (1975) Results of a Phenytoin Quality Control Scheme, in *Clinical Pharmacology of Antiepileptic Drugs* (Schneider H., Janz D., and Gardner-Thorpe C., eds.), Springer, Berlin.

Riess W. (1974a) Drug Assay by Radioactive Reagents, in *The Poisoned Patient. The Role of the Laboratory* (Ciba Foundation Symposium 26), Elsevier, Amsterdam.

Riess W. (1974b) The double radioisotope derivative technique for the assay of drugs in biological material. I. The determination of maprotiline. *Anal. Chim. Acta* **68,** 363–376.

Riess W. (1977) The double radioisotope derivative technique for the assay of drugs in biological material. II. The determination of opipramol. *Anal. Chim. Acta.* **88,** 109–115.

Riess W., Brechbuhler S., and Dubois J. P. (1979) Rational analysis of drugs in biological fluids with particular reference to the tricyclic antidepressants. *Prog. Drug. Metab.* **3,** 115–151.

Risch S. C., Huey L. Y., and Janowsky D. S. (1979a) Plasma levels of tricyclic antidepressants and clinical efficacy: Review of the literature. I. *J. Clin. Psychiat.* **40,** 4–16

Risch S. C., Huey L. Y., and Janowsky D. S. (1979b) Plasma levels of tricyclic antidepressants and clinical efficacy: Review of the literature. II. *J. Clin. Psychiat.* **40,** 58–69.

Riva E., Hrdina P. D., and Morselli P. L. (1975) Measurement of desipramine in brain tissue by a radioisotope derivative technique. *J. Pharm. Pharmacol.* **27,** 797–799.

Rivera-Calimlim L. and Hershey L. (1984) Neuroleptic concentrations and clinical response. *Ann. Rev. Pharmacol. Toxicol.* **24,** 361–368.

Rubinstein G., McIntyre I., Burrows G. D., Norman T. R., and Maguire K. P. (1983) Metabolism of Tricyclic Antidepressant Drugs, in *Drugs in Psychiatry* vol I *Antidepressants* (Burrows G. D., Norman T. R., and Davies B. M., eds.), Elsevier, Amsterdam.

Scoggins B. A. and Maguire K. P. (1977) Measurement of Tricyclic Antidepressant Drugs in Plasma, in *Handbook of Studies on Depression* (Burrows G. D., ed.), Excerpta Medica, Amsterdam.

Scoggins B. A. Maguire K. P., Norman T. R., and Burrows G. D. (1980) Measurement of tricyclic antidepresants. 1. A review of methodology. *Clin. Chem.* **25,** 5–17.

Sjoqvist F., Hammer W., Idestrom C. M., Lind M., Tuck D., and Asberg M. (1968) Plasma Level of Monomethylated Tricyclic Antidepressants and Side-Effects in Man, in *Proceedings of the European Society for the Study of Drug Toxicity* Excerpta Medica, Amsterdam.

Tompsett S. L. (1974) The Phenothiazines, in *Progress in Chemical Toxicology* vol. 5 (Stolman A., ed.), Academic, New York.

Turner W. J., Turano P., and Badzinski S. (1976) An Attempt To Establish Quality Control in Determination of Plasma Chlorpromazine by a Multi-Laboratory Collaboration, in *Pharmacokinetics of Psychoactive Drugs* (Gottschalk L. and Merlis S., eds.), Spectrum, New York.

Zuleski F. R., Loh A., and DiCarlo F. J. (1977) Assay of human plasma for nortriptyline by radio-acetylation and thin-layer chromatography. *J. Chromatogr.* **132,** 45–49.

Gas–Liquid Chromatography of Antidepressant, Antipsychotic, and Benzodiazepine Drugs in Plasma and Tissues

Thomas B. Cooper

1. Introduction

This review attempts to cover a broad range of literature on the analysis of psychotropic drugs in tissues and body fluids. A considerable emphasis on the quantification of plasma drug levels will be found because this is an area that has generated most of the published analytical procedures. Concentrations of these drugs in tissues, urine, and spinal fluid samples are easily quantifiable using very minor modifications of these procedures. A rationale for assuming that plasma level concentrations of these drugs and their metabolites reflect the concentration of these compounds in the biophase surrounding the CNS receptor follows.

Since the introduction of neuroleptic drugs in the 1950s and the tricyclic antidepressant drugs in the late 1950s and early 1960s, considerable investigation and refinement of the mode of administration of these drugs have been made (e.g., efficacy of once-a-day medication and of long-acting medication). In terms of the quantity of drug to be administered to each individual patient, however, until the last few years the clinician was forced to treat individual patients with drugs at his or her disposal simply by starting the patient on what is generally accepted as a reasonable dosage of medication and then, if the clinical response is not obtained, increasing the amount of medication given until the patient shows either clinical improvement or troublesome side effects, or the actual quantity of medication given exceeds the sometimes arbitrarily determined upper limit. The latter is particularly pertinent in the case of the neuroleptic drugs in which it is still common practice to "titrate" the patient until extrapyramidal side effects are observed and then to reduce the medication slightly, hoping the patient will improve at this new level. The recognition that tardive dyskinesia is a major unwanted side effect of neuroleptic medication has prompted much work in ways to minimize drug exposure of the psychotic patient.

In some branches of medicine the value of plasma levels of drugs required to optimize patient treatment or to avoid toxicity has been well documented (Cooper, 1978). In psychiatry the levels of drug and metabolite present in human plasma on "therapeutic" doses of antidepressant and antipsychotic medication are of the order of 10^{-7} to 10^{-12} gm/mL. It is only in the past 15 yr that technology has developed to the point at which many of these drugs can be quantitated.

In psychiatry considerable effort has been expended in the examination of plasma drug concentration, clinical efficacy, and side effect relationships. It has been clearly shown that patients treated with identical doses of tricyclic antidepressant or antipsychotic medication show marked interindividual differences in their pseudo-steady-state plasma concentrations of both the parent compound and its various metabolites (Fig. 1). The correlation, therefore, between dose ingested and drug plasma level is usually weak, and it is probable that plasma drug levels will be more strongly correlated with drug concentration at the receptor site. There is considerable evidence indicating that much of this plasma level variation is genetically determined. It has been suggested that such constitutional variation in the rate of drug metabolism may explain therapeutic failures and side effects. In other branches of medicine this variation in dosage requirement can often be controlled by dosage titration to physiologically quantifiable end points such as bleeding time, clotting time, and so on. As early as 1964 Brodie suggested that in psychiatry the administration of antipsychotic medication in a dose just sufficient to elicit the extrapyramidal system side effects mentioned above was an attempt by the skilled physician to correct for the known 10–20-fold variation in the interindividual metabolism of such drugs. Such procedures have not, however, gained wide acceptance. May and VanPutten 1978 stated that "at best, antipsychotic drugs are prescribed in a pragmatic trial-and-error basis; at worst they are prescribed by rote." It is possible, and indeed probable, that many patients deemed "nonresponders" are receiving too little or too much medication.

First, however, one must justify the assumption that the blood drug concentration is correlated to the drug concentration in the biophase surrounding the receptor site. Animal studies have shown a strong correlation between plasma level and brain tissue concentration of these drugs, with the brain concentration being

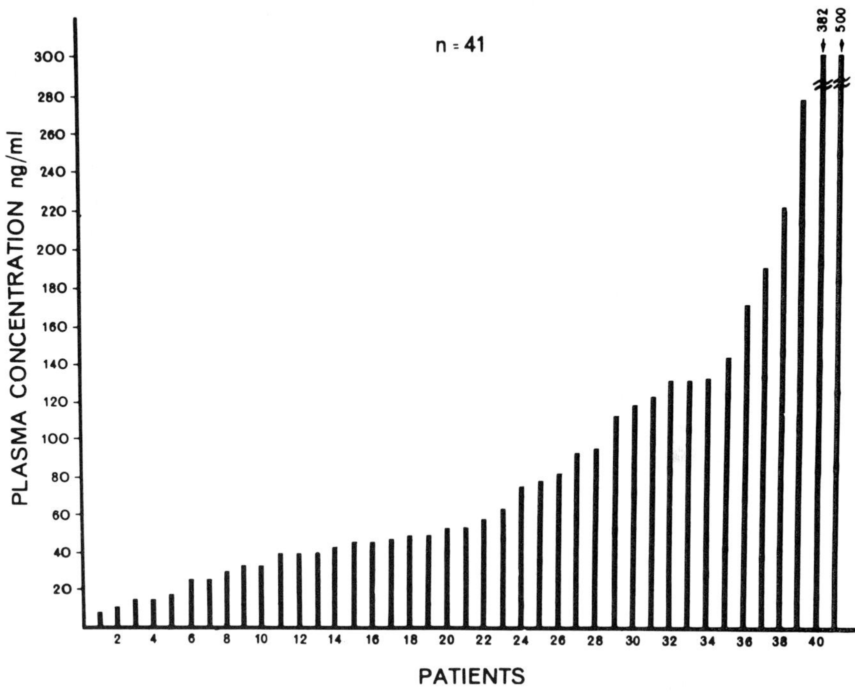

Fig. 1. Variation observed in steady-state plasma level of desmethyl-imipramine when patients received 50 mg tid reprinted with the kind permission from Pfeiffer and Smythies (1976).

5–20 times that of plasma (Nagy, 1977; Dingell et al., 1964; Jori et al., 1971). In man, two studies have demonstrated similar strong correlations between plasma and CSF levels of drugs and major metabolites using gas chromatography/mass spectrometry (GC/MS) with deuterated internal standards (Muscettolla et al., 1978; Alfredsson et al., 1976; Hanin et al., 1985). The inhibition of norepinephrine (noradrenaline, NA) uptake in isolated adrenergic neurons incubated in the plasma of patients treated with various antidepressant drugs has been shown to be highly correlated with the plasma concentration, but not the dose, of the drug (Sjoqvist, 1979; Borga et al., 1969; Tuck et al., 1972, 1973; Hamberger and Tuck, 1973). Tyramine, an indirect-acting sympathomimetic amine, when injected in an intraveneous bolus causes a transient increase in blood pressure. This action can be blocked by the administration of a tricyclic antidepressant, presumably by blockade of the reuptake mechanism of the noradrenergic neurons (Borga et al., 1969; Tuck et al., 1972, 1973; Hamberger and Tuck,

Table 1
Effect of Metabolites of Tricyclic Antidepressants
on the Neuronal Uptake of Norepinephrine[a]

Drug or metabolite	EC_{50}, mol/L	Relative to DMI	Relative to NT
Imipramine (IMI)	250	0.22	—
2-OH-IMI	180	0.31	—
Desipramine (DMI)	55	1.00	—
2-OH-DMI	61	0.90	—
Amitriptyline (AT)	380	—	0.20
Nortriptyline (NT)	75	—	1.00
E-10-OH-AT	670	—	0.11
Z-10-OH-AT	720	—	0.10
E-10-OH-NT	160	—	0.47
Z-10-OH-NT	160	—	0.47
Desmethyl-NT	1000	—	0.08
AT-N-oxide	7300	—	0.01

[a]Imipramine series data from Potter et al. (1979); amitriptyline series data from Bertilsson et al. (1979).

1973). Blockade of this tyramine pressor response in patients chronically treated with antidepressants has been shown to be strongly correlated to the plasma concentration of the drug (Freyschuss et al., 1970; Ghose et al., 1976; Ghose, 1980a; Mulgirigama et al., 1977). In a novel approach, Ghose et al. (1980b) have used this observation to measure the pharmacodynamic "half-life" of amitriptyline and its metabolites; they measured the decrease in pharmacological blockade rather than the decrease in blood concentration. More recent interest has been focused on the role of various metabolites of the tricyclic antidepressant drugs and the neuroleptic drugs because they are now known to be psychoactive, to cross the blood–brain barrier, and to be present in plasma in quantities of the same order of magnitude as the parent compound. The relative potencies of NA blockade of some of the antidepressant drugs and metabolites are shown in Table 1. The hydroxy metabolite of the prototypic antipsychotic chlorpromazine has been shown "in vitro" to have dopamine-blocking activity approximately equal to that of the parent compound, and "in vivo" it releases prolactin. The preceding data give a reasonable justification for the assumption that the blood or plasma concentration of the drug in steady state is strongly correlated with the drug concentration in the biophase surrounding the receptor, and that quantitation of pharmacologically active metabolites is warranted. Based on these data, studies investigating clinical efficacy and side effect relationships have scientific validity.

The methodologies developed for the assay of plasma drug concentrations are equally applicable to quantitation in human or animal tissues with only minor modification. The concentration of these compounds is usually considerably higher in tissues, as indicated by volume of distribution values of 10–40 L/kg.

2. Sample Collection

2.1. Blood Samples

Until quite recently it was generally assumed that plasma and serum drug concentrations from the same patient were not statistically different (Stiller et al., 1980). Hullett et al. (1984) have demonstrated, however, that there are considerable differences between the drug concentration in plasma and serum collected at the same time from individual patients. This manuscript was open to some criticism because of the use of blood-collection systems, all of which at one time or another have been implicated in artifactual lowering of plasma concentrations of the psychotropic drugs (Orsulak et al., 1984). We have repeated this study in our laboratories using collection tubes prepared by us and have demonstrated that the plasma-to-serum difference is real and warrants consideration in terms of interpretation of experiments in which both of these matrices have been analyzed and the data used interchangeably. Prior to 1982, the use of 'Vacutainors (Becton, Dickinson and Co.) caused a spurious lowering of plasma drug concentrations. The plasticizer (tris-butoxyethylphosphate) in the container stoppers displaced the basic drug from the α_1 acid glycoprotein in plasma, thus increasing the "free" fraction of the drug in plasma. This increased free fraction then redistributed between erythrocytes and plasma, with a resultant spurious lowering of the plasma level of the drug (Borga et al., 1977). This artifact has been observed with several basic lipophilic drugs including the antidepressants and antipsychotics (Fremstad and Bergerund, 1976; Brunswick and Mendels, 1977; Veith et al., 1978).

Since this observation, Beckton, Dickinson and Co. have prepared tubes specifically for the collection of such basic drug assay samples, and these apparently do not in any way interfere with quantitation of these compounds (Thoma et al., 1979). It is emphasized that only tubes specifically manufactured for therapeutic drug monitoring are free of TBEP. Orsulak et al. (1984) have reported that serum separator and serum transport tubes cause spurious lowering of tricyclic antidepressant (TCA) serum levels, and Missen and Gwyn (1976) report that an antioxidant (*N*-isopropyl-*N'*-phenyl-*p*-phenylenediamine) gave rise to a spurious peak with a retention time similar to that of methadone and, therefore, several antidepressant drugs. This commercial antioxidant was found to be present in tubes used by the police for the collection of blood alcohol levels. The second possible problem is the use of a heparin lock for multiple blood collections. Recent studies have demonstrated that as little as 50 U of heparin can cause a 30% increase in "free" drug concentration. The mechanism of action is thought to be the induction of lipoprotein lipase activity by heparin with consequent degradation of lipoprotein (Krauss et al., 1974). Basic drugs bind to lipoproteins, and the resultant increased catabolism of these compounds increases levels of unbound drug. This increased level of unbound drug will cause changes in a number of the parameters derived from pharmacokinetic analysis (Wood et al., 1979; Desmond et al., 1980). However, this observation has been challenged by Naranjo et al. (1980), who suggest that the above observation resulted at least in part from the fact that subjects were in a nonfasting state, which may be the major determinant of activation of the lipoprotein lipase. Whether it be heparin activation or food ingestion, clearly the change in free drug concentration must be controlled in pharmacokinetic studies.

The author strongly recommends that investigators carefully evaluate such collection systems at frequent intervals against an all-glass system to guard against such pitfalls. Disposable syringes are often used for the collection of blood samples, but again these must be tested against an all-glass system.

Thoma et al. (1979) have reported that split TCA samples collected and separated within 2h or kept in contact with red cells for 6d at 4°C yielded similar results. Orsulak and Gerson (1980) state that tissues and biofluids can be stored at –20°C for at least 6 mo. In our laboratory we have demonstrated that tertiary and secondary amine TCAs are stable for more than 12 mo if stored in tightly stoppered glass tubes. Thus, although it is obviously preferable to process samples as soon as possible, TCA samples can be collected after normal working hours, placed in the refrigerator (not a freezer), and separated the next morning. The long-term stability of samples kept frozen enables batch processing, thus increasing cost effectiveness and decreasing variance in clinical studies by analysis of all samples of a given patient on the same day in the same system.

In our experience, for long-term storage it is advisable to put samples in glass tubes with screw caps. We have observed a freeze-drying effect when using snap-cap polypropylene tubes. This is caused presumably because the occasional tube's cap is not airtight. For this reason it is also recommended that "frost-free" freezers not be used for long-term storage because this increases the potential for freeze-drying.

2.2. Tissue Samples

Dissected samples should be placed in a preweighed vial, reweighed, and then either immediately frozen or the sample homogenized in 0.1N hydrochloric acid and stored frozen. We have, in our laboratory, analyzed various tissues using this procedure and found it to be highly satisfactory (Cooper et al., 1981; Jason et al., 1981; Friedman et al., 1983; Barkai et al., 1984). It is emphasized that using a biological matrix identical to that of the matrix to be analyzed for standard curve generation is essential in all studies. Thus, drug-free plasma, serum, or tissue is used in each case to construct a standard curve. For example, in our rat experiments we used brain tissue from untreated animals. In general, we have found that the extraction procedures used for plasma are equally applicable to these tissues. In the situation in which a tissue contains a great deal of lipid material, we have found that prior extraction of the acid homogenate, after addition of the internal standard or standards, with heptane or hexane before making the sample basic and extracting the drug helps clean up the extract considerably without loss of drug. Again, in this situation the standard curve must be treated in identical fashion.

In the case in which separate quantitation of free and conjugated drug is needed, then some preliminary work should be performed to ascertain the stability of the conjugate in an acid media. Few psychotropic conjugates can be readily hydrolyzed at low pH when kept cold (frozen). Acid hydrolysis usually requires harsh conditions of pH < 2.0 and temperatures 60–95°C.

However, recent experience with the antidepressant nomifensine, which forms a highly labile conjugate, requires this cautionary note.

3. Storage of Samples

As previously stated, tissue and biofluids containing TCA drugs can be stored frozen for at least 1 yr at –20°C. We routinely use 0.1N hydrochloric acid or sulfuric acid for tissue homogenate preparations and store aliquots of these homogenates at –20°C. Plasma samples do not need to be acidified, but urine samples should be made acidic before freezing. Tissues stored whole prior to homogenization are kept at –80°C.

Antipsychotic drugs may present more difficult problems in that some are photosensitive and may require precautions such as collection in light-shielded containers (May et al., 1978) and protection from fluorescent lighting during extraction and preparation for assay using gas–liquid chromatography (GLC) (Rivera-Calimlim and Siracusa, 1977). Studies on the stability of these drugs has usually been done in aqueous solution and only rarely in the presence of blood plasma. Javaid et al. (1982) suggest the use of an antioxidant in sample collection and storage of antipsychotic drugs samples because of their experience with trifluperazine. Curry (1978) has suggested that storage of chlorpromazine (CPZ) samples may result in increased levels of CPZ brought about by reduction of CPZ N-oxide. Friedel (1976) suggested that storage of CPZ may result in increased levels of CPZ because of reduction of the sulfoxide to CPZ. Traficante et al. (1979) demonstrated that rapid in vitro sulfoxidation of CPZ by a hemecatalyst in hemoglobin is completely inhibited by the presence of a protein-like factor in human plasma. The in vitro sulfoxidation of perphenazine in the presence of erythrocytes when samples were kept in a refrigerator overnight has been reported (Larsen and Naestoft, 1973). This would seem, to some extent, to be contrary to the findings of Traficante, although the observations were made with different compounds. Other investigators have mentioned the protection afforded by plasma both in preventing oxidation and absorption to glassware during extraction procedures (Cooper, 1978; Midha et al., 1979). Very clearly, plasma at physiological pH has a large capacity for scavenging free radicals and therefore reducing such reactions. Finally, Hawes et al. (1984) and Hubbard et al. (1984)

have suggested that the findings of Traficante and perhaps others may in actual fact be artifacts caused by the extraction of CPZ at a highly basic pH, where CPZ can be converted to CPZ sulfoxide and CPZ-N-oxide converted to CPZ. Extraction at physiological pH prevents this conversion.

Davis et al. (1977) state that samples from patients receiving CPZ showed increasing amounts of CPZ with time even when the samples were stored at below 0°C. They postulated that elevation of CPZ content resulted from reversion of a metabolite to the parent compound. However, these samples were extracted at a strongly basic pH and therefore conversion of the CPZ *N*-oxide may have been an artifact of the extraction procedure rather than the storage procedure. Certainly in our own experience using CPZ with dichlorpromazine as an internal standard and extraction at pH 8.0 we have not been able to demonstrate significant change of CPZ concentration with time when stored at –15°C.

Of particular interest in terms of stability of samples, however, is the observation that the new antidepressant drug nomifensine forms a glucuronide conjugate that is highly unstable (Heptner et al., 1978). This has subsequently been shown by Dawling et al. (1979) to be unstable at room temperature; this rapid hydrolysis was markedly inhibited by storage at 4°C, where a 1-wk incubation resulted in nomifensine levels 1.5 times the original concentration. Thus no specific recommendations that will cover storage of all drugs for assay can be made.

Until carefully controlled studies are performed, it is obviously necessary to take great care in the handling of these samples. We routinely recommend that samples be collected and centrifuged and plasma transferred to a tightly stoppered glass tube and stored frozen within 2 h of collection. There is a clear need for detailed examination of the claims cited above, so that a more rational collection and storage proposal can be devised. Until such experiments are available, however, we are forced to insist upon collection, rapid freezing, and assay as soon as feasibly possible. Shipment of samples in the unfrozen state is a routine procedure in many studies and is widely utilized by commercial laboratories. We have studied the stability of nortriptyline plasma samples shipped via regular US mail in the nonfrozen state when compared to split samples shipped in dry ice. With this compound, results with samples shipped unfrozen were comparable to those from split samples shipped in dry ice. Other data on the stability of TCAs at room temperature indicate that shipment in the unfrozen state is satisfactory. Shipment of samples containing phenothiazine-type drugs, however, needs thorough evaluation before utilizing similar transportation procedures.

4. Extraction Procedure

The antidepressant and antipsychotic drugs, i.e., tricyclic drugs, are highly lipophilic bases with pk_a values of 8.5–10 or higher, and therefore at basic pH the free base can be extracted into a relatively nonpolar organic phase and at pH < 3 can be reextracted from this organic phase into an aqueous phase. Plasma and tissue extraction procedures generally involve either one or three steps, depending upon the method of analysis. The single basic extraction step into an organic solvent followed by a concentration procedure can be used in GLC, but the majority of procedures involve a three-step method in which the intact moeity can be further purified from endogenous impurities by utilizing a backwash step into an acid followed by reextraction from the aqueous phase by appropriate pH adjustment and addition of a suitable organic solvent. This solvent residue can then be derivatized (silylation, esterification, alkylation, and so on), if necessary, prior to GLC analysis using highly selective detectors such as electron-capture or thermionic detectors.

Bonded phase extraction procedures are gaining favor in the assay of psychotropic drugs. These procedures are simple, with little time and handling required because the single-tube procedure eliminates centrifugation and transfer of extracts. In its simplest form, all that is required is a small vacuum manifold, capable of accommodating approximately ten bonded phase columns simultaneously. The columns are washed with methanol then with water or buffer, and then the plasma sample together with internal standards applied. The proteins are then washed from the column with a weak buffer solution and the drug eluted from the column using methanol or methanolic acid solution. The eluate can then be taken to dryness for GLC injection or in some cases can be injected without evaporation on a high-pressure liquid chromatography (HPLC) apparatus. Such procedures have been described by Thoma et al. (1979), Narasimhachari (1981), and Kwong et al. (1982). The use of methanol that has been acidified by hydrogen chloride gas is a useful procedure that forms the hydrochloride salt of the drug, helping prevent absorption of the drug on glass surfaces.

A scientific instrument manufacturer (Varian Associates, Sunnyvale, CA 94089) has recently marketed a device that uses this procedure; cassettes containing ten columns are processed to the point at which the drug is retained on the column ready for elution and the cassette is then loaded into an instrument that automatically sequentially injects the contents of each column onto a HPLC. The small final volume (<0.3 mL) used to elute the sample from the column has obvious direct implication for GLC assay work. Preliminary data generated utilizing this system suggest that the major time-consuming extraction procedures may well be unnecessary for a variety of drugs.

5. Chromatographic Procedures

5.1. Column Selection

Glass columns are popular in biological studies because of their inertness. Packed and capillary columns are used, the latter permitting much better resolution of compounds of interest, but at considerably higher cost (Rosseel et al., 1978; Van Brunt, 1983; Gupta et al., 1983). Literally dozens of stationary phases are used in GLC assay, with a wide range of stationary phase loading (0.5–30%). The most common stationary phase used for the assay of the psychotropic drugs is the silicone phase OV-17. This phase has been used with as little as 0.75% (Borga and Garle, 1972) and as high as 10% load (Burch et al., 1979). One manufacturer (Supelco Inc., Bellefonte Park, PA) has prepared a column packing (SP-1767) specifically tested and guaranteed to perform satisfactory separations of tricyclic antidepressant drugs. Other less polar stationary phases commonly used in these separations include OV-1, OV-101, and SE 30 (Weder and Bickel, 1968; Bailey et al., 1977; Javaid et al., 1981, 1982; Dubois et al., 1976; Lachartre et al., 1982). More polar phases have been used OV-225 (O'Brien and Hinsvark, 1976) and Carbowax 20 M (Eicholtz, 1975; Nyberg and Martensson, 1977). Mixtures of stationary phases have been reported by some investigators (Braithwaite and Widdop, 1971; Corona and Bonferoni, 1976; Dorrity et al., 1977), but seem to offer little advantage over a carefully optimized procedure utilizing a single phase.

5.2. Internal Standards

The use of an internal standard in gas chromatography is, in this author's opinion, mandatory, and considerable care must be taken in the selection of each internal standard. Clearly the ideal standard is a compound with essentially identical structure, partition coefficient, vapor pressure, and so on. The use of deuterium or ^{13}C-labeled compounds, especially the latter (the isotope-dilution method), meets all of the above requirements, but requires a mass spectrometer to determine the ratio of the labeled to unlabeled compound. In GLC it is common practice to use an analog or closely related compound as internal standard. It is recommended that when mixtures of secondary and tertiary amine compounds are to be analyzed, then ideally two internal standards reflecting each structure should be used. Thus, if a derivative of the secondary amine is needed, then the secondary amine internal standard will control the derivative step. When a derivative step is not included and the method has demonstrated linearity using either a tertiary or secondary amine internal standard, then this additional refinement, though desirable, is not absolutely necessary. When two internal standards are used, it is strongly recommended that the ratio between these internal standards be routinely calculated and recorded. This simple step guards against unusual matrix effects in which the tertiary amine and secondary amine partition coefficients may change.

5.3. Flame Ionization Detection

The flame ionization detector (FID) is severely limited because of lack of sensitivity in terms of routine monitoring of therapeutic drug levels. Several investigators have published methods using this detector (Braithwaite and Widdop, 1971; Hucker and Stauffer, 1974; Corona and Bonferoni, 1976; Norman et al., 1977; O'Brien and Hinsvark, 1976; Brunderlein et al., 1977; Nyberg and Martensson, 1977; Burch et al., 1979), and when these procedures are used carefully they are adequate for the measurement of steady-state plasma level of patients treated with normal therapeutic doses of psychotropic drugs. They have the disadvantage of requiring relatively large volumes of plasma (3–10 mL/assay), and none are suitable for the analysis of single-dose pharmacokinetic studies unless very large plasma volumes are utilized. Such methods are suitable in animal experimentation in which pharmacological dosages in the micromolar range are utilized (Weder and Bickel, 1968). However, the high sensitivity and selectivity of other GLC detection systems means that the extensive clean-up procedures needed for the FID are not necessary in many applications.

5.4. Electron-Capture Detection

The electron-capture detector (ECD) is a selective detector that can detect as little as a picogram (10^{-12} g) of a compound containing an electrophyllic substituent. The compound elutes from the GLC column into the detector and absorbs electrons, thereby reducing the electron population. The electrons are themselves produced by ionization of the carrier gas with β emission from a ^{3}H, ^{63}NI, or other suitable isotope source. If a constant (dc mode) or intermittant (pulsed-mode) potential is applied to the electrodes, the decrease in the standing current can be used to quantify the eluting compound. The original ECD system had the major drawback of a very narrow linear range; this has been overcome by the introduction of a "constant-current" device. In the former, once the pulse characteristics had been chosen they were kept constant and the reduction of the standing current by the electron-capturing material passing through the detector was measured as the detector's response. Thus the passage of a substance through the detector is signaled by a change of current at constant pulse frequency. Electronic modifications to this system have allowed the standing current to be maintained at a constant level by increasing the pulse frequency as the electrophyllic material passes through the detector. A frequency-to-voltage converter transmits this information to a recorder or data acquisition system. The pulse frequency in such a system is proportional to the amount of substance introduced over more than four decades (Maggs et al., 1971). This electronically simple modification has alleviated one of the most vexing problems associated with the use of this detector. The detector system is extremely sensitive to oxygen and other contaminants, and considerable care must be taken to ensure the purity of the carrier gas entering the system. The writer strongly recommends the use of the highest-purity carrier gas, gas driers, and oxygen traps as a routine procedure. The choice of derivatives to be formed can have considerable impact upon the sensitivity of the assay procedure both from the point of view of electron-capturing ability of the derivative and whether the compound captures electrons in a dissociative manner or a nondissociative manner. Therefore selection of the derivative must be carefully reviewed. An excellent chapter on acylation is to be found in the book by Blau and King (1977) and in a seminar paper by Walle and Ehrsson (1970). The differentiation between dissociative and non-dissociative properties of the molecule is discussed in detail by Wentworth and Chen (1967). To make a gross simplification, a molecule can capture electrons in a dissociative manner to produce a negative ion and a radical or a nondissociative manner to form a negative ion only. The former is favored by a high temperature to facilitate bond breaking, whereas the latter is favored by a low temperature. The molar response of a compound can vary by a factor of 1000 over a 100°C range of detector oven temperature. It is therefore a worthwhile effort to determine optimal conditions for each compound to increase sensitivity via one or other of these mechanisms (Pettit et al., 1967; Poole, 1976). In a review of papers on assay methodology in this field, it is unfortunate to have to report that this optimization is rarely reported and presumably little attention is paid to these phenomena.

5.4.1. Antidepressant Drugs

These compounds must be derivatized to form a compound with electrophyllic properties. The secondary amines can be readily isolated using acylating agents such as trifluoroacetic anhydride (TFAA), pentafluoropropionic anhydride (PFPA), heptafluorobutyric anhydride (HFBA), and a variety of other derivatizing reagents. Representative methods are: desmethylimipramine (Ervik et al., 1970); nortriptyline (Borga and Garle, 1972); maprotiline (Geiger et al., 1975); nomifensine (Vereczkey et al., 1976); and zimelidine (Larsen and Marinelli, 1978). The procedure described by Walle and Ehrsson (1970), in which excess HFBA is removed by washing the organic solution of the derivatized compound with water and ammonia solution, yields extremely clean chromatograms and has proven highly satisfactory in our own laboratory work with electron-capture detection systems.

The tertiary amines have proven more resistant to assay using this detector, in that an electrophilic derivative is more difficult to form. Wallace et al. (1975), Hamilton et al. (1975), Hartvig et al. (1976), and Hartvig and Naslund (1977) have developed methods using a novel approach in which tertiary amines can be analyzed by electron-capture detection after oxidation to the anthroquinone or by conversion to a carbamate by use of trichloroethylchloroformate. The method requires the prior separation of the secondary amine because it also forms the same derivative. The separation procedure described by Hartvig et al. (1976) is, however, rapid and practical and has much to recommend it. One suspects that this procedure would have gained much greater recognition and widespread use if it had not been introduced at approximately the same time that the nitrogen-phosphorous detector became available. The latter, which is to be discussed later in this chapter, essentially revolutionized the assays of psychotropic drugs. The monoamine oxidase-inhibiting drugs phenelzine and tranylcypromine have been quantitated utilizing their respective hydrazine and primary amine functions for derivatization (Cooper et al., 1978; Baselt et al., 1977; Anggard and Hankey, 1969; Calverley et al., 1981; Hampson et al., 1984).

5.4.2. Neuroleptic Drugs

The electron-capture detector was first used by Curry (1968) to analyze CPZ in plasma. This publication essentially started the in-depth investigation of plasma levels/clinical efficacy, and its use has been widespread. The introduction of the internal standard dichlorpromazine by Flint et al. (1971) improved precision and accuracy, and several other modifications have been introduced since the original procedure was described. Thus Whelpton and Curry (1975) and Davis et al. (1977) have used 3% OV-225 as the liquid phase; this enables separation of diazepam, which had previously been shown to interfere with the assay. Hubbard et al. (1984) have demonstrated that CPZ *N*-oxide is converted to CPZ if the extract is carried out above pH 8.0, whereas CPZ can be adequately extracted at physiological pH, thus avoiding this artifact. Losses at low ng/mL concentration have been addressed by Davis et al. (1977), who advocate the use of microgram quantities of promazine as a carrier. Christoph et al. (1972) used trifluperidol as an internal standard and a high-volume sample injection port that increased sensitivity by allowing injection of the total final extract (100 μL) onto the GLC column. The metabolites of CPZ, specifically the sulfoxides, 7-hydroxyCPZ and desmethylCPZ, have also been quantitated using this detector (Mackay et al., 1974; Curry and Evans, 1975), as have haloperidol (Zingales, 1971; Marcucci et al., 1971; Forsman et al., 1974), perphenazine (Larsen and Naestoft, 1973; Hansen and Larsen, 1974), penfluridol (Cooper, S. F. et al., 1975), fluphenazine (Rivera-Calimlim and Siracusa, 1977; Airoldi et al., 1974), loxapine (Cooper and Kelly, 1979; Cooper et al., 1979), clozapine (Simpson and Cooper, 1978), and pipotiazine (Cooper and Lapierre, 1981).

5.4.3. Benzodiazepines

The majority of the GLC assays reported for the determination of specific benzodiazepines in biological fluids use electron-capture detection because of the presence of an electronegative group (usually a halogen or nitro group) in the 7 position of the molecule. A halogen in the 2' position of the 5 phenyl ring and a carbonyl group in position 2 of the 1,4 benzodiazepine ring enhance this response. Detailed reviews on benzodiazepines, with extensive references, are to be found in the chapters by Greenblatt and Shader (1974), deSilva (1981, 1983), and Norman and Burrows (1981). With the exception of chlordiazepoxide and its two major metabolites, benzodiazepines are volatile and thermostable to the extent that separation can be achieved at the high temperatures required for GLC assays of these compounds. The benzodiazepines extract readily into a suitable organic solvent (e.g., benzene, toluene, benzene : methyl chloride), and this solvent extract can be evaporated to dryness and the sample reconstituted in a small volume and injected directly. The selectivity of the ECD eliminates the need for any additional sample clean-up or preparation. There are more than 100 publications describing the assay of benzodiazepines. Representative methods are for diazepam and metabolites (deSilva et al., 1964, 1966; Weinfeld, 1977; Greenblatt, 1978), oxazepam, lorazepam (Knowles et al., 1971; Knowles and Ruelius, 1972; Vessman et al., 1972), temezepam (Fucella et al., 1972; Belvedere et al., 1972), bromazepam, (deSilva et al., 1974a), flurazepam and metabolites (deSilva et al., 1974b), clonazepam, flunitrazepam, and metabolites (deSilva and Bekersky 1974; deSilva et al., 1974c), nitrazepam (Ehrsson and Tilly, 1973), and triazolobenzodiazepines (Greenblatt et al., 1981).

5.5. Nitrogen-Phosphorous Detector (NPD)

The nitrogen and phosphorous alkali flame detector, also known as the thermionic detector, was first described by Karmen and Giuffrida (1964), Karmen (1964), and Giuffreda (1964). The basic composition of this detector involved a flame system with the addition of an alkali source (K, Rb, or Cs) brought close to the flame tip. This source has been variously metal wire, porous ceramic impregnated with alkali salt, a compressed bead of alkali, or a cylinder. The heated alkali salt evaporates and results in increased sensitivity for nitrogen- and phosphorous-containing compounds. The NPD is approximately 30,000 times more sensitive to nitrogen and 60,000 times more sensitive to phosphorous than it is to carbon. Despite this enhanced sensitivity, the detector proved highly erratic and unpredictable in performance with long-term stability problems.

In 1974 Kolb and Bischoff reported a novel NPD with low background current, high sensitivity and selectivity, and excellent long-term stability. Their novel arrangement consisted of a standard flame detector, a rubidium silicate bead on an electrically-heated platinum wire (negative polarity), a collector electrode (positive polarity), and a very low hydrogen flow (3–6 cm^2/min), which means that a flame cannot be formed; rather, a low-temperature plasma is formed, with the hydrogen burning only at the surface of the heated glass bead. The temperature of the bead can be controlled independently of the hydrogen flow. The major advantages of this system over the original detector are its very low background current (picoamps compared to nanoamps) and the minimal loss of the alkali content of the rubidium glass, thus maintaining efficiency for a long period (Hartigan and Purcell, 1974). In our laboratory, NPD beads have been used daily for as long as 8 mo before requiring replacement.

The mechanism by which this detector acquires such high sensitivity and selectivity is not completely understood (Brazhnikov et al., 1970). Kolb and Bischoff (1974) and Kolb et al. (1977) have suggested that the vaporized Rb atoms react with the radicals in the flame, resulting in Rb$^+$ ions and electrons. The electrons impinging upon the collector give the background current, and the Rb$^+$ ions return to the bead surface, preventing Rb loss. Ionization of nitrogen-containing compounds to CN-radicals disturbs this vapor phase equilibrium, which in turn results in increased alkali vaporization to restore the equilibrium. This in turn results in an increase in the electron current. Olah et al. (1979) disagree with this theory and suggest that the highly efficient ionization of the alkali atoms required for the above hypothesis is not met at the operating temperatures of the detector, nor does it readily explain the good sensitivity of nonvolatile alkali salt sources. Their alternative suggestion is that the features of the signal-temperature functions indicate a surface catalytic ionization mechanism in which the alkali atoms do not leave the surface of the bead; rather, the electron transfer takes place on the bead surface. This phenomena could explain equally well the long-term stability of the detector and, in addition, explain the problems of interpretation in the case of nonvolatile alkali salts.

5.5.1. Selection of Internal Standards

5.5.1.1. TRICYCLIC ANTIDEPRESSANTS.   A majority of the published methods utilize other chemically related TCAs as internal standards. Use of TCAs has a distinct advantage of meeting the criteria set forth for internal standards. A problem of co-ingestion of two or more TCAs, e.g., in an overdose situation, can be guarded against by using TCAs not available in the country where the analysis is being carried out. For example, in the USA butriptyline and chlorimipramine and its desmethyl metabolite are excellent choices because they are not marketed in this country (Cooper, 1981). Diethyl analogs of amitriptyline were used by Hucker and Stauffer (1977), and propyl and butyl analogs of nomifensine by Bailey et al. (1977). The structural analog of amitriptyline (N 7084) (Gupta et al., 1976; Jorgensen, 1975), Ciba 34276 (Borga and Garle, 1972), and 2-chloramitriptyline (Hartvig et al., 1976; Hartvig and Naslund, 1977) have also been utilized. These possess the structural and physicochemical properties of the target compound, but are research pharmacueticals not available for human use. Although excellent choices for internal standards, these analogs are difficult to synthesize and are not readily available from pharmaceutical manufacturers.

Other internal standards that belong to different chemical classes have also been used, and these include promazine (Bruderlein et al., 1977; Belvedere et al., 1975; O'Brien and Hinsvark, 1976) and maprotiline (Norman et al., 1977; Bailey and Jatlow, 1976; Connor et al., 1977; Bertrand et al., 1978).

5.5.1.2. BICYCLIC AND TETRACYCLIC ANTIDEPRESSANTS.   Methods for the tetracyclic antidepressants utilize cyproheptadine (Lachatre et al., 1982) and 7-methylmianserin (Vink and Van Hal, 1980). For the bicyclic antidepressants, zimelidine, CPK-191 (Larsen and Marinelli, 1978), viloxazine, and imipramine (Fazio et al., 1984) have been employed.

5.5.2. Techniques of Application

The introduction of the NPD made feasible the measurement of many psychotropics that heretofore had proven difficult or impossible with the then "state-of-the-art" technology (Cooper, 1981). This was particularly true for the tertiary amines (e.g., amitriptyline, imipramine, doxepin, and so on). The subsequent plethora of publications using this technology attests to its simplicity of operation, specificity, reliability, and cost effectiveness. An additional fortunate aspect of the NPD is that methods developed for flame ionization detection systems can usually be transferred to the NPD with little modification. A sensitivity gain of 10–40-fold can thus be achieved, and specificity is markedly enhanced. A major problem is that a halogenated solvent cannot be used because this causes marked deterioration of the rubidium bead and platinum wire. At least one manufacturer has utilized a time-delay switch in which the current of the bead is switched off for a predetermined time interval and then turned on when the solvent front has passed through the detector. Gifford et al. (1975) reported on the use of a six-pot port and a back flush valve that vented the solvent front to the atmosphere before connecting the detector. The use of silylating agents to form less polar derivatives with good chromatographic properties is unfortunately unsuitable for the NPD in that very large solvent fronts are observed that simply swamp the detector, obscure the compound of interest, and rapidly decrease the sensitivity of the bead. Despite this limitation, at least one method, using silylating agents without removal of excess reagent, has been published with conditions such that the peak of interest eluted on the tail portion of the solvent front. The reported accuracy and precision of this assay were excellent (Franklin et al., 1978). The use of imidazole and amide derivatizing agents is not recommended unless excess reagent can be removed before injection. In our hands the procedure first described by Walle and Ehrsson (1970) for use with the ECD has extensive application in the utilization of this detector. We have found that many compounds can be derivatized and then "cleaned up" in this fashion before injection (Cooper, 1981). The majority of published methods utilizing the NPD involve a three-step extraction procedure, but others have reported that only a single-step extraction is necessary (Jorgensen, 1975; Rosseel et al., 1978; Bredesen and Ellingsen, 1981).

Other improvements in the general technique have been reduction of the volume of the final extraction solvent to 50 μL or less (Cooper, T. B. et al., 1975, 1976; Hucker and Stauffer, 1977; Dorrity et al., 1977; Dawling and Braithwaite, 1978; Abernethy et al., 1981; Corona et al., 1983). The latter group also describes an apparatus for obtaining adequate solvent volume from such a small volume for injection using a "bubble cap capillary collector." Using this apparatus, the investigators reported that they were able to obtain 10-µML aliquots from a 25-µML total volume added for extraction. In our own laboratory we used tapered 3-mL tubes for this final extraction. After centrifugation, the aqueous lower layer is removed using a fine-bore capillary pipet; the remaining sample plus a small amount of aqueous phase is then recentrifuged. The organic supernatant can be easily sampled from the tapered tip of the tube and a practiced technician can obtain four 5-µML injections from this phase. Several attempts have been made, with varying success, to produce a single method for all tricyclic antidepressants (Gifford, 1976; Witts and Turner, 1977; Kristinsson, 1981). The last-named author introduced a concept of back-extraction with methanolic formic acid followed by evaporation that, it was claimed, minimized drug loss. Certainly one extraction procedure can be used for all antidepressant drugs in common use, but internal standards must still be varied according to the medication given (Volin, 1981).

5.5.2.1. Tricyclic Antidepressant Assays.   The literature on the assay of antidepressant drugs using the NPD is very extensive and in many cases repetitive. Recent reviews of this literature, which are recommended as source material, include Scoggins et al. (1980), Scoggins and Maguire (1981), Gupta and Molnar (1979), and Van Brunt (1983).

Some of the earliest publications on TCA quantitation using the NPD were those by Jorgensen (1975), Cooper, T. B. et al. (1975, 1976), Bailey and Jatlow (1976), Vasiliades and Bush (1976), Gupta et al. (1976), Gifford (1976), Hucker and Stauffer (1977), Dorrity et al. (1977), Dawling and Braithwaite (1978), and Bertrand et al. (1978). All demonstrated high sensitivity (5–20 ng/mL) and precision (CVs generally less than 10%). Perhaps the simplest way to present these data is to dichotomize the methods into those that require derivatization of the secondary amine functions and those that do not.

Representative methods that do not require a derivative step include those of Cooper, T. B. et al. (1975, 1976), Gifford et al. (1975), Bailey and Jatlow (1976), Vasiliades and Bush (1976), Hucker and Stauffer (1977), Dorrity et al. (1977), Dawling and Braithwaite (1978), Bertrand et al. (1978), Abernathy et al. (1981), and Hals et al. (1982).

Methods that do require a derivative step include those of Jorgensen (1975), Gupta et al. (1976), Connor et al. (1977), Bailey et al. (1977), Dhar and Kutt (1979), and Midha et al. (1980).

Other antidepressant drugs of different structure, e.g., tetracyclic and bicyclic compounds, have also been quantitated using the NPD. These include mianserin (Vink and Hal, 1980; Lachatre et al., 1982; Kristinsson, 1981; Gupta et al., 1977; Volin, 1981) and viloxazine (Fazio et al., 1984).

5.5.2.2. ANTIPSYCHOTIC DRUG ASSAY.   The NPD has not been quite so extensively used in the assay of the antipsychotic drugs, although it is adequately sensitive and selective for those compounds (Cooper, 1983). Chlorpromazine, the prototypic phenothiazine, and some of its metabolites (Bailey and Guba, 1979; Linnoila and Dorrity, 1978), haloperidol (Bianchetti and Morselli, 1978), trifluoperazine (Javaid et al., 1982; Roscoe et al., 1982), butaperazine (Javaid et al., 1979), and fluphenazine (Franklin et al., 1978; Javaid et al., 1981) have been assayed using this detector.

5.5.2.3. BENZODIAZEPINE ASSAY.   As stated, these compounds are easily quantified using electron-capture techniques with high precision and sensitivity in simple organic extracts of plasma without further work-up being required. The NPD can be used for this purpose with many of these compounds, but offers little advantage over the electron-capture procedures. Examples are to be found in publications by Dhar and Kutt (1979), Verebey et al. (1982), and Abernathy et al. (1983).

6. Conclusion

Gas–liquid chromatography has proven a powerful analytical tool in the assay of CNS-active drugs at concentrations of the order 10^{-6} to 10^{-12} g/g of tissue or fluid. The introductions of the constant-current electron-capture detector and the thermionic detector were major advances in this area that made possible routine clinical laboratory assay of such compounds.

The relatively recent introduction of HPLC with its many advantages has at present eclipsed GLC in terms of analytical procedures published during the past 5 yr. Some have even suggested that GLC may become obsolete in these applications. Certainly the cost factor involved in having both instruments available is high, but in this reviewer's opinion it is necessary. A good analytical laboratory must utilize a variety of technologies depending upon the physicochemical properties of the compound of interest and the biological matrix from which it must be extracted before analysis. Gas-liquid chromatography instrumentation is cost effective, highly reliable, and capable of high capacity and complements other analytical procedures in a busy laboratory.

References

Abernathy D. R., Arendt R. M., Lauven P. M., and Greenblatt D. J. (1983) Determination of Ro 15-1788, a benzodiazepine antagonist, in human plasma by gas–liquid chromatography with nitrogen-phosphorus detection. *Pharmacology* **26**, 285–289.

Abernathy D. R., Greenblatt D. J., and Shader R. I. (1981) Tricyclic antidepressant determination in human plasma by gas–liquid chromatography using nitrogen-phosphorous detection: Application to single-dose pharmacokinetic studies. *Pharmacology* **23**, 57–63.

Airoldi L., Marcucci F., Mussini E., and Garattini S. (1974) Distribution of penfluridol in rats and mice. *Eur. J. Pharmacol.* **25**, 291–295.

Alfredsson G., Wode-Helgodt B., and Sedvall G. (1976) A mass fragmentographic method for the determination of chlorpromazine and two of its active metabolites in human plasma and CSF. *Psychopharmacology* **48**, 123–131.

Anggard E. and Hankey A. (1969) Derivatives of sympathomimetic amines for gas chromatography with electron capture detection and mass spectrometry. *Acta Chem. Scand.* **23**, 3110–3119.

Bailey D. N. and Guba J. J. (1979) Gas-chromatographic analysis for chlorpromazine and some of its metabolites in human serum, with use of a nitrogen detector. *Clin. Chem.* **25**, 1211–1215.

Bailey D. N. and Jatlow P. I. (1976) Gas-chromatographic analysis for therapeutic concentrations of amitriptyline and nortriptyline in plasma, with use of a nitrogen detector. *Clin. Chem.* **22**, 777–781.

Bailey E., Fenoughty M., and Richardson L. (1977) Automated high resolution gas chromatographic analysis of psychotropic drugs in biological fluids using open-tubular glass capillary columns. 1. Determination of nomifensine in human plasma. *J. Chromatogr.* **131**, 347–355.

Barkai A. I., Suckow R. F., and Cooper T. B. (1984) Imipramine and its metabolites: Relationship to cerebral catecholamines in rats in vivo. *J. Pharmacol. Exp. Ther.* **230**, 330–335.

Baselt R. C., Stewart C. B., and Shaskan E. (1977) Determination of serum and urine concentrations of tranylcypromine by electron-capture gas chromatography. *J. Anal. Toxicol.* **1**, 215–217.

Belvedere G., Burti L., Frigerio A., and Pantarotto C. (1975) Gas chromatographic–mass fragmentographic determination of 'steady-state' plasma levels of imipramine and desipramine in chronically treated patients. *J. Chromatogr.* **111**, 313–321.

Belvedere G., Tognoni G., Frigerio A., and Morselli P. L. (1972) Specific, rapid and sensitive method for gas-chromatographic determination of methyloxazepam in small samples of blood. *Anal. Lett.* **5**, 531–541.

Bertilsson L., Mellstrom B., and Sjoquist F. (1979) Pronounced inhibition of noradrenaline uptake by 10-hydroxy-metabolites of nortriptyline *Life Sci.* **25**, 1285–1292.

Bertrand M., Dupuis C., Gagnon M., and Dugal R. (1978) Nanogram-range determination of plasma imipramine by gas–liquid chromatography using a selective nitrogen/phosphorus detector. *Clin. Biochem.* **11**, 117–120.

Bianchetti G. and Morselli P. L. (1978) Rapid and sensitive method for determination of haloperidol in human samples using nitrogen-phosphorus selective detection. *J. Chromatogr.* **153**, 203–209.

Blau K. and King G. (1977) *Handbook of Derivatives for Chromatography* Heydon, London.

Borga O. and Garle M. (1972) A gas chromatographic method for the quantitative determination of nortriptyline and some of its metabolites in human plasma and urine. *J. Chromatogr.* **68**, 77–88.

Borga O., Azarnoff D. L., Forshell G. P., and Sjoqvist F. (1969) Plasma protein binding of tricyclic antidepressants in man. *Biochem. Pharmacol.* **18**, 2135–2143.

Borga O., Piafsky K. M., and Nilsen O. G. (1977) Plasma protein binding of basic drugs. 1. Selective displacement of α_1-acid glycoprotein by tris (2-butoxyethyl) phosphate. *Clin. Pharmacol. Ther.* **22**, 539–544.

Braithwaite R. A. and Widdop B. (1971) A specific gas-chromatographic method for the measurement of "steady-state" plasma levels of amitriptyline and nortriptyline in patients. *Clin. Chim. Acta* **35**, 461–472.

Brazhnikov M. V., Gurev M. V., and Sakodynsky K. I. (1970) Thermionic detectors in gas chromatography. *Chromatogr. Rev.* **12**, 1–41.

Bredesen J. E. and Ellingsen O. F. (1981) Rapid isothermal gas–liquid chromatographic determination of tricyclic antidepressants in serum with use of a nitrogen-selective detector. *J. Chromatogr.* **204**, 361–367.

Brodie B. B. (1964) Distribution and Fate of Drugs, in *Absorption and Distribution of Drugs* (Binns, T. B. ed.) William and Wilkins, Baltimore.

Bruderlein H., Kraml M., and Dvornik D. (1977) A gas–liquid chromatographic method for the determination of butriptyline in serum. *Clin. Biochem.* **10**, 3–7.

Brunswick D. J. and Mendels J. (1977) Reduced levels of tricyclic antidepressants in plasma from vacutainers. *Commun. Psychopharmacol.* **1**, 131–134.

Burch J. E., Raddats M. A., and Thompson S. G. (1979) Reliable routine method for the determination of plasma amitriptyline and nortriptyline by gas chromatography. *J. Chromatogr.* **162,** 351–356.

Calverley D. G., Baker G. B., Coutts R. T., and Dewhurst W. G. (1981) A technique for measurement of tranylcypromine in rat brain regions using gas chromatography with electron-capture detection. *Biochem. Pharmacol.* **30,** 861–867.

Christoph G. W., Schmidt D. E., Davis J. M., and Janowsky D. S. (1972) A method for determination of chlorpromazine in human blood serum. *Clin. Chim. Acta* **38,** 265–270.

Connor J. N., Johnson G. F., and Solomon H. M. (1977) Quantitation of amitriptyline and nortriptyline in human serum. *J. Chromatogr.* **143,** 415–421.

Cooper S. F. and Lapierre Y. D. (1981) Gas–liquid chromatographic determination of pipotiazine in plasma of psychiatric patients. *J. Chromatogr.* **222,** 291–296.

Cooper S. F., Albert J.-M., and Dugal R. (1975) Gas-liquid chromatographic determination of penfluridol in plasma. *Int. Pharmacopsychiat.* **10,** 78–88.

Cooper S. F., Dugal R., and Bertrand M. J. (1979) Determination of loxapine in human plasma and urine and identification of three urinary metabolites. *Xenobiotica* **9,** 405–414.

Cooper T. B. (1978) Plasma level monitoring of antipsychotic drugs. *Clin. Pharmacokinet.* **3,** 14–38.

Cooper T. B. (1981) Nitrogen Phosphorous Detector: Experiences in Psychotropic Drug Level Monitoring, in *Clinical Pharmacology in Psychiatry* (Usdin E., ed.) Elsevier, New York.

Cooper T. B. (1983) Plasma Level Monitoring of Antipsychotic drugs, in *Handbook of Clinical Pharmacokinetics* (Gibaldi M. and Prescott L., eds.) ADIS Health Science Press, Australia.

Cooper T. B. and Kelly R. G. (1979) GLC analysis of loxapine, amoxapine and their metabolites in serum and urine. *J. Pharm. Sci.* **68,** 216–219.

Cooper T. B., Allen D., and Simpson G. M. (1975) A sensitive GLC method for the determination of imipramine and desmethylimipramine using a nitrogen detector. *Psychopharmacol. Commun.* **1,** 445–454.

Cooper T. B., Allen D., and Simpson G. M. (1976) A sensitive method for the determination of amitriptyline and nortriptyline in human plasma. *Psychopharmacol. Commun.* **2,** 105–116.

Cooper T. B., Bost R., and Sunshine I. (1981) Postmortem blood and tissue levels of loxapine and its metabolites. *J. Anal. Toxicol.* **5,** 99–100.

Cooper T. B., Robinson D. S., and Nies A. (1978) Phenelzine measurement in human plasma: A sensitive GLC-ECD procedure. *Psychopharmacol. Commun.* **2,** 505–512.

Corona G. L. and Bonferoni B. (1976) Simultaneous determination of therapeutic levels of amitriptyline and nortriptyline in plasma by gas–liquid chromatography. *J. Chromatogr.* **124,** 401–404.

Corona G. L., Bonferoni B., Frattini P., Cucchi M. L., and Santagostino G. (1983) Gas chromatographic routine analysis of five tricyclic antidepressants in plasma. *J. Chromatogr.* **277,** 347–351.

Curry S. H. (1968) Determination of nanogram quantities of chlorpromazine and some of its metabolites in plasma using gas–liquid chromatography with an electron capture detector. *Anal. Chem.* **40,** 1251–1255.

Curry S. H. (1978) Pharmacokinetics and psychotropic drugs. *Psychol. Med.* **8,** 177–180.

Curry S. H. and Evans S. (1975) Assay of 7-hydroxychlorpromazine, and failure to detect more than small quantities, in plasma of responding schizophrenics. *Psychopharmacol. Commun.* **1,** 481–490.

Davis C. M., Meyer C. J., and Fenimore D. C. (1977) Improved gas chromatographic analysis of chlorpromazine in blood serum. *Clin. Chim. Acta* **78,** 71–77.

Dhar A. K. and Kutt H. (1979) Monitoring diazepam and desmethyldiazepam concentrations in plasma by gas–liquid chromatography, with use of a nitrogen-sensitive detector. *Clin. Chem.* **25,** 137–140.

Dawling S. and Braithwaite R. A. (1978) Simplified method for monitoring tricyclic antidepressant therapy using gas–liquid chromatography with nitrogen detection. *J. Chromatogr.* **146,** 449–456.

Dawling S., Braithwaite R., and Crome P. (1979) Nomifensine overdose and plasma drug concentration. *Lancet* **1,** 56.

de Silva J. A. F. (1981) Benzodiazepine Analysis: Electron Capture-GLC and Ancillary Techniques, in *Clinical Pharmacology in Psychiatry.* (E. Usdin, ed.) Elsevier, New York.

de Silva J. A. F. (1983) Analytical strategies for therapeutic monitoring of drugs and metabolites in biological fluids. *J. Chromatogr.* **273,** 19–42.

de Silva J. A. F. and Bekersky I. (1974) Determination of clonazepam and flunitrazepam in blood by electron-capture gas–liquid chromatography. *J. Chromatogr.* **99,** 447–460.

de Silva J. A. F., Bekersky I., Brooks M. A., Weinfeld R. E., Glover W., and Puglisi C. V. (1974) Determination of bromazepam in blood by electron-capture GLC and of its major urinary metabolites by differential pulse polarography. *J. Pharm. Sci.* **63,** 1440–1445.

de Silva J. A. F., Koechlin B. A., and Bader G. (1966) Blood level distribution patterns of diazepam and its major metabolite in man. *J. Pharm. Sci.* **55**, 692–702.

de Silva J. A. F., Puglisi C. V., Brooks M. A., and Hackman M. R. (1974a) Determination of flurazepam (Dalmane) and its major metabolites in blood by electron-capture gas–liquid chromatography and in urine by differential pulse polography. *J. Chromatogr.* **99**, 461–483.

de Silva J. A. F., Puglisi C. V., and Munno N. (1974b) Determination of clonazepam and flunitrazepam in blood and urine by electron capture GLC. *J. Pharm. Sci.* **63**, 520–527.

de Silva J. A. F., Schwartz M. A., Stefanovic V., Kaplan J., and D'Arconte L. (1964) Determination of diazepam (Valium) in blood by gas liquid chromatography. *Anal. Chem.* **36**, 2099–2105.

Desmond P. V., Roberts R. K., Wood A. J. J., Dunn G. D., Wilkinson G. R., and Schenker S. (1980) Effect of heparin administration on plasma binding of benzodiazepines. *Br. J. Clin. Pharmacol.* **9**, 171–175.

Dhar A. K. and Kutt H. (1979) An improved gas liquid chromatographic procedure for the determination of amitriptyline and nortriptyline levels in plasma using nitrogen sensitive detectors. *Ther. Drug Monit.* **1**, 209–216.

Dingell J. V., Sulser F., and Gillette J. R. (1964) Species differences in the metabolism of imipramine and desmethylimipramine (DMI). *J. Pharmacol. Exp. Ther.* **143**, 14–22.

Dorrity F. Jr., Linnoila M., and Habig R. L. (1977) Therapeutic monitoring of tricyclic antidepressants in plasma by gas chromatography. *Clin. Chem.* **23**, 1326–1328.

Dubois J. P., Kung W., Theobald W., and Wirz B. (1976) Measurement of clomipramine, *N*-desmethylclomipramine, imipramine and de-hydroimipramine in biological fluids by selective ion monitoring, and pharmacokinetics of clomipramine. *Clin. Chem.* **22**, 892–897.

Ehrsson J. and Tilly A. (1973) Electron-capture gas chromatography of nitrazepam and in human plasma as methyl derivative. *Anal. Lett.* **6**, 197–210.

Eicholtz P. C. N. (1975) Determination of therapeutic levels of amitriptyline in serum by gas liquid chromatography. *J. Chromatogr.* **111**, 456–458.

Ervik M., Walle T., and Ehrsson J. (1970) Quantitative gas chromatographic determination of nanogram levels of desipramine in serum. *Acta Pharm. Suec.* **7**, 625–634.

Fazio A., Crisafulli P., Primerano G., D'Agostino A. A., Oteri G., and Pisani F. (1984) A sensitive gas chromatographic assay for the determination of serum viloxazine concentration using a nitrogen-phosphorous-selective detector. *Ther. Drug Monit.* **6**, 484–488.

Flint D. R., Ferullo C. R., Levandoski P., and Hwang B. (1971) More sensitive gas-chromatographic measurement of chlorpromazine in plasma. *Clin. Chem.* **17,** (letter to the editor), 830–831.

Forsman A., Martensson E., Nyberg G., and Ohmann R. (1974) A gas chromatographic method for determining haloperidol. *Naunyn Schmiedebergs Arch. Pharmacol.* **286,** 113–124.

Franklin M., Wiles D. H., and Harvey D. J. (1978) Sensitive gas chromatographic determination of fluphenazine in human plasma. *Clin. Chem.* **24,** 41–44.

Fremstad D. and Bergerund K. (1976) Reduced binding of quinidine in plasma from vacutainers. *Clin. Pharmacol. Ther.* **20,** 120.

Freyschuss U., Sjoqvist F., and Tuck D. (1970) Tyramine pressor effects in man before and during treatment with nortriptyline or ECT. Correlation between plasma level and effect of nortriptyline. *Pharmacologia Clinica* **2,** 72–78.

Friedel R. O. (1976) Relationship of pharmacological models of bioavailability of chlorpromazine with drug and metabolite blood levels. *Psychopharmacol. Bull.* **12,** 63–64.

Friedman E., Cooper T. B., and Dallob A. (1983) Effects of chronic antidepressant treatment on serotonin receptor activity in mice. *Eur. J. Pharmacol.* **89,** 69–76.

Fucella L. M., Tosolini G., Moro E., and Tamassia V. (1972) Study of physiological availability of temazepam in man. *Int. J. Clin. Pharmacol.* **6,** 303–309.

Geiger U. P., Rajagopalan T. G., and Riess W. (1975) Quantitative assay of maprotiline in biological fluids by gas–liquid chromatography. *J. Chromatogr.* **114,** 167–173.

Ghose K. (1980a) Sympathomimetic amines and tricyclic antidepressant drugs. *Neuropharmacology* **19,** 1251–1254.

Ghose K. (1980b) Decreased tyramine sensitivity after discontinuation of amitriptyline therapy. *Eur. J. Clin. Pharmacol.* **18,** 151–157.

Ghose K., Gifford L., Turner P., and Leighton M. (1976) Studies of the interaction of desmethylimipramine with tyramine in man after a single oral dose, and its correlation with plasma concentration. *Br. J. Clin. Pharmacol.* **3,** 334–337.

Gifford L. A. (1976) Progress towards a comprehensive plasma assay for the tricyclic and tetracyclic antidepressants. *Postgrad. Med. J.* **52** (3 suppl.), 69–71.

Gifford L. A., Turner P., and Pare C. M. B. (1975) Sensitive method for the routine determination of tricyclic antidepressants in plasma using a specific nitrogen detector. *J. Chromatogr.* **105,** 107–113.

Giuffrida L. (1964) A flame ionization detector highly selective and sensitive to phosphorus. *J. Assoc. Offic. Agric. Chem.* **47,** 293–300.

Greenblatt D. J. (1978) Simultaneous gas-chromatographic analysis for diazepam and its major metabolite, desmethyldiazepam, with use of double internal standardization. *Clin. Chem.* **24,** 1838–1841.

Greenblatt D. J. and Shader R. I. (1974) Benzodiazepines (first of two parts). *N. Eng. J. Med.* **291,** 1011–1015.

Greenblatt D. J., Divoll M., Moschitto L. J., and Shader R. I. (1981) Electron capture gas chromatographic analysis of the triazolobenzo-diazepines alprazolam and triazolam. *J. Chromatogr.* **225,** 202–207.

Gupta R. N. and Molnar G. (1979) Measurement of therapeutic concentrations of tricyclic antidepressants in serum. *Drug Metab. Rev.* **9,** 79–97.

Gupta R. N., Molnar G., and Gupta M. L. (1977) Estimation of maprotiline in serum by gas chromatography, with use of nitrogen specific detector. *Clin. Chem.* **23,** 1849–1852.

Gupta R. N., Molnar G., Hill R. E., and Gupta M. L. (1976) Estimation of amitriptyline and its metabolites in serum and urine by GLC using nitrogen-specific detector. *Clin. Biochem.* **9,** 247–251.

Gupta R. N., Stefanec M., and Eng F. (1983) Determination of tricyclic antidepressant drugs by gas chromatography with the use of a capillary column. *Clin. Biochem.* **16,** 94–97.

Hals P.-A., Lundgren T.-I., and Aarbakke J. (1982) A sensitive gas chromatographic assay for amitriptyline and nortriptyline in plasma. *Ther. Drug Monit.* **4,** 365–369.

Hamberger B. and Tuck J. R. (1973) Effect of tricyclic antidepressants on the uptake of noradrenaline and 5-hydroxytryptamine by rat brain slices incubated in buffer or human plasma. *Eur. J. Clin. Pharmacol.* **5,** 229–235.

Hamilton H. E., Wallace J. E., and Blum K. (1975) Spectrophotometric and gas–liquid chromatographic determination of amitriptyline. *Anal. Chem.* **47,** 1139–1143.

Hampson D. R., Baker G. B., Nazarali A. J., and Coutts R. T. (1984) A rapid and sensitive gas chromatographic method for the analysis of tranylcypromine in brain tissue using acetylation and pentafluorobenzoylation. *J. Biochem. Biophys. Meth.* **9,** 85–87.

Hanin I., Koslow S. H., Kocsis J. H., Bowden C. L., Brunswick D., Frazer A., Carl J., and Robins E. (1985) Cerebrospinal fluid levels of amitriptyline, nortriptyline, imipramine and desmethylimipramine *J. Affec. Disord.* **9,** 69–78.

Hansen C. E. and Larsen N.-E. (1974) Perphenazine concentrations in human whole blood. *Psychopharmacologia* (Berl.) **37,** 31–36.

Hartigan M. J. and Purcell J. E. (1974) Analytical performance of a novel nitrogen-selective detector and its applications with glass open tubular columns. *J. Chromatogr.* **99,** 339–348.

Hartvig P. and Naslund B. (1977) Simultaneous determination of plasma amitriptyline and nortriptyline as trichloroethyl carbamates by electron-capture gas chromatography. *J. Chromatogr.* **133,** 367–371.

Hartvig P., Strandberg S., and Naslund B. (1976) Determination of plasma amitriptyline by electron-capture gas chromatography after oxidation to anthraquinone. *J. Chromatogr.* **118,** 65–74.

Hawes E. M., Cowper M. J., McKay G., Hubbard J. W., and Midha K. K. (1984) Minimal Interconversion of Chlorpromazine and Metabolites in Blood Under Physiological Conditions, in *Proceedings of the Third International Symposium on Biological Oxidation of Nitrogen Inorganic Molecules,* London.

Heptner W., Hornke I., Cavagna F., Fehlhaber H. W., Rupp W., and Neubauer H. P. (1978) Metabolism of nomifensine in man and animal species. *Arzneimittelforsch.* **28,** 58–64.

Hubbard J. N., Cowper M. J., McKay G., Hawes E. M., and Midha K. K. (1984) Spuriously High Plasma or Blood Chlorpromazine Due to Reduction of *N*-Oxide During Analysis, in *Proceedings of the Third International Symposium on Biological Oxidation of Nitrogen in Organic Molecules,* London.

Hucker H. B. and Stauffer S. C. (1974) GLC method for quantiative determination of amitriptyline in human plasma. *J. Pharm. Sci.* **63,** 296–297.

Hucker H. B. and Stauffer S. C. (1977) Rapid, sensitive gas-liquid chromatographic method for determination of amitriptyline and nortriptyline in plasma using a nitrogen sensitive detector. *J. Chromatogr.* **138,** 437–442.

Hullett F. J., Levy A. B., and Tachiki K. H. (1984) Serum imipramine concentrations lower than plasma values. *J. Clin. Psychiatry* **45,** 516–518.

Jason K. M., Cooper T. B., and Friedman E. (1981) Prenatal exposure to imipramine alters early behavioral development and beta adrenergic receptors in rats. *J. Pharmacol. Exp. Ther.* **217,** 461–466.

Javaid J. I., Dekirmenjian H., and Davis J. M. (1982) GLC analysis of trifluoperazine in human plasma. *J. Pharm. Sci.* **71,** 63–66.

Javaid J. I., Dekirmenjian H., Liskevych U., and Davis J. M. (1979) Determination of butaperazine in biological fluids by gas chromatography using nitrogen specific detection system. *J. Chromatogr. Sci.* **17,** 666-670.

Javaid J. I., Dekirmenjian H., Liskevych U., Lin R.-L., and Davis J. M. (1981) Fluphenazine determination in human plasma by a sensitive gas chromatographic method using nitrogen detector. *J. Chromatogr. Sci.* **19,** 439–443.

Jorgensen A. (1975) A gas chromatographic method for the determination of amitriptyline and nortriptyline in human plasma. *Acta Pharmacol. Toxicol.* **36,** 79–90.

Jori A., Bernardi D., Muscettola G., and Garattini S. (1971) Brain levels of imipramine and desipramine after combined treatment with these drugs in rats. *Eur. J. Pharmacol.* **15,** 85–90.

Karmen A. (1964) Specific detection of halogens and phosphorus by flame ionization. *Anal. Chem.* **36,** 1416–1421.

Karmen A. and Giuffrida L. (1964) Enhancement of the response of the hydrogen flame ionization detector to compounds containing halogens and phosphorus. *Nature* **201,** 1204–1205.

Knowles J. A. and Ruelius H. W. (1972) Absorption of 7-chloro-1,3-dihydro-3-hydroxy-5-phenyl-2H-l,4-benzodiazepin-2-one (oxazepam) in humans. Determination of the drug by gas–liquid chromatography with electron capture detection. *Arzneimittelforsch.* **22,** 687–692.

Knowles J. A., Comer W. H., and Ruelius H. W. (1971) Disposition of 7-chloro-5-(0-chlorophenyl)1,3-dihydro-3-hydroxy-2H-1,4-benzodiazepin-2-one (lorazepam) in humans. Determination of the drug by electron capture gas chromatography. *Arzneimittelforsch.* **21,** 1055–1059.

Kolb B. and Bischoff J. (1974) A new design of a thermionic nitrogen and phosphorus detector for GC. *J. Chromatogr. Sci.* **12,** 625–629.

Kolb B., Auer M., and Pospisil P. (1977) Reaction mechanism in an ionization detector with tunable selectivity for carbon, nitrogen and phosphorus. *J. Chromatogr. Sci.* **15,** 53–63.

Krauss R. M., Levy R. I., and Fredrickson D. S. (1974) Selective measurement of two lipase activities in post heparin plasma from normal subjects and patients with hyperlipoproteinemia. *J. Clin. Invest.* **54,** 1107–1124.

Kristinsson J. (1981) A gas chromatographic method for the determination of antidepressant drugs in human serum. *Acta Pharmacol. Toxicol.* **49,** 390–398.

Kwong T. C., Martinez R., and Keller J. M. (1982) Bonded phase extraction of plasma tricyclic antidepressant drugs for gas chromatographic analysis. *Clin. Chim. Acta* **126,** 203–208.

Lachartre G. F., Nicot G. S., Merle L. J., and Valette J. P. (1982) Determination of mianserin in human plasma by gas–liquid chromatography. *Ther. Drug Monit.* **4,** 359–364.

Larsen N.-E. and Marinelli K. (1978) Determination of zimelidine and its demethylated metabolite in human plasma by gas chromatography. *J. Chromatogr.* **156,** 335–339.

Larsen N.-E. and Naestoft J. (1973) Determination of perphenazine and fluphenazine in whole blood by gas chromatography. *Med. Lab. Technol.* **30,** 129–132.

Linnoila M. and Dorrity F. (1978) Measurement of plasma and erythrocyte chlorpromazine and N-monodesmethylchlorpromazine levels by gas chromatography with a nitrogen sensitive detector. *Acta Pharmacol. Toxicol.* **42**, 264–270.

Mackay A. V. P., Healey A. F., and Baker J. (1974) The relationship of plasma chlorpromazine to its 7-hydroxy and sulphoxide metabolites in a large population of chronic schizophrenics. *Br. J. Clin. Pharmacol.* **1**, 425–430.

Maggs R. J., Joynes P. L., Davies A. J., and Lovelock J. E. (1971) The electron capture detector—A new mode of operation. *Anal. Chem.* **43**, 1966–1971.

Marcucci F., Airoldi L., Mussini E., and Garattini S. (1971) A method for the gas chromatographic determination of butyrophenones. *J. Chromatogr.* **59**, 174–177.

May P. R. A. and Van Putten T. (1978) Plasma levels of chlorpromazine in schizophrenia. *Arch. Gen. Psychiatry* **35**, 1081–1087.

May P. R. A., Van Putten T., Jenden D. J., and Cho A. K. (1978) Test dose response in schizophrenia. *Arch. Gen. Psychiat.* **35**, 1091–1097.

Midha K. K., Charette C., Cooper J. K., and McGilveray I. J. (1980) Comparison of a new GLC-AFID method with a GLC–MS selected ion monitoring technique and a radioimmunoassay for the determination of plasma concentrations of imipramine and desipramine. *J. Anal. Toxicol.* **4**, 237–243.

Midha K. K., Loo J. C. K., Hubbard J. W., Rowe M. L., and McGilveray I. J. (1979) Radioimmunoassay for chlorpromazine in plasma. *Clin. Chem.* **25**, 166–168.

Missen A. W. and Gwyn S. A. (1976) Another source of contamination from blood-sample containers. *Clin. Chem.* **24**, 2063–2064.

Mulgirigama L. D., Pare C. M. B., Turner P., Wadsworth J., and Witts D. J. (1977) Tyramine pressor responses and plasma levels during trycyclic antidepressant therapy. *Postgrad. Med. J.* **53** (suppl. 4), 30–34.

Muscettola G., Goodwin F. K., Potter W. Z., Claeys M. M., and Markey S. P. (1978) Imipramine and desipramine in plasma and spinal fluid. *Arch. Gen. Psychiat.* **35**, 621–625.

Nagy A. (1977) Blood and brain concentrations of imipramine, clomipramine and their monomethylated metabolites after oral and intramuscular administration in rats. *J. Pharm. Pharmacol.* **29**, 104–107.

Naranjo C. A., Abel J. G., Sellers E. M., and Giles H. G. (1980) Unaltered diazepam plasma binding using indwelling heparinized cannulae for sampling. *Br. J. Clin. Pharmacol.* **9**, 103–105.

Narasimhachari N. (1981) Evaluation of C_{18} Sep-Pak cartridges for biological sample clean-up for tricyclic antidepressant assays. *J. Chromatogr.* **225**, 189–195.

Norman T. R. and Burrows G. D. (1981) Methods for the Measurement of Psychotropic Drugs. Antipsychotics and Antianxiety Agents, in *Psy-*

chotropic Drugs: Plasma Concentrations and Clinical Response (Burrows G. D. and Norman T. R., eds.) Marcel Dekker, New York.

Norman T. R., Maguire K. P., and Burrows G. D. (1977) Determination of therapeutic levels of butriptyline in plasma by gas–liquid chromatography. *J. Chromatogr.* **134,** 524–528.

Nyberg G. and Martensson E. (1977) Quantitative analysis of tricyclic antidepressants in serum from psychiatric patients. *J. Chromatogr.* **143,** 491–497.

O'Brien J. E. and Hinsvark O. N. (1976) GLC determination of doxepin plasma levels. *J. Pharm. Sci.* **65,** 1068–1069.

Olah K., Szoke A. and Vajta Z. S. (1979) On the mechanism of Kolb's N-P selective detector. *J. Chromatogr. Sci.* **17,** 497–502.

Orsulak P. J. and Gerson B. (1980) Therapeutic monitoring of tricyclic antidepressants: Quality-control considerations. *Ther. Drug Monit.* **2,** 233–242.

Orsulak P. J., Sink M., and Weed J. (1984) Blood collection tubes for tricyclic antidepressant drugs: A reevaluation. *Ther. Drug Monit.* **6,** 444–448.

Pettit B. C., Simmonds P. G., and Zlatkis A. (1967) Influence of detector temperature on sensitivity of electron absorbing derivatives. *J. Chromatogr. Sci.* **7,** 645–650.

Pfeiffer C. C. and Smythies J. R., eds. (1976) *International Review of Neurobiology* vol. 19, Academic, New York.

Poole C. F. (1976) Temperature dependence of electron capture response. *J. Chromatogr.* **118,** 280–281.

Potter W. Z., Calil H. M., Manian A. A., Zavadil A. P., and Goodwin F. K. (1979) Hydroxylated metabolites of tricyclic antidepressants: preclinical assessment of activity. *Biol. Psychiat.* **14,** 601–613.

Rivera-Calimlim L. and Siracusa A. (1977) Plasma assay of fluphenazine. *Commun. Psychopharmacol.* **1,** 233–242.

Roscoe R. M. H., Cooper J. K., Hawes E. M., and Midha K. K. (1982) A GLC-nitrogen phosphorous detector assay for trifluoperazine in plasma. *J. Pharm. Sci.* **71,** 625–627.

Rosseel M. T., Bogaert M. G., and Claeys M. (1978) Quantitative GLC determination of cis and trans-isomers of doxepin and desmethyl-doxepin. *J. Pharm. Sci.* **67,** 802–805.

Scoggins B. A. and Maguire K. P. (1981) Methods for the Measurement of Psychotropic Drugs Antidepressants, in *Psychotropic Drugs: Plasma Concentration and Clinical Response* (Burrows G. D. and Norman T. R., eds.) Marcel Dekker, New York.

Scoggins B. A., Maguire K. P., Norman T. R., and Burrows G. D. (1980) Measurement of tricyclic antidepressants. 1. A review of methodology. *Clin. Chem.* **26,** 5–17.

Simpson G. M. and Cooper T. B. (1978) Clozapine plasma levels and convulsions. *Am. J. Psychiat.* **135,** 99–100.

Sjoqvist F. (1979) Monitoring of antidepressant drug plasma levels: The next ten years. *Prog. Neuropsychopharmacol.* **3,** 201–210.

Stiller R. C., Perel J. M., Lin F. C., and Narayanan S. (1980) A comparative study of serum versus plasma for tricyclic antidepressant drugs. *Clin. Chem.* **26,** 1000.

Thoma J. J., Bondo P. B., and Kozak C. M. (1979) Tricyclic antidepressants in serum by a Clin-Elut™ Column extraction and high pressure liquid chromatographic analysis. *Ther. Drug Monit.* **1,** 335–358.

Traficante L. J., Sakalis G., Siekierski J., Rotrosen J., and Gershon S. (1979) Rapid in vitro sulfoxidation of chlorpromazine by human blood: Inhibition by an endogenous plasma protein factor. *Life Sci.* **24,** 337–346.

Tuck J. R., Hamberger B., and Sjoqvist F. (1972) Uptake of ^{3}H-noradrenaline by adrenergic nerves of rat iris incubated in plasma from patients treated with various psychotropic drugs. *Eur. J. Clin. Pharmacol.* **4,** 212–216.

Tuck J. R., Kahan E., and Siwers B. (1973) Biological and pharmacokinetic evidence for generic equivalence of three imipramine preparations: Comparison with a new imipramine analogue. *Acta Pharmacol. Toxicol.* **32,** 304–313.

Van Brunt N. (1983) Application of new technology for the measurement of tricyclic antidepressants using capillary gas chromatography with a fused silica DB5 column and nitrogen phosphorus detection. *Ther. Drug Monit.* **5,** 11–37.

Vasiliades J. and Bush K. C. (1976) Gas liquid chromatographic determination of therapeutic and toxic levels of amitriptyline in human serum with a nitrogen-sensitive detector. *Anal. Chem.* **48,** 1708–1711.

Veith R. C., Raisys V. A., and Perera C. (1978) The clinical impact of blood collection methods on tricyclic antidepressants as measured by GC/MS-SIM. *Commun. Psychopharmacol.* **2,** 491–494.

Verebey K., Jukofsky D., and Mule S. J. (1982) Confirmation of EMIT benzodiazepine assay with GLC/NPD. *J. Anal. Toxicol.* **6,** 305–308.

Vereczkey L., Bianchetti G., Rovei V., and Frigerio A. (1976) Gas chromatographic method for the determination of nomifensine in human plasma. *J. Chromatogr.* **116,** 451–456.

Vessman J., Freij G., and Stromberg S. (1972) Determination of oxazepam in serum and urine by electron capture gas chromatography. *Acta Pharm. Suec.* **9,** 447–456.

Vink J. and Van Hal H. J. M. (1980) Simplified method for determination of the tetracyclic antidepressant mianserin in human plasma using gas chromatography with nitrogen detection. *J. Chromatogr.* **181,** 25–31.

Volin P. (1981) Therapeutic monitoring of tricyclic antidepressant drugs in plasma or serum by gas chromatography. *Clin. Chem.* **27**, 1785–1787.

Wallace J. E., Hamilton H. E., Goggin L. K., and Blum K. (1975) Determination of amitriptyline at nanogram levels in serum by electron capture gas–liquid chromatography. *Anal. Chem.* **47**, 1516–1519.

Walle T. and Ehrsson H. (1970) Quantitative gas chromatographic determination of picogram quantities of amino and alcoholic compounds by electron capture detection. I. Preparation and properties of the heptafluorobutyryl derivatives. *Acta Pharm. Suec.* **7**, 389–406.

Weder H. J. and Bickel H. M. (1968) Separation and determination of imipramine and its metabolites from biological samples by gas–liquid chromatography. *J. Chromatogr.* **37**, 181–189.

Weinfeld R. E., Posmaster H. N., Khoo K. C., and Puglisi C. V. (1977) Rapid determination of diazepam and nordiazepam in plasma by electron capture gas–liquid chromatography: Application in clinical pharmacokinetic studies. *J. Chromatogr.* **143**, 581–595.

Wentworth W. E. and Chen E. (1967) Electron attachment mechanisms. Analytical importance of temperature and concentration variable. *J. Gas Chromatogr.* **5**, 170–179.

Whelpton R. and Curry S. H. (1975) Gas-chromatographic separation of chlorpromazine, diazepam and N-desmethyldiazepam. *J. Pharm. Pharmacol.* **27**, 970–971.

Witts D. J. and Turner P. (1977) A single comprehensive gas chromatographic assay for tricyclic and tetracyclic antidepressant drugs and their major metabolites in plasma. *Br. J. Clin. Pharmacol.* **4**, 249–252.

Wood M., Shand D. G., and Wood A. J. J. (1979) Altered drug binding due to the use of indwelling heparinized cannulas (heparin lock) for sampling. *Clin. Pharmacol. Ther.* **25**, 103–107.

Zingales I. A. (1971) A gas chromatographic method for the determination of haloperidol in human plasma. *J. Chromatogr.* **54**, 15–24.

Gas Chromatographic–
Mass Spectrometric Analysis
of Antidepressants, Neuroleptics,
and Benzodiazepines

Kenton L. Reed

1. Introduction

Gas chromatography–mass spectrometry (GC–MS) has been used for the identification of drugs since 1968 when Hammar and coworkers (1968) employed this technique for the analysis of chlorpromazine and some of its metabolites isolated from human blood. Since that time this analytical procedure has been successfully adapted for the analysis of many psychotropic drugs. Clinical applications of GC–MS have been reviewed (Hill and Whelan, 1984; Harvey, 1984). This review will be concerned with the application of MS to the identification and quantitation of antidepressants, neuroleptics, benzodiazepines, and their metabolites isolated from a complex biological matrix. When used in combination with the excellent separating power of GC, the combined GC–MS provides an analytical technique that is highly selective and sensitive and has the versatility to quantitatively determine levels of psychotropic drugs and their relevant metabolites. These properties make MS the method of choice for the analysis of these compounds. Disadvantages of this technique, however, include the initial high cost of equipment, the sophisticated level of skill required for the operation and maintenance of the instrumentation, and the difficulty in interpretation of the data. For these reasons MS is not used for the routine analysis of these drugs and metabolites, but is used in those cases in which low levels are expected and high sensitivity is required, positive identification of drugs is mandatory, such as in drug screens or overdoses, and studies elucidating the structure of the metabolites of these compounds. In addition, MS is used as a reference method for validating other analytical procedures, in bioavailability studies, in testing the purity of various preparations, and in verifying the structure of various chemical derivatives. The GC–MS technique has been used in both research and clinical investigations to determine levels of these drugs and metabolites from the low picogram to the microgram range.

This chapter will present a brief description of mass spectroscopic systems used in the analysis of these drugs followed by a review of the applications of this methodology to the analysis of antidepressants, neuroleptics, benzodiazepines, and their naturally occurring metabolites. The major portion of the work reported here involves the combined GC–MS technique with either electron impact (EI) or chemical ionization (CI) used as the source of ionization. Other methodologies, however, including fast atom bombardment (FAB), negative ion MS, and combined high-performance liquid chromatography–mass spectrometry (HPLC–MS) will be discussed.

2. Instrumentation

Comprehensive reviews of mass spectrometer instrumentation, especially as it is used in combination with GC, have already been published (*see* Todd, 1986; Rose, 1984; Rose and Johnstone, 1982; Craig et al., 1979). Essentially, a MS system consists of (1) an interface through which the sample to be analyzed is introduced, (2) an ion source that ionizes the sample, (3) a mass analyzer that separates the ions according to their mass/charge ratio *(m/z)*, (4) an ion detector, and (5) a data system used for interpretation of the mass spectral information. The ion detector, which is normally either a discrete or continuous dynode electron multiplier, and data system will not be discussed in this section.

2.1. Interfaces

2.1.1. Gas Chromatograph

The majority of works published on the analysis of the drugs discussed in this chapter uses a mass spectrometer interfaced with a gas chromatograph. McFadden (1979) has reviewed many of the available GC–MS interfaces. With a packed column the carrier gas flow rate is much higher than the high vacuum pumping system can remove. The interface most commonly used for this type of column is a single-stage jet separator. This device uses the principle that gases of different molecular weight will have different rates of diffusion in an expanding supersonic jet stream to preferentially remove the carrier gas. This separator, which is usually made of glass in order to minimize catalytic decomposition of the sample, will transfer approximately 25% of the drug being analyzed to the mass spectrometer. This yield will depend on experimental conditions and must be established empirically.

For open tubular or capillary columns, the open-split/direct coupling interface can be used for either the rigid glass column or the flexible fused silica column. The capillary column is connected to the MS inlet line via an open coupling tube that is held at atmospheric pressure. This avoids the partial evacuation of the column, which would reduce its efficiency, as well maintaining retention times that are comparable to those found in conventional GC. The coupling tube fits loosely over both the column end and the inlet of the mass spectrometer so that both ends are close together. The inlet line draws the outflow from the capillary column and transfers it directly into the mass spectrometer. Air is excluded from the system by enclosing the coupling tube and flowing helium through the enclosing device.

The simplest means of transferring the effluent from the gas chromatograph to the mass spectrometer is to use a fused silica column inserted directly into the ion source. This latter method of coupling the gas chromatograph to the mass spectrometer has a number of advantages, including no adsorption or decomposition on the surface of the interface, no dead volume, and total transfer of the sample. Cramers et al. (1981) calculated that the maximum decrease in column efficiency caused by placing the outlet of the column in a vacuum is 12.5%. In practice, the effects on column efficiency resulting from operating with the column end in a vacuum seem to be minimal, and this mode of interfacing is rapidly gaining in popularity (LeClerq et al., 1982).

2.1.2. Direct Insertion

Probably the second most commonly used interface is the direct-insertion probe. This means of introducing the sample into the mass spectrometer is normally used to analyze samples that have been separated by some other technique such as thin-layer chromatography (TLC) or HPLC. This technique can also be used to obtain reference spectra of pure compounds and in compound identification. It is not useful for analyzing samples that are mixtures. Advantages of this procedure include high sensitivity, no losses at the interface, and the ability to be used with both EI and CI.

2.1.3. High-Performance Liquid Chromatography

Interfacing HPLC with MS (HPLC–MS) is becoming important in studies of drug conjugation and in studies of drugs that are thermally labile or have low volatility. At present, however, the interfaces used in HPLC–MS must still be considered to be in a developmental stage since the number of analytical investigations is limited. McFadden (1980) and Arpino (1985) have reviewed some of the current developments in HPLC–MS interfacing, whereas Karger and Vouras (1985) have discussed chromatographic aspects of the HPLC–MS interface. Authors in many of the more recent papers using this technique have coupled their mass spectrometer to an HPLC instrument that uses a microbore column. Using these narrower columns increases the compatability with MS by reducing the amount of solvent used to carry out the separation of the compounds.

The moving belt interface mechanically transfers the effluent from an HPLC instrument into the mass spectrometer via an endless moving belt. The solvent is evaporated from the belt, and in the ion source the sample is flash evaporated. This system of sample introduction is useful with either CI or EI. Initially, this technique was limited to volatile solvents that could easily be evaporated. The development of the microbore column, however, decreased the quantity of solvent to be evaporated and permitted the use of aqueous solvents. For many compounds, thermal desorption of the sample from the belt has proven to be a problem.

Direct liquid introduction (DLI) uses a flow restrictor to fraction off a small portion of the column effluent. The restrictor is often a pin hole that produces a jet of small droplets that are evaporated as they pass into the ion source. One of the major drawbacks of this system is the frequent blockage of the pinhole.

The thermospray (TS) interface heats the eluent from the liquid chromatograph rapidly to vaporize it as it flows through a nozzle. The rapid vaporization of the sample as it passes through the nozzle causes the formation of a supersonic jet of superheated fine droplets. The majority of the solvent is removed by diffusion of these droplets to a vacuum system while the ions are deflected into the mass analyzer. This method for sampling not only provides a method of introducing the HPLC eluent into the mass spectrometer, but also provides a new means of ionizing the sample. Ionization takes place as a result of the interaction of the sample with the solvent ions. In cases in which nonpolar solvents are used, an electron beam is used to provide ions in a manner similar to CI.

2.2. Ion Source

The two most commonly used means of ionizing the compounds discussed in this chapter are EI and CI. Electron impact is the dominant means of ionization in the studies of drugs and their metabolites. CI is rapidly catching up as the method of choice, however.

2.2.1. Electron Impact Ionization

In the EI source, the gaseous sample molecules are impacted by a beam of electrons that have been accelerated by a potential of the order of 70 eV. These impacted molecules often lose an electron, thereby forming excited positively charged ions that often undergo further reactions to form a series of fragment ions as well as neutral fragments. This ionization procedure usually leads to a high degree of fragmentation, yielding considerable information about the molecular structure of the compound. Unfortunately, in many cases very few of the high-mass fragments are detected. These high-mass fragments are less subject to interference from contaminants, and for routine monitoring it is advantageous to have an abundance of these. The degree of fragmentation can be reduced by decreasing the accelerating potential of the electrons. This normally reduces the total ion efficiency, however, causing a decrease in overall sensitivity.

2.2.2. Chemical Ionization

Chemical ionization is a procedure to reduce the degree of fragmentation by using a reagent gas that is ionized by a high-energy electron beam (150–500 eV). The ionized reagent gas molecules then react with the sample molecules to produce an ionized sample species. In many cases, CI produces a reaction that either leads to a transfer of a proton to the sample molecule to form an $(M + 1)^+$ ion, or an abstraction of a hydride ion from the sample molecule to form an $(M - 1)^+$ ion. The energy imparted during the ion molecule reaction is normally less than that transferred during EI ionization, and the resulting ion undergoes less fragmentation.

Negative ion MS is an especially useful technique for compounds containing atoms with high electron affinities. There are three important processes for the generation of negative ions: electron capture, dissociative electron capture, and ion pair production. The most important mechanisms are the electron capture and dissociative electron capture, both processes requiring low-energy electrons, i.e., less than 15 eV. These low-energy electrons are produced in CI reactions in which the high-energy electrons, 100 eV or greater, interact with the reagent gas. The low-energy electrons are reaction products from the formation of the positive ions of reagent gas, and in addition, are formed from the nonionizing collisions of the electron beam with reagent gas molecules. The negative ion spectra are dominated by quasi-molecular anions, as well as some structurally significant fragment ions. The structural information obtained from the negative ion spectra is similar to that obtained from positive ion CI spectra, primarily identification of the molecular ion. If the analyte molecules possess good electron-capturing properties, however, the intensity of the negative ion species can be orders of magnitude higher than the intensity of the positive ion species. The high yields of negative ions result in a corresponding increase in sensitivity (Hunt et al., 1976; Hunt and Crow, 1978). Negative ion CI has been used to detect extremely low levels of the benzodiazepines.

2.2.3. Fast-Atom Bombardment Ionization

Fast-atom bombardment (FAB) (Barber et al., 1982) is an extremely useful technique for the characterization of involatile metabolites of drugs and underivatized conjugates. This process uses a beam of high-energy atoms, usually argon or xenon. The beam is focused on the sample, which is dispersed in a low-vapor-pressure liquid such as glycerol. This procedure produces an intense flux of even electron, positive, or negative quasi-molecular ions. This procedure is advantageous because it provides high sensitivity, the analyte need not be volatile, and there is no heating (therefore it is effective for the analysis of heat-labile compounds). Fast atom bombardment would appear to be an ideal ionizing technique for use with an HPLC system.

2.3. Mass Analyzer

The characteristic that distinguishes mass spectrometers is the method used to separate ions of different masses. For the analyses of the compounds to be discussed in this work, there are only three types of mass analyzers extensively used. These are: (1) the single-focusing magnetic sector mass analyzer, (2) the double-focusing mass analyzer, and (3) the quadrupole mass filter. In discussing the mass analyzers, an important concept to understand is that of mass resolution. Resolution is defined by the equation:

$$R = \Delta M \tag{1}$$

where R is the resolution, M is the mean m/z mass of two ions of approximately equal intensity separated by a valley whose height is 10% of the peak height of either ion, and Δ is the difference in mass between the two peaks. The quadrupole instruments are normally operated at unit mass discrimination, i.e., the ion peak width at the base is set to unity across the entire mass range. The single-focusing instruments are capable of a resolution of up to 7000, whereas the double-focusing instruments are capable of a resolution of 100,000 or greater. Most applications discussed here use resolution to the nearest integer value, with only a few examples using high-resolution information. For most analyses of drugs and metabolites, the low resolution is more than adequate to resolve any interference from compounds coeluting with the drug or metabolite. In addition the low-resolution equipment is less costly and easier to operate. The principal use of high-resolution spectrometers is in the identification of metabolites of drugs.

2.3.1. Single-Focusing Magnetic Sector Mass Analyzer

The single-focusing magnetic sector mass analyzer typically consists of a homogeneous magnetic field and two ion slits. The ions formed in the ion source are accelerated through the source slit and into the magnetic field. The field, which is at right angles to the trajectory of the ion beam, causes a force on the beam perpendicular to both the field and the direction of the trajectory of the beam. This force causes the ion beam to assume a radial flight path, with the radius being directly proportional to the ion momentum and inversely proportional to the magnetic field strength. The collector slit is positioned such that only ions traveling in a radial path of fixed radius will pass through the slit. Ions with different momentum can be focused on the collector slit by varying the strength of the magnetic field. The term single-focusing refers to the ability of the magnetic field to correct for directional divergence of the entering ion beam. Directional divergence is a result of a number of variables, including differences in the velocity vector of the molecule before ionization, differences in the position of molecules when they are ionized, and charge-repulsion forces. Thus, directional focusing eliminates the need to have a collimated ion beam entering the magnetic field. The resolution of this type of instrument is determined by the radius of curvature of the instrument and the width of the source and collector slits.

2.3.2. Double-Focusing Mass Analyzer

The ions accelerated through the source slit do not contain identical amounts of kinetic energy. The energy differences result from variations in the kinetic energy that the molecules contained at the time of ionization. This difference in kinetic energy causes a broadening of each of the ion peaks since the magnetic sector separates ions according to momentum. Therefore, ions of constant mass, but with differing velocities, will be focused over a range of the applied magnetic field. Corrections for the variation in the kinetic energy can be made by the addition of an electric field sector. This sector is composed of an entrance slit, an ion lens, and an exit slit. The ion lens is generally a radial electric field generated by two cylindrical electrodes on a common axis. The ion beam arriving at the entrance slit will experience an electrostatic force such that at a given potential, ions with identical m/z values, but heterogeneous velocities, will diverge. The width of the exit lens can be adjusted to accept a wider or narrower energy spread in the beam. The collector slit is positioned such that ions of similar m/z values, but differing slightly in velocity, will be refocused at a single point after emerging from the magnetic field. When the collector slit is located such that both the velocity focusing and the directional convergence occur at the same position, the instrument is said to be double focusing.

2.3.3. Quadrupole Mass Filters

Quadrupole mass filters are devices consisting of four rods spatially arranged to lie at the corners of a square. One pair of rods opposite one another receives a positive direct current (dc) and radio-frequency (RF) voltage, while the other receives a negative dc and an RF voltage equal to the other, but 180° out of phase. Positive ions are extracted from the ion source and focused along the longitudinal axis of the four rods. This arrangement causes an oscillating field within the quadrupole, and, for a given RF/dc ratio, only ions of a given mass-to-charge ratio will pass through the field and reach the detector. Ions with other m/z ratios will have unstable oscillations and strike the rods. The theoretical basis of the quadrupole mass filter has been reviewed by Todd (1981, 1984), Feser and Kogler (1979), and Dawson (1976). This filter device is normally operated at unit resolution. The quadrupole mass analyzers suffer from the phenomenon known as mass discrimination, which is manifested by a decrease in the transmission efficiency of ions of higher masses. For most applications discussed in this chapter, mass discrimination is not of particular importance since few drugs, metabolites, or derivatives of these compounds have ions with masses above m/z 500. At an m/z of 500 or less, mass discrimination effects are minimal.

2.3.4. Ion Trap Analyzer

The ion trap analyzer was introduced at about the same time as the conventional quadrupole analyzer. It did not become a commercially available instrument, however, until 1984. The theory and operation of this type of instrument have been reviewed by Todd (1986) and Stafford and coworkers (1984). In this device, the ion formation and the mass analysis take place in the same physical space with a high background pressure of helium resulting in an enhancement of the mass resolution and instrument sensitivity. The relatively low vacuum requirements make this an ideal analyzer for interfacing with a gas chromatograph. Physically, the instrument is small, the electrode tolerances are less critical than in a quadrupole mass filter, and the vacuum requirements are less than in a conventional mass spectrometer. These factors make this analyzer less expensive to purchase and easier to operate, and result in reduced maintenance compared to other mass spectroscopic instrumentation. At the same time it offers analytical capabilities similar to much larger conventional instruments.

2.4. Selected Ion Monitoring

The two modes for collecting mass spectral data are the scan mode and the selected-ion monitoring (SIM) mode. In the scan mode the mass range to be examined is continuously swept and a mass spectrum is obtained. The data obtained from the chromatographic run will consist of a series of mass spectra. The spectral data can then be used to identify the peaks obtained in the chromatographic separation. For quantitative purposes it is more efficient to monitor a specific ion fragment associated with the compound to be determined. In a given time period, the proportion of time spent monitoring the specific ion (100%) is far greater than it would be if during that same period the entire mass range was monitored. In the latter case the fraction of time spent monitoring the specific ion would be 1/(number of mass units swept). This ion-monitoring procedure results in an increase in the sensitivity of the assay for a given compound. The procedure can be extended such that several selected ions of different *m/z* can be monitored simultaneously. This procedure will be less sensitive than monitoring a single ion, however. When high sensitivity is required for a particular application, it is important to monitor as few ions as is practical to ensure specificity of the compound.

3. Sample Preparation

Most analyses of drugs and their metabolites are carried out on samples isolated from whole blood, plasma, or serum. In addition, there are a number of reports in which the assays are performed on samples isolated from cerebrospinal fluid, urine, saliva, or tissue samples. Artifacts may be introduced into the analysis from the sample storage, contaminants from the drug itself, contaminants introduced during the sample preparation, or contaminants isolated from the biological matrix. Ende and Spiteller (1982) have reviewed contaminants of mass spectrometric analyses from sources other than sample impurities or artifacts produced in the sample workup.

3.1. Sample Collection

It is important to select a device for the collection of biological samples that does not adversely affect the quantitative procedure or give rise to erroneous results caused by an interaction between the biological matrix and the collection device. Plasma obtained from human subjects is the most common source of biological samples, with the blood for these analyses normally being collected in a heparinized glass or polystyrene container. The effects of the plasticizer, tris(2-butoxyethyl)phosphate, found in the stoppers of these tubes is well documented (Borga et al., 1977; Brunswick and Mendels, 1977). The plasticizer causes a decrease in the binding of the tricyclic antidepressants to the alpha-1-glycoprotein. This inhibition of binding causes a redistribution of the antidepressant, resulting in an increase in the amount of drug bound to the erythrocytes. Tubes are now available without this plasticizer. A number of investigators who use the heparinized tubes for the collection of antidepressants, neuroleptics, or benzodiazepines take the precaution of not allowing the blood to come in contact with the stopper. This is advisable, especially when one is analyzing a new drug whose properties have not been well documented. There have also been questions arising from the use of an indwelling needle flushed with heparin for the collection of blood samples. Some drugs have been reported to show an increase in the free fraction after heparin injection (Scoggins et al., 1980; Wood et al., 1979; Routledge et al., 1980; Desmond et al., 1980; Wiegand and Levy, 1979), although other investigators have not observed this phenomena (Silber et al., 1980; Naranjo et al., 1980). Until this question is resolved, it is probably wise to avoid using heparin in the indwelling needle for sample collection. It is important to evaluate the sample collection procedure used for a specific drug or metabolite for any effect the procedure may have on the quantitation or any bias in the biological data that may result from the methodology.

3.2. Sample Storage

Once the sample is collected, it is common practice to store the sample at –20°C or colder if the assay is not to be carried out within 2 wk. Although some tricyclics have been shown to be stable over a period of 1 mo when stored at 4°C, unless this stability has been shown for the specific compound under investigation, it is advisable to freeze the samples as quickly as possible. Some neuroleptics and benzodiazepines have been shown to be photosensitive and susceptible to photodegradation. One should assume the neuroleptic or benzodiazepine to be studied is photosensitive unless the photostability of the compounds has been demonstrated. Sample and standards should be stored in the dark or at least in a light-impenetrable container.

3.3. Extraction

3.3.1. Antidepressants

The antidepressants discussed in this review are mostly lipophilic strong bases. Most extraction procedures take advantage of these properties to separate the drug from the biological matrix. A three-step liquid-liquid extraction procedure is the most commonly used method for the initial purification. This would involve first making the sample alkaline using sodium hydroxide. It is important that the final pH of the mixture be greater than pH 10. To the alkaline sample, a hydrophobic solvent such as hexane (5–10 mL) is added. The hexane will often have a small amount of isoamyl alcohol mixed with it (1–5%). The isoamyl alcohol increases the polarity of the extraction solvent, decreases emulsion formations, and is reported to reduce adsorption on glassware. The sample and hexane are mixed for approximately 15 min using a sample shaker or rocker, and the phase separation is accomplished by centrifugation. To the hexane, which now contains the analyte, is added approximately 1 mL of acid, usually dilute hydrochloric or sulfuric. The mixing again takes place using either a sample shaker or rocker, followed by centrifugation for phase separation. The acidic aqueous phase is made alkaline by the addition of the base, ammonium hydroxide, and a small amount of hydrophobic solvent such as toluene to which has been added isoamyl alcohol (15%). The analyte is extracted into the organic solvent using vigorous mixing with a vortex mixer, and the solvent is separated from the aqueous phase by centrifugation. At this point there is some variation in the method of sample treatment. Approximately half of the studies reported inject the sample extracted at this stage, whereas the others evaporate the solvent using a stream of nitrogen and then add a derivatizing agent to react with the secondary amines (the sample is then blown dry and reconstituted in a solvent suitable for injecting). Antal and coworkers (1980) studied a number of pitfalls in the extraction of desipramine. These investigators found that there was considerable adsorptive loss of desipramine when the sample was blown dry. Neither polypropylene nor polystyrene containers improved the recovery. The recovery was improved by silanizing the glassware, but the silanizing procedure gave rise to a contaminant that interfered with the determination. A method of cleaning the glassware was found that reduced the problem of adsorption, but the cleaning technique was too lengthy and complex to be used on a routine basis. If the sample contains one of the hydroxylated metabolites of the tricyclic antidepressants, the derivatization step is necessary for the analysis.

Solid-phase extraction techniques have recently been proposed for extracting antidepressants from their biological matrix (Narasimhachari et al., 1981; Skrinska et al., 1984) for determination by GC–MS. This procedure would greatly simplify the sample clean-up, as well as decrease the amount of organic solvents used. The technique involves the use of a small column containing a column-packing material such as C_{18} bonded silica. These small columns are first washed with methanol, then an acidic buffer containing approximately 5% methanol, followed by the sample, to which has been added the internal standard. The sample is then collected by washing it off with a small amount of organic solvent such as methanol. The sample can now be injected or blown dry, derivatized, and then reconstituted for injection.

3.3.2. Neuroleptics

The extraction procedure used for the antidepressants (i.e., extraction into solvent, back-extraction into acid, and a final extraction into solvent) can be used for the extraction of neuroleptics. It is more common with these compounds, however, to use a large volume of solvent in the final extraction step, concentrate the sample by taking it to dryness, then reconstitute using a small volume of an appropriate solvent for injection. A large number of different solvents have been used in the final extraction step for neuroleptics; this is different from the situation with the antidepressants, in which nearly half of the studies reviewed used the toluene-isoamyl alcohol combination. Additional problems arise in the extraction of neuroleptics as a result of their tendency to adsorb onto glass surfaces. Some investigators have tried to minimize this problem with the use of silanized glassware or polypropylene tubes. A procedure for the isolation and extraction of a neuroleptic and its metabolites was reported by Gruenke and coworkers (1985). These investigators devised a procedure to extract chlorpromazine and five of its metabolites.

3.3.3. Benzodiazepines

The extraction procedure for the benzodiazepines is similar to that of the neuroleptics, i.e., extracting into solvent, back-extracting into acid, extracting into solvent, taking the sample to dryness, followed by reconstituting in an appropriate solvent for the analysis. There is much more heterogeneity in the choice of solvents in this class of compounds than in the other two. There is a tendency to use less polar extraction solvents in the first step. A combination of benzene and 1,2-dichloroethane is probably most often used. There seems to be no preference of solvents in the final extraction, and ethyl acetate, diethyl ether, and benzene have been used. No problems with adsorption has been reported for this class of compounds, and the use of silanized glassware appears to be the exception rather than the rule in this case.

3.4. Derivatization

Detection of the intact molecular compound is the preferred method of quantitative determination. As a result of phenomena such as column adsorption or poor chromatographic characteristics, however, derivatization is often necessary. Before one decides to derivatize, it is prudent to examine whether a more suitable choice of column or a different method of treating the glassware might obviate the need for this procedure. The addition of this chemical reaction step adds more time to the assay, as well as increasing the amount of sample manipulation. Each added step to the procedure increases the likelihood of contamination and other sources of error. The choice of the derivatizing agent and the conditions used for carrying out this procedure are important. A judicious choice of these parameters can greatly enhance the chromatographic properties as well as improve the specificity and detectability of the drug or metabolite.

3.4.1. Antidepressants

For the antidepressants the necessity of derivatizing the secondary amines is debatable. Investigators are divided on whether this procedure is required. For those who do prefer derivatization, the fluoroacyl analogs are the preferred products because of the simplicity of the derivatizing procedure and the high volatility of the products. The derivatizing agent most commonly used is the trifluoroacetic anhydride (TFAA), with pentafluoroproprionic anhydride (PFPA) and heptafluorobutyric anhydride (HFBA) also being employed. Less commonly used for derivatizing these compounds is acetic anhydride.

3.4.2. Neuroleptics

Unlike the antidepressants, most investigators derivatize the secondary amine metabolites of the neuroleptics. Derivatizations of these secondary amines are usually carried out with TFAA. Alfredsson and coworkers (1976) found that derivatization with TFAA caused the sulfoxide metabolites of both chlorpromazine and norchlorpromazine to be converted to their corresponding parent compounds, chlorpromazine and norchlorpromazine. This conversion could cause erroneously high results for these two compounds unless the two sulfoxides are separated from the chlorpromazine and norchlorpromazine. Gruenke and coworkers (1985) avoided this problem by analyzing chlorpromazine and its sulfoxide metabolite without derivatization. The sample was then derivatized with *N*-methyl-*bis*(trifluoroacetamide) (MBTFA) to form the trifluoroacyl derivative and re-injected to quantitate the norchlorpromazine metabolites. The hydroxy metabolites of chlorpromazine were derivatized with *N,O-bis*(trimethylsilyl)trifluoroacetamide (BSTFA). Fluphenazine was derivatized by the addition of *N,O-bis*(trimethylsilyl)acetamide (BSA) (McKay et al., 1983).

3.4.3. Benzodiazepines

Derivatization of the benzodiazepines is far less common than for the antidepressants or neuroleptics. Most of the parent compounds can be chromatographed without any modification. An investigation of imidazobenzodiazepine-3-carboxamide by Rubio and coworkers (1982b) used PFPA to derivatize the amide functional group of this compound. This reaction does not lead to a fluoroacyl derivative; rather, the reaction product formed is a nitrile. These workers have investigated the reaction and claim it to be quantitative even at very low levels. Cautreels and Jeanniot (1980) studied ethyl loflazepate and some of its biotransformation products. The parent drug and one of its major metabolites were found to have very poor chromatographic characteristics. The secondary amine groups of both compounds were derivatized with 1-iodobutane using the procedure of De Gier and 't Hart (1979). The reaction was carried out by the addition of *N,N*-dimethylacetamide, tetrabutylammonium hydroxide, and 1-iodobutane. This resulted in butylation at the 1-position of the 1,4-diazepine ring. In addition, the proton in the 3-position of the parent compound was sufficiently activated to be substituted by a methoxy group.

4. Quantitation

Quantitative analysis using a mass spectrometer as the detector requires the use of an internal standard. This standard is required to correct for sample loss during the extraction process, variable yields in derivatization procedures, and variations in the mass spectral detection system itself. In addition, the internal standard may act as a carrier to improve the extraction efficiency of analytes present at very low concentrations. Ideally, the chemical characteristics of the standard should be identical to those of the analyte to be determined, and it should be added to the biological sample before the extraction procedure. In general, ions of high mass are preferred for quantitative purposes since they are more selective and less subject to interference. When the intensity of high-mass ions is low, the investigator must make a decision as to whether to sacrifice sensitivity by monitoring the high-mass fragment, or to sacrifice selectivity by using a more abundant, lower-mass fragment. Problems with interference when using ions of either low or high mass may require the use of alternative ionization procedures such as CI, or alternative detection methodology such as negative-ion detection. Quantitative results are normally obtained using SIM and comparing the peak height or area of the analyte with that of the internal standard. The optimization of the quantitative procedure by the choice of parameters such as the internal standard, the ions monitored, and extraction procedure has been reviewed (Vink et al., 1980; Falkner, 1981; De Leenheer and Nelis, 1981).

4.1. Stable-Isotope Internal Standard

Analogs of the parent compounds that are labeled with stable isotopes are the ideal internal standard. Chemical and physical properties such as adsorption, extraction, derivatization, and ionization should be nearly identical to those of the parent compound. The carrier effect of the analog should be maximal, thus facilitating the quantitation of extremely low levels of the analyte. Deuterium-labeled analogs are most often used. The labeled compound is normally produced by a hydrogen exchange reaction carried out in ^2H_2O. The exchange procedure may produce a mixture of products with varying degrees of substitution. It is important that the degree of substitution be such that interference between the monitored ion fragments of the analyte and internal standard is minimal. The spectra of an acetylated derivative of fluphenazine and a deuterated analog of fluphenazine is shown in Fig. 1. Ion fragments at either m/z 280 or 419 are useful for the quantitation of fluphenazine, since they are both specific and relatively abundant. From the spectra of the acetylated deuterium-labeled fluphenazine, one would monitor the m/z 284 or 423 peak as the standard. The ion contribution of the deuterated compound to the m/z 280 peak is minimal, whereas there is no measureable contribution to the m/z 419 peak. The contribution of the fluphenazine derivative to the deuterated internal standard is also seen to be minimal, i.e., the m/z 284 and 423 peaks of the fluphenazine will not significantly contribute to the monitored peaks of the standard. Contributions of the analog to the monitored peaks of the analyte will limit the lower level of detectability of the compound. If the ion fragment contribution of the analyte to the monitored peaks of the internal standard is significant, one can use a nonlinear procedure to quantitatively determine the concentration of the drug. Other

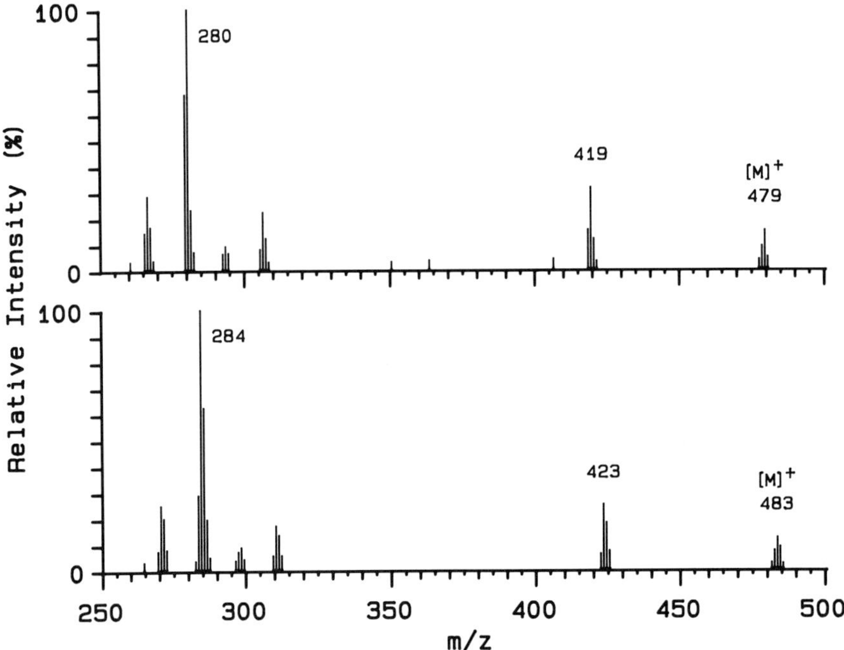

Fig. 1. EI mass spectra (70 eV) of the acetylated derivatives of fluphenazine (top) and tetradeutero-fluphenazine (bottom).

stable isotope analogs containing ^{15}N, ^{18}O, or ^{13}C have been prepared. These latter compounds have also been used in bioavailability studies (Colburn et al., 1980). It is important that the label is in a
nonlabile position; otherwise it may be lost during ionization. If the
label is lost during the ionization process, CI with monitoring of the
pseudo-molecular ion may be required. The use of stable isotope
internal standards for quantitative procedures has been reviewed
(Millard, 1978; Haskins, 1982).

4.2. Chemical-Analog Internal Standards

In cases in which the stable-isotope analog is not readily available or is difficult to synthesize, a chemical analog may be used as
the internal standard. This analog should be chosen such that the
extraction, adsorption, and derivatization properties are similar to
those of the analyte. For many compounds it is suggested that a
tertiary and secondary amine be used as the internal standard for
the tertiary and secondary amine analytes; this appears to be

especially true for the tricyclic antidepressants. If the same ion fragment is to be monitored for both the analyte and the standard, it is then necessary that the two compounds be separated chromatographically. For standards that do not produce similar ion fragments, the chromatographic separation is not necessary.

5. Antidepressants

Antidepressants have become the treatment of choice for most types of depressive illness. They are used in the treatment of both endogenous depression and reactive depression. The largest class of antidepressant drugs is represented by the tricyclic antidepressants. In addition to depression, the tricyclic compounds have been used in the treatment of a number of other disorders, including chronic pain, eating disorders, enuresis, and migraine. Both animal studies and indirect evidence from investigations on humans have suggested that the principal modes of action of tricyclic antidepressants are through interactions with the noradrenergic and serotonergic systems. Many of the undesirable side effects associated with these compounds are related to their anticholinergic actions. Studies have demonstrated wide interindividual variations in steady-state levels of antidepressants, with 5- to 10-fold differences commonly being found in patients receiving similar doses. It is clear from these studies that patients must be individually dosed, and basing the drug treatment on a standard dosage regimen may lead to inadequate treatment for some patients and excessive treatment for others. Numerous studies investigating the relationship between plasma levels of antidepressants and clinical outcome have been carried out. Although the results are not totally unambiguous, there is a strong suggestion that the measurement of plasma levels of antidepressants can provide the clinician with valuable information concerning the pharmacological treatment of the patient. Plasma levels can also be used to determine patient compliance, differentiate between toxicity and effects of the illness itself, and assess the clinical situation when a patient fails to respond to a particular medication.

5.1. Tricyclics

The tricyclic antidepressants are characterized by their three-ring structure. The molecular structure of several tricyclic anti-

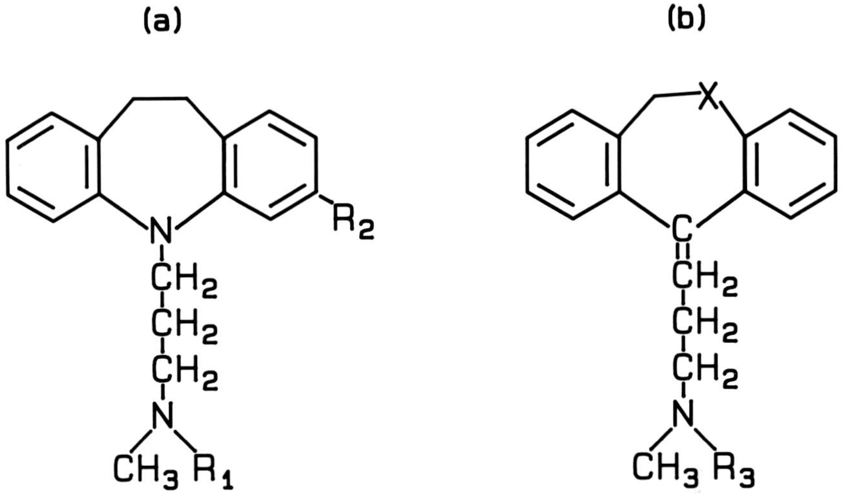

(a)   **(b)**

(a) Imipramine $R_1 = CH_3$, $R_2 = H$
   Desipramine $R_1 = H$, $R_2 = H$
   Clomipramine $R_1 = CH_3$, $R_2 = Cl$
   Desmethylclomipramine $R_1 = H$, $R_2 = Cl$
(b) Amitriptyline $R_3 = CH_3$, $X = CH_2$
   Nortriptyline $R_3 = H$, $X = CH_2$
   Doxepin $R_3 = CH_3$, $X = O$
   Desmethyldoxepin $R_3 = H$, $X = O$
   Dothiepin $R_3 = CH_3$, $X = S$
   Northiaden $R_3 = H$, $X = S$

Fig. 2. Structures of several tricyclic antidepressants.

depressants and metabolites are shown in Fig. 2. This class of compounds has been used in the treatment of depression since Kuhn (1958) reported imipramine to be effective in the treatment of endogenous depression. The demethylated secondary amine derivatives of the tertiary tricyclics are metabolic products that are themselves therapeutically active. Several secondary amine tricyclics, such as desipramine, which is a metabolic product of imipramine, are themselves marketed for the treatment of depression. Studies of imipramine have strongly suggested that a minimum combined imipramine plus desipramine level of 150 ng/mL is necessary for therapeutic response. Nortriptyline has been shown to have a curvilinear response, with the therapeutic window being defined by levels between 50 and 150 ng/mL. The therapeutic ranges for the other tricyclic antidepressants are less clearly defined and await more studies to eliminate the ambiguities.

5.1.1. Imipramine and Desipramine

Imipramine is metabolized to the therapeutically active desipramine and to the hydroxy metabolite, 2-hydroxyimipramine, which is suspected of having therapeutic activity. In addition, desipramine can be metabolized to form the 2-hydroxydesipramine, which also may be therapeutically active. Most studies of imipramine in plasma or serum by GC–MS have measured only the two known therapeutically active components, imipramine and desipramine. The gas chromatographic conditions are very similar, with most separations being carried out on 1–2 m glass columns packed with OV-17 (Frigerio, et al., 1972; Biggs et al., 1975; Claeys et al., 1976; Alkalay et al., 1979; Midha et al., 1980; Narasimhachari et al., 1981). Other column packings used for the separation include OV-2250 (Jenkins and Friedel, 1978), OV-225 (Wilson et al., 1977), OV-1 (Belvedere et al., 1976b), OV-101 (Narasimhachari and Friedel, 1979), and Carbowax 20M (Chinn et al., 1980). The EI spectrum of imipramine has a base peak at m/z 234 or 235 with a relatively strong molecular ion at m/z 280, whereas the underivatized desipramine has a base peak at m/z 234 or 235 and a relatively strong molecular ion peak at m/z 266. The CI spectrum is dominated by the base peak (which represents the protonated molecular ion) at m/z 281 or 267 for imipramine and desipramine, respectively. Several reports include the synthesis of deuterated analogs of these compounds for use as internal standards (Shaw and Markey, 1977; Heck et al., 1978; Claeys et al., 1976; Narasimhachari et al., 1981). Stable isotopes are the ideal internal standards since they will have nearly identical extraction, chromatographic, and mass spectroscopic properties. Narasimhachari and coworkers (1981) have developed a procedure for the measurement of imipramine, desipramine, 2-hydroxyimipramine, and 2-hydroxydesipramine. The hydroxy compounds and secondary amines are derivatized using N-trifluoroacetamide. This procedure used EI ionization with deuterated analogs of these compounds as internal standards. The limits of detection and reproducibility were not reported. Alkalay and coworkers (1979b) reported an EI ionization technique using clomipramine and desmethylclomipramine as the internal standards for imipramine and desipramine, respectively. The secondary amines were derivatized using HFBA. The coefficient of variation was reported to be 19% at a concentration level of 1–3 ng/mL using 0.2–2 mL of plasma in the extraction procedure. They reported a minimum level of detection of 0.2 ng/mL for imipramine and 0.1 ng/mL for desipramine. Chemical ionization MS was developed by Jenkins and Friedel (1978), with methane being used as the carrier gas as well as the reactant gas. They reported the procedure to be quantitative at 1 ng/mL, with a coefficient of variation typically less than 5% when using 2 mL of plasma. This procedure used a two-part analysis for the imipramine and desipramine and would be too time-consuming and complex for routine use. Chinn and coworkers (1980) developed a similar methane CI procedure, but they did not require derivatization and used the deuterated analogs as the internal standards. Extracting 2 mL of plasma, they reported that the inter-assay and intra-assay coefficients of variation were less than 11%. The lower limit of quantitation was reported to be less than 10 ng/mL. Covey and coworkers (1985) used thermospray HPLC–MS to identify imipramine, desipramine, 2-hydroxyimipramine, 2-hydroxydesipramine, and the N-oxide metabolite of imipramine. Chromatography was carried out on a 3-μm C_{18} reverse-phase column with a mobile phase consisting of methanol:acetonitrile:0.1M ammonium acetate (36:36:28), and employed at a flow rate of 1.4 mL/min. The sample was ionized using the thermospray ionization technique. The sensitivity of the assay was not reported. These investigators pointed out the ease of obtaining useful spectra from very labile compounds, such as the N-oxide metabolite of imipramine, with minimal sample preparation and no chemical derivatization. The methods described for imipramine and desipramine all have sufficient sensitivity for routine monitoring of plasma samples and have the additional advantage of being extremely selective, thereby eliminating most problems arising from contamination from other metabolites.

5.1.2. Amitriptyline and Nortriptyline

Amitriptyline is biotransformed into the therapeutically active nortriptyline as well as the biologically active 10-hydroxyamitriptyline. The nortriptyline undergoes biotransformation to form the biologically active 10-hydroxynortriptyline. Both 10-hydroxy compounds have *cis* (Z) and *trans* (E) stereoisomeric forms. As with the imipramine assays previously described, the chromatography is normally carried out on a 1–2-m glass column packed with OV-17.

The EI ionization spectra of amitriptyline has a strong base peak at
m/z 58, with very weak intensities for the molecular ion. Nortriptyline has a base peak at m/z 44 and also has an extremely weak
molecular ion peak. Since neither the m/z 58 nor the m/z 44 peaks is
very selective, these compounds are not good candidates for EI
ionization. Chemical ionization spectra of these compounds are all
dominated by their protonated molecular ions, making amitriptyline, nortriptyline, and their 10-hydroxy metabolites ideal candidates for CI analysis. Figure 3 shows a comparison of amitriptyline
spectra obtained using EI and CI techniques. Assays have been
developed for analysis of all four compounds, i.e., amitriptyline,
nortriptyline, and their corresponding 10-hydroxy metabolites
(Garland et al., 1979; Ishida et al., 1984), for nortriptyline and
10-hydroxynortriptyline (Ziegler et al., 1976; Alvan et al., 1977),
and for amitriptyline and nortriptyline (Biggs et al., 1976; Garland,
1977; Wilson et al., 1977; Jenkins and Friedel, 1978; Chinn et al.,
1980). The EI method of Ishida and coworkers (1984) utilized two
analogs as internal standards, one for amitriptyline and 10-
hydroxyamitriptyline and the other for nortriptyline and 10-
hydroxynortriptyline. The method included dehydration of the
two hydroxy metabolites to form the corresponding 10,11-dienes
and derivatization of the nortriptyline and 10-hydroxynortriptyl-
ine with TFAA. The lower limit of detection was reported as 2
ng/mL for all compounds except the 10-hydroxynortriptyline,
which had a lower limit of detection of 5 ng/mL. These values
were obtained using 1 mL of plasma in the extraction procedure.
The m/z 58 peak was monitored for the detection of both amitriptyline and 10-hydroxyamitriptyline with considerably less than
full baseline separation between the two peaks. This will cause
problems, particularly in situations in which the concentration
of either of these compounds is significantly higher than that of
the other. Garland and coworkers (1979) synthesized the deuterated analogs of the four compounds for use as internal standards and determined the levels by means of CI, with isobutane
being used as both the carrier gas and the CI reagent gas. Their
procedure extracted 1 mL of plasma, and the derivatization
was carried out by the addition of TFAA to the crude extraction
material. Under their conditions, the nortriptyline and 10-hy-
droxynortriptyline formed the trifluoroacyl derivatives, whereas
both the 10-hydroxyamitriptyline and 10-hydroxynortriptyline
were dehydrated to form the corresponding 10,11-dienes. The
assay had a reported lower limit of quantitation of 0.5 ng/mL

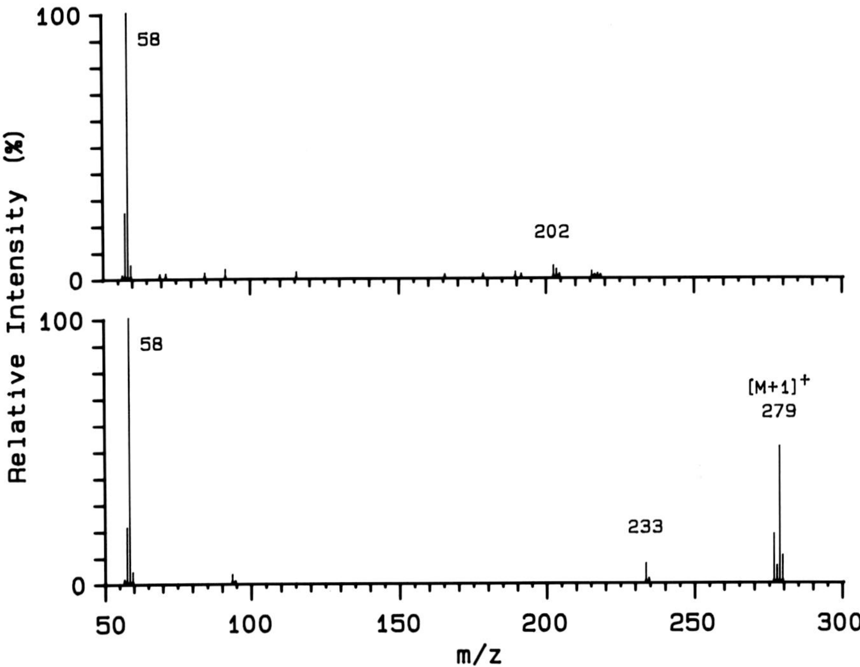

Fig. 3. EI mass spectra (70 eV) of amitriptyline (top) and the CI (methane) mass spectra of amitriptyline (bottom).

for all compounds except 10-hydroxynortriptyline, which had a lower limit of 1 ng/mL. Assay precision and accuracy were given as less than 5% for all compounds. This assay has ample sensitivity for plasma level studies as well as having the high specificity associated with monitoring the protonated molecular ions. Hassan and Pesce (1986) used thermospray HPLC–MS for the detection and identification of amitriptyline in samples of human gastric lavage. In this procedure, 1 mL of sample was mixed with 1 mL of methanol, and the mixture was shaken for 30 s, centrifuged, and analyzed directly by thermospray ionization. No column was used for chromatographic separation. The protonated molecular ion at m/z 278, the base peak, was used to identify amitriptyline.

5.1.3. Clomipramine and Desmethylclomipramine

Clomipramine is a chlorinated analog of imipramine and, like imipramine, is biotransformed by demethylation. Both clomipramine and desmethylclomipramine can undergo further biotransformation to form the 8-hydroxy metabolites. None of the presently available GC–MS assays were developed to analyze all four compounds simultaneously. Gaskell (1980) developed a method that will quantitate only clomipramine. This procedure required only 0.5 mL of plasma and used high-resolution mass spectrometry with EI ionization. Although the assay demonstrates good sensitivity and reproducibility (a coefficient of variation of 2.5% at a concentration of 5 ng/mL), it is not particularly useful since most studies require the analysis of both clomipramine and desmethylclomipramine. Several EI ionization procedures have been reported (Dubois et al., 1976; Alfredsson et al., 1977; Alkalay et al., 1979b). Two of these investigators (Dubois et al., 1976; Alfredsson et al., 1977) synthesized the deuterium analogs of clomipramine and desmethylclomipramine for use as internal standards. Desmethylclomipramine was derivatized by the addition of PFPA, and the molecular ions of the clomipramine and derivatized desmethylclomipramine were monitored to maximize selectivity. Dubois and coworkers (1976) reported that their procedure, which uses 1 mL of whole blood in the extraction, can quantitatively determine levels as low as 0.3 ng/mL. At a concentration of 4.8 ng/mL of both compounds, the coefficient of variation was given as 3.5% for clomipramine and 3.3% for desmethylclomipramine. Lapin and Karobath (1980) also used the deuterated analogs to determine clomipramine and desmethylclomipramine. Their procedure used CI mass spectrometry with methanol as the reactant gas. The secondary amine, desmethylclomipramine, was not derivatized, and the protonated molecular ions of both compounds were monitored for quantitative analysis. They reported the sensitivity to be greater than that required for routine determination of therapeutic concentrations; they did not indicate the reproducibility or the lower limit of quantitation, however.

5.1.4. Doxepin and Desmethyldoxepin

Doxepin is structurally similar to amitriptyline, with an oxygen atom substituted in one of the bridging methylene groups. The major metabolite with known pharmacological activity is desmethyldoxepin. Both doxepin and desmethyldoxepin can exist in either the *cis* (Z) or *trans* (E) isomeric form. Several methods have been reported for the determination of doxepin and desmethyldoxepin using EI ionization (Biggs et al., 1976; Frigerio et al., 1977; Davis et al., 1983). Two procedures derivatized the desmethyldoxepin with TFAA and used glass columns packed with OV-17 for the chromatography (Biggs et al., 1976; Davis et al., 1983). The other method used acetic anhydride for the derivatization and OV-1 for the chromatography (Frigerio et al., 1977). The spectrum of doxepin has a base peak at *m/z* 58 and no other peak of significant intensity, whereas both derivatives of desmethyldoxepin have a base peak at *m/z* 234. All of these procedures reported a sensitivity of approximately 1 ng/mL, but, doxepin was determined by monitoring the relatively nonspecific ion fragment of *m/z* 58. Vasiliades (1979) reported a method for the determination for doxepin only. This procedure will not be of much interest, since most studies would require the determination of all active compounds. Chemical ionization MS has been carried out on doxepin and desmethyldoxepin (Wilson et al., 1977; Jenkins and Friedel, 1978) using methane as both reactant gas and carrier gas. Wilson and coworkers (1977) used a 1.5-m glass column packed with OV-225 and derivatized desmethyldoxepin with TFAA, whereas Jenkins and Friedel (1978) performed the chromatography on a 1.5-m glass column packed with SP-2250DB, with the derivatizing agent being heptafluorobutyramide. Both procedures monitored the protonated molecular ions, i.e., doxepin and the derivatized desmethyldoxepin, which greatly increases the specificity of the assay. Neither of these research groups indicated the lower limits of detectability. All of the procedures were capable of separately quantitating the *cis* and *trans* isomers of doxepin and desmethyldoxepin.

5.1.5. Dothiepin and Northiaden

Dothiepin is a tricyclic antidepressant structurally similar to doxepin. This compound has a sulfur atom substituted for the ring oxygen atom of doxepin. It is metabolized in the body to form the desmethyl metabolite northiaden. Both dothiepin and northiaden are biotransformed to the corresponding dothiepin S-oxide and northiaden S-oxide. All three metabolites of dothiepin are suspected of being biologically active. Maguire and coworkers (1981) reported a technique for measuring dothiepin, northiaden, and dothiepin S-oxide. This procedure used EI ionization and performed the chromatography on a 2-m glass column packed with OV-101. They synthesized a deuterium analog of dothiepin as the internal standard for both dothiepin and dothiepin S-oxide and used protriptyline as the internal standard for northiaden. The spectra of dothiepin and dothiepin S-oxide yielded a strong ion fragment at *m/z* 58 for both compounds. Nothiaden, which was derivatized with TFAA, yielded a spectrum with a base peak at *m/z*

217. In their procedure, they monitored the *m/z* 58 peak for dothiepin and dothiepin S-oxide and the m/z 217 peak for northiaden. Extracting either 1 or 2 mL of plasma, they obtained coefficients of variation of 11 and 10% at a concentration of 5 ng/mL for dothiepin and northiaden, respectively, and a coefficient of variation of 10% at a concentration of 10 ng/mL for dothiepin S-oxide. Their extraction procedure was not capable of extracting out the northiaden S-oxide. As has been previously mentioned in discussions concerning the EI spectra of compounds of this basic structure, i.e. amitriptyline and doxepin, the *m/z* 58 ion fragment is not sufficiently selective to give reliable measurements of subnanogram quantities of these compounds, and caution must be used in interpreting low levels with this procedure. Crampton and coworkers (1980) reported a technique for measuring dothiepin using CI with methane as the carrier gas and reagent gas, and synthesized a deuterium-labeled analog of dothiepin as the internal standard; the chromatography was performed on a 1-m glass column packed with OV-17. The CI spectrum had a base peak at *m/z* 296 corresponding to the protonated molecular ion and a strong peak at *m/z* 58. Monitoring the pseudo-molecular ion, they reported a minimum measurable concentration of 0.5 ng/mL. The utility of this method is limited since they did not include the active metabolites in their procedure.

5.1.6. Metapramine

Bougerolle and coworkers (1985) developed a technique for the analysis of metapramine, a tricyclic antidepressant differing from imipramine in that it contains a methyl group at the azepine nitrogen atom and a methylamino group at the C-10 position. Several metabolites of this compound are also suspected of possessing biological activity and were determined in this procedure. The extraction used 2 mL of plasma, with the isolated compounds being derivatized using TFAA. The separation of the compounds was conducted on a 25-m SE-30 fused silica column. Detection of the metapramine was carried out using EI ionization and SIM. The base peak and at least one other ion were measured for each of the compounds, with the ion fragments ranging from *m/z* 192 to 362. At a concentration of 1 ng/mL, metapramine had a coefficient of variation of 9.4%, whereas at a concentration of 5 ng/mL, the three metabolites measured had coefficients of variation of 8.2, 7.6, and 8.0%.

5.2. Tetracyclics

5.2.1. Maprotiline

Maprotiline is a tetracyclic with pharmacological actions similar to those of the tricyclic antidepressants. It differs from some tricyclics, however, in that it has sedative and antiaggressive properties, is less anticholinergic, and has negligible effects on blood pressure and heart rate. There are reports suggesting that this drug has a more rapid onset of action than the tricyclics. It has been analyzed using both EI (Jindal et al., 1980) and CI (Skrinska et al., 1984; Alkalay et al., 1980, 1979a). Jindal and coworkers (1980) quantitated both maprotiline and desmethylmaprotiline using a synthesized trideuterated analog of maprotiline as the internal standard. After extracting 1 mL of plasma, the sample was derivatized by the addition of TFAA. Chromatography was carried out on a 1.8-m glass column packed with OV-17, and the effluent analyzed using EI ionization. The mass spectrum of the trifluoroacetamide derivative of maprotiline has a molecular ion at m/z 373, a base peak at m/z 345, and another very intense peak at m/z 191, whereas the derivatized desmethylmaprotiline has a molecular ion at m/z 359, a base peak at m/z 331, and an intense peak at m/z 191. The base peaks at m/z 345 for maprotiline, m/z 331 for desmethylmaprotiline, and m/z 348 for the trideuterated analog of maprotiline were monitored. At a concentration of 10 ng/mL of maprotiline and desmethylmaprotiline, these investigators reported a reproductibility of approximately 2% for both compounds with an assay sensitivity of 2 ng/mL for each compound. Alkalay and coworkers (1979a) used a trideuterated maprotiline analog as an internal standard and derivatized the samples with HFBA. The chromatography was carried out using a 1-m glass column packed with Poly S-179. The analysis was performed with CI using methane as both the reactant and carrier gas. Starting with 0.5–2.0 mL of whole blood, these investigators reported an averaged coefficient of variation of 7.4% at concentrations of 5 and 50 ng/mL. They were able to determine maprotiline levels over a range of 0.5–150 ng/mL. Alkalay and coworkers (1980) used this same procedure to determine the bioavailability of oral and tablet forms of the drug. In this research, a hexadeuterated maprotiline analog was used as the internal standard, whereas the aqueous medication contained the trideuterated analog of maprotiline. The aqueous and tablet forms of the medication were taken simultaneously, and the maprotiline and trideuterated analog levels of the drug in whole blood were determined. Skrinska and coworkers (1984) used a disposable C_{18} extraction column to extract the maprotiline from the biological matrix. Using 1 mL of plasma, these investigators reported an absolute recovery of 93.2%. The analysis of maprotiline was carried out using trideuterated nortriptyline as the internal standard, derivatizing with TFAA, and using CI with methane as the reactant gas. The chromatography was performed on a 1.2-m glass column packed with OV-17. The protonated molecular ion of the derivatized maprotiline and nortriptyline were monitored for quantitation. At 200 ng/mL, the coefficient of variation was reported to be 5.6%.

5.2.2. Mianserin

Mianserin is an antidepressant with clinical indications similar to those of the tricyclics. This compound is less anticholinergic and is considered to have fewer side effects. Mianserin has been analyzed using EI ionization (De Ridder et al., 1977; Jindal et al., 1982; Maguire et al., 1982). All three studies used deuterated internal standards and monitored the molecular ion at *m/z* 264, which is a very strong ion fragment. The procedure used by De Ridder and coworkers (1977) used a purification by HPLC before the sample was injected into the GC–MS, making this technique quite slow and cumbersome to use on a routine basis. Jindal and coworkers (1982) used 1 mL of plasma for the extraction, separated the compounds on a 1.8-m glass column filled with OV-1, and obtained a coefficient of variation of 8% at a concentration of 5 ng/mL. Maguire and coworkers (1982) used a 2-m glass column packed with OV-101 and reported a sensitivity of the assay of 1 ng/mL with a coefficient of variation of 10% at a concentration of 5 ng/mL.

5.3. Other Antidepressants

5.3.1. Trazodone

Trazodone is a triazolopyridine derivative used as an antidepressant. Patients receiving the drug for therapeutic purposes have plasma concentrations that are considerably higher than plasma levels of those receiving tricyclic antidepressants. Gammans (1985) developed an assay for the determination of trazodone and its tetra-deuterated analog. They were attempting to differentiate between the rates of release of a liquid and capsule formulation of trazodone. A solution of the stable isotope-labeled analog of trazodone and the unlabeled drug in the solid form was simultaneously administered to the patient. The procedure used EI ionization with SIM. The molecular ions at m/z 371 and 375 for trazodone and the deuterated analog were quite weak, and the base peaks were at m/z 205 and 209, respectively. The base peaks could not be used, however, since the undeuterated trazodone also had a relatively strong ion at m/z 209. The compounds were analyzed using the peaks at m/z 231 and 235 for the trazodone and deuterated trazodone, respectively. The extraction used 1 mL of plasma, etoperidone was used as the internal standard, and the chromatographic separation was carried out on a 4-m DB-1 fused-silica capillary column. The inter-assay coefficient of variation for this procedure was 5% at a concentration of 60 ng/mL. No attempts were made to determine the lower limit of quantitation since 60 ng/mL was more than sufficient sensitivity for the bioavailability study. Belvedere and coworkers (1975a) reported a methodology for the determination of trazodone in rat plasma. Their procedure used an external standard, and no indications of the reproducibility were given.

5.3.2. Nomifensine

Nomifensine is a tetrahydroisoquinoline derivative whose pharmacological profile is similar to that of imipramine and amitriptyline. Structurally this compound is dissimilar to any of the tricyclics. It appears to have antidepressant actions similar to imipramine, but has fewer side effects and a more rapid onset of therapeutic action. Bagchi and coworkers (1985) developed a procedure for the quantitative determination of this compound in rat plasma or rat brain extract. One milliliter of the biological material was extracted with benzene and derivatized using TFAA. The internal standard was a deuterated nomifensine analog synthesized by these investigators. The chromatography was carried out using a 1.8-m glass column packed with OV-17, whereas the MS was performed using EI ionization with SIM. The spectrum of the TFAA derivative of this compound yielded a molecular ion at m/z 334 and a base peak at m/z 222. The base peak was used for quantitation to increase the sensitivity of the assay. At a concentration of nomifensine of 10 ng/mL, the coefficient of variation was 3.6%, whereas the limit of determination was reported as 1–2 ng/mL.

6. Neuroleptics

Compounds that have the ability to cause psychomotor slowing and affective indifference to environmental stimuli without producing sleep are defined as neuroleptics (Delay and Deniker, 1952). The primary use of neuroleptics is in the treatment of schizophrenia, organic psychoses, and the manic phase of manic-depressive illness (Goodman and Gilman, 1975). Clinically these drugs are also used in the treatment of many other disorders, and it is estimated that by the end of 1970 over 250 million people had been treated with these compounds (Crane, 1973). The three most important classes of drugs with neuroleptic properties are the phenothiazines, thioxanthenes, and butyrophenones. The large variation in the individual dosages of neuroleptics required for the treatment of schizophrenia has led many researchers to investigate the relationship between clinical efficacy and plasma levels of these drugs. At present, the monitoring of plasma levels of neuroleptic drugs and their active metabolites has not been developed to an extent where it is a useful clinical procedure. The very low levels of these compounds found in plasma (0.05–150 ng/mL), as well as the numerous metabolites of these drugs found in vivo, have caused analytical difficulties in attempts to determine a relationship between plama levels and clinical outcome. Structures of some of the neuroleptics are shown in Fig. 4.

6.1. Phenothiazines

6.1.1. Chlorpromazine

Chlorpromazine was one of the first neuroleptics used in the treatment of schizophrenia and is still widely used. Numerous studies of this compound have been carried out to determine its metabolism, pharmacokinetics, and relationship of plasma concentration to clinical efficacy. Determinations of this drug and its metabolites by MS have been carried out using EI excitation (Essien et al., 1975; Alfredsson and Sedvall, 1976; Alfredsson et al., 1976; May et al., 1978; Taulov et al., 1980a,b; McKay et al., 1982; Maurer and Pfleger, 1984). Electron impact ionization is well suited for the determination of chlorpromazine since the mass spectrum yields a very strong molecular ion at m/z 318. Sensitivities in the subnanogram range can be obtained when this molecular ion is monitored using SIM techniques. Other major fragments are found at m/z 86 and 58 (base peak). Alfredsson and coworkers (1976) determined

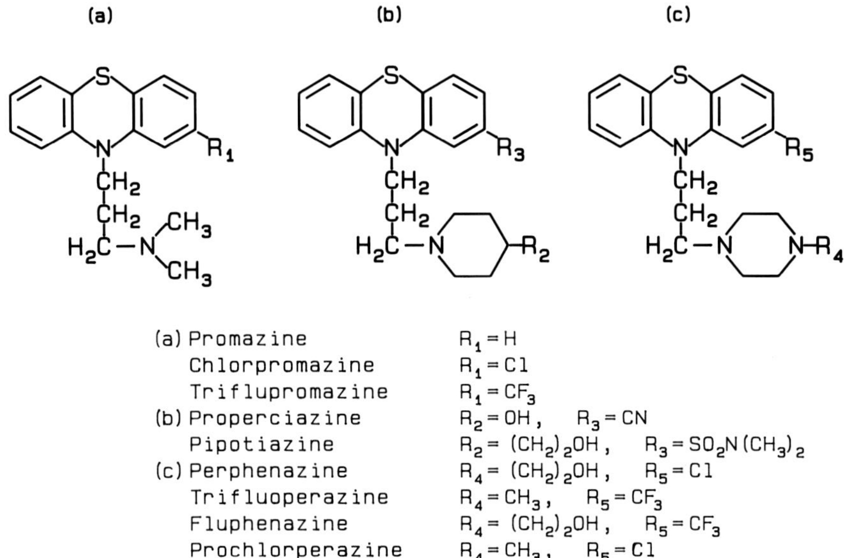

(a) Promazine $R_1 = H$
   Chlorpromazine $R_1 = Cl$
   Triflupromazine $R_1 = CF_3$
(b) Properciazine $R_2 = OH$, $R_3 = CN$
   Pipotiazine $R_2 = (CH_2)_2OH$, $R_3 = SO_2N(CH_3)_2$
(c) Perphenazine $R_4 = (CH_2)_2OH$, $R_5 = Cl$
   Trifluoperazine $R_4 = CH_3$, $R_5 = CF_3$
   Fluphenazine $R_4 = (CH_2)_2OH$, $R_5 = CF_3$
   Prochlorperazine $R_4 = CH_3$, $R_5 = Cl$

Fig. 4. Structures of several phenothiazine neuroleptics.

levels of chlorpromazine, monodesmethylchlorpromazine, and 7-hydroxychlorpromazine in plasma, cerebrospinal fluid, and tissue using the deuterated analogs of these compounds as the internal standards. Chromatography was carried out using a 1.5-m glass column packed with 3% SP-2100. Monodesmethylchlorpromazine was derivatized using TFAA to form the trifluoroacetyl derivative, whereas 7-hydroxychlorpromazine was derivatized with BSA to form the trimethylsilyl derivative. In their study they found that the sulfoxide metabolite of chlorpromazine was converted to chlorpromazine by the derivatizing reagents. There was no interference, however, since the extraction procedure separated chlorpromazine and monodesmethylchlorpromazine from their sulfoxide metabolites. The sulfoxide metabolites and 7-hydroxychlorpromazine were isolated together. Possible interference from the metabolites was eliminated by using BSA as the derivatizing reagent since it did not cause any of the sulfoxides to be reduced. Interference from other known metabolites of chlorpromazine was not detected. Using these procedures, the investigators found the lower limit of detection of chlorpromazine, monodesmethylchlorpromazine, and 7-hydroxychlorpromazine to be approximately 1 ng/mL in plasma and the limit of detection of chlorpromazine in cerebrospinal fluid to be approximately 200 pg/mL. In both cases these lower limits were determined using 3 mL of plasma or cerebrospinal fluid. Plasma levels of 10 patients who had been receiving chlorpromazine for at least 2 wk ranged from 1.7 to 122 ng/mL for chlorpromazine, undetected to 12.2 ng/mL for monodesmethylchlorpromazine; and 0.6–19.5 ng/mL for 7-hydroxychlorpromazine. The levels of chlorpromazine in cerebrospinal fluid were found to be approximately 3% of the levels found in plasma. A method similar to that of Alfredsson and Sedvall (1976) was used to determine the relationship between plasma and saliva levels of chlorpromazine (May et al., 1978). McKay and coworkers (1982) determined levels of chlorpromazine in volunteers who were given a single 50-mg dose of chlorpromazine. Prochlorperazine was used as the internal standard, and the chromatography was carried out on a 1.22-m glass column packed with 3% OV-1. As in the previous methods, EI excitation was used and the molecular ion of chlorpromazine, m/z 318, was monitored. Using 2 mL of plasma, they were able to determine levels of chlorpromazine as low as 0.25 ng/mL with a coefficient of variation of 5.1%. Temperature effects on the mass spectra of N-oxide and N-oxide sulfoxide metabolites of chlorpromazine, promazine, and methotrimeprazine were studied using a direct insertion probe (Essian et al., 1975). These spectra were compared with those obtained when the corresponding compounds were introduced by the combined GC–MS technique. The study showed that, in general, the N-oxide and N-oxide sulfoxides are thermally labile. For instance, the promazine-N-oxide and promazine-N-oxide sulfoxide broke down under their gas chromatographic conditions to phenothiazine. Gruenke and coworkers (1985) reported a procedure for the determination of chlorpromazine and its major metabolites from biological fluids. Deuterium-labeled analogs of chlorpromazine and each of the metabolites were synthesized for use as internal standards. Approximately 10–100 times more labeled compound than the expected amount of unlabeled compound was added as the internal standard. This high concentration of the labeled material acted as a carrier and improved the efficiency of the extraction. The extraction procedure was devised to enable the investigators to isolate distinct groups of the metabolites in different fractions. Chromatographic separation was performed on a 2-m glass column packed with OV-17. Electron-impact ionization was used, with the molecular ion complex of each compound being monitored. The only exception was chlorpromazine sulfoxide; for this compound the base peak was used for quantitation.

Using 2 mL of plasma in the extraction procedure, coefficients of variation for chlorpromazine were less than 5% at 272 ng/mL and 7% at a concentration of 18 ng/mL. Details of the EI-induced fragmentation of some of the phenothiazine *S*-oxides and *S,S*-dioxides were studied by Taulov and coworkers (1980a,b). Maurer and Pfleger (1984) have developed a GC–MS technique for the rapid identification of phenothiazines and their metabolites in urine. The chromatographic separation was carried out on a 60-cm nickel column packed with 5% OV-101, and the ionization energy for the EI excitation was 90 eV. This procedure is designed to give a rapid, unequivocal identification of phenothiazines for clinical or forensic estimation of intoxication.

6.1.2. Trifluperazine

Trifluperazine is another phenothiazine widely used in the treatment of schizophrenia. This drug undergoes extensive metabolism (Breyer et al., 1974) and, like other piperazine side chain phenothiazines, is considerably more potent than chlorpromazine. As a result of its high potency, trifluperazine is administered in much lower doses than chlorpromazine; thus the plasma drug concentrations are too low to measure by many of the standard analytical techniques. Midha and coworkers (1982) studied the pharmacokinetics of trifluperidol using EI excitation and SIM. Under these excitation conditions the molecular ion peak of trifluperazine at m/z 407 is ideal for quantitative studies since it is specific and the relative abundance of the ion is high. The chromatography was carried out on a 1.22-m glass column packed with OV-1. Either prochlorperazine or a deuterated analog of trifluperidol was used as the internal standard. The average overall recoveries for the trifluperazine and prochlorperazine were 102 and 91%, respectively, and these were not concentration-dependent over a range of trifluperazine concentrations of 1–10 ng/mL. Using 2 mL of plasma, the coefficient of variation of their method was 6.8% at a concentration of 0.078 ng/mL and 1.5% at a concentration of 5.0 ng/mL. In examining the pharmacokinetic properties of trifluperidol in volunteers, they found plasma concentrations ranging from approximately 0.1 to 3 ng/mL. Whelpton and coworkers (1982) developed a procedure for the analysis of trifluperazine employing EI excitation with deuterated trifluperidol as the internal standard. The chromatography was carried out utilizing a 1-m glass column packed with 3% OV-225. Using 2–4 mL of plasma, the coefficient of variation at 20, 2, and 0.2 ng/mL of trifluperidol was found to be 1.6, 3.3, and 13%, respectively. The addition of the deuterated internal standard was found to greatly enhance the recovery of the drug, especially at very low levels.

6.1.3. Fluphenazine

Fluphenazine is another potent phenothiazine with a piperazine side chain. It is available as both the fluphenazine dihydrochloride salt, which is administered orally, and as the decanoate or heptanoate ester, which is administered intramuscularly. Both of these formulations are used extensively in the treatment of schizophrenia. McKay and coworkers (1983) have described a GC–MS procedure that monitors the m/z 406 ion fragment of the trimethylsilyl derivative of fluphenazine and the m/z 372 ion fragment of the trimethylsilyl derivative of the perphenazine, which was used as the internal standard. The base peak of the derivatized fluphenazine at m/z 281 could not be used for quantitation since there was considerable interference from the column bleed. The chromatography was performed on a 1.8-m glass column packed with 3% OV-1. The coefficient of variation for fluphenazine was 4.6% at a concentration of 0.078 ng/mL and 3.5% at a concentration of 10.0 ng/mL. In pharmacokinetic studies using volunteers, plasma levels of fluphenazine ranged from 0.078 ng/mL to approximately 0.77 ng/mL. Brooks and Heyes (1982) studied the N-oxides and S-oxides of fluphenazine by desorption chemical ionization (DCI) and fast-atom bombardment (FAB). These modes of excitation were utilized in an attempt to develop a procedure in which the oxides of these thermally labile compounds could be identified. Electron impact and normal CI spectra of these compounds have been reported to yield extremely weak or no identifiable molecular ion (Melikian et al., 1977). Using either DCI or FAB, Brooks and Heyes (1982) were able to obtain spectra with strong molecular or pseudo-molecular ion peaks.

6.1.4. Mequitazin

Mequitazin is a phenothiazine that has a potent pharmacological effect at very low dosages. Like other potent neuroleptics, the plasma levels of mequitazin are low, and the determination of plasma levels requires an assay of high sensitivity. The first chromatographic procedure for the measurement of this drug was published by Fourtillan and coworkers (1984) using EI excitation. The spectrum of mequitazin yields a base peak at m/z 124 with a strong molecular ion at m/z 322, whereas the spectrum of the chemical analog that was used as the internal standard, IBF-28145, yielded a base peak at m/z 110 and a strong molecular ion at m/z 308. Quantitation was carried out by monitoring the molecular ion of both mequitazin and the internal standard. The extraction procedure required 2 mL of plasma or 1 mL of urine for the analysis. The chromatography was carried out using a 25-m fused silica capillary column that was wall coated with CP Sil 5. The lower limit of detection was reported to be 0.5 ng/mL of mequitazin, whereas the coefficient of variation was reported to be 9% at a plasma level of 1 ng/mL.

6.1.5. Promazine

Promazine was investigated by Covey and coworkers (1985) using thermospray ionization. The thermally labile N-oxide and N-oxide sulfoxide metabolites of promazine were also determined using their procedure. Chromatographic separation took place on a 3-μm C_{18} reverse-phase column using a mixture of methanol, acetonitrile, and 0.1M ammonium acetate (36:36:28) flowing at a rate of 1.4 mL/min. The thermospray ionization is extremely mild, resulting in spectra containing a major peak at the protonated molecular ion with few other fragments to aid in structural identification. No quantitative determinations for these drugs or metabolites were reported.

6.1.6. Propericiazine

Propericiazine is a piperidyl phenothiazine that is used at low doses as an antipsychotic. As a result of the low dosage and extensive first-pass metabolism, the plasma concentration of this compound is low. De Leenheer and coworkers (1985) reported a procedure to analyze for propericiazine using fluphenazine as the internal standard. Both the drug to be studied and the internal standard were acylated using TFAA. The chromatographic separation was carried out on a 12-m OV-101 fused silica capillary column coupled directly to the mass spectrometer. Spectra were obtained using both EI and CI ionization conditions. For CI, isobutane was used as the reagent gas. These investigators found that although the CI spectra were dominated by the pseudo-molecular ions, there was no increase in sensitivity using this procedure. Quantitation was performed using EI ionization. The coefficient of variation was 3.2% at 5 ng/mL and 2.8% at 20 ng/mL. The between-run coefficient of variation was 4.8% at 5 ng/mL and 3.8% at 20 ng/mL. The detection limit was 1 ng/mL; at this level the signal-to-noise ratio was 3:1. They found that the amount of isopropanol in the extraction solvent strongly affected the recovery.

6.2. Butyrophenones

The butyrophenones are a class of drugs with high neuroleptic potency. Haloperidol was the first butyrophenone to be introduced clinically and is still one of the most widely used antipsychotic drugs. Unlike the phenothiazines, haloperidol is not extensively metabolized in vivo, and only one pharmacologically active metabolite is known. The lack of pharmacologically active metabolites makes this an ideal compound with which to assess the interrelationship between plasma drug concentrations and clinical response. The structure of several of the butyrophenones are shown in Fig. 5.

6.2.1. Haloperidol

The EI spectrum of haloperidol (Marcucci et al., 1971; Moulin et al., 1979; Pape, 1981) yields a molecular ion of extremely low

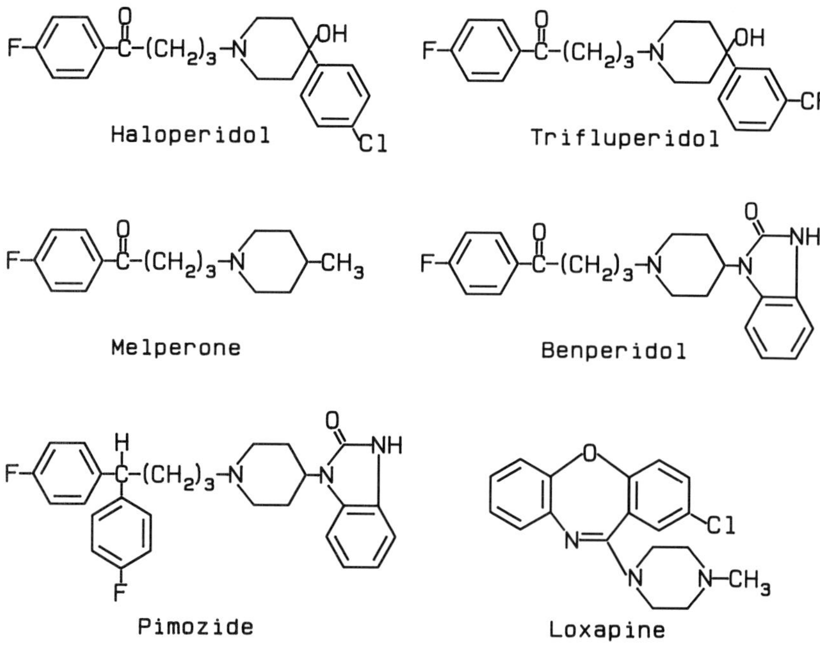

Haloperidol

Trifluperidol

Melperone

Benperidol

Pimozide

Loxapine

Fig. 5. Structures of several butyrophenone neuroleptics and other neuroleptics.

intensity at *m/z* 375 with major ions at *m/z* 237 and 224 (base peak). Moulin and coworkers (1979) used EI in combination with SIM to determine plasma levels of haloperidol in patients receiving this drug. Chlorine-substituted haloperidol was used as the internal standard and, like haloperidol, has two major ion fragments at *m/z* 237 and 224. Chromatographic separation was carried out on an OV-17 WCOT column with the common fragments at *m/z* 237 and 224 being monitored. The authors pointed out a number of deficiencies that needed to be overcome before this procedure could be used as a definitive reference method for haloperidol. In a toxicological investigation, Pape (1981) used EI as well as CI to obtain a mass spectral identification of an active metabolite of haloperidol, reduced haloperidol or hydroxyhaloperidol. Chemical ionization has been used to quantitate haloperidol in plasma by several investigators (Hornbeck et al., 1979; Szczepanik-Van Leeuwen, 1985). Hornbeck and coworkers (1979) described a procedure using trifluperidol as the internal standard and methane-ammonia as the reactant gas. The CI spectrum was dominated by the protonated molecular ion at *m/z* 376. The gas chromatography was carried out on a 2-ft glass column packed with 3% SP 2100. Their extraction procedure was about 2 h long and required 2 mL of patient plasma. The reproducibility of this method was found to be ±15% at a concentration level of 0.5 ng/mL. These authors report that at levels of less than 0.5 ng/mL, the reproducibility of the method was too variable to allow for quantitation. The analysis described by Szczepanik-Van Leeuwen (1985) utilized CI with ammonia as the reactant gas and tetra-deuterated haloperidol as the internal standard. The deuterated internal standard helps minimize the losses in the extraction process and decreases the adsorption on the column, as well as acting as a carrier to increase the overall recovery. The extraction and chromatographic procedures were similar to those used by Hornbeck et al. (1979), with the exception that 100 ng of thioridazine was added to the sample, which they claimed greatly improved the peak shape and recovery. At a concentration of 2.0 ng/mL, the procedure had a coefficient of variation of 10.1% with a demonstrated detection level as low as 280 pg/mL.

6.2.2. Melperone

Melperone, an experimental butyrophenone, has been quantitated by GC–MS (Chan and Okerholm, 1983), with the mass spectrometer operated in the EI mode and the chromatography carried out on a 1.5-m glass column packed with SE-30. The molecular ion at m/z 263 is only a minor ion, with major ions at m/z 125, 123, and 112 (base peak). Utilizing SIM, these workers had a coefficient of variation of 17% at a concentration of 2.5 ng/mL. During the course of this study, they noticed a drug-related peak that was tentatively identified as the reduced alcohol of melperone.

6.2.3. Other Butyrophenones

Marcucci and coworkers (1971) utilized MS as a reference technique to verify methodology being developed to analyze several butyrophenones by GC with electron-capture detection. The butyrophenones were introduced using direct insertion techniques as well as GC. Chromatographic separation was carried out utilizing a 2-m glass column packed with 3% OV-17, whereas the mass spectrometer was operated in an EI mode with the ionization potential maintained at 70 eV. The mass spectra indicated that during the gas chromatographic procedure both haloperidol and trifluperidol remained structurally intact, whereas both benperidol and fluoropipamide were modified. Maurer and Pfleger (1983) have presented a method for the identification of benperidol, bromperidol, droperidol, fluanisone, haloperidol, melperone, moperone, penfluridol, pipamerone, and trifluperidol and their major basic metabolites in urine by computerized GC–MS. Their procedure was developed to allow for the rapid identification of butyrophenones for toxicological purposes and does not attempt to quantitate the drugs or their metabolites.

6.3. Other Neuroleptics

6.3.1. Loxapine

Loxapine is an antipsychotic used in the treatment of schizophrenia. Lutz and coworkers (1982) developed a technique for analyzing biological samples from humans using GC–MS with a deuterated analog of loxapine as the internal standard. Chromatography was carried out using a 1.8-m glass column packed with 1% OV-1. The extraction procedure used 3 mL of urine, 1 mL of plasma, or 1 g of tissue, and the quantitation was performed using SIM. The EI spectra of loxapine yielded strong ion fragments at m/z 257, 83 (base peak), and 70, with a less intense molecular ion at m/z 327. To increase the specificity of the assay, the molecular ion peak was monitored, although the use of this less-intense peak limited the sensitivity of the procedure. The interrun precision was determined over a 2-mo period and was reported to have a coefficient of variation of 3% for urine and 6% for plasma at a concentration level of 10 ng/mL. The assay sensitivity was 2 ng/mL.

6.3.2. Flutroline

Flutroline is a gamma-carboline that has exhibited neuroleptic properties and useful clinical antipsychotic activity. A GC–MS assay was developed (Falkner et al., 1984) for the determination of flutroline in plasma using SIM and capillary GC. The extraction was carried out using 1 mL of plasma. The underivatized flutroline was reported to exhibit poor chromatographic characteristics, and several derivatization procedures were attempted. The trimethyl-silyl derivative of flutroline was prepared and chromatographed using a 4–5-m Durabond DB-1 or DB-5 WCOT capillary column, which was directly interfaced to the ion source. Quantitation was carried out by monitoring the *m/z* 266 peak of both the flutroline and the flutroline analog. These investigators reported the lower limit of reproducible quantitation to be 3 ng/mL.

6.3.3. Pimozide

Pimozide is a diphenylbutylpiperidine that is structurally unrelated to the tricyclic or butyrophenone psychotropics. Janssen (1971) has suggested there exists some topographical similarity between this compound and the classical neuroleptics. This compound has been used in the treatment of Gilles de la Tourette's syndrome, a disorder whose symptoms begin in childhood. Determination of this compound in adolescents presents several difficulties. The dose used in the treatment of this disorder is normally very low, an average of 1–2 mg/d, resulting in low plasma concentrations, and the amount of blood available for study is limited. Reed and coworkers (unpublished observations) developed a procedure for the analysis of pimozide in plasma of adolescents being treated for Tourette's syndrome with this compound. A tetradeuterated analog of pimozide was synthesized for use as an internal standard. The sample was prepared by extracting 1 mL of plasma and derivatizing with acetic anhydride. The chromatography was carried out using a 1-m DB-5 fused silica capillary column directly inserted into the ion source. The GC effluent was analyzed using EI excitation. The spectrum of the derivatized compound has a base peak at *m/z* 272 with a small molecular ion at *m/z* 503. For quantitation, the *m/z* 272 for pimozide and *m/z* 276 for the tetra-deuterated analog were monitored. The coefficient of variation was

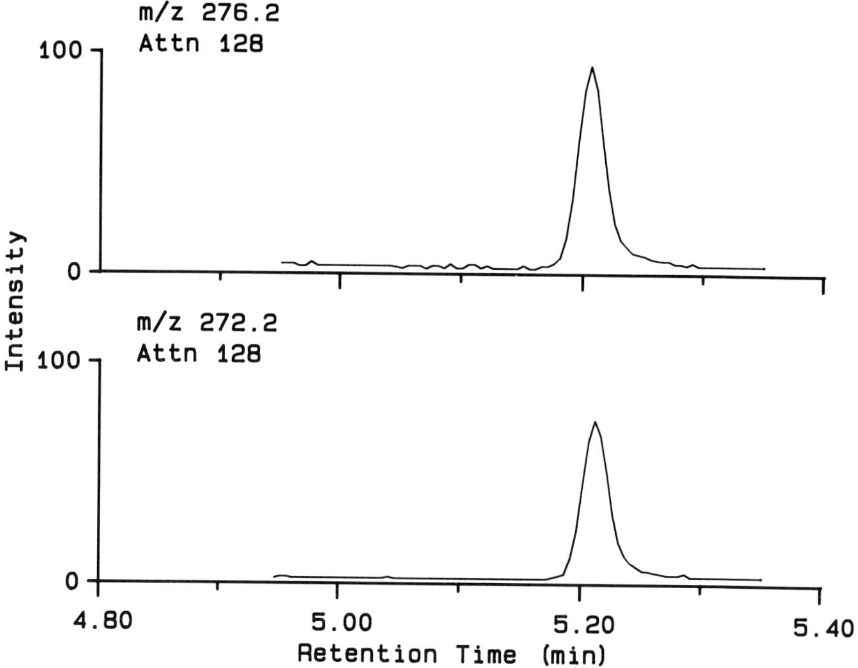

Fig. 6. Selected ion chromatograms of the acetate derivatives of pimozide (*m/z* 272.2) and tetra-deutero-pimozide (*m/z* 276.2) obtained from extracted patient plasma (1 mL). The data presented in this figure correspond to a plasma concentration of 3.2 ng/mL.

3.9% at a concentration of 3.5 ng/mL and 6.5% at a concentration of 1.2 ng/mL. The lower limit of quantitation was reported to be 0.1 ng/mL. An SIM chromatogram for this compound is shown in Fig. 6. Patient levels ranged from 0.3 to 12 ng/mL.

7. Benzodiazepines

The benzodiazepines are the most commonly prescribed drugs in the world, being extensively used as anxiolytics, anti-convulsants, and hypnotic agents. One reason for the widespread use of these drugs is their high therapeutic index (ratio of lethal dose:therapeutic dose), which makes them relatively safe. For example, in a study of 27 cases of overdoses by the hypnotic nitrazepam, Matthews and coworkers (1969) found no incidence of loss of conciousness or respiratory depression, even though the

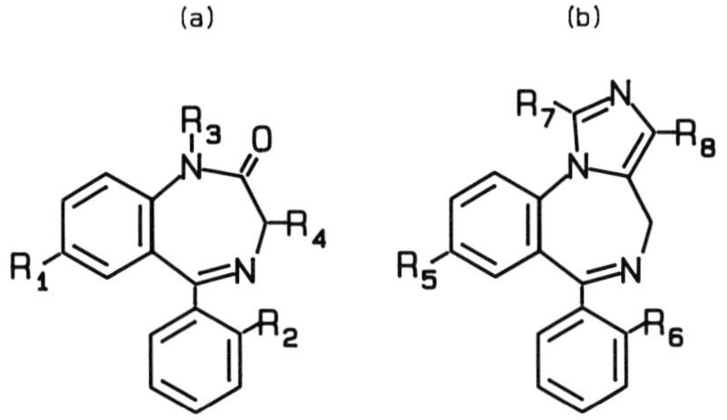

(a) (b)

(a) Diazepam $R_1 = Cl$, $R_2 = H$, $R_3 = CH_3$, $R_4 = H$
   Nordiazepam $R_1 = Cl$, $R_2 = H$, $R_3 = H$, $R_4 = H$
   Clonazepam $R_1 = NO_2$, $R_2 = Cl$, $R_3 = H$, $R_4 = H$
   Flurazepam $R_1 = Cl$, $R_2 = F$, $R_3 = CH_2CH_2N(CH_2CH_3)_2$, $R_4 = H$
   Oxazepam $R_1 = Cl$, $R_2 = H$, $R_3 = H$, $R_4 = OH$
   Lorazepam $R_1 = Cl$, $R_2 = Cl$, $R_3 = H$, $R_4 = H$
   Ethyl Loflazepate $R_1 = Cl$, $R_2 = F$, $R_3 = H$, $R_4 = CO_2CH_2CH_3$
(b) Midazolam $R_5 = Cl$, $R_6 = F$, $R_7 = CH_3$, $R_8 = H$
   Imidazobenzodiazepine- $R_5 = Cl$, $R_6 = Cl$, $R_7 = H$, $R_8 = CONH_2$
   3-carboxamide

Fig. 7. Structures of several benzodiazepines.

overdoses ranged to up to 40 times the hypnotic dose. Benzodiazepines undergo extensive biotransformation, often with the metabolites themselves being pharmacologically active. Correlations between plasma concentrations and clinical efficacy of these compounds and their active metabolites have not been established. The majority of reports published on the use of MS in the determination of benzodiazepines and their metabolites are studies of absorption, distribution, biotransformation, identification of metabolites, and identification for the purpose of clinical and forensic toxicology. The structures of several benzodiazepines are shown in Fig. 7.

7.1. Anxiolytics

7.1.1. Tofizopam

Tofizopam and its metabolites were determined in biological samples (Tomori et al., 1982) using a GC–MS assay. The mass spectrometer was operated in the EI excitation mode with the chromatography being carried out either on a 6-ft glass column containing 2% XE-60 or a 2-ft glass column packed with 2% OV-101. The compounds were isolated from 10-mL aliquots of urine.

After an initial extraction, the samples were purified using TLC. The major peaks from the TLC purification were then reconstituted and analyzed on the GC–MS. The reconstituted material was either injected directly or silylated and then injected. A number of de-methylated metabolites were identified with the aid of MS. The high abundance of the molecular ion species, as well as the presence of structurally significant fragments, made this method practical for the identification of the metabolites.

7.1.2. Ethyl Loflazepate

Ethyl loflazepate is a benzodiazepine characterized by its potent antianxiety activity. Cautreels and Jeanniot (1980) developed an assay for this drug as well as two of its metabolites, the carboxylic acid metabolite and the decarboxylated metabolite. The acid metabolite proved to be quite labile, converting rapidly to the decarboxylated metabolite, and no quantitative data could be obtained on the individual compounds separately. Quantitative data was obtained for the sum of the carboxylic acid plus decarboxylated metabolites. The chromatography was carried out using a 1.5-m glass column packed with OV-1. Two chemical analogs were used as internal standards, and the analysis was carried out using CI with ammonia as the reactant gas. The extracted compounds and internal standards were derivatized with 1-iodobutane. The EI spectra of these compounds were difficult to monitor since only low-intensity ion fragments were observed. The CI spectra gave strong pseudo-molecular ions for all compounds used in the quantitation. The coefficient of variation was 2.4% at a concentration of 50 ng/mL for the combined metabolites. Although the assay could quantitate 1 ng/mL of the parent compound, none was detected in plasma obtained from volunteers receiving the drug.

7.1.3. Imidazobenzodiazepine-3-carboxamide

Rubio and coworkers (1982b) described a method for measuring imidazobenzodiazepine-3-carboxamide in plasma using a tetradeuterated analog of the compound as the internal standard. The quantitation was carried out using negative-ion CI mass spectrometry. The amide moiety of this compound was converted to a nitrile using PFPA and triethylamine. This derivatization was reported to be quantitative even when using picogram amounts of material. The mass spectrum of this compound did not show the usual strong molecular ion peak associated with negative ion detection of benzodiazepines; in fact, no significant molecular ion peak was detected. The spectrum obtained was an intense single peak at m/z 316 corresponding to the $(M\text{-}HCl)^-$ fragment ion. Using 1 mL of plasma in the extraction, the coefficient of variation was 17% at a concentration level of 0.1 ng/mL.

7.2. Anticonvulsants

7.2.1. Clonazepam

Clonazepam is primarily used as an anticonvulsant agent for controlling "petit mal" minor motor seizures. It is extensively metabolized, with the principle metabolite being its nitro bioreduction product, the 7-amino metabolite. Min and Garland (1977) developed an assay to simultaneously determine both clonazepam and its principal metabolite using GC–MS, with the internal standards being the ^{15}N-substituted analog of both the parent compound and metabolite. A high degree of specificity for the assay was obtained utilizing CI with ammonia as the reactant gas. The spectra of both compounds were very simple, with the molecular ion-proton extraction product dominating. The protonated molecular ion for both the parent clonazepam and its metabolite were monitored using SIM. The compounds were separated chromatographically using a 1-m glass column packed with 1% OV-25, with methane being used as the carrier gas. The extraction procedure used either 0.5 or 1 mL of plasma. The limit of detection for the assay was reported to be 1 ng/mL for clonazepam and 2 ng/mL for the metabolite. A further investigation by these workers (Garland and Min, 1979) studied the use of negative ion CI for the determination of clonazepam, using a stable isotope analog of this compound as the internal standard. The gas chromatographic and ionization conditions were the same as previously described, with the exception that ammonia was not used as the reagent gas; methane was used as both the carrier gas and reactant gas. The spectrum consisted principally of the M^- molecular ion with a small $(M\text{-}Cl)^-$ fragment ion. Selected ion monitoring was used for quantitation, with the M^- ion being monitored for both clonazepam and the internal standard. This procedure was demonstrated to be 15–25 times more sensitive than the previously reported positive ion CI technique used by these investigators (Min and Garland, 1977). The sensitivity of the assay was estimated to be 0.1 ng/mL using 0.5 mL of plasma. The sensitivity was limited by the amount of isotope contamination from the internal standard that caused interference at the M⁻ ion of clonazepam. The ion chromatograms were virtually free of noise, with samples containing 0.5 ng/mL yielding a signal-to-noise ratio of greater than 100. Colburn and coworkers (1980) used this negative ion CI procedure and several stable isotopic variants of clonazepam to quantitatively assess the contribution of gut content, intestinal wall, and liver to the first-pass metabolism of clonazepam following simultaneous oral and intravenous dosing. The high sensitivity of the negative-ion assay procedure allowed these workers to carry out their analysis using only 0.15 mL of blood.

7.3. Hypnotics

7.3.1. Midazolam

Vasiliades and Sahawneh (1982) developed a GC–MS assay using EI excitation and SIM. Using 1–2 mL samples of serum, a sensitivity of 2 ng/mL was reported. These investigators found the sensitivity of the assay to be limited because of the poor chromatographic characteristics of the underivatized midazolam. Rubio and coworkers (1982a) used the analytical procedure of Garland and Miwa (1980) for determining the plasma levels of midazolam and its hydroxymethyl and desmethyl metabolites. The deuterated analogs of each of the compounds was used as the respective internal standard. To improve the chromatographic resolution of hydroxymethylmidazolam, this compound was reacted with BSA to form the trimethylsilyl derivative. The GC was carried out on a 1.52-m glass column packed with 3% Poly-S. The spectra of each of these compounds showed a pattern similar to the other negative-ion spectra reported, i.e., a dominant molecular ion peak M⁻. The trimethylsilyl derivative also showed a small peak corresponding to a $(M-(CH_3)_3SiOH)^-$ ion. This assay used 1-mL aliquots of plasma for the extraction procedure. The limit of quantitation was reported to be 1 ng/mL for all three compounds, with the coefficient of variation being less than 6% for each at a plasma level of 5 ng/mL. In their investigation no quantifiable amount of the desmethylmidazolam was found.

7.3.2. Flurazepam

Flurazepam is one of the most widely used drugs for the treatment of insomnia. This compound is extensively metabolized, and in pharmacokinetic studies the plasma levels of the parent drug have rarely exceeded several ng/mL. Miwa and Garland

(1981) have reported a highly specific and sensitive assay for flurazepam using a deca-deuterated analog of the compound as an internal standard and assaying by means of negative ion CI MS. The chromatography was carried out on a 4-ft micropacked column filled with OV-17. Methane was used as the carrier gas for the chromatography as well as the reactant gas for the CI. The negative ion spectra of flurazepam is quite simple, consisting only of the m/z 387 (M^-) molecular ion. The assay was carried out using 2 mL of plasma in the extraction procedure. In order to improve the efficiency of the extraction procedure, a structural analog of flurazepam was added. At a plasma concentration of 0.135 ng/mL, the coefficient of variation was reported to be 4%, whereas the lower limit of quantitation was 0.012 ng/mL.

7.4. Other Benzodiazepines

Garland and Miwa (1980) used negative ion CI for the analysis of several benzodiazepines and their metabolites. The spectra of these compounds contained one major peak, either the molecular (M^-) or the $(M-H)^-$ ion. The only exception reported was the spectrum of bromazepam, which was dominated by the Br^- peak and had only a minor $(M-H)^-$ peak. Methane was used as the carrier gas for the chromatographic separation as well as the reactant gas for the CI. The chromatography was carried out on a 4-ft column packed with 3% OV-17. Nordiazepam was assayed from either 1 mL of plasma or 1 mL of the brain homogenate using a deuterated analog of nordiazepam as the internal standard. The minimum measurable concentration was reported to be 1 ng/mL for nordiazepam in either plasma or brain homogenate.

The thermal stability of 16 benzodiazepines was studied (Joyce et al., 1984). Chromatographic separation was carried out using a 25-m fused silica column coated with SE-30. The capillary column was directly interfaced to the ion source. The analysis was carried out using either EI ionization or CI with ammonia as the reagent gas. The positive ion CI was found to be no more sensitive than the EI determination. These investigators found that seven of the compounds studied were thermally stable, whereas the other nine showed varying degrees of thermal decomposition. The nine unstable compounds were divided into five groups according to their mechanism of decomposition. These results suggest that care must be taken in the interpretation of any gas chromatographic data unless the thermal stability of the compound being studied is demonstrated.

Maurer and Pfleger (1981) developed a procedure for the determination of benzodiazepines and their primary metabolites in urine. Their approach was developed for clinical and toxicological screening of benzodiazepines. The methodology used EI ionization, with the chromatography being carried out on a 0.6-m nickel column packed with 20% UCC-W. The urine samples were acid hydrolyzed and then extracted. The extracts were then reconstituted in methanol and injected. The identification of the drug and metabolite was made by a computerized matching of known spectra with those obtained experimentally.

8. Concluding Remarks

In this review the role of MS in the analysis of three classes of psychotropic drugs, namely antidepressants, neuroleptics, and benzodiazepines, has been examined. The highly specific and sensitive technique offered by combined GC–MS provides the definitive method for the quantitation of these drugs, as well as for the characterization of metabolites of these compounds. The high cost of the instrumentation precludes this technique from being widely used for the routine analysis of these drugs. With the advent of newer and less costly apparatus such as the ion trap detector, however, this technique may become far more accessible.

References

Alfredsson G. and Sedvall G. (1976) Mass Fragmentographic Analysis of Chlorpromazine in Human Plasma, in *Antipsychotic Drugs: Pharmacodynamics and Pharmacokinetics* Sedvall G., Urnas B., and Zotterman Y., eds.) Pergamon, Oxford.

Alfredsson G., Wode-Helgodt B., and Sedvall G. (1976) A mass fragmentographic method for the determination of chlorpromazine and two of its active metabolites in human plasma and CSF. *Psychopharmacology* **48,** 123–131.

Alfredsson G., Wiesel F., Fyro B., and Sedvall G. (1977) Mass fragmentographic analysis of chlomipramine and its mono-demethylated metabolite in human plasma. *Psychopharmacology* **52,** 25–30.

Alkalay D., Carlsen S., Khemani L., and Bartlett M. F. (1979a) Selected ion monitoring assay for the antidepressant maprotiline. *Biomed. Mass Spectrom.* **6,** 435–438.

Alkalay D., Volk J., and Carlsen S. (1979b) A sensitive method for the simultaneous determination in biological fluids of imipramine and desipramine or clomipramine and *N*-desmethylclomipramine by gas chromatography mass spectrometry. *Biomed. Mass Spectrom.* **6**, 200–204.

Alkalay D., Wagner Jr., W. E., Carlsen S., Khemani L., Volk J., Bartlett M. F., and LeSher A. (1980) Bioavailability and kinetics of maprotiline. *Clin. Pharmacol. Ther.* **27**, 697–703.

Alvan G., Borga O., Lind M., Palmer L., and Siwers B. (1977) First pass hydroxylation of nortriptyline: Concentrations of parent drug and major metabolites in plasma. *Eur. J. Clin. Pharmacol.* **11**, 219–224.

Antal A., Mercik S., and Kramer P. A. (1980) Technical considerations in the gas chromatographic analysis of desipramine. *J. Chromatogr. Biomed. Appl.* **183**, 149–157.

Arpino P. J. (1985) Ten years of liquid chromatography–mass spectrometry. *J. Chromatogr.* **323**, 3–11.

Bagchi S. P., Lutz T., and Jindal S. P. (1985) Gas chromatographic–mass spectrometric determination of nomifensine using a stable isotope-labeled analogue as an internal standard. *J. Chromatogr. Biomed. Appl.* **344**, 362–366.

Barber M., Borsoli R. S., Elliot G. J., Sedgewick R. D., and Tyler A. N. (1982) Fast atom bombardment mass spectrometry. *Anal. Chem.* **54**, 645A–657A.

Belvedere G., Frigerio A., and Pantarotto C. (1975a) Gas chromatographic–mass fragmentographic determination of trazodone in rat plasma. *J. Chromatogr.* **112**, 631–636.

Belvedere G., Burti L., Frigerio A., and Pantarotto C. (1975b) Gas chromatographic–mass fragmentographic determination of "steady-state" plasma levels of imipramine and desipramine in chronically treated patients. *J. Chromatogr.* **111**, 313–321.

Biggs J. T., Holland W. H., Chang S., Hipps P. P., and Sherman W. R. (1976) Electron beam ionization mass fragmentographic analysis of tricyclic antidepressants in human plasma. *J. Pharm. Sci.* **65**, 261–268.

Borga O., Piafsky K. M., and Nilson O. G. (1977) Plasma protein binding of basic drugs. *Clin. Pharmacol. Ther.* **22**, 539–544.

Bougerolle A. M., Chabard J. L., Bargnoux H., Petit J., Berger J. A., and Dordain G. (1985) Simultaneous determination of metapramine and its demethylated metabolites in plasma by gas chromatography–mass spectrometry. *J. Chromatogr. Biomed. Appl.* **345**, 59–66.

Breyer U., Gaertner H. J., and Prox A. (1974) Formation of identical metabolites from piperazine- and dimethylamino-substituted phenothiazine drugs in man, rat, and dog. *Biochem. Pharmacol.* **23**, 313–322.

Brooks P. and Heyes W. F. (1982) Desorption chemical ionization and fast atom bombardment mass spectrometry of fluphenazine oxide derivatives. *Biomed. Mass Spectrom.* **9,** 522–526.

Brunswick D. J. and Mendels J. (1977) Reduced levels of tricyclic antidepressants in plasma from vacutainers. *Commun. Psychopharmacol.* **1,** 131–134.

Cautreels W. and Jeanniot J. P. (1980) Quantitative analysis of CM 6912 (ethyl loflazepate) and its metabolites in plasma and urine by chemical ionization gas chromatography mass spectrometry. Application to pharmacokinetic studies in man. *Biomed. Mass Spectrom.* **7,** 565–571.

Chan K. Y. and Okerholm R. A. (1983) Quantitative analysis of melperone in human plasma by gas chromatography–mass spectrometry-selected ion monitoring. *J. Chromatogr. Biomed. Appl.* **274,** 121–127.

Chinn D. M., Jennison T. A., Crouch D. J., Peat M. A., and Thatcher G. W. (1980) Quantitative analysis for tricyclic antidepressant drugs in plasma or serum by gas chromatography-chemical-ionization mass spectrometry. *Clin. Chem.* **26,** 1201–1204.

Claeys M., Muscettola G., and Markey S. P. (1976) Simultaneous measurement of imipramine and desipramine by selected ion recording with deuterated internal standards. *Biomed. Mass Spectrom.* **3,** 110–116.

Colburn W. A., Bekersky I., Min B. H., Hodshon B. J., and Garland W. A. (1980) Contribution of gut contents, intestinal wall and liver to the first-pass metabolism of clonazepam in the rat. *Res. Commun. Chem. Path. Pharmacol.* **27,** 73–90.

Covey T. R., Crowther J. B., Dewey E. A., and Henion J. D. (1985) Thermospray liquid chromatography–mass spectrometry determination of drugs and their metabolites in biological fluids. *Anal. Chem.* **57,** 474–481.

Craig R. D., Bateman R. H., Green B. N., and Millington D. S. (1979) Mass spectrometry instrumentation for chemists and biologists. *Philos. Trans. Roy. Soc. London.* **A293,** 135–155.

Cramers C. A., Scherpenzeel G. J., and LeClerq P. A. (1981) Increased speed of analysis in directly coupled gas chromatographic–mass spectrometry systems: Capillary columns at sub-atmospheric outlet pressures. *J. Chromatogr.* **203,** 207–216.

Crampton E. L., Glass R. C., Marchant B., and Rees J. A. (1980) Chemical ionization mass fragmentographic measurement of dothiepin plasma concentrations following a single oral dose in man. *J. Chromatogr. Biomed. Appl.* **183,** 141–148.

Crane G. E. (1973) Clinical psychopharmacology in its 20th year. *Science* **181,** 124–128.

Davis T. P., Veggeberg S. K., Hameroff S. R., and Watts K. L. (1983) Sensitive and quantitative determination of plasma doxepin and desmethyl-doxepin in chronic pain patients by gas chromatography and mass spectrometry. *J. Chromatogr. Biomed. Appl.* **273**, 436–441.

Dawson P. H. (1976) *Quadrupole Mass Spectrometry and Its Applications.* Elsevier, New York.

De Gier J. J. and Hart B. J. (1979) Sensitive gas chromatographic method for the determination of diazepam and *N*-desmethyldiazepam in plasma. *J. Chromatogr.* **163**, 304–309.

Delay J. and Deniker P. (1952) Trente-huit Cas de Psychoses Traites par la Cure Prolongée et Continué de 4560 RP, in *Compte Rendu du Congres* Masson et Cie, Paris.

De Leenheer A. P. and Nelis H. J. C. F. (1981) Development and evaluation of selected assays for drugs and drug metabolites in biological materials. *Analyst* **106**, 1025–1035.

De Leenheer A. P., Jonckheere J. A., Lefevere M. F., De Broe M. E., and Verpooten G. A. (1985) Measurement of propericiazine in plasma by capillary gas chromatography and selected ion monitoring detection. *Biomed. Mass Spectrom.* **12**, 25–29.

De Ridder J. J., Koppens P. C. J. M., and Van Hal H. J. M. (1977) Mass fragmentographic assay of nanogram amounts of the antidepressant drug mianserin hydrochloride (Org GB 94) in human plasma. *J. Chromatogr. Biomed. Appl.* **143**, 289–297.

Desmond P. V., Roberts R. K., Wood A. J., Dunn D. G., Wilkinson G. R., and Schenker S. (1980) Effect of heparin administration on plasma binding of benzodiazepines. *Br. J. Clin. Pharmacol.* **9**, 171–175.

Dubois J. P., Kung W., Theobald W., and Wirz B. (1976) Measurement of clomipramine, *N*-desmethyl-clomipramine, imipramine, and dehydroimipramine in biological fluids by selective ion monitoring, and pharmacokinetics of clomipramine. *Clin. Chem.* **22**, 892–897.

Ende M. and Spiteller G. (1982) Contaminants in mass spectrometry. *Mass Spectrom. Rev.* **1**, 29–62.

Essien E. E., Cowan D. A., and Beckett A. H. (1975) Metabolism of phenothiazines: Identification of N-oxygenated products by gas chromatography and mass spectrometry. *J. Pharm. Pharmacol.* **27**, 334–342.

Falkner F. C. (1981) Comments on some common aspects of quantitative mass spectrometry. *Biomed. Mass Spectrom.* **8**, 43–46.

Falkner F. C., Fouda H. G., and Mullins F. G. (1984) A gas chromatographic/mass spectrometric assay for flutroline, a gamma-carboline antipsychotic agent, with direct derivation on a moving needle injector. *Biomed. Mass Spectrom.* **11**, 482–485.

Feser K. and Kogler W. (1979) The quadrupole mass filter for gc–ms applications. *J. Chromatogr. Sci.* **17**, 57–63.

Fourtillan J. B., Girault J., Bouquet S., and Lefebvre M. A. (1984) Determination of mequitazin in human plasma and urine by capillary column gas-liquid chromatography–mass spectrometry. *J. Chromatogr. Biomed. Appl.* **309**, 391–396.

Frigerio A., Belvedere G., De Nadai F., Fanelli R., Pantarotto C., Riva E., and Morselli P. L. (1972) A method for the determination of imipramine in human plasma by gas-liquid chromatography–mass fragmentography. *J. Chromatogr.* **74**, 201–208.

Frigerio A., Pantarotto C., Franco R., Gomeni R., and Morselli P. L. (1977) Quantitative determination of doxepin and desmethyldoxepin in rat plasma by means of gas-liquid chromatography–mass fragmentography. *J. Chromatogr.* **130**, 354–360.

Gammans R. E., Kerns E. H., Bullen W. W., Covington R. R., and Russell J. W. (1985) Gas chromatographic–mass spectrometric method for trazodone and a deuterated analogue in plasma. *J. Chromatogr. Biomed. Appl.* **339**, 303–312.

Garland W. A., Muccino R. R., Min B. H., Cupano J., and Fann W. E. (1979) A method for the determination of amitriptyline and its metabolites nortriptyline, 10-hydroxyamitriptyline, and 10-hydroxynortriptyline in human plasma using stable isotope dilution and gas chromatography–chemical ionization mass spectrometry (GC–CIMS). *Clin. Pharmacol. Ther.* **25**, 844–856.

Garland W. A. (1977) Quantitative determination of amitriptyline and its principal metabolite, nortriptyline, by GLC–chemical ionization mass spectrometry. *J. Pharm. Sci.* **66**, 77–81.

Garland W. A. and Min B. H. (1979) Determination of clonazepam in human plasma by gas chromatography–negative ion chemical ionization mass spectrometry. *J. Chromatogr.* **172**, 279–286.

Garland W. A. and Miwa B. J. (1980) Methane negative chemical ionization analysis of 1,3-dihydro-5-phenyl-1,4-benzodiazepin-2-ones. *Environ. Health Perspect.* **36**, 69–76.

Gaskell S. J. (1980) Gas chromatography/high-resolution mass spectrometry as a reference method for clomipramine determination. *Postgrad. Med. J.* **56**, 90–93.

Goodman L. S. and Gilman A. (1975) *The Pharmacological Basis of Therapeutics* Macmillan, New York.

Gruenke L. D., Craig J. C., Klein F. D., Nguyen T. L., Hitzemann B. A., Holaday J. W., Loh H. H., Braff L., Fischer A., Glick I. D., Hartmann F., and Bissell D. M. (1985) Determination of chlorpromazine and its major metabolites by gas chromatography–mass spectrometry: Application to biological fluids. *Biomed. Mass Spectrom.* **12**, 707–713.

Hammar C. G., Holmstedt B., and Ryhage, R. (1968) Mass fragmentography. Identification of chlorpromazine and its metabolites in human blood by a new method. *Anal. Biochem.* **25**, 532–548.

Harvey D. J. (1984) The use of mass spectrometry in pharmacokinetic and drug-metabolism studies. *Mass Spectrom.* **7**, 293–352.

Haskins N. J. (1982) The application of stable isotopes in biochemical research. *Biomed. Mass Spectrom.* **9**, 269–277.

Hassan M. and Pesce A. J. (1986) The use of thermospray LC/MS for drug identification in human gastric lavage. *Vestec Thermospray Newslett.* **2**, 1–2.

Heck H. d'A., Flynn N. W., Buttrill Jr., S. E., Dyer R. L., and Anbar M. (1978) Determination of imipramine in plasma by high pressure liquid chromatography and field ionization mass spectrometry: Increased sensitivity in comparison with gas chromatography mass spectrometry. *Biomed. Mass Spectrom.* **5**, 250–257.

Hill R. E. and Whelan D. T. (1984) Mass spectrometry and clinical chemistry. *Clin. Chim. Acta* **139**, 231–294.

Hornbeck C. L., Griffiths J. C., Nebrosky R. J., and Faulkner M. A. (1979) A gas chromatographic mass spectrometric chemical ionization assay for haloperidol with selected ion monitoring. *Biomed. Mass Spectrom.* **6**, 427–430.

Hunt D. F. and Crow F. W. (1978) Electron capture negative ion chemical ionization mass spectrometry. *Anal. Chem.* **50**, 1781–1784.

Hunt D. F., Stafford G. C., Crow F. W., and Russell J. W. (1976) Pulsed positive negative ion chemical ionization mass spectrometry. *Anal. Chem.* **48**, 2098–2105.

Ishida R., Ozaki T., Uchida H., and Irikura T. (1984) Gas chromatographic–mass spectrometric determination of amitriptyline and its major metabolites in human serum. *J. Chromatogr. Biomed. Appl.* **305**, 73–82.

Janssen P. A. J. (1971) A new series of neuroleptic drugs. The 4,4-diphenylbutyl-piperidines and their relationship with other neuroleptics. *Clin. Trials J.* **8** (suppl. II), 7–23.

Jenkins R. G. and Friedel R. O. (1978) Analysis of tricyclic antidepressants in human plasma by GLC–chemical–ionization mass spectrometry with selected ion monitoring. *J. Pharm. Sci.* **67**, 17–23.

Jindal S. P., Lutz T., and Vestergaard P. (1980) GLC-Mass spectromeric determination of maprotiline and its major metabolite using stable isotope labeled analog as internal standard. *J. Pharm. Sci.* **69**, 684–687.

Jindal S. P., Lutz T., and Vestergaard P. (1982) Selected ion monitoring assay for the antidepressant mianserin in human plasma with stable isotope labeled analog as internal standard. *J. Anal. Toxicol.* **6**, 34–37.

Joyce J. R., Bal T. S., Ardrey R. E., Stevens H. M., and Moffat A. C. (1984) The decomposition of benzodiazepines during analysis by capillary gas chromatography/mass spectrometry. *Biomed. Mass Spectrom.* **11,** 284–289.

Karger B. L. and Vouras P. (1985) A chromatographic perspective of high-performance liquid chromatography–mass spectrometry. *J. Chromatogr.* **323,** 13–32.

Kuhn R. (1958) The treatment of depressive states with G22355 (imipramine hydrochloride). *Am. J. Psychiat.* **115,** 459–464.

Lapin A. and Karobath M. (1980) A sensitive and specific method for the determination of chlorimipramine and desmethylchlorimipramine in plasma using selected ion monitoring with chemical ionization. *Biomed. Mass Spectrom.* **7,** 588–591.

LeClerq P. A., Scherpenzeel G. J., Vermeer E. A. A., and Cramers C. A. (1982) Increased speed of analysis in directly coupled gas chromatography–mass spectrometry systems. II. Advantages of vacuum outlet operation of thick-film capillary columns. *J. Chromatogr.* **241,** 61–71.

Lutz T., Jindal S. P., and Cooper T. B. (1982) GLC/MS assay for loxapine in human biofluids and tissues with deuterium labeled analog as an internal standard. *J. Anal. Toxicol.* **6,** 301–304.

Maguire K. P., Norman T. R., Burrows G. D., and Scoggins B. A. (1982) A pharmacokinetic study of mianserin. *Eur. J. Clin. Pharmacol.* **21,** 517–520.

Maguire K. P., Norman T. R., Burrows G. D., and Scoggins B. A. (1981) Simultaneous measurement of dothiepin and its major metabolites in plasma and whole blood by gas chromatography–mass fragmentography. *J. Chromatogr. Biomed. Appl.* **222,** 399–408.

Marcucci F., Mussini E., Airoldi L., Fanelli R., Frigerio A., De Nadai F., Bizzi A., Rizzo M., Morselli P. L., and Garattini S. (1971) Analytical and pharmacokinetic studies on butyrophenones. *Clin. Chim. Acta.* **34,** 321–332.

Matthews H., Proudfoot A. T., Aitkin R. C. B., Raeburn J. A., and Wright N. (1969) Nitrazepam—a safe hypnotic. *Br. Med. J.* **3,** 23–25.

Maurer H. and Pfleger K. (1981) Determination of 1,4- and 1,5-benzodiazepines in urine using a computerized gas chromatographic–mass spectrometric technique. *J. Chromatogr. Biomed. Appl.* **222,** 409–419.

Maurer H. and Pfleger K. (1983) Screening procedure for detecting butyrophenone and bisfluorophenyl neuroleptics in urine using a computerized gas chromatographic–mass spectrometric technique. *J. Chromatogr. Biomed. Appl.* **272,** 75–85.

Maurer H. and Pfleger K. (1984) Screening procedure for detection of phenothiazine and analogous neuroleptics and their metabolites in urine using a computerized gas chromatographic–mass spectrometric technique. *J. Chromatogr. Biomed. Appl.* **306,** 125–145.

May P. R. A., Van Putten T., Jenden D. J., and Cho A. K. (1978) Test dose response in schizophrenia: Chlorpromazine blood and saliva levels. *Arch. Gen. Psychiat.* **35,** 1091–1097.

McFadden W. H. (1979) Interfacing chromatography and mass spectrometry. *J. Chromatogr. Sci.* **17,** 2–16.

McFadden W. H. (1980) Liquid chromatography–mass spectrometry systems and applications. *J. Chromatogr. Sci.* **18,** 97–115.

McKay G., Hall K., Cooper J. K., Hawes E. M., and Midha K. K. (1982) Gas chromatographic–mass spectrometric procedure for the quantitation of chlorpromazine in plasma and its comparison with a new high-performance liquid chromatographic assay with electrochemical detection. *J. Chromatogr. Biomed. Appl.* **232,** 275–282.

McKay G., Hall K., Edom R., Hawes E. M., and Midha K. K. (1983) Subnanogram determination of fluphenazine in human plasma by gas chromatography mass spectrometry. *Biomed. Mass Spectrom.* **10,** 550–555.

Melikian A. P., Flynn N. W., Petty F., and Wander J. D. (1977) Chemical ionization mass spectra of phenothiazine derivatives and their oxygenated analogues. *J. Pharm. Sci.* **66,** 228–232.

Midha K. K., Roscoe R. M. H., Hall K., Hawes E. M., Cooper J. K., McKay G., and Shetty H. U. (1982) A gas chromatographic mass spectrometric assay for plasma trifluoperazine concentrations following single doses. *Biomed. Mass Spectrom.* **9,** 186–190.

Midha K. K., Charette C., Cooper J. K., and McGilveray I. J. (1980) Comparison of a new GLC–AFID method with a GLC–MS selected ion monitoring technique and a radioimmunoassay for the determination of plasma concentrations of imipramine and desipramine. *J. Anal. Toxicol.* **4,** 237–243.

Millard B. J. (1978) *Quantitative Mass Spectrometry* Heyden, London.

Min B. H. and Garland W. A. (1977) Determination of clonazepam and its 7-amino metabolite in plasma and blood by gas chromatography–chemical ionization mass spectrometry. *J. Chromatogr.* **139,** 121–133.

Miwa B. J. and Garland W. A. (1981) Determination of flurazepam in human plasma by gas chromatography–electron capture negative chemical ionization mass spectrometry. *Anal. Chem.* **53,** 793–797.

Moulin M. A., Camsonne R., Davy J. P., Poilpre E., Morel P., Debruyne D., Bigot M. C., Dedieu M., and Hardy M. (1979) Gas chromatography–electron-impact and chemical-ionization mass spectrometry of haloperidol and its chlorinated homologue. *J. Chromatogr.* **178,** 324–329.

YNaranjo C. A., Abel J. G., Sellers E. M., and Giles H. G. (1980) Unaltered diazepam plasma binding using indwelling heparinized cannulae for sampling. *Br. J. Clin. Pharmacol.* **9**, 103–105.

Narasimhachari N. and Friedel R. O. (1979) N-Alkylation of secondary amine tricyclic antidepressants as a general method for their quantitation by GC–MS–SIM technique. Preparation of N-ethyl derivatives of desimipramine, nortriptyline, protriptyline and desmethyldoxepin. *Anal. Lett.* **12**, 77–86.

Narasimhachari N., Saady J., and Friedel R. O. (1981) Quantitative mapping of metabolites of imipramine and desipramine in plasma samples by gas chromatographic–mass spectrometry. *Biol. Psychiat.* **16**, 937–944.

Pape B. E. (1981) Isolation and identification of a metabolite of haloperidol. *J. Anal. Toxicol.* **5**, 113–117.

Reed K. L., Sandor P., and Lee M. Determination of pimozide in plasma by electron impact gas chromatography-mass spectrometry (manuscript in preparation).

Rose M. E. (1984) Gas chromatography–mass spectrometry and high performance liquid chromatography–mass spectrometry. *Mass Spectrom.* **7**, 196–292.

Rose M. E. and Johnstone R. A. W. (1982) *Mass Spectrometry for Chemists and Biochemists* Cambridge University Press, Cambridge, UK.

Routledge P. A., Kitchell B. B., Bjornsson T. D., Skinner T., Linnoila M., and Shand D. G. (1980) Diazepam and N-desmethyldiazepam redistribution after heparin. *Clin. Pharmacol. Ther.* **27**, 528–532.

Rubio F., Miwa B. J., and Garland W. A. (1982a) Determination of midazolam and two metabolites of midazolam in human plasma by gas chromatography–negative chemical–ionization mass spectrometry. *J. Chromatogr. Biomed. Appl.* **233**, 157–165.

Rubio F., Miwa B. J., and Garland W. A. (1982b) Determination of imidazobenzodiazepine-3-carboxamide, a new anxiolytic agent, in human plasma by gas chromatography–negative chemical–ionization mass spectrometry. *J. Chromatogr. Biomed. Appl.* **233**, 167–173.

Scoggins B. A., Maguire K. P., Norman T. R., and Burrows G. D. (1980) Measurement of tricyclic antidepressants. 1. A review of methodology. *Clin. Chem.* **26**, 5–17.

Shaw G. J. and Markey S. P. (1977) The synthesis of N-deuteromethylated desipramine and imipramine. *J. Labelled Compounds Radiopharmaceut.* **13**, 315–322.

Silber B., Lo M., and Riegelman S. (1980) The influence of heparin administration on the plasma protein binding and disposition of propranolol. *Res. Commun. Pathol. Pharmacol.* **27**, 419–429.

Skrinska V., Ohman J., and Wellstead C. (1984) Gas chromatography–mass spectrometry of maprotiline in serum. *Clin. Chem.* **30**, 1276–1277.

Stafford Jr., G. C., Kelley P. E., Syka J. E. P., Reynolds W. E., and Todd J. F. J. (1984) Recent improvements in and analytical applications of advanced ion trap technology. *Int. J. Mass Spectro. Ion Process.* **60**, 85–98.

Szczepanik-Van Leeuwen P. A. (1985) Improved gas chromatographic–mass spectrometric assay for haloperidol utilizing ammonia chemical ionization and selected-ion monitoring. *J. Chromatogr.* **339**, 321–330.

Taulov I. G., Tamas J., Hegedus-Vajda J., and Simov D. (1980a) Mass spectrometric studies of some phenothiazine-S,S-dioxides. *Acta Chim. Acad. Sci. Hung.* **105**, 117–126.

Taulov I. G., Tamas J., Hegedus-Vajda J., and Simov D. (1980b) Mass spectrometric studies of some phenothiazine-S-dioxides. *Acta Chim. Acad. Sci. Hung.* **105**, 109–116.

Todd J. F. J. (1981) A survey of the current state of quadrupole mass spectrometry. *Dyn. Mass Spectrom.* **6**, 3–13.

Todd J. F. J. (1984) A survey of the current state of quadrupole mass spectrometry. *Int. J. Mass Spectrom. Ion Process.* **60**, 3–13.

Todd J. F. J. (1986) Instrumentation in Mass Spectrometry, in *Advances in Mass Spectrometry 1985* John Wiley, New York.

Tomori E., Horvath G., Elekes I., Lang T., and Korosi J. (1982) Investigation of the metabolites of tofizopam in man and animals by gas-liquid chromatography–mass spectrometry. *J. Chromatogr.* **241**, 89–99.

Vasiliades J. and Sahawneh T. (1982) Midazolam determination by gas chromatography, liquid chromatography and gas chromatography–mass spectrometry. *J. Chromatogr. Biomed. Appl.* **228**, 195–203.

Vasiliades J., Sahawneh T. M., and Owens C. (1979) Determination of therapeutic and toxic concentrations of doxepin and loxapine using gas-liquid chromatography with a nitrogen-sensitive detector, and gas chromatography–mass spectrometry of loxapine. *J. Chromatogr. Biomed. Appl.* **164**, 457–470.

Vink J., Van Hal H. J. M., and Koppens P. C. J. M. (1980) From biological sample to final result of analysis: A long way to go using GCMS in drug research. *Adv. Mass Spectrom.* **8B**, 1251–1260.

Whelpton R., Curry S. H., and Watkins G. M. (1982) Analysis of plasma trifluoperazine by gas chromatography and selected ion monitoring. *J. Chromatogr. Biomed. Appl.* **228**, 321–326.

Wiegand U. W. and Levy G. (1979) Effect of heparin injection on plasma protein binding of bilirubin and salicylate in rats. *J. Pharm. Sci.* **68**, 1483–1486.

Wilson J. M., Williamson L. J., and Raisys V. A. (1977) Simultaneous measurement of secondary and tertiary tricyclic antidepressants by GC/MS chemical ionization mass fragmentography. *Clin. Chem.* **23**, 1012–1017.

Wood M., Shand D. G., and Wood A. J. J. (1979) Altered drug binding due to the use of indwelling heparinized cannulas (heparin lock) for sampling. *Clin. Pharmacol. Ther.* **25**, 103–107.

Ziegler V. E., Fuller T. A., and Biggs J. T. (1976) Nortriptyline and 10-hydroxynortriptyline plasma concentrations. *J. Pharm. Pharmacol.* **28**, 849–850.

High-Performance Liquid Chromatographic Analysis of Antidepressants, Neuroleptics, and Benzodiazepines

Donald F. LeGatt

1. Introduction

Whether one is a toxicologist attempting to confirm drug overdose or establish cause of death, a physician/clinical chemist seeking to correlate clinical response with a serum level, or a research scientist collecting pharmacokinetic data, precise accurate analytical methods for drugs are required. Such a method, high-performance liquid chromatography (HPLC), has become an invaluable analytical tool not only in the research environment, but in clinical and forensic laboratories as well. This technique has been applied extensively to the qualitative and quantitative analysis of three groups of widely prescribed and abused mood-altering drugs—the antidepressants, neuroleptics, and benzodiazepines. This chapter will review the types of HPLC procedures available for the three drug groups, concentrating on methods published within the last 7 yr. Considerations pertaining to specimen collection, storage, and handling and specimen preparation prior to chromatography will be addressed. Detailed evaluations of individual methods will not be done, and the appropriate articles should be searched for more specific information.

2. Theory

Since the theory of HPLC has been discussed in previous volumes, this portion of the chapter will be brief. HPLC is a separation technique causing mixtures of compounds to be resolved by exploiting differences in the physical and chemical properties of the compounds. The rate of migration of each compound in a mixture passing through a bed or column of stationary phase under the influence of the moving liquid, the mobile phase, is dependent on the differing affinities each compound has for the two phases. Compounds are then directed individually into a detection system that records the elution of the compounds from the stationary phase.

Liquid chromatographic separations of antidepressants, neuroleptics, and benzodiazepines utilize liquid-solid (adsorption) chromatography and liquid-liquid (partition) chromatography, the latter of which includes reverse-phase chromatography, normal-phase- chromatography, and ion-pair chromatography.

Adsorption chromatography describes a system in which the solid stationary phase (silica gel) is more polar than the mobile phase. The mobile phase in a typical adsorption system utilizes a nonpolar solvent (e.g., hexane) as the prinicple solvent to which a second polar solvent (e.g., water, an alcohol) may be added in small concentrations ($\leq 5\%$). Separation involves competition between drug molecules in the sample and molecules of the mobile phase for sites on the active adsorbent surface.

Partition chromatography encompasses a wide, often confusing range of column types and operating conditions. This method involves the coating or chemical bonding of a liquid stationary phase on the surface of the support, usually silica. Reverse-phase chromatography, a form of partition chromatography, describes a system in which the mobile phase is more polar than the stationary phase. Stationary phases are nonpolar, usually hydrophobic alkyl- or aryl-silane compounds. Mobile phases are aqueous with water-miscible organic solvents (methanol, acetonitrile) added as elution modifiers. The major advantage of nonpolar stationary phases over other liquid chromatography systems is rapid attainment of equilibrium after alterations in mobile phase composition, allowing for the operation of gradient elution separations. Elution characteristics of compounds in reverse-phase chromatography are critically dependent on mobile-phase composition, with compounds eluting more rapidly as the organic solvent proportion in the mobile phase is increased.

Polar stationary phases [e.g., cyano (CN) functionality] chemically bonded to silica particles represent another form of partition chromatography often referred to as normal-phase chromatography. As with adsorption chromatography, mobile phases are generally organic in composition. However, CN-bonded columns are versatile, also capable of operation in the reverse-phase mode. This type of column is usually preferred to aryl-/alkyl-silane-bonded (e.g., C_{18}) and adsorption-phase columns because problems associated with peak tailing on C_{18}-type columns and equilibration/retention time shifting on adsorption columns are eliminated. In addition, ion pairing is usually unnecessary.

Ion-pair partition chromatography has been employed to eliminate peak tailing of basic compounds. For elution of the basic drugs to be discussed in this chapter, a negatively charged ion (e.g., sulfate, sulfonate, perchlorate) is added to the mobile phase. The mechanism involves either formation of a neutral ion pair, which is partitioned between the stationary and mobile phases, or loading of the surface of the stationary phase with the added ionic solution, which then functions as an ion exchanger. Advantages claimed for ion-pair chromatography over the conventional methods of liquid chromatographic separation include enhanced column efficiency and improved peak symmetry. However, the use of ion pairing has been challenged by some investigators (Bannister et al., 1981; Breutzmann and Bowers, 1981) who have claimed the above-mentioned benefits are offset by the problem of slow column equilibration.

Ultraviolet (UV), fluorescence, or electrochemical detectors are utilized for analysis of the antidepressants, neuroleptics, and benzodiazepines, with fixed- or variable-wavelength UV detectors representing the majority. Fluorescence and electrochemical detectors are more restrictive, being suitable only for compounds possessing fluorescence or oxidizing potential, respectively. In addition, the electrochemical detector is potentially less stable than the more dependable UV detector. However, the latter two detector types generally possess greater sensitivity and increased selectivity compared to the UV detector.

3. Antidepressants

Measurement of serum or plasma levels of antidepressants, particularly imipramine, desipramine, and nortriptyline, is clinically useful in certain situations (Glassman et al., 1985) and may prove beneficial in treatment of overdose or in establishing cause of death. HPLC at present is a method of choice for analysis of this drug group.

3.1. Preanalytical Concerns

Prior to extraction and chromatography, proper collection procedures and sample-handling techniques must be followed to ensure analysis is performed on a representative specimen. Timing of specimen collection in relationship to drug administration is also important, but will not be discussed here in detail.

Most analytical techniques for antidepressants have utilized plasma collected in heparinized glass or polystyrene tubes, or serum as the biological specimens of choice. However the plasma/erythrocyte ratios for antidepressants are variable between individuals (Linnoila et al., 1978), indicating whole blood determinations may be more acceptable.

Heparin has been implicated in increasing the free fraction of various drugs after injection (Wood et al., 1979; Desmond et al., 1980; Routledge et al., 1980). However, other investigators have not observed this effect (Narango et al., 1980; Silber et al., 1980). The mechanism supposedly involves stimulation of lipoprotein lipase, causing inactivation of lipoproteins to which the drugs are highly bound. Resultant increases in free drug concentration could lead to decreased plasma or serum levels because of redistribution into erythrocytes. Investigations with antidepressants, which are also highly lipoprotein bound, have not been done. The solution is to avoid the use of heparin in indwelling needles or cannuli used for sample collection.

More of a historical note than of current practical importance is the documented reduction in serum or plasma antidepressant concentrations when blood was collected in Vacutainer tubes (Borga et al., 1977). The lower levels were caused by inhibition of antidepressant binding to alpha-1-glycoprotein by tris(2-butoxyethyl)phosphate (TBEP), a stopper plasticizer, resulting in drug redistribution to the erythrocyte fraction. However, a current concern with regard to blood collection tubes involves the serum separator blood collection tubes. This tube type with a gel material providing a physical barrier between the serum and the erythrocyte fractions following centrifugation is very popular. Sample pour-off is eliminated, resulting in both saved time and a reduced possibility of specimen misidentification. However, is the gel layer inert? Unfortunately it appears not. Significant concentration decreases for lidocaine, pentobarbital, and phenytoin in serum of specimens collected in Becton-Dickson serum separator tubes have been found (Quattrocchi et al., 1983). Serum concentration decreases have also been documented for amitriptyline, imipramine, and desipramine in specimens collected in similar tubes (Orsulak et al., 1984). To date, however, interference with total drug serum concentrations in specimens collected in Corvac® serum separator tubes has not been found in our laboratory. Jewesson and Remick (1986) found no difference between standard evacuated tubes and serum separator tubes, (Vacutainer®, Monoject®) used when sampling blood for serum tricyclic antidepressant analysis. Nevertheless, all collection containers should be properly evaluated in vitro and in vivo.

Once collected in a suitable container, plasma or serum samples may be stored at 4°C for up to 1 mo or at –20°C if time between collection and analysis is greater. Particular precautions should be taken processing blood samples for nomifensine analysis. To prevent in vitro degradation of the pharmacologically inactive, acid-labile, temperature-sensitive conjugate of the active free form (Dawling et al., 1979; McIntyre et al., 1981), serum or plasma should be rapidly processed and frozen at –20°C until analyzed. If specimens are to be mailed or transported to a referral laboratory, plasma or serum frozen on dry ice is the most acceptable procedure. Transport of whole blood is unacceptable.

3.2. Extraction

Direct liquid chromatography of an intact serum or plasma sample for antidepressant analysis is generally not feasible for three reasons: the expected lifetime of the chromatographic column would be in jeopardy, the resultant chromatograms would be virtually uninterpretable because of the multitude of peaks, and the analyses would not provide the required sensitivity. Consequently, some form of extraction and concentration is required. The Cole-Parmer Instrument Co. (Chicago, IL) has overcome two of the previously mentioned limitations by introducing the internal surface reverse-phase (ISRP) concept, which allows for direct injection of serum without prior protein removed (Cole-Parmer Instrument Company, 1986). The protein-excluding pores on the hydrophilic surface force the larger protein molecules to pass through the column unretained. The analytes of interest are resolved by the interior surface, which is chemically bonded hydrophobic reverse phase. Application to analysis of drugs with serum concentration in the microgram per liter range was not demonstrated. The direct injection of filtered urine specimens for the analysis of the antidepressant citalopram and associated metabolites has been accomplished (Oyehaug et al., 1984). Adsorption to glassware, particularly in the final evaporation or concentration stage, accounts for substantial drug loss. Most researchers acid wash, silanize, or utilize both processes for glass-ware treatment in an attempt to overcome this problem. For acid washing, hydrochloric or chromic acid followed by a deionized water or ammonium hydroxide/water rinse in used. Silanization is accomplished with various reagents, one being dimethyldichloro-silane in toluene coupled with an organic solvent rinse. Siliconiza-tion with commercial solutions such as Serva® siliconizing solution or Surfasil® (Pierce Chemical Co.) is a viable alternative applied in our laboratory to minimize drug adsorption. Rinsing of clean glass-ware with the solution is followed by baking at 100°C for 1 h to bond the alkylsiloxane film to the surface. Pretreatment of glass-ware with *N*-butylamine has been done to deactivate glassware adsorption sites (Dehaas-Vermeulen and Thompson, 1984). Di-ethylamine (0.5%) added to hexane extraction solvent has also been used to minimize adsorption (Johnson et al., 1982). The definitive solution to the adsorption problem is to avoid the use of glassware. Dixon and Martin (1981) found neglible adsorp-tion using inexpensive disposal polypropylene tubes for extraction and concentration steps in an assay for simultaneous amitripty-line and imipramine determination. However, significant ad-sorption was found by Thoma and coworkers (1979), who used polypropylene tubes for extraction and evaporation. A novel ap-proach to reduce adsorption was proposed by Edelbroek and coworkers (1982) in a procedure for amitriptyline. The tetracyclic maprotiline was added to the organic phase prior to evaporation, and reduced adsorption of amitriptyline and metabolites was observed.

Antidepressants in general are lipophilic strong bases and therefore are isolated as free bases by adjustment of the biological specimen to pH 9–12, followed by extraction with an organic sol-vent or organic solvent mixture. Organic solvent quality should be high (i.e., HPLC or glass-distilled grade) to ensure optimal chromatography. An alcohol such as isoamyl alcohol is sometimes added (1–5%) to enhance extraction and minimize emulsion forma-tion. For quantitative purposes, an internal standard possessing structural similarity to the compound(s) in question is incorporated into the sample prior to pH adjustment. This is done to minimize variable recovery caused by adsorption. Methods published to date require serum or plasma volumes ranging from 1 to 4 mL. The latter volume is inconvenient and in some cases impractical for routine clinical analysis. After extraction the nonaqueous phase is either evaporated to dryness followed by reconstitution with mobile phase or aqueous buffer (reverse-phase systems only) or is back-extracted to eliminate interference from endogenous constituents. For back-extraction, a mineral acid such as hydrochloric or sulfuric acid is satisfactory. However, oxidizing acids such as perchloric or nitric acid should be avoided (Norman and Maguire, 1985). Unlike gas chromatography, derivatization or structural modification of the antidepressants is usually unnecessary for HPLC. However, formation of fluorescent derivatives of maprotiline and oxaprotiline from dansyl chloride (Breyer-Pfaff et al., 1984) and prechromatographic alkylation and oxidation of diclofensine and its major plasma metabolites (Strojny and de Silva, 1985) have been described. These two manual methods, however, are labor-intensive and would not be readily applicable to the clinical laboratory.

For determination of hydroxylated antidepressant metabolites in plasma, a hydrolysis step is not required, since only the free or unconjugated metabolites possess clinical or toxic activity. However, in the pharmacokinetic studies, plasma or urine are hydrolyzed with β-glucuronidase enzyme (Bock et al., 1982; Strojny and de Silva, 1985) or acid, usually hydrochloric (Biggs et al., 1979), to liberate the hydroxylated metabolites from conjugation to glucuronide, allowing for analysis of total metabolite.

Although serum and plasma are the specimens of choice, whole blood has also been used for antidepressant quantitation (Smith et al., 1982). Hemolysis of the sample is required prior to initiation of the extraction process.

Manual extraction procedures for antidepressants and associated metabolites have recoveries ranging from 53 to 93%.

Gaining in popularity for antidepressant analyses are disposable extraction columns. Columns packed with diatomaceous earth (Clin Elut®, Analytichem International, Harbor City, CA) or C_{18}- and cyanopropyl-bonded silica (Bond Elut®, Analytichem International, Harbor City, CA; Sep-Pak®, Waters Assoc., Milford MA) provide acceptable to good recoveries (75–100%) of antidepressants from plasma or serum with small solvent (eluate) volumes (Thoma et al., 1979; Tasset and Hassan, 1982; Beierle and Hubbard, 1983a,b; Kobayashi et al., 1984; Lensmeyer and Evenson, 1984). However, recoveries of only 65% for imipramine and desipramine were realized from serum using Sep-Pak® C_{18} cartridges (Waters Assoc.) (Kobayashi et al., 1984). Elimination of mixing, centrifugation, and extract transferring reduces technical time and minimizes losses caused by adsorption.

The Prep I® automated sample processor (Dupont) has been utilized for extraction of antidepressants from serum (Koteel et al., 1982). This automated instrument is a reversible centrifuge, in which solvents pass through disposable resin cartridges packed with Styrene divinylbenzene copolymer into two separate collection cups. One cup collects unadsorbed solutes and wash solution, and the second, subsequent to a change in rotation, collects the eluate solvent containing the analytes of interest. Recoveries were from 72 to 97%. A second automated system, the Technicon "Fast LC" system (Technicon Instruments Corp, Tarrytown, NY) has been described for antidepressant quantitation (Bannister et al., 1981). With pipeting of the serum sample into the sample tray being the only manual manipulation, this system provides extraction and HPLC analysis under microprocessor control. Recoveries are respectable (95% and 76% for tertiary and secondary antidepressants, respectively). Varian Instruments (Sunnyvale, CA) markets an automated HPLC sample preparation/injection system called the AASP®. The AASP® automates sorbent extraction of eight tricyclic antidepressants from serum and injects the isolated sample into the liquid chromatograph portion of the instrument (Varian Instruments, Sample Preparation Digest no. 2). The major advantages of automated systems compared to manual procedures are reduction in dedicated technologist's time, sample saving, and decreased organic solvent requirement. System cost need not be a deterrent of automated systems, since the AASP® may be purchased for less than $20,000 (Canadian).

3.3. Analysis

3.3.1. Tricyclics

The largest portion of those drugs categorized as antidepressants has a common structural nucleus: a three-membered ring. Some of the more commonly prescribed and analyzed tricyclic antidepressants are displayed in Fig. 1. Published analytical procedures for tricyclic antidepressants describe analysis of parent drug and associated metabolites or simultaneous determination of two or more drugs. The former method type will be discussed first.

Amitriptyline is the most frequently administered drug of this class of antidepressants. The methods to be discussed apply to serum and plasma, the exceptions being for analysis in whole blood (Smith et al., 1982) and urine (Biggs et al., 1979). Some procedures only discuss analysis of the drug and the active de-

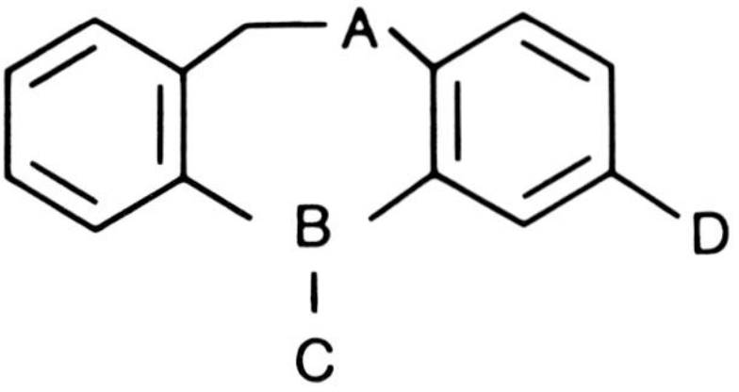

Fig. 1. Structures of selected tricyclic antidepressants.

	A	B	C	D
Amitriptyline	CH_2	C	$=CHCH_2CH_2N(CH_3)_2$	H
Nortriptyline	CH_2	C	$=CHCH_2CH_2NHCH_3$	H
Imipramine	CH_2	N	$-CH_2CH_2CH_2N(CH_3)_2$	H
Desipramine	CH_2	N	$-CH_2CH_2CH_2NHCH_3$	H
Clomipramine	CH_2	N	$-CH_2CH_2CH_2N(CH_3)_2$	Cl
Trimipramine	CH_2	N	$-CH_2\underset{CH_3}{CH}CH_2N(CH_3)_2$	H
Doxepin	O	C	$=CHCH_2CH_2N(CH_3)_2$	H

methylated metabolite nortriptyline (Biggs et al., 1977; Brodie et al., 1977a; Watson and Stewart, 1977; Sonsalla et al., 1982b), whereas others describe simultaneous determination of various hydroxylated metabolites as well (Kraak and Bijster, 1977; Mellstrom and Braithwaite, 1978; Biggs et al., 1979; Bock et al., 1982; Edelbroek et al., 1982; Smith et al., 1982; Suckow and Cooper, 1982). The latter methods are more acceptable because quantitation of the pharmacologically active hydroxylated metabolites is possible, and hydroxylated metabolite noninterference in quantitating the major active substances (amitriptyline, nortriptyline) is guaranteed. The instrument systems applied for amitriptyline analysis are based on adsorption (Watson and Stewart, 1977; Edelbroek et al., 1982; Smith et al., 1982; Sonsalla et al., 1982b), reverse-phase (Brodie et al., 1977a; Kraak and Bijster, 1977; Biggs et al., 1979; Bock et al., 1982), and ion-pair chromatography operated in the reverse-phase (Suckow and Cooper, 1982) and adsorption (Mellstrom and Braithwaite, 1978) modes. The majority of columns of varying lengths (10–30 cm) were packed with 5-μm particle diameter packing, which provides both increased sensitivity and resolution compared to the 10-μm material. Although the undisputed superiority of one chromatographic system cannot be demonstrated, Suckow and Cooper (1982) accomplished resolution of amitriptyline and five metabolites [*cis* (Z) and *trans* (E) 10-hydroxynortriptyline, *cis* (Z) and *trans* (E) 10-hydroxyamitriptyline, nortriptyline] in 20 min with a reverse-phase method using heptanesulfonate as an ion-pairing reagent (Fig. 2). A more rapid analytical method using an adsorption system resolved the above-mentioned compounds, with the exception of *cis* (Z) and *trans* (E) 10-hydroxyamitriptyline, in 10 min (Edelbroek et al., 1982). Although the discipline of chiral pharmacology is receiving more attention, the clinical significance of monitoring optically active isomers of the same amitriptyline metabolite remains to be proven. Furthermore, the ability of a method to resolve the *cis* (Z) and *trans* (E) isomers of 10-hydroxyamitriptyline may be unimportant, since concentrations of the latter are usually less than 5 μg/L in patients receiving amitriptyline. Analysis of amitriptyline and metabolites is limited to instruments with UV detectors, since response to fluorometric or electrochemical systems is poor or nonexistent. The most widely used UV wavelength is 254 nm. Although variable wavelength detectors set at the far UV region (220–200 nm) have been used, increased interference from concurrently administered drugs, particularly the benzodiazepines, may result. Sensitivity of assays for amitriptyline ranged from 2 to 25 μg/L using varying amounts of serum or plasma for extraction (1–4 mL). As with all chromatographic techniques, interference by other prescribed or over-the-counter medications that coextract can present a serious problem in HPLC analysis of amitriptyline. Therefore, disclosure of all medications currently prescribed is required to ensure accurate quantitation.

Imipramine, like amitriptyline, is biotransformed into a therapeutically active demethylated metabolite (desipramine) and biologically active hydroxylated metabolites (2-hydroxyimipramine, 2-hydroxydesipramine). Consequently, an assay capable of resolving all four compounds will be beneficial for pharmacokinetic studies and potentially useful in a clinical environment. However, at present most clinicians request imipramine and desipramine only. Some HPLC assays discuss only imipramine and desipramine determination (Reece et al., 1979; Kobayashi et al., 1984), whereas others provide resolution of the four specified compounds (Sutfin and Jusko, 1979; Godbillon and Gauron, 1981; Suckow and Cooper, 1981). Fekete and coworkers (1981) resolved imipramine, desipramine, didesmethylimipra-

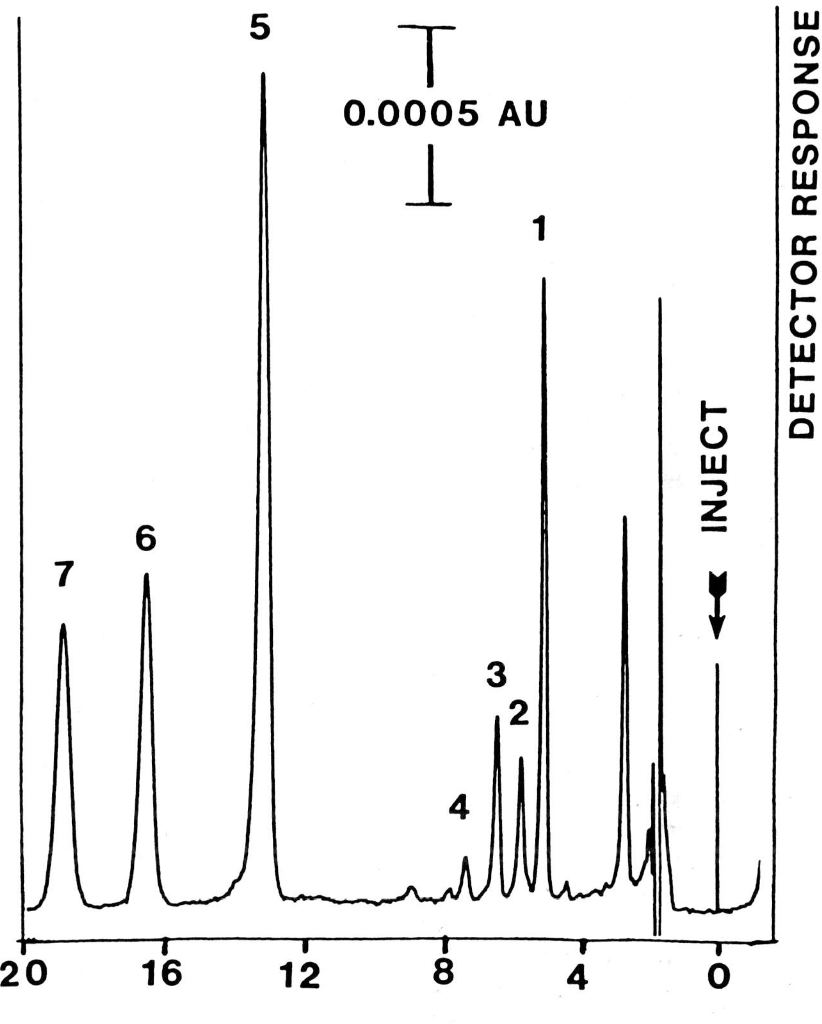

Fig. 2. Sample chromatogram of a plasma sample (1 mL) from a patient receiving amitriptyline analyzed on a 5-μm reverse-phase (trimethylsilyl) column, using the counter-ion heptanesulfonate. Compounds resolved include (1) *trans*-10-OH-nortriptyline, (2) *trans*-10-OH-amitriptyline, (3) *cis*-10-OH-nortriptyline, (4) *cis*-10-OH-amitriptyline, (5) loxapine (internal standard), (6) nortriptyline, and (7) amitriptyline (from Suckow and Cooper, 1982).

mine, 2-hydroxyimipramine, and imipramine-N-oxide in 15 min. However, quantitation of the latter two compounds in biological samples would be difficult because of a significant solvent front. Furthermore, reliable quantitation is jeopardized by the absence of internal standards. Chromatography is performed using adsorption (Sutfin and Jusko, 1979; Godbillon and Gauron, 1981), reverse-phase (Reece et al., 1979; Fekete et al., 1981; Kobayashi et al., 1984), and ion-pair reverse-phase (Suckow and Cooper, 1981) systems. Fluorescence, electrochemical, and UV detectors have been utilized in instruments for imipramine determination. Sensitivity of the assays ranged from 0.5 to 10 µg/L with manual extraction of 1–2 mL of serum or plasma. The most applicable system appears to be one described by Suckow and Cooper (1981). Manual extraction of plasma (1 mL), chromatography on a µ-Bondapack C_{18} reverse-phase column with an acetonitrile/acetate mobile phase (40:60) containing heptanesulfonate as the ion-pairing agent, and electrochemical detection resulted in resolution of imipramine, desipramine, and the associated 2-hydroxylated metabolites within 15 min. Sensitivity of 5 µg/L was established.

Clomipramine, the chlorinated imipramine analog, and its monodemethylated metabolite desmethylclomipramine, have been resolved and quantitated by adsorption (Westenberg et al., 1977a,b; Godbillon and Gauron, 1981; Diquet et al., 1982), adsorption ion-pair (Mellstrom and Tybring, 1977), and reverse-phase (Linnoila et al., 1982) chromatography. Although Linnoila and coworkers (1982) have quantitated the eight hydroxy metabolites of clomipramine and desmethylclomipramine, these metabolites, which may possess biological activity, have not been extensively investigated by HPLC. Another minor metabolite, di-desmethylclomipramine, has been analyzed simultaneously with clomipramine and desmethylclomipramine (Mellstrom and Tybring, 1977). Ultraviolet detection at 250–254 nm has been used almost exclusively for clomipramine determination. Although both clomipramine and the demethylated metabolite exhibit higher absorbances at 220 nm, a wavelength in the near UV range was selected to eliminate potential interference from solvent impurities (Westenberg et al., 1977b). An electrochemical detector has also been applied to detection of this tricyclic drug (Linnoila et al., 1982). Assay sensitivities range from 2 to 10 µg/L and 5 to 18 µg/L for clomipramine and desmethylclomipramine, respectively, when 1–4 mL of serum or plasma were extracted.

HPLC procedures dedicated solely to the analysis of doxepin are few. Whall and Dokladalova (1979) describe an assay for the separation of the *cis* (Z) and *trans* (E) isomers of the drug. This assay, however, does not have clinical application, since a pharmacologically active metabolite desmethyldoxepin was not considered. Faulkner and Lee (1983) have described an assay for the simultaneous quantitation of doxepin and desmethyldoxepin in plasma. Using a 5-μm reverse-phase column and an UV detector, the two compounds and internal standard (desipramine) were resolved in 9 min. The lower detection limit was 5 μg/L for both compounds. The assay appears suitable for routine monitoring, although a wide solvent front in the depicted chromatograms may be potentially problematic for desmethyldoxepin analysis. Thoma and coworkers (1979) achieved good chromatography of doxepin, desmethyldoxepin, and two internal standards (trimipramine, protriptyline) within 6 min on a CN-bonded column. The short analysis time was facilitated by maintenance of column temperature at 60°C. Interference from the solvent front did not occur, and although sensitivities were not stated, calibration curves were linear from 25 to 1000 μg/L with extraction of 1-mL samples. This assay is versatile with applications to amitriptyline/imipramine analysis.

Trimipramine, an imipramine analog, and three major metabolites (desmethyltrimipramine, 2-hydroxytrimipramine, 2-hydroxydesmethyltrimipramine) have been analyzed on a reverse-phase system with electrochemical detection (Suckow and Cooper, 1984). Good sensitivity (3 μg/L) and rapid analysis (10 min) make this assay applicable to routine monitoring of trimipramine and its metabolites. However, interference did occur in plasma from patients receiving imipramine, desipramine, clomipramine, chlorpromazine, amoxapine, loxapine, trazodone, and haloperidol. Other tricyclic and tetracyclic antidepressants and most benzodiazepines did not interfere because they do not respond electrochemically.

Brodie and coworkers (1977b) developed an HPLC procedure for dothiepin and the potentially active monodesmethyl metabolite northiaden. With a reverse-phase system (Partisil 10-ODS column) and UV detection (231 nm), 10 μg/L of each component could be detected on extraction of 2 mL of plasma or serum. Analysis time was approximately 12 min. Noninterference from other metabolites (S-oxides and 2-hydroxydothiepin) was confirmed.

Cianopramine, a cyano analog of imipramine, has been quantitated by Hojabri and Glennon (1985) utilizing adsorption chromatography with either UV or fluorescence detection. Fluoroscence detection was preferred, since both increased sensitivity (12.5 μg/L compared to 50 μg/L for UV detection—1 mL plasma extracted) and improved chromatography were realized.

The assays just described for tricylic antidepressants are for the most part dedicated or designed for quantitation of one drug and its associated metabolites. Procedures for analysis of more than one tricyclic antidepressant have also been described. There are two types. The first applies to analysis of a group of antidepressants (tricyclics, tricyclics/bicyclic), with the group members analyzed individually (Moyes and Moyes, 1977; Hackett and Dusci, 1979; Dixon and Martin, 1981 Blanc et al., 1983; Lagerstrom et al., 1983; Rop et al., 1985; Rop et al., 1986). The second type of procedure achieves simultaneous resolution and analysis of tricyclic, tricyclic/tetracyclic, or tricyclic/novel drug combinations (Proelss et al., 1978; Vandermark et al., 1978; Thoma et al., 1979; Streator et al., 1980; Bannister et al., 1981; Breutzmann and Bowers, 1981; Kabra et al., 1981; Wallace et al., 1981a; Johnson et al., 1982; Koteel et al., 1982; Sonsalla et al., 1982a; Beierle and Hubbard, 1983a; Yang and Evanson, 1983; Lensmeyer and Evanson, 1984; Sutfin et al., 1984; Visser et al., 1984; Yufu et al., 1984). Adsorption, reverse-phase, and reverse-phase/ion-pair chromatography have been utilized. Most of these procedures have routine applicability. The most suitable depends on the particular application required. However, some of the more impressive separations have been achieved with CN-bonded columns operated in the reverse-phase mode. Yang and Evanson (1983) and Lensmeyer and Evanson (1984), using a Xorbax cyanopropylsilane (CN) column (Dupont, Wilmington, DE), resolved nine and ten antidepressants, respectively in 16 min (Fig. 3). Extraction of 2-mL serum specimens utilizing a single solvent/freezing extraction step (Yang and Evanson, 1983) or Bond Elut® disposable extraction columns (Lensmeyer and Evanson, 1984) yielded detection limits of 5–10 μg/L. Interference from hydroxylated antidepressant metabolites was not significant, although thioridazine and chlorpromazine metabolites did cochromatograph with some analytes. Procedures designed for analysis of more than one antidepressant are ideally suited for the routine high-volume clinical laboratory. The laboratory interested in implementing such a method must test as many drugs/drug metabolites as possible for potential chromatographic

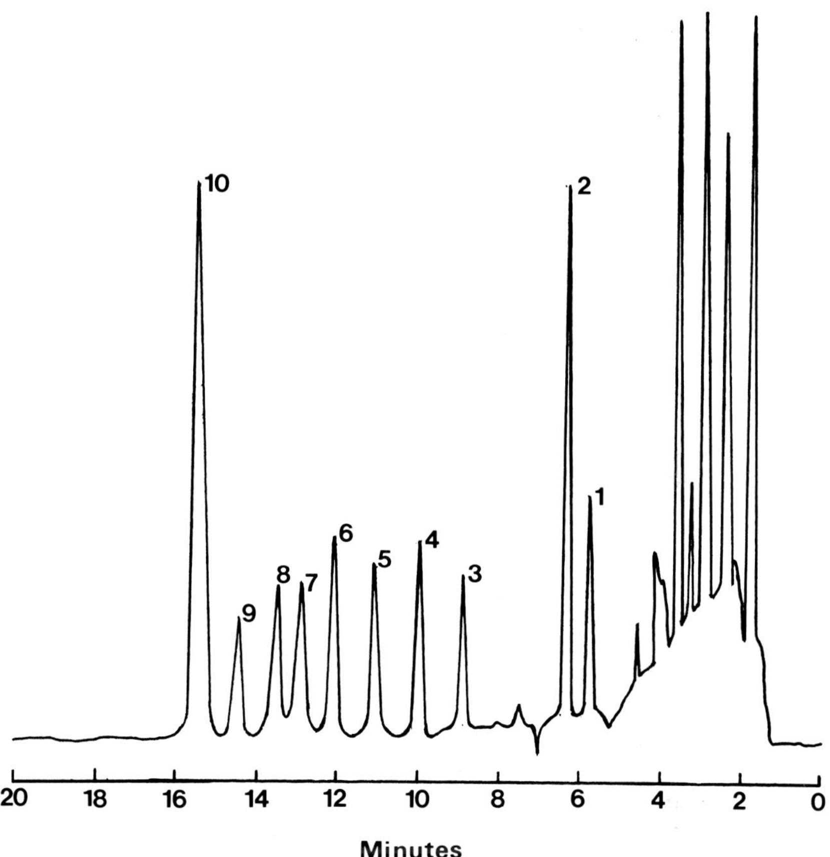

Fig. 3. Chromatogram of extract from serum supplemented with each of the following compounds to 100 µg/L: (1) 8-OH-amoxapine; (2) 7-OH-loxapine; (3) amoxapine; (4) desmethyldoxepin; (5) doxepin; (6) desipramine; (7) nortriptyline; (8) imipramine; (9) amitriptyline; (10) trimipramine (from Lensmeyer and Evenson, 1984).

interference to ensure credible quantitative analyses of the antidepressants concerned.

3.3.2. Tetracyclics

Tetracylic antidepressants are relatively novel and less frequently used drugs compared to the tricyclic congeners. Nevertheless, the tetracyclics have well-documented clinical efficacy. The structures of the more frequently encountered tetracyclic antidepressants are displayed in Fig. 4.

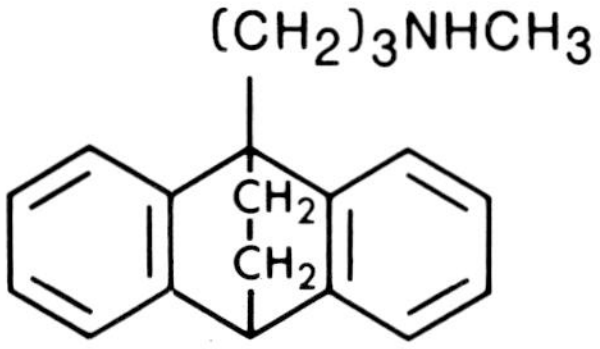

(CH$_2$)$_3$NHCH$_3$

CH$_2$
CH$_2$

MAPROTILINE

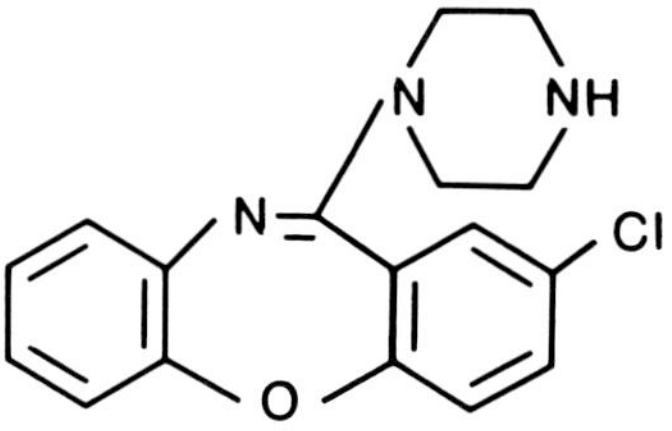

N   NH

N

Cl

O

AMOXAPINE

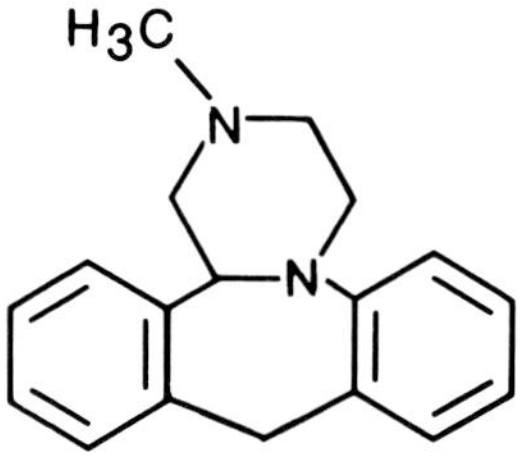

H$_3$C

N

N

MIANSERIN

Fig. 4. Structures of selected tetracyclic antidepressants.

Although maprotiline, a noradrenaline-uptake inhibitor, has been analyzed most often by gas chromatography, HPLC procedures have been published for the drug involving determination of either maprotiline alone (Beierle and Hubbard, 1983b; Ketchum et al., 1983), in combination with an active hydroxylated congener oxaprotiline (Kuss and Feistenauer, 1981; Breyer-Pfaff et al., 1984), or in combination with its demethylated metabolite, desmethylmaprotiline (Salonen and Scheinin, 1983; Wong and Waugh, 1983; Yufu et al., 1984). The clinical benefit of codetermination of the metabolite remains questionable. Both adsorption chromatography with detection of fluorescence derivatives (Breyer-Pfaff et al., 1984) or UV detection (Beierle and Hubbard, 1983b) and reverse-phase methods utilizing UV detection (Kuss and Feistenauer, 1981; Ketchum et al., 1983; Salonen and Scheinin, 1983; Wong and Waugh, 1983; Yufu et al., 1984) have been applied. Limits of detection of 2–15 μg/L for extraction of 1 mL serum/plasma samples were achieved with detection specifically within the far UV region. However, the increased sensitivity (as compared to near UV region detection) was offset by increased interference from tricyclic antidepressants and benzodiazepines. Although the method of Breyer-Pfaff and coworkers (1984) for maprotiline and oxaprotiline in plasma was sensitive, accurate, and reproducible, it is a laborious technique not applicable to a routine clinical laboratory. Salonen and Scheinin (1983) and Wong and Waugh (1983) offer perhaps the most suitable methods, with baseline resolution of maprotiline, the desmethylated metabolite, and internal standards in 10 min.

The tetracyclic amoxapine undergoes hydroxylation to inactive 7-hydroxy and active 8-hydroxy metabolites. The ability to separate the 7- and 8-hydroxy metabolites may not be required, since serum/plasma concentrations of the former are virtually negligible (less than 5 µg/L). Nevertheless, the most suitable HPLC procedure should successfully resolve all three compounds. The majority of techniques for amoxapine determination has utilized reverse-phase systems (Kimball and Lampert, 1982; Tasset and Hassan, 1982; Ketchum et al., 1983; Wong and Waugh, 1983; Johnson et al., 1984; Yufu et al., 1984; Kobayashi et al., 1985; Suckow and Cooper, 1985). Beierle and Hubbard (1983b) used adsorption chromatography, but were unable to resolve the 7- and 8-hydroxylated metabolites. Ultraviolet detection at 250–254 nm has been employed the most frequently. However, amoxapine and its metabolites are oxidizable, and Suckow and Cooper (1985) have exploited this characteristic and utilized an electrochemical detector. Amoxapine, the two hydroxylated derivatives, and internal standard were resolved within 15 min. Limit of detection was 5 µg/L for 1 mL of plasma. Tasset and Hassan (1982) and Johnson and coworkers (1984), using reverse-phase/UV detection systems, achieved acceptable resolution within 8 and 10 min, respectively. However, a decrease in sensitivity results (Tasset and Hassan, 1982: 50 µg/L) or a temperature-controlled column is required (Johnson et al., 1984: 60°C). Kobayashi and coworkers (1985) subjected sera to enzyme hydrolysis for liberation of the conjugates of the two hydroxylated metabolites and subsequent simultaneous quantitation with the parent drug. Carbamazepine, selected as internal standard, was not a practical choice, since this drug is often prescribed to depressed individuals. Most methods discussed potentially interfering compounds.

Mianserin possesses two pharmacologically active metabolites, desmethyl- and 8-hydroxy mianserin. Suckow and coworkers (1982), using reverse-phase ion-pair chromatography with electrochemical detection, achieved simultaneous resolution and quantitation of the three compounds within 12 min. Sensitivities were excellent (5 µg/L for 1 mL plasma). However, the involved extraction technique and potentially unreliable chromatographic/ detection system may limit the usefulness of this method in the busy clinical laboratory. Wong and coworkers (1984) described a reverse-phase/UV detection method for mianserin. Although this method provided comparable sensitivity, analysis of mianserin metabolites was not addressed.

Aptazapine, a potential antidepressant structurally related to mianserin, has been assayed on an adsorption system with fluorometric detection (DeHaas-Vermeulen and Thompson, 1984). Achieving a sensitivity of 2.5 µg/L on extraction of 1 mL plasma, aptazapine and the internal standard mianserin were resolved within 8 min. Although resolution was achieved and peak tailing was minimal, the chromatographic peaks were broad. Chromatography could potentially be improved by substituting the 10-µm silica packing with a 5-µm material. No mention of interference from coadministered drugs was made.

3.3.3. Miscellaneous Antidepressants

Nontricylic, second generation, novel; these names have been applied to a potpourri of compounds bound by a common thread— possession of antidepressant activity. The structures of three of the more common miscellaneous antidepressants are shown in Fig. 5.

HPLC procedures for analysis of trazodone, a triazolopyridine derivative, have utilized reverse-phase chromatography with either UV (Ankier et al., 1981; Root and Ohlson, 1984; Wong et al., 1984; Gerson et al., 1986; Miller and Devane, 1986a), fluorescence (Gupta and Lew, 1985), or electrochemical (Suckow, 1983) detection and adsorption chromatography with fluorescence detection (Caldwell and Flanagan, 1985). Ankier and coworkers realized a sensitivity of 20 µg/L with extraction of 1 mL plasma. Ether extraction preceded the 8-min chromatographic analysis. Although use of an internal standard was claimed, an external standard in fact was utilized. Addition of an internal standard prior to extraction should have been done to compensate for any losses incurred during the procedure. Wong and coworkers (1984) achieved greater sensitivity (5 µg/L, 1 mL plasma extracted), but chromatographic analysis was more lengthy (15 min). An extensive patient evaluation was performed with trazodone plasma concentrations determined in 26 patients (73–1678 µg/L). Root and Ohlson (1984), upon extraction of 1-mL specimens of serum, plasma, whole blood, urine, gastric contents, and tissue homogenates, achieved sensitivity of 50 µg/L. Gupta and Lew (1985), utilizing fluorescence

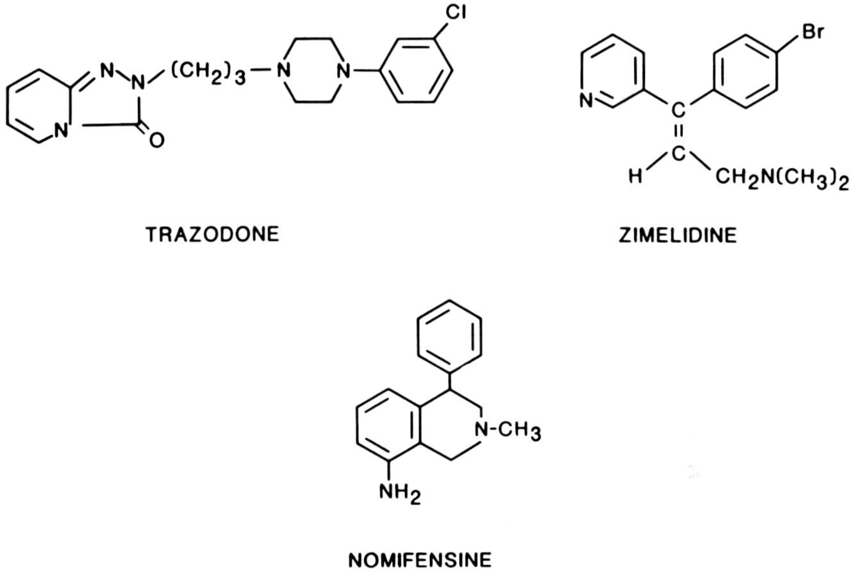

TRAZODONE

ZIMELIDINE

NOMIFENSINE

Fig. 5. Structures of selected miscellaneous antidepressants.

detection, could quantitate 1 ng of trazodone injected on-column with harmine as internal standard. A hydrolytic product of trazodone, 1-*m*-chlorophenylpiperazine (*m*-CPP), which may possess antidepressant activity, was quantitated simultaneously with the parent drug utilizing a reverse-phase system coupled with an electrochemical detector (Suckow, 1983) or a UV detector (Miller and Devane, 1986a). Minimum quantifiable levels of detection from 1-mL plasma samples were 5 and 5–10 µg/L for *m*-CPP and trazodone, respectively. Caldwell and Flanagan (1985) and Gerson and coworkers (1986) detailed techniques applicable to routine monitoring of trazodone in the clinical laboratory. Although metabolites were not discussed, 6 and 4 min analysis times with sensitivities of 250 and 100 µg/L, respectively, were realized. Since the method described by Caldwell and Flanagan requires only 20 µL of patient specimen, assay sensitivity could easily be enhanced by increasing sample size.

Zimelidine, a selective inhibitor of serotonin uptake, has demonstrated antidepressant activity. The metabolic profile includes a major active demethylated product, norzimelidine, and a group of pharmacologically inactive compounds, including di-demethylated zimelidine and zimelidine-*N*-oxide. Therefore, an HPLC procedure must ensure simultaneous determination of zimelidine and norzimelidine and guarantee noninterference from the inactive metabolites. Emanuelsson and Moore (1978) utilized a unique combination of a silica gel column and aqueous mobile phase to resolve zimelidine, norzimelidine, and the two respective E-isomers (internal standards) within 10 min. However, the sensitivity (25 μg/L; 1 mL plasma extracted) may not be adequate for pharmacokinetic studies. Chromatographic conditions resulted in peak tailing, and the inactive metabolites were not discussed. Westerlund and coworkers (1979), using adsorption ion-pair chromatography with UV detection, resolved zimelidine, norzimelidine, and two internal standards in a 10-min chromatographic analysis. Sensitivity was excellent (0.15 μg/L; 1 mL plasma extracted), and interference by inactive metabolites was ruled out. Applications to quantitation in whole blood, urine, and rat brain were also described. The disadvantages of this method were the laborious column preparation and equilibration techniques required prior to analysis. Westerlund and Erixson (1979) described a reverse-phase system with UV detection. Although the sensitivity (1.5 μg/L; 1 mL sample extracted) was not as impressive as the method just described, column preparation was much less sophisticated. A 9-min chromatographic time allowed for baseline resolution of two internal standards, zimelidine, norzimelidine, zimelidine-*N*-oxide, and zimelidine primary amine metabolite. Inclusion of aliphatic tertiary amine, dimethyloctylamine, in the mobile phase optimized the chromatography.

Nomifensine, an isoquinoline derivative with antidepressant properties, has three major metabolites: 4-hydroxynomifensine, 4-hydroxy-3-methoxynomifensine, and 3-hydroxy-4-methoxynomifensine. Although resolution of the three metabolites and the parent drug is a prerequisite for pharmacokinetic investigation, clinical application may only require resolution of nomifensine from the group of metabolites, since clinical efficacy of the later compounds remains to be established. Lindberg and coworkers (1983) described a reverse-phase/UV detection procedure for nomifensine in human serum. An 8-min chromatographic analysis successfully resolved nomifensine and the internal standard *p*-chlorodisopyramide. Excellent sensitivity was achieved (2 μg/L; 1 mL serum extracted), and interference from metabolites and other coadministered medications was negligible. A reverse-phase method for quantitative analysis of the three nomifensine metabolites in human plasma has been described (Lindberg, 1985). Utilizing UV detection, determination of 2 μg/L of each metabolite has been claimed for extraction of 1 mL of plasma. Although chromatography was completed by 8 min, baseline irregularity may contribute to quantitative difficulties on routine application.

HPLC has been applied to several other miscellaneous antidepressants, including citalopram (Oyehaug et al., 1982, 1984), indalpine (Jozefczak et al., 1982), viloxazine (Gillilan and Mason, 1981), indeloxazine (Kamimura et al., 1985), and diclofensine (Strojny and de Silva, 1985). Adsorption or reverse-phase chromatography coupled with UV or fluorescence detection have been applied in these methods.

4. Neuroleptics

Therapeutic drug monitoring of neuroleptics has lagged behind antidepressant monitoring. The lack of sensitive, specific methodologies and the absence of correlation between plasma/serum drug concentrations and clinical effect has restricted quantitation of neuroleptics, until recently, to the confines of the research laboratory. Although monitoring of neuroleptics is still controversial, the clinical laboratory is experiencing increased pressure to expand its analytical repertoire to include this drug group. HPLC measurement of neuroleptics in biological specimens is relatively new in comparison to gas chromatographic procedures and, at present, is not the major analytical tool for neuroleptic analysis. Nevertheless, HPLC instrumentation has improved to the point where this technique will eventually play a more significant role in therapeutic drug monitoring of neuroleptics.

4.1. Preanalytical Concerns

As with antidepressants, precautions must be taken in the collection and storage of biological samples for neuroleptic analysis. In fact, the neuroleptics are even more susceptible to loss prior to analysis. Blood collection tubes with TBEP in the stoppers must be avoided to prevent plasma/RBC redistribution of the neuroleptic. Adsorption phenomena must be prevented. Avoidance of plastic containers and Teflon-lined screw caps for storage, and silanization/siliconization of glassware designated for specimen storage has been recommended (Fekete et al., 1981). Use of polypropylene tubes and caps for sample preparation and storage has also been suggested (Stoll et al., 1984). Furthermore, these compounds are particularly susceptible to photodecomposition. In addition, in vitro isomerization of the active *cis* (Z) isomer of thiothixene to the inactive *trans* (E) isomer can occur in biological specimens exposed to UV light (Bogema et al., 1982). All stock standard solutions, biological samples, and crystalline compounds should be stored in the dark or in light-impenetrable containers. Since neuroleptics are susceptible to oxidation in alkaline media, sodium metabisulfate or ascorbic acid may be added to prevent oxidation loss. Decomposition by heavy metals can pose a problem, and acid washing may therefore be required (Bell, 1985). Chlorpromazine, fluphenazine, and thioridazine demonstrated stabilities of 1 wk (Gupta et al., 1981), 6 mo (Wills and Gelder, 1980) and 3 wk (Wells et al., 1983) in frozen plasma. Chlorpromazine, chlorpromazine sulfoxide, and chlorpromazine N-oxide were found to be stable in whole blood under physiological conditions (Hawes et al., 1986).

4.2. Extraction

Prolongation of column life, reduction in the number of extraneous chromatographic peaks, and improved sensitivity permitting quantitation in the μg/L and ng/L ranges—these are the primary reasons for separating the particular neuroleptic from the associated biological medium prior to chromatographic analysis.

Separation and concentration of the neuroleptics, however, spawns a serious problem—drug adsorption. As with antidepressant analysis, various precautions alone or in combination have been taken to avoid or at least minimize this problem. Silanization of glassware or utilization of siliconized tubes have been the most popular solutions. Polypropylene tubes have been suggested for replacement of glassware. Another remedy has involved elimination of evaporation steps in the extraction procedure. Addition of isopropyl or isopentyl alcohol (3–5%) or isopropylamine to the extractant solvent has also reduced adsorption losses. Supplementation of the original specimen with either a constant amount of the neuroleptic to be quantitated ("carrier" solution) (Bogema et al., 1982) or a particular internal standard found to prevent adsorption (Midha et al., 1981) have also been done.

Extraction of chlorpromazine and associated metabolites from plasma and whole blood has presented analytical problems. In the presence of alkali, chlorpromazine *N*-oxide is reduced to chlorpromazine, resulting in falsely elevated concentrations of the parent drug (Hubbard et al., 1985; McKay et al., 1985). In whole blood, chlorpromazine was oxidized to chlorpromazine sulfoxide. With the exception of alkalinization of plasma with sodium carbonate, alkaline extraction methods for chlorpromazine must be avoided.

Virtually all published procedures for neuroleptics involve manual solvent extraction. The drugs and metabolites are partitioned into an organic solvent or organic solvent mixture from alkalinized biological specimen. However, the similarity ends here. The organic extract is either evaporated to dryness and reconstituted for analysis, back-extracted into an acidic medium, which is then injected directly onto the chromatograph, or back-extracted as above into acid, followed by basification, organic solvent reextraction, evaporation, and reconstitution in mobile phase. Although removal of endogenous interfering substances by back-extraction techniques is a prerequisite for detection of picogram and nanogram neuroleptic levels, promethazine is not a suitable candidate for such techniques because of instability in weakly acidic solutions (Leelavathi et al., 1985). Recoveries of neuroleptics extracted by manual solvent techniques ranged from 33 to 98%.

Direct injection of diluted, deproteinized serum or plasma has been done in an attempt to circumvent laborious extraction procedures (Brinkman et al., 1981; Smith, 1981). Sensitivity is compromised (10–20 µg/L for extraction of unspecified sample volumes) and in some clinical situations may not be sufficient for reliable quantitation. This author feels extra time spent with extraction techniques will not only optimize sensitivity, but ensure better chromatography and longer column life.

Derivatization of compounds is generally regarded as the phenomenon unique to gas chromatography. However, Wallace and coworkers (1981) employed trichloroethylchloroformate derivatization of an organic extract of promethazine for liquid chromatography using UV detection. This procedure has limited use, since it is only applicable to phenothiazine neuroleptics with a tertiary amine group containing at least one methyl group. Furthermore, implementation of a derivatization step negates any advantage HPLC possesses over GLC for analysis of neuroleptics.

As stated previously, analytical procedures for neuroleptics have employed manual solvent extraction almost exclusively. However, Bogema and coworkers (1982) have utilized reusable C_{18} Sep-PAK cartridges (Waters Associates, Milford, MA) for extraction of *cis* (Z) and *trans* (E) isomers of thiothixene from human plasma. The cartridges were activated with methanol and deionized water prior to addition of plasma, internal standard solution, and NaOH. Elution of the cartridges with hexane/isopropanol followed by evaporation and reconstitution in mobile phase precedes HPLC analysis. Optimal recovery was 62%. This procedure not only reduced the analysis time, but eliminated chromatographic interferences found with the coevaluated procedure of manual solvent extraction.

4.3. Analysis

4.3.1. Phenothiazines

Within the neuroleptic family of drugs, the phenothiazine group is undoubtedly the largest. The structures of some selected phenothiazine neuroleptics are displayed in Fig. 6. The analytical challenge facing the individual initiating an HPLC assay for these compounds involves not only achieving sensitivity, but guaranteeing chromatographic integrity or resolution. Phenothiazines undergo extensive metabolism in vivo: ring sulfur atom oxidation, aromatic hydroxylation with conjugation, N-dealkylation, and N-oxidation of the ring, along with combinations of these processes. Chlorpromazine, for example, has 168 postulated metabolites, some of which may be pharmacologically active. Development of an HPLC assay capable of resolving such a large number of compounds is virtually impossible. Therefore, development of an assay that allows for simultaneous determination of the parent drug and a select number of important detectable metabolites must suffice.

The prototype of the phenothiazine group is chlorpromazine, and as previously indicated, this drug is extensively metabolized. Most of the HPLC procedures for chlorpromazine have not detected metabolites in biological samples of patients receiving the drug, although some investigators have demonstrated metabolite noninterference in analysis of the parent drug by chromatography of solutions spiked with metabolites. Allender and coworkers (1983), however, have achieved simultaneous detection of chlorpromazine, 7-hydroxychlorpromazine sulfoxide, chlorpromazine sulfoxide, norchlorpromazine, and norchlorpromazine sulfoxide in liver and blood extracts from a chlorpromazine overdose case. Reverse-phase (Fekete et al., 1981; Smith, 1981; Murakami et al., 1982; McKay et al., 1983), adsorption (Allender et al., 1983; Przy-

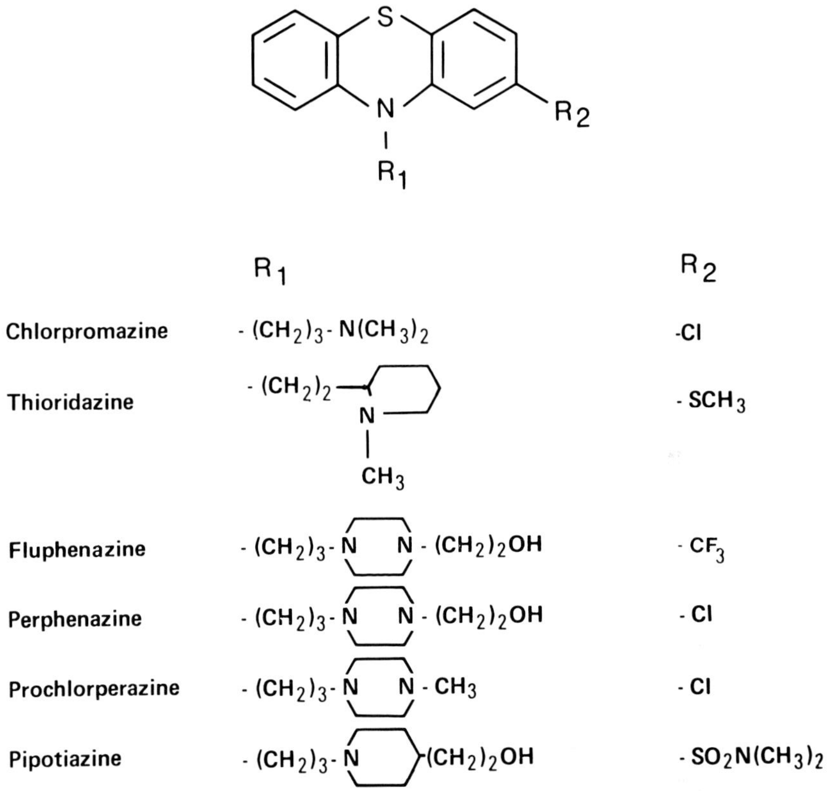

Fig. 6. Structures of selected phenothiazine neuroleptics.

	R_1	R_2
Chlorpromazine	$-(CH_2)_3-N(CH_3)_2$	$-Cl$
Thioridazine	$-(CH_2)_2-$ (2-methylpiperidinyl)	$-SCH_3$
Fluphenazine	$-(CH_2)_3-N$(piperazinyl)$N-(CH_2)_2OH$	$-CF_3$
Perphenazine	$-(CH_2)_3-N$(piperazinyl)$N-(CH_2)_2OH$	$-Cl$
Prochlorperazine	$-(CH_2)_3-N$(piperazinyl)$N-CH_3$	$-Cl$
Pipotiazine	$-(CH_2)_3-N$(piperidinyl)$(CH_2)_2OH$	$-SO_2N(CH_3)_2$

borowska et al., 1985), and normal phase (Midha et al., 1981; Cooper et al., 1983; Hubbard et al., 1985) chromatography have been applied. Operation in the reverse-phase mode would be the most suitable, since elution of the polar metabolites prior to the relatively nonpolar parent would optimize detection of these low concentration compounds. This point may be purely academic, for in reality, chlorpromazine metabolites appear undetectable by HPLC in the plasma/serum of patients receiving therapeutic doses of the drug. Electrochemical or UV detection have been utilized, with a sensitivity as low as 0.25 µg/L achieved with the former detector type (Cooper et al., 1983). This degree of sensitivity, although essential for pharmacokinetic and bioavailability studies, is not required for routine therapeutic monitoring. A therapeutic range of 50–500 µg/L has been proposed for chlorpromazine in plasma of patients on chronic therapy (Bell, 1985). Consequently, the less-sensitive UV detector with detection limits as low as 1 μg/L should suffice. Two limitations of electrochemical detection are the presence of wide solvent fronts, possibly caused by endogenous material response, and a requirement for routine cleaning of detector components, again a consequence of endogenous material. Unfortunately, potential interferences from coadministered medications were not discussed.

Thioridazine, a widely used phenothiazine antipsychotic, also undergoes extensive transformation in humans, primarily by S-oxidation to two active metabolites [thioridazine-2-sulfoxide (mesoridazine) and thioridazine-2-sulfone (sulforidazine)] and various inactive products. Unlike other phenothiazines, hydroxylation and N-demethylation routes are virtually nonexistent. Therefore an HPLC method must be able to simultaneously determine thioridazine and the two active metabolites, and also guarantee noninterference from the inactive products. With the exception of one procedure, which did not include discussion of the 2-sulfone metabolite (McCutcheon, 1979), all three compounds have been analyzed simultaneously. Interferences from other metabolites were ruled out in most procedures. Either adsorption (Skinner et al., 1981; Kilts et al., 1982) or reverse-phase (McCutcheon, 1979; Stoll et al., 1984) systems have been described, employing either UV or electrochemical detectors. Sensitivities ranged from 0.1 μg/L for extraction of 1 mL of plasma (electrochemical) (Stoll et al., 1984) to 250 μg/L for 2 mL of whole blood (UV) (McCutcheon, 1979). Bell (1985) has proposed a plasma therapeutic range of 1000–1500 μg/L for thioridazine. Consequently, all procedures discussed herein possess sufficient sensitivity for detection of thioridazine in patients being treated chronically with the drug. However, nonresponders with low plasma levels and researchers pursuing pharmacokinetic and bioavailability studies will be best served by an assay with a sensitivity of ≤10 μg/L. Rapid chromatography (8–10 min) is obtained by all three research groups, with resolution of thioridazine and the two active metabolites. Interference from coadministered compounds is documented for most procedures. Lack of interference from benzodiazepines, antidepressants, and antimuscarinic-antiparkinsonian agents, which are frequently administered concomitantly with neuroleptics, is particularly important. A reverse-phase procedure using UV detection has been described for the determination of thioridazine and its major metabolites in postmorten tissues and fluids (Allender, 1985). This procedure, which utilizes enzyme digestion, is intended for toxicological application only, and does not possess sufficient sensitivity for chronic therapeutic applications. Two additional thioridazine metabolites, the diastereoisomers of thioridazine 5-sulfoxide, may be important in assessing electrocardiographic abnormalities in thioridazine-treated patients (Gottschalk et al., 1978). Two methods have been described for quantitative analysis of the 5-sulfoxide isomers. Wells and coworkers (1983) resolved thioridazine, mesoridazine, sulforidazine, and the 5-sulfoxide isomers, utilizing adsorption chromatography and postcolumn oxidation prior to fluorometric detection. Chromatographic time is lengthy (25 min) and the extraction and post-column oxidation procedures appear cumbersome. The procedure described by Hale and Poklis (1984) utilizes adsorption chromatography with UV detection, but applies to analysis of the 5-sulfoxide isomers only. Utilization of this method in human investigation is not feasible, since mesoridazine, a thioridazine metabolite was selected as internal standard. Sensitivities of the two methods were 2 and 5 µg/L, respectively (2 mL specimen extracted).

Perphenazine, a low-dose neuroleptic, is a demanding candidate for HPLC analysis. A therapeutic range of 2–6 nmol/L (0.8–2.4 µg/L) has been proposed (Larsen et al., 1985)—substantially lower than ranges for chlorpromazine and thioridazine. Although perphenazine possesses a known pharmacologically active metabolite (7-hydroxyperphenazine), Larsen and coworkers (1985) did not consider quantitation because of its questionable ability to cross the blood–brain barrier. However, simultaneous quantitation of the inactive, dealkylated perphenazine metabolite has been proposed, not only to ensure chromatographic noninterference, but to indicate a high metabolising capacity. Reverse-phase chromatography has been utilized exclusively for determination of perphenazine (Tjaden et al., 1976; Larsson and Forsman, 1983; Larsen et al., 1985), with the latter two references having accomplished simultaneous analysis of the dealkylated metabolite and the parent drug. Both UV and electrochemical detection have been applied, with sensitivity as low as 0.2 µg/L reported. To achieve this sensitivity, serum volumes of 2.5–3.0 mL were required—a potential problem for drug monitoring in the pediatric patient. Times for chromatographic analysis ranged from approximately 7 min using a 3-µm column packing (Larsen et al., 1985) to 15 min using a 10-µm support (Larsson and Forsman, 1983). Because of the high HPLC sensitivity settings required to detect this drug, baseline noise is more evident when compared to analyses for the previously discussed neuroleptics. With a substantial background, particular note must be made of coadministered medications to ensure noninterference. Larsen and coworkers (1985) appear to have undertaken the most comprehensive evaluation of analytical selectivity, with other neuroleptics, antidepressants, and antiparkinsonian drugs tested for interference. Only diazepam and methotrimeprazine interfered.

Fluphenazine, like perphenazine, presents a formidable challenge to the analyst. A postulated therapeutic range of 0.2–4.0 μg/L (Bell, 1985) has severely restricted the development of HPLC methods. Consequently only two such procedures were found. Tjaden and coworkers (1976), using reverse-phase chromatography with electrochemical detection, achieved a sensitivity of 1 μg/L with extraction of 3 mL of serum. With adsorption chromatography and UV detection, Viala and coworkers (1983) could detect 0.5 μg/L when 2 mL of serum or whole blood were extracted. The only researchers to achieve a limit of sensitivity for fluphenazine of less than 0.2 μg/L (0.16 μg/L) upon extraction of a practical serum volume (1 mL) were Goldstein and Van Vunakis (1981), using a combined HPLC-RIA (radioimmunoassay) technique. The HPLC component was only used to resolve fluphenazine from the sulfoxide metabolite and other phenothiazines. Detection was accomplished solely by RIA. The sophisticated nature of this method excludes it from routine clinical application. The primary metabolites of fluphenazine are the sulfoxide and 7-hydroxy derivatives in either free or conjugated forms. With the exception of Goldstein and Van Vunakis (1981), who resolved the sulfoxide metabolite from the parent drug, the quantitation of metabolites was not addressed.

Singular HPLC procedures have been described for analysis of methotrimeprazine in plasma and urine (Murakami et al., 1982), prochlorperazine in plasma (Sankey et al., 1982), and pipotiazine in plasma and urine (Le Roux et al., 1982). Extraction of 1 mL of plasma yielded a sensitivity of 2 μg/L for methotrimeprazine. This sensitivity appears sufficient for routine clinical monitoring. Chromatography was accomplished within 12 min. Unfortunately, metabolite and comedication interferences (with the exception of chlorpromazine) were not discussed. The reverse-phase/electrochemical detection system permitted successful chromatography of extracted urine specimens. For prochlorperazine analysis in plasma (CN-bonded column in reverse-phase mode/electrochemical detection), 0.2 μg/L was detectable upon extraction of 5 mL of specimen. Although this sensitivity may allow for chronic therapy monitoring, the large specimen volume required is impractical for clinical application, particularly for the pediatric population. Metabolite determination was not discussed, and chlorpromazine was the only noninterferant mentioned. Sensitivities of 0.25 μg/L and 2 μg/L in plasma and urine, respectively (2 mL specimen extracted), were obtained for pipotiazine upon application of an adsorption/fluorescence detection system. A 15-min analysis guaranteed selectivity, with no interference found from the 7-hydroxysulfide and N-oxide metabolites. Other potential interferants were not discussed.

Two comprehensive articles have described assay strategy for singular and simultaneous determinations of phenothiazine, thioxanthene, and butyrophenone neuroleptics and antihistamines (Curry et al., 1982) and liquid chromatographic analysis of benzodiazepines and selected phenothiazine drugs (Brinkman et al., 1981). Curry and coworkers (1982), utilizing conventional and radial compression CN-bonded columns in the reverse-phase mode with UV or electrochemical detection, have established numerous chromatographic separations with sensitivity as low as 0.1 μg/L with a 10-mL sample extracted. Clinical applications were discussed. With a CN-bonded column in the reverse-phase mode and a unique detection system, Brinkman and coworkers (1981) analyzed selected parent phenothiazines at concentrations to 10 μg/L subsequent to deproteinization of unspecified serum volumes and direct supernatant injection. After resolution on the HPLC column, the compounds of interest passed into a postcolumn photochemical reactor. Upon UV irradiation for approximately 2 min, conversion into fluorescence products was realized. The newly formed fluorescence products passed into a fluorometer for detection. Application of this technique to a routine clinical setting would be limited because of inherent sophistication.

Although classified as an antihistamine and antinauseant, promethazine is worthy of brief mention. Promethazine is extensively metabolized by S-oxidation, demethylation, N-oxidation, hydroxylation, and conjugation. Consequently one would think metabolite determination or noninterference should be considered. In reality, however, the metabolites are usually not detected in the serum or plasma of patients receiving the drug, and only after extraction of 10 mL of whole blood have they been found (Taylor and Houston, 1982). Promethazine instability in weakly acidic solutions has hindered analytical development because back-extraction into acid, a technique so important for removing endogenous interference for low level quantitation, must be avoided. Nevertheless, successful HPLC analysis of the drug in biological samples has been realized. Reverse-phase chromatography utilizing either a conventional hydrophobic (Wallace et al., 1981b; Taylor and Houston, 1982) or CN-bonded (Wallace et al., 1981c; Allender and Archer, 1984; Leelavathi et al., 1985) stationary phase has been coupled with UV or electrochemical detection to realize limits of sensitivity as low as 0.1 μg/L with 2 mL of plasma extracted. For routine monitoring, sensitivity should approach 1.0 μg/L. Three promethazine methodologies applying unique techniques such as HPLC valve switching to prevent accumulation of electroactive impurities at an electrochemical detector (Leelavathi et al., 1985), trichloroethyl chloroformate derivatization (Wallace et al., 1981a), and enzyme digestion of tissue homogenates (Allender and Archer, 1984) have been described.

A novel phenothiazine derivative, 10-[3-(3-hydroxypyrrolidinyl)propyl]-2-trifluoromethyl phenothiazine, has been quantitated in mice brain and blood using reverse-phase chromatography with electrochemical detection (Shibanoki et al., 1986). Application to human studies was suggested.

4.3.2. Thioxanthenes

The thioxanthene class of neuroleptics, although small in size relative to the phenothiazine group, is nevertheless therapeutically important. The structures of some thioxanthene neuroleptics are displayed in Fig. 7. Members of this drug group, particularly thiothixene and clopenthixol, serve as excellent examples of the chiral aspect of pharmacology. Consequently, an HPLC procedure for a thioxanthene neuroleptic must resolve the active and inactive isomers of the drug.

Thiothixene, a low-dose thioxanthene neuroleptic, exists in the *cis* (Z) and *trans* (E) forms, with only the *cis* (Z) isomer possessing pharmacological activity. Published HPLC procedures have successfully resolved the thiothixene isomers using CN-bonded columns operated in the reverse-phase mode coupled with UV detection at 229 nm (Bogema et al., 1982; Dorey et al., 1983; Narasimhachari et al., 1984). The lowest sensitivity limit achieved was 0.5 μg/L with 1 mL of plasma extracted (Narasimhachari et al., 1984). Sufficient sensitivity has been achieved since plasma concentrations of 4–20 μg/L (Bogema et al., 1982) and 10–100 μg/L

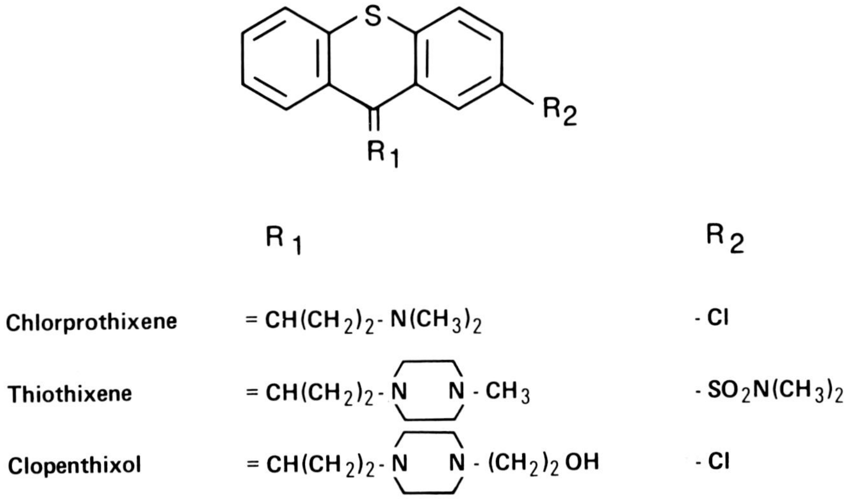

Fig. 7. Structures of selected thioxanthene neuroleptics.

(Bell, 1985) have been reported in patients receiving thiothixene on a chronic basis. Bogema and coworkers (1982) also achieved resolution of a demethylated metabolite, N-desmethylthiothixene. However, detection of this metabolite in plasma of patients receiving thiothixene was not demonstrated. The two earlier methods required addition of a constant volume of carrier thiothixene solution to all samples prior to extraction, consequently increasing assay sensitivity to an undefined level. In addition, the chromatographic column had to be saturated prior to analysis with three injections of a high-concentration thiothixene standard (1 mg/L). The quantitation of *cis* (Z)-thiothixene was placed in jeopardy by the large solvent front, particularly at low plasma concentrations (less than 5 µg/L) (Bogema et al., 1982). A 10-µm stationary phase was probably a contributing factor to the presence of *cis* (Z)-thiothixene on the shoulder of the solvent peak. Narasimhachari and coworkers (1984), utilizing a 5-µm CN-bonded column, alleviated the solvent peak interference problem. With *trans* (E)-thiothixene added in excess to serve as an internal standard and carrier, chromatography was realized in 6 min with no interfering drugs or metabolites found. Furthermore, utilization of a carrier *cis* (Z)-thiothixene solution or presaturation of the analytical column were not required. All three methods demonstrated no biotransformation of active *cis* (Z)-thiothixene to inactive *trans* (E)-thiothixene in humans. However, in vitro isomerization to the *trans* (E) isomer occurred on exposure to UV light (Bogema et al., 1982), necessitating proper specimen storage and handling. The converse, isomerization of *trans* (E)- to *cis* (Z)-thiothixene upon repeated freeze/thaw cycles was demonstrated by Narasimhachari and coworkers (1984). This latter phenomenon can be prevented by freezing individual aliquots of a large stock standard solution.

Chlorprothixene, marketed as Taractan®, is biotransformed in humans via oxidation and N-demethylation to the sulfoxide, *N*-desmethylsulfoxide, and sulfoxide-*N*-oxide metabolites. Although the sulfoxide metabolite is the predominant species in plasma following therapeutic administration and overdosage, pharmacological activity of this metabolite and others is questionable. Brooks and coworkers (1985) have achieved simultaneous resolution of chlorprothixene, chlorprothixene sulfoxide, and *N*-desmethylchlorprothixene in a 10-min chromatographic analysis using a CN-bonded column operated in the reverse-phase mode. Ultraviolet and electrochemical detectors placed in series with the latter detector downstream of the former were utilized. Sensitivity limits for chloroprothixene and the sulfoxide metabolite were 5.0 µg/L or less for extraction of 1 mL of plasma. The N-desmethyl metabolite was not quantitated because of low, variable recovery. Although electrochemical detection demonstrated comparable sensitivity and precision to UV detection, an endogenous interferent prevented measurement of the sulfoxide metabolite.

Clopenthixol, like thiothixene, exists as two geometric isomers: an active *cis* (Z) and an inactive *trans* (E) isomer. N-Dealkylation, an important route of biotransformation, results in the formation of a second pair of geometric products, the *cis* (Z) and *trans* (E) isomers of N-desalkylclopenthixol, which like the *trans* (E) isomer of the parent compound, possess minimal neuroleptic activity. However, an HPLC assay for clopenthixol should resolve all four compounds, not only for specificity assurance, but for evaluation of the metabolic capacity of the individual in question. Two similar procedures have been described for the simultaneous determination of the *cis* (Z) and the *trans* (E) isomer of clopenthixol and desalkylclopenthixol in serum (Aaes-Jorgensen, 1980) and plasma or whole blood (Viala et al., 1983) utilizing adsorption chromatography and UV detection. Sensitivities for the parent drug and metabolite were 0.5–0.8 µg/L and 2.5 µg/L, respectively for extraction of a 2-mL specimen. Chromatographic analysis time was 15 min. Unfortunately, most of the benzodiazepines and other neuroleptics (flupenthixol, droperidol, fluphenazine) interfered with the assay. No interference was found from tricyclic anti-depressants and antiparkinsonian drugs. As with thiothixene, endogenous transformation of the *cis* (Z) isomer of clopenthixol to the *trans* (E) isomer is virtually nonexistent (Aaes-Jorgensen, 1980).

4.3.3. Butyrophenones

The butyrophenone neuroleptics, although structurally dissimilar to the phenothiazines, show many of their pharmacological properties. The structures of some butyrophenone compounds are displayed in Fig. 8.

Haloperidol, the prototype of the butyrophenone class, is one of the most potent neuroleptics. Although the drug was considered to possess no pharmacologically active metabolites, a reduced (dihydroxylated) metabolite is reportedly 10% as potent as haloperidol (Morselli et al., 1981). Although Browning and coworkers (1982) suggested a high reduced haloperidol/haloperidol ratio correlates with poor clinical response, routine monitoring of the reduced metabolite simultaneously with the parent drug remains controversial. Reverse-phase (Miyazaki et al., 1981; Jatlow et al., 1982; Kogan et al., 1983; Korpi et al., 1983; Dhar and Kutt, 1984; Hequet et al., 1985) and adsorption (McBurney and George, 1984) chromatography with UV or electrochemical detection have been utilized for haloperidol determination in serum or plasma. Parkinson (1985) has described an HPLC assay for the simultaneous determination of three butyrophenone neuroleptics (haloperidol, spiroperidol, trifluperidol) in rat brain homogenates. Since haloperidol serum/plasma concentrations found after acute or chronic dosing vary from 2 to 245 µg/L, an HPLC assay must possess sufficient sensitivity to accommodate this range. With two exceptions (Miyazaki et al., 1981; Kogan et al., 1983), HPLC assays for haloperidol discussed here offer such sensitivity. Some of the methods present disadvantages restricting their routine use. Miyazaki and coworkers (1981) used an external rather than an internal standard, and method selectivity was not discussed. Jatlow and associates (1982) required two internal standards for haloperidol quantitation, one of them the commonly prescribed antidepressant desipramine, to circumvent selectivity problems. The chromatographic analysis presented by McBurney and George (1984) displays troublesome peak tailing by both haloperidol and the internal standard fenethazine. Finally, quantitation appears cumbersome in the method of Hequet and coworkers (1985), since haloperidol and an internal standard, chlorohaloperidol, elute on

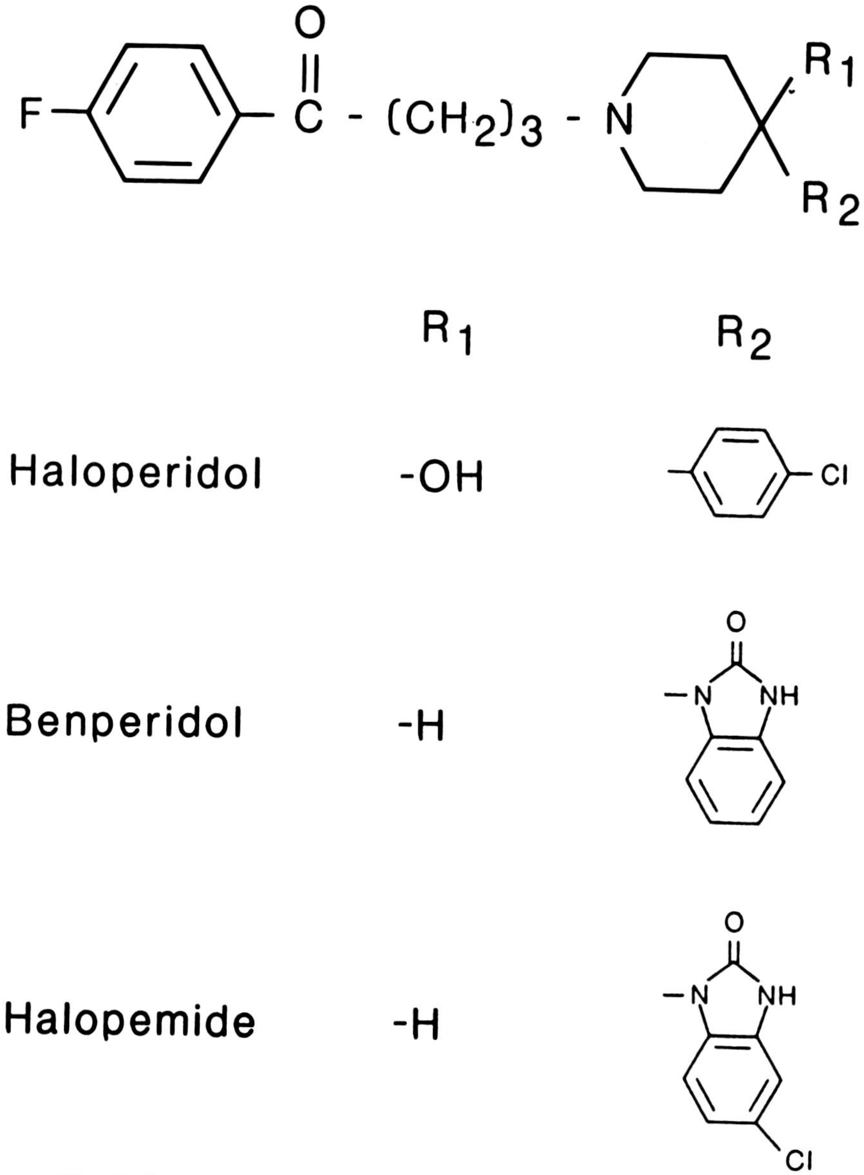

Fig. 8. Structures of selected butyrophenone neuroleptics.

the tail of a huge preceding peak. Korpi and coworkers (1983) accomplished simultaneous quantitation of haloperidol and the reduced metabolite in a 15–20-min chromatographic analysis using a CN-bonded column and electrochemical detection (Fig. 9). Miller and Devane (1986b) have described a modification of a previously

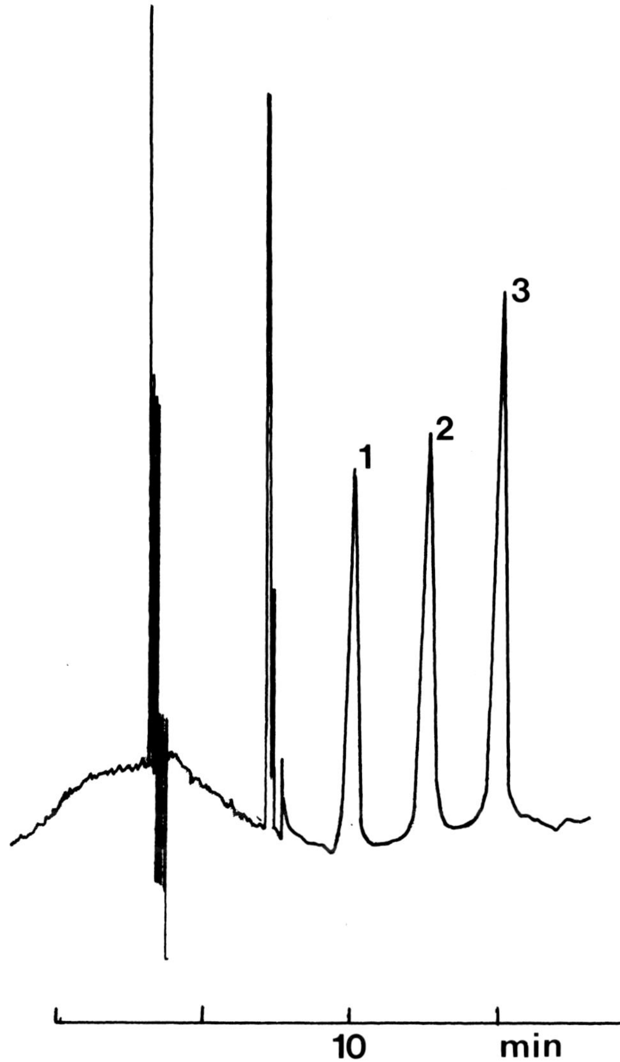

Fig. 9. Chromatogram of patient's extracted serum after haloperidol treatment. (1) reduced haloperidol; (2) haloperidol; (3) chlorohaloperidol (internal standard) (from Korpi et al., 1983).

published method (Jatlow et al., 1982), which also permits such simultaneous quantitation. Using reverse-phase chromatography with UV detection, extraction of a 1-mL specimen realizes a sensitivity of 2 μg/L for both compounds in a 10-min chromatographic run. Good chromatography was also shown by Dhar and Kutt (1984), who achieved resolution of haloperidol and the internal standard chlorohaloperidol within 10 min on a C_8-bonded reverse-phase column coupled to a UV detector. Potential interferences from coadministered medications were discussed for most of the haloperidol HPLC methods. Particular notice should be taken of interferences from benzodiazepines and anticholinergic drugs, for they are used commonly in combination with haloperidol.

Benperidol, structurally related to haloperidol, is eight times as potent. Kruger and coworkers (1984) have developed an assay utilizing a reverse-phase column (C_{18}) and UV detection. With haloperidol as internal standard, limits of sensitivity ranging from 0.5 to 1.0 µg/L for extraction of 2 to 4 mL of plasma volumes were realized. Chromatographic analysis was accomplished within 15 min, and the assay can be applied to the determination of other butyrophenones. Potential interference from coadministered medications was not discussed.

Halopemide, a benperidol congener, has been determined by reverse-phase ion-exchange chromatography with UV detection (Van Rooij et al., 1979). Sodium dodecyl sulfate (SDS) acting as a cation exchanger was adsorbed to the stationary phase and halopemide in a proteinated form (pH, 6.5) interacted with the SDS. A cleavage product of halopemide previously identified in rats was coanalyzed with the parent drug, but this product was not found in plasma samples of patients receiving halopemide. Limits of sensitivity for halopemide and the cleavage product were 7.4 and 12.9 µg/L, respectively. Unfortunately, no internal standard was utilized in the assay, and potential interferences from other drugs were not discussed.

4.3.4. Miscellaneous Neuroleptics

Pimozide, a potent long-acting neuroleptic with a diphenylbutylamine nucleus, has been quantitated in plasma by reverse-phase chromatography combined with fluorescence detection (Miyao et al., 1983). With a methylated analog as internal standard, pimozide could be detected at concentrations as low as 0.3 µg/L from extraction of 1 mL of specimen. Ultraviolet detection could only realize a sensitivity limit of 5.0 µg/L with a similar specimen volume extracted. This exceptional sensitivity is required, for plasma concentrations were below 3.3 µg/L at all times after oral administration of a 3 mg tablet. Chromatography was accomplished within 10 min, and no endogenous interferants were found. Interferences from other coadministered medications were not mentioned. Assay recovery (90–100%) was optimized using silanized extraction tubes.

Isofloxythepin and oxyprothepin, neuroleptics of the tricyclic dibenzothepin class, have been quantitated in pharmaceutical preparations using an ion-pair reverse-phase system (Pacakova et al., 1984). Detection was accomplished using a UV and electrochemical detector connected in series.

5. Benzodiazepines

Benzodiazepines are the most widely used and abused drugs in the Western world. Marketed as having a variety of functions (anti-anxiety, muscle relaxant, hypnotic, anticonvulsant), these drugs are similar both structurally and in their pharmacological actions. The structures of some selected benzodiazepines are displayed in Fig. 10. Routine therapeutic drug monitoring of benzodiazepines, unlike the case for antidepressants and neuroleptics, is not commonplace. Although minimum therapeutic levels for diazepam in treatment of acute anxiety and epilepsy have been quoted (Raisys et al., 1980), no concrete relationships between benzodiazepine biological fluid levels and clinical effect have been formulated. The only exception is clonazepam in the treatment of epilepsy, and even in this case, the legitimacy of a therapeutic range has been questioned. Consequently, the majority of analytical techniques have been developed specifically for pharmacokinetic and toxicological applications. Chromatographic techniques, particularly GLC and HPLC, are the major analytical tools for qualitative and quantitative analysis of benzodiazepines. Because of the heat liability of some of these compounds (chlordiazepoxide, oxazepam), HPLC is the preferred analytical method.

5.1. Preanalytical Concerns

Although benzodiazepines are less susceptible to degradation and adsorption than antidepressants and neuroleptics, storage conditions must still be optimized to ensure stability. Since photodegradation, particularly of the nitrobenzodiazepines (nitrazepam, clonazepam), may occur in either biological specimens or standard solutions (Kelly et al., 1982), storage in the dark or in amber bottles, preferably at 4°C, is recommended. Samples stored at room temperature, but not exposed to light, also showed con-

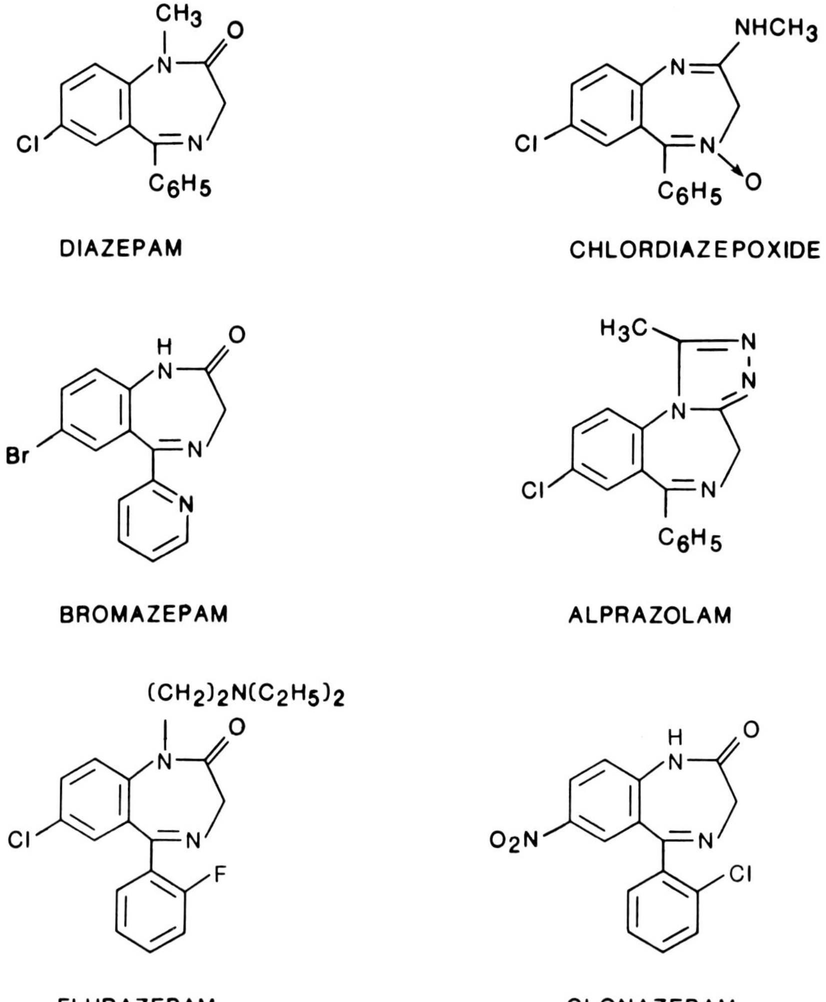

DIAZEPAM

CHLORDIAZEPOXIDE

BROMAZEPAM

ALPRAZOLAM

FLURAZEPAM

CLONAZEPAM

Fig. 10. Structures of selected benzodiazepines.

siderable decomposition, which suggests temperature is also an important factor. Desmethylchlordiazepoxide, a primary metabolite of chlordiazepoxide, is unstable in solution (Foreman et al., 1980), even when frozen (Divoll et al., 1982). Therefore, solutions for calibration samples should be prepared weekly. Wong (1983) demonstrated good stability for diazepam and nordiazepam in solution (6 mo), but limited stability for chlordiazepoxide, desmethylchlordiazepoxide, and nitrazepam (10 d). Hydrolysis of benzodiazepines to the respective benzophenones may occur in acidic media. Consequently, storage in or exposure to acid conditions for a prolonged period should be avoided.

5.2. Extraction

As for antidepressants and neuroleptics, extraction procedures are required to optimize the detection of benzodiazepines in biological fluids. Although methanol deproteinization of serum followed by direct injection of supernatant has been done for determination of demoxepam, a chlordiazepoxide metabolite (Brinkman et al., 1981), this type of procedure is the exception rather than the rule.

The majority of HPLC procedures for benzodiazepines are manual, involving sample basification, organic solvent extraction, evaporation of the organic phase, and reconstitution prior to HPLC analysis of intact drug. Back-extraction has been incorporated into several methods to eliminate endogenous interferants. Because these drugs are weak bases, other procedures in place of an initial basification either use a neutral buffer (pH 7.0–7.4) or extract a sample directly with no prior pH manipulation. Although silanized glassware has been employed for benzodiazepine detection (Adams et al., 1984; Ratnaraj et al., 1984), the majority of methods have not found adsorption to be a problem and, hence, have utilized untreated glassware. Polypropylene tubes have been employed in place of glassware for quantitation of nitrazepam in plasma (Kelly et al., 1982). For determination of hydroxylated metabolites in urine, enzymatic hydrolysis of the conjugates, primarily glucuronides, is required prior to extraction. Recoveries of benzodiazepines and respective metabolites from biological media for manual extraction procedures ranged from 34 to 100%. Sajgo and coworkers (1981) for an unknown reason deproteinized serum prior to extracting the vacuum-dried residue. This seemingly unnecessary procedure resulted in poor recovery (34%).

Extraction columns (Bond Elut®, Extrelut 1®) have been utilized as alternatives to manual extraction for the determination of benzodiazepines (Good and Andrews, 1981; Rao et al., 1982; Heizmann et al., 1984; Kozu, 1984; Kabra and Nzekwe, 1985). Although the speed of analysis and minimal specimen handling should imply more efficient extraction (80–100%), manual extraction procedures remain the choice of most investigators.

A sophisticated extraction instrument, the Prep I® (Dupont, U.S.A.), previously described for antidepressant analysis, has been utilized for extraction of serum for determination of clonazepam (Taylor et al., 1984).

5.3. Analysis

5.3.1. Anxiolytics

The prototype benzodiazepine anxiolytic is diazepam. The three major metabolites, all pharmacologically active, are 3-hydroxydiazepam (temazepam), nordiazepam, and oxazepam. Associated conjugates are also formed. An HPLC method, depending on the particular application, should be capable of achieving resolution of most of the above-mentioned compounds. In serum, plasma, or whole blood, only diazepam and nordiazepam are present in detectable amounts and resolution of oxazepam and temazepam is unimportant. However, if urine is to be analyzed for a pharmacokinetic project, resolution of diazepam and all three major metabolites should be realized. With the exception of one procedure utilizing adsorption chromatography with UV detection (Perchalski and Wilder, 1978), reverse-phase chromatography with UV detection has been applied exclusively to HPLC analysis of diazepam (Brodie et al., 1978; Kabra et al., 1978; MacKichan et al., 1979; Vree et al., 1979; Raisys et al., 1980; Tjaden et al., 1980; Cotler et al., 1981; Ratnaraj et al., 1981; Rao et al., 1982; Tada et al., 1985). Wavelength selection ranged from 230 to 254 nm, with 245 nm the most popular. Chromatographic analysis times are highly variable, ranging from 5 to 22 min. Minimum therapeutic levels in plasma/serum for diazepam in the treatment of anxiety and epilepsy have been quoted as 400 and 500 μg/L, respectively (Raisys et al., 1980). The HPLC procedures referred to in this paragraph offered sensitivities in serum/plasma/whole blood ranging from 1 to 50 μg/L and 2 to 50 μg/L for diazepam and nordiazepam, respectively, upon extraction of 0.1–2.0 mL of sample. Therefore, these assays are not only suitable for clinical monitoring, but possess sufficient sensitivities for pharmacokinetic investigations. Since assay development was often only for this latter purpose, potential interferences from coadministered medications are not discussed in a large number of the methods. One of the more impressive procedures has been described by Rao and coworkers (1982), who developed a rapid HPLC method combining Bond Elut® column extraction with a 10-min reverse-phase chromatographic analysis of diazepam and the three major metabolites (Fig. 11). Minute samples (50–100 μL) of whole blood, serum, or plasma

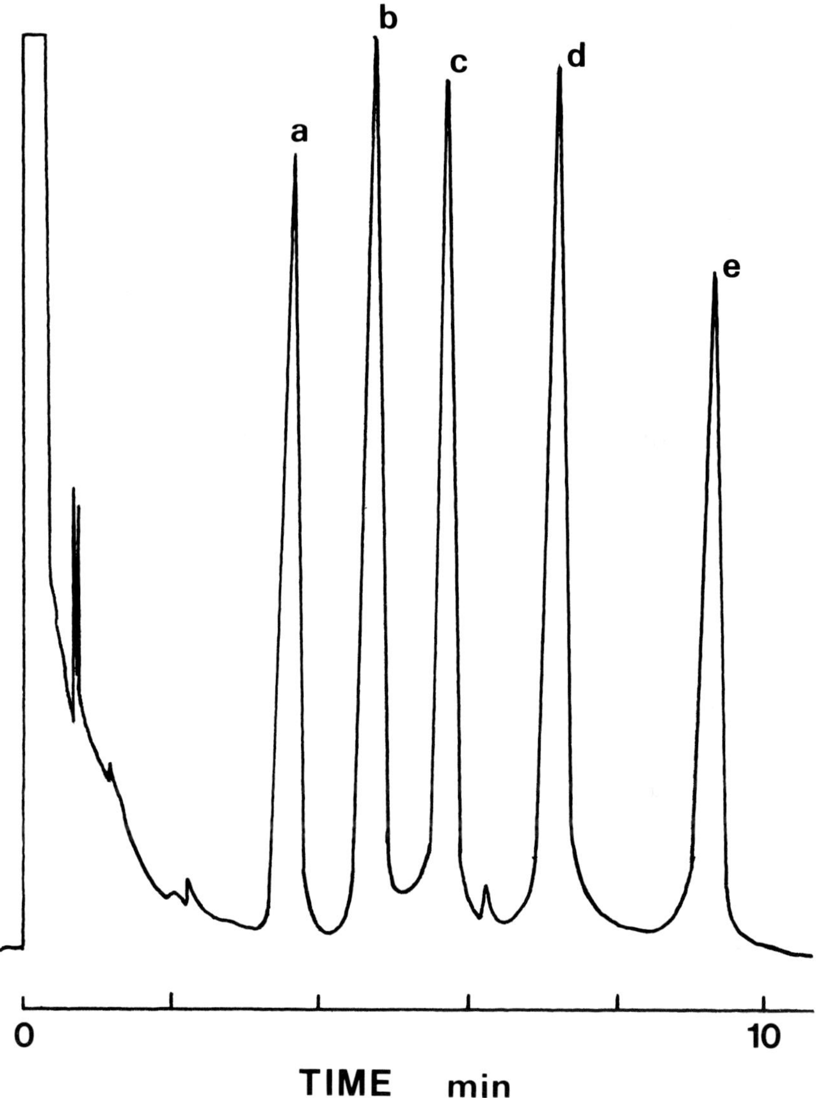

Fig. 11. Chromatogram of diazepam and its metabolites. Blood containing 100 ng each of (a) methylnitrazepam (internal standard), (b) oxazepam, (c) 3-hydroxydiazepam, (d) *N*-desmethyldiazepam, and (e) diazepam extracted through a Bond Elut C_{18} column (from Rao et al., 1982).

yielded 25 μg/L sensitivities for all compounds. The method has particular application to the study of diazepam pharmacokinetics in a small animal model as well as in clinical monitoring of pediatric patients.

Chlordiazepoxide, another extensively used benzodiazepine anxiolytic, is biotransformed by oxidation to two principle active metabolites, desmethylchlordiazepoxide and demoxepam. Two additional metabolites, nordiazepam and oxazepam, are usually not detected in serum or plasma unless overdosage has occurred. Consequently, any HPLC procedure for this drug should at least resolve the first three mentioned compounds, and the majority of procedures mentioned here accomplish that. However, absolute resolution of all five compounds would be required for urine analysis. As for diazepam, reverse-phase chromatography coupled with UV detection has been the primary HPLC analytical combination (Greizerstein and Wojtowicz, 1977; Peat et al., 1979; Peat and Kopjak, 1979; Ascalone, 1980; Divoll et al., 1982). Sensitivities ranged from 30 μg/L to an unspecified concentration of less than 200 μg/L upon extraction of 0.5–2 mL of biological sample. Times for chromatographic analyses for most methods were within 15 min. Although the methods were developed for pharmacokinetic studies in animals and humans, monitoring during long-term therapy was occasionally suggested as a potential application. However, the lack of references to possible interference from coadministered medications precludes such assays from clinical evaluations. Skellern and coworkers (1978) described an HPLC procedure utilizing an adsorption-phase column and UV detection. Sensitivity limits and possible interference were not discussed. Furthermore, the 40-min chromatographic analysis is not acceptable for either clinical or research applications. Chlordiazepoxide and the N-desmethyl metabolite were determined by reverse-phase chromatography and electrochemical detection (Hackman and Brooks, 1981). The limit of sensitivity was not improved compared to the methods employing UV detection (50 μg/L; 1 mL of plasma extracted), and the other metabolites (demoxepam, nordiazepam, and oxazepam) were poorly detected under the detector operating conditions selected. Brinkman and coworkers (1981) have described a method of questionable utility, which involves post-column photooxidation of demoxepam into a highly fluorescent product. Application of the technique to chlordiazepoxide and other metabolites was not pursued.

Bromazepam, a low-dose anxiolytic benzodiazepine, has been determined by reverse-phase HPLC coupled with UV detection in human plasma (Hirayama and Kasuya, 1983; Heizmann et al., 1984). Two major metabolites, 3-hydroxybromazepam and a cleavage product, 2-(2-amino-5-bromo-3-hydroxybenzoyl)pyridine, were determined in urine following enzymatic deconjugation (Heizmann et al., 1984). Limits of sensitivity for bromazepam in plasma were 5–6 μg/L for extraction of 1-mL volumes. Heizmann and coworkers (1984) utilized a rapid column extraction procedure (Extrelut 1®), whereas Hirayama and Kasuya (1983) employed a lengthy extraction procedure involving freezing steps. However, the chromatographic analysis of the latter procedure was 12 min compared to 20 min for the former method.

Alprazolam, another low-dose benzodiazepine, has been determined in human serum and plasma by both adsorption (Adams et al., 1984) and reverse-phase (McCormick et al., 1984; Edinboro and Backer, 1985) chromatography with UV detection. Only the α-hydroxy and 1-demethyl metabolites have demonstrated pharmacological activity, but are present at concentrations of less than 5% of corresponding alprazolam concentrations. Therefore, quantitation of these compounds is not relevant. Although Adams and coworkers (1984) achieved a 0.5-μg/L limit of sensitivity upon extraction of a 0.5-mL specimen, as compared to a 2.5-μg/L limit for a 2-mL specimen (McCormick et al., 1984), the chromatography demonstrated by the former research group is relatively poor, with a huge solvent front masking the internal standard (triazolam) peak. The method of Edinboro and Backer (1985) was designed specifically for analysis of alprazolam in postmorten whole blood specimens. A sensitivity of 5 μg/L was established upon extraction of 1-mL specimens. Steady-state concentrations of 25–55 μg/L have been found in six patients receiving 1.5–6.0 mg of drug (McCormick et al., 1984). Consequently, all three methods have sufficient sensitivity for potential routine monitoring.

Reverse-phase chromatography with UV detection has been applied to the determination of various miscellaneous anxiolytic benzodiazepines and associated metabolites in biological fluids, including clobazam (Brachet-Liermain et al., 1982; Ratnaraj et al., 1984; Tomasini et al., 1985), adinazolam (Peng, 1984), pinazepam (Grassi et al., 1977), and clorazepate (Colin et al., 1983). Clorazepate deserves particular mention because of its spontaneous decarboxylation to nordiazepam in vitro and in vivo. Therefore, an extrapolation kinetic method was suggested for estimation of clorazepate concentration in the study of Colin and coworkers (1983). However, clinical evaluation would only be concerned with nordiazepam estimation. Ethyl loflazepate has been detected in baboon plasma using reverse phase chromatography coupled with an on-line radioactivity monitor (Davi et al., 1985). Application to analysis in humans is restricted by the required administration of ^{14}C-labeled drug. Quantification of the dopamine receptor antagonist R-(+)-7-chloro-8-hydroxy-1-phenyl-2,3,4,5,-tetrahydro-1H-3-methyl-3-benzazepine in an animal model was accomplished with reverse-phase chromatography and electrochemical detection (Kilts et al., 1985). Adsorption chromatography with UV detection has been utilized for quantification of tofisopam in serum (Sajgo et al., 1981).

The previously discussed publications on anxiolytic benzodiazepines were developed for determination of specific drugs and their associated metabolites. Procedures have been developed for the simultaneous analysis of various combinations of intact benzodiazepines (Foreman et al., 1980; Good and Andrews, 1981; Lensmeyer et al., 1982; Wong, 1983; Gill et al., 1986) or benzophenone derivatives (Violon et al., 1982) in biological fluids. Particular emphasis is on diazepam and chlordiazepoxide and the respective metabolites. Violon and coworkers (1982), by hydroylsis with hydrochloric acid, converted parent benzodiazepines into benzophenone derivatives. This procedure, although designed for clinical toxicology, has limited application. Hydrolysis of several of the benzodiazepines resulted in formation of the same benzophenone. Consequently, the specific drug cannot be readily identified. The simultaneous quantitative analyses of intact benzodiazepine combinations were accomplished with reverse-phase/UV detection combinations. All extractions were manual solvent procedures, with the exception of the method of Good and Andrews (1981), which employed Bond Elut® extraction columns. Assay interferences from other drugs were discussed in most of these procedures. Sensitivities were sufficient for clinical and toxicological applications (25–100 µg/L) upon extraction of practical serum volumes (0.5–1.0 mL). Implementation of one of these quantitative procedures may be potentially more useful for a service-oriented laboratory. Gill and coworkers (1986) have examined the retention characteristics of 21 benzodiazepines on both adsorption and reverse-phase systems with UV detection. Extraction procedures were not applied, since dissolved drug sample were chromatographed directly. This method was intended for qualitative determination of benzodiazepines in toxicology specimens.

5.3.2. Hypnotics

Nitrazepam, a prominent benzodiazepine hypnotic, is largely converted to inactive metabolites, primarily 7-amino-nitrazepam and 7-acetamidonitrazepam. These metabolites and several minor cleavage, hydroxylation, and conjugation products are excreted in the urine. Both reverse-phase (Ho et al., 1983) and ion-pair adsorption (Kelly et al., 1982) chromatography with UV detection have been applied to measurement of nitrazepam in plasma. The former procedure could also be adapted to quantification of temazepam, another benzodiazepine hypnotic, in plasma. Both procedures utilized manual solvent extraction procedures. Sensitivities were identical (5.0 μg/L), although the method employing the ion-pair adsorption system required only 0.5 mL of sample, compared to 1.0 mL for the alternate procedure. Chromatographic analysis times were comparable (approximately 16 min), and metabolite noninterference in the ion-pair adsorption method was guaranteed. Although metabolite identification was not discussed by Ho and coworkers (1983), the two compounds have not been identified in plasma extracts of patients receiving nitrazepam (Kelly et al., 1982). Therefore, resolution of all three compounds for plasma analysis is not required. However, urine determination of nitrazepam and its metabolites requires complete resolution of nitrazepam and the 7-amino and 7-acetamido metabolites. Kozu (1984), in a procedure developed specifically for urine analysis, identified the two primary metabolites by reverse-phase chromatography and UV detection in a 20-min chromatographic analysis. Although limits of sensitivity were not provided, 1.5 mg/L metabolite concentrations were determined upon extraction of 10-mL specimen volumes.

Midazolam, which is pharmacologically similar to diazepam, has been indicated for induction of anesthesia. Reverse-phase chromatography with UV detection has been utilized for quantitation of midazolam and its major 1-hydroxymethyl metabolite in human serum (Vasiliades and Sahawneh, 1981), analysis of midazolam and four metabolites in dog plasma (Puglisi et al., 1985), and determination of the parent drug and 3-hydroxylated metabolites in plasma and urine of man and dog (Vree et al., 1981). Free hydroxymethylmidazolam was not detected in serum of patients who had received midazolam (Vasiliades and Sahawneh, 1981), and only after enzyme hydrolysis of the sample could the metabolite be determined (Vree et al., 1981). Although Puglisi and coworkers (1985) resolved midazolam and four metabolites (1-hydroxymethylmidazolam, 4-hydroxymidazolam, 4-hydroxy-1-hydroxymethylmidazolam, 1-desmethylmidazolam) in 12 min, only midazolam and the 1-hydroxymethyl metabolite were detected in an oral toxicity study in the dog. Although designed for pharmacokinetic applications, Vasiliades and Sahawneh (1981) discussed potential interference from coadministered medications. Limits of sensitivity quoted for midazolam were 50 µg/L (Puglisi et al., 1985), 30 µg/L (Vree et al., 1981), and 15 µg/L (Vasiliades and Sahawneh, 1981) for extraction of 1.0 mL, 0.2 mL, and an unspecified serum volume, respectively. Vree and coworkers (1981) admitted 30 µg/L was not sensitive enough for elucidating a complete kinetic profile of the drug in man. Furthermore, a suitable internal standard was not found, and loss during extraction was not compensated for. Puglisi and de Silva (1981) have described an adsorption chromatography-UV detection HPLC procedure for determination of a midazolam analog in whole blood, plasma, and urine.

Quantitation of estazolam in biological fluids (serum, urine, gastic contents) using reverse-phase chromatography coupled with UV detection has been described (di Tella et al., 1986). The method was designed for toxicological, not clinical, specimens, since concentrations below 0.6 mg/L were not tested.

Methods have been described for determination of the hydroxyethyl metabolite of flurazepam in human urine by adsorption chromatography with UV detection (Weinfield and Miller, 1981) and for the quantitation of two flunitrazepam metabolites, the 7-amino and 7-aminodesmethyl compounds in plasma by reverse-phase chromatography with UV detection (Sumirtapura et al., 1982). Unfortunately, HPLC procedures for the two parent drugs, flurazepam and flunitrazepam, were not found.

5.3.3. Anticonvulsants

Although clonazepam is not the only benzodiazepine compound possessing anticonvulsant activity, it is the primary member of this benzodiazepine class. Clonazepam undergoes extensive biotransformation to relatively inactive metabolites, the primary products being the 7-amino and 7-acetamido compounds. Therefore, routine clinical monitoring of the metabolites is not required. Most of the methods discussed here do not undertake metabolite determination or even discuss potential interference from these compounds. However, Heazlewood and Lemass (1984) guaranteed complete resolution of clonazepam from the metabolites on the reverse-phase system applied. Furthermore, these investigators selected a wavelength at which the metabolites were virtually undetectable (306 nm). Since the other methods for clonazepam quantitation in serum or plasma used reverse-phase chromatography (Rovei and Sanjuan, 1980; Shaw et al., 1983; Taylor et al., 1984; Petters et al., 1984; Kabra and Nzekwe, 1985), the polar metabolites probably were resolved at earlier retention times than clonazepam and the selected internal standards. The first three references listed in the former statement also employed a clonazepam-specific wavelength (306–311 nm). Therefore, assay specificity was ensured for all methods. Petters and coworkers (1984) reported the quantitation of clonazepam and the 7-amino and 7-acetamido metabolites in human plasma. Unfortunately, two extraction and chromatographic procedures are required— one for the parent drug and the second for the two metabolites. A therapeutic range of 20–70 µg/L has been assigned in my institution for clonazepam in chronic treatment of seizure disorders. All HPLC clonazepam assays described here achieved adequate sensitivity for this range, with limits as low as 2 µg/L reported. Serum/plasma volumes required for analysis ranged from 0.1 to 2 mL. Any larger requirements would be impractical for the pediatric population, where this drug is extensively used. Interferences were discussed for most of the methods. Particular notice should be taken of the methods' ability to resolve carbamazepine, an anticonvulsant often coadministered with clonazepam.

6. Summary

Antidepressants, neuroleptics, benzodiazepines—the µg/L concentrations of these compounds found in vivo present a formidable challenge to any analytical method. HPLC has met this challenge and has proven itself to be a valuable methodology for determination of these compounds in biological fluids. Nevertheless, if one is interested in applying one of the HPLC procedures discussed to his or her own laboratory, specific questions must be answered. Can the procedure by applied to the biological fluid of interest? Is the method practical for routine monitoring in pediatrics? Is the assay sufficiently sensitive, accurate, specific, and rapid for the chosen application? Is the cost of the extraction and analytical system within the constraint of the laboratory budget? These are but a few of the questions to be addressed.

Fortunately, from the plethora of literature on HPLC analysis of antidepressants, neuroleptics, and benzodiazepines, procedures answering "yes" to the previously posed questions can be found.

References

Aaes-Jorgensen T. (1980) Specific high-performance liquid chromatographic method for estimation of the *cis*(Z)- and *trans*(E)-isomers of clopenthixol and a N-dealkyl metabolite. *J. Chromatogr. Biomed. Appl.* **183,** 239–245.

Adams W. J., Bombardt P. A., and Brewer J. E. (1984) Normal-phase liquid chromatographic determination of alprazolam in human serum. *Anal. Chem.* **56,** 1590–1594.

Allender W. J. (1985) High-pressure liquid chromatographic determination of thioridazine and its major metabolites in biological tissues and fluids. *J. Chromatogr. Sci.* **24,** 541–545.

Allender W. J. and Archer A. W. (1984) Liquid chromatographic analysis of promethazine and its major metabolites in human postmortem material. *J. Forens. Sci.* **29,** 515–526.

Allender W. J., Archer A. W., and Dawson A. G. (1983) Extraction and analysis of chlorpromazine and its major metabolites in post mortem material by enzymic digestion and HPLC. *J. Anal. Toxicol.* **7,** 203–206.

Ankier S. I., Martin B. K., Rogers M. S., Carpenter P. K., and Graham C. (1981) Trazodone—a new assay procedure and some pharmacokinetic parameters. *Br. J. Clin. Pharmacol.* **11,** 505–509.

Ascalone V. (1980) Determination of chlordiazepoxide and its metabolites in human plasma by reversed-phase high-performance liquid chromatography. *J. Chromatogr. Biomed. Appl.* **181,** 141–146.

Bannister S. J., van der Wal Sj., Dolan J. W., and Synder L. R. (1981) Liquid-chromatographic analysis for common tricyclic antidepressant drugs and their metabolites in serum or plasma with the Technicon "FAST-LC" system. *Clin. Chem.* **27,** 849–855.

Beierle F. A. and Hubbard R. W. (1983a) Liquid chromatographic separation of antidepressant drugs. I. tricyclics. *Ther. Drug Monit.* **5,** 279–292.

Beierle F. A. and Hubbard R. W. (1983b) Liquid chromatographic separation of antidepressant drugs. II. Amoxapine and maprotiline. *Ther. Drug Monit.* **5,** 293–301.

Bell J. L. (1985) Methods for monitoring neuroleptic drugs. *Am. Assoc. Clin. Chem. TDM-T* **6**, 1–6.

Biggs S. R., Brodie R. R., Hawkins D. R., and Midgley I. (1977) The Use of High Pressure Liquid Chromatography and Mass Spectrometry for the Analysis of Amitriptyline and Its Metabolites in Plasma, in *Clinical Toxicology* vol. 18 (Leonard B. J. and Duncan W. A. M., eds.) Excerpta Medica, Elsevier, Amsterdam.

Biggs S. R., Chasseaud L. F., Hawkins D. R., and Midgley I. (1979) Determination of amitriptyline and its major basic metabolites in human urine by high-performance liquid chromatography. *Drug Metab. Dispos.* **7**, 233–236.

Blanc M., Tran M.-A., and Cotonat J. (1983) Technique de dosage de trois antidépresseurs tricycliques par chromatographie liquide haute performance. *Therapie* **38**, 17–20.

Bock J. L., Giller E., Gray S., and Jatlow P. (1982) Steady-state plasma concentrations of *cis*- and *trans*-10-OH amitriptyline metabolites. *Clin. Pharmacol. Ther.* **31**, 609–616.

Bogema S. C., Narasimhachari N., Mumtaz M., Goldin S., and Friedel R. O. (1982) Separation and quantitation of *cis*- and *trans*-thiothixene in human plasma by high-performance liquid chromatography. *J. Chromatogr. Biomed. Appl.* **233**, 257–267.

Borga O., Piafsky K. M., and Nilsen O. G. (1977) Plasma protein binding of basic drugs. *Clin. Pharmacol. Ther.* **22**, 539–544.

Brachet-Liermain A., Jarry Ch., Faure O., Guyot M., and Loiseau P. (1982) Liquid chromatography determination of clobazam and its major metabolite N-desmethylclobazam in human plasma. *Ther. Drug Monit.* **4**, 301–305.

Breutzmann D. A. and Bowers L. D. (1981) Reversed-phase liquid chromatography and gas chromatography/mass fragmentography compared for determination of tricyclic antidepressant drugs. *Clin. Chem.* **27**, 1907–1911.

Breyer-Pfaff U., Wiatr R., and Nill K. (1984) Measurement of maprotiline and oxaprotiline in plasma by high-performance liquid chromatography of fluorescent derivatives. *J. Chromatogr. Biomed. Appl.* **309**, 107–114.

Brinkman U. A., Welling P. L. M., De Vries G., Scholten A. H. M. T., and Frei R. W. (1981) Liquid chromatography of demoxepam and phenothiazines using a post-column photochemical reactor and fluorescence detection. *J. Chromatogr. Biomed. Appl.* **217**, 463–471.

Brodie R. R., Chasseaud L. F., and Hawkins D. R. (1977a) Separation and measurement of tricyclic antidepressant drugs in plasma by high-performance liquid chromatography. *J. Chromatogr. Biomed. Appl.* **143**, 535–539.

Brodie R. R., Chasseaud L. F., Crampton E. L., Hawkins D. R., and Risdall P. C. (1977b) High performance liquid chromatographic determination of dothiepin and northiaden in human plasma and serum. *J. Int. Med. Res.* **5,** 387–390.

Brodie R. R., Chasseaud L. F., and Taylor T. (1978) High-performance liquid chromatographic determination of benzodiazepines in human plasma. *J. Chromatogr.* **150,** 361–366.

Brooks M. A., DiDonato G., and Blumenthal H. P. (1985) Determination of chlorprothixene and its sulfoxide metabolite in plasma by high-performance liquid chromatography with ultraviolet and amperometric detection. *J. Chromatogr. Biomed. Appl.* **337,** 351–362.

Browning J. L., Harrington C. A., Burch N. R., and Davis C. M. (1982) Quantification of haloperidol and reduced haloperidol metabolites in plasma of psychiatric patients. *Fed. Proc. Fed. Am. Soc. Exp. Biol.* **41,** 1635.

Caldwell R. and Flanagan R. J. (1985) Measurement of trazodone in biological fluids by high performance liquid chromatography. *J. Chromatogr. Biomed. Appl.* **345,** 187–191.

Cole-Parmer Instrument Co. (1986) New HPLC column for fast drug analysis. Information brochure CHR-NP, Chicago, Illinois.

Colin P., Sirois G., and Lelorier J. (1983) High-performance liquid chromatography determination of dipotassium clorazepate and its major metabolite nordiazepam in plasma. *J. Chromatogr. Biomed. Appl.* **273,** 367–377.

Cooper J. K., McKay G., and Midha K. K. (1983) Subnanogram quantitation of chlorpromazine in plasma by high-performance liquid chromatography with electrochemical detection. *J. Pharm. Sci.* **72,** 1259–1262.

Cotler S., Puglisi C. V., and Gustafson J. H. (1981) Determination of diazepam and its major metabolites in man and in the cat by high-performance liquid chromatography. *J. Chromatogr. Biomed. Appl.* **222,** 95–106.

Curry S. H., Brown E. A., Hu O. Y.-P., and Perrin J. H. (1982) Liquid chromatographic assay of phenothiazine, thioxanthene and butyrophenone neuroleptics and antihistamines in blood and plasma with conventional and radial compression columns and UV and electrochemical detection. *J. Chromatogr. Biomed. Appl.* **231,** 361–376.

Davi H., Guyonnet J., Necciari J., and Cautreels W. (1985) Determination of circulating ethyl loflazepate metabolites in the baboon by radio-high-performance liquid chromatography with injection of crude plasma samples: Comparison with solvent extraction and thin-layer chromatography. *J. Chromatogr. Biomed. Appl.* **342,** 159–165.

Dawling S., Braithwaite R., and Crome P. (1979) Nomifensine overdose and plasma drug concentration. *Lancet* **1,** 56.

Dehass-Vermeulen J. and Thompson T. A. (1984) Assay for the determination of the tetracyclic antidepressant compound aptazapine in plasma by high performance liquid chromatography. *J. Chromatogr. Biomed. Appl.* **306**, 412–416.

Desmond P. V., Roberts R. K., Wood A. J., Dunn G. D., Wilkinson G. R., and Schenker S. (1980) Effect of heparin administration on plasma binding of benzodiazepines. *Br. J. Clin. Pharmacol.* **9**, 171–175.

Dhar A. K. and Kutt H. (1984) Improved liquid-chromatographic determination of haloperidol in plasma. *Clin. Chem.* **30**, 1228–1230.

Diquet B., Gaudel G., Colin J. N., and Singlas E. (1982) Dosage plasmatique de la clomipramine et de la demethylclomipramine par chromatographie liquide à haute performance. *Ann. Biol. Clin.* (Paris) **40**, 321–324.

di Tella A. S., Ricci P., di Nunzio C., and Cassandro P. (1986) A new method for the determination in blood and urine of a novel triazolobenzodiazepine (estazolam) by HPLC. *J. Anal. Toxicol.* **10**, 65–67.

Divoll M., Greenblatt D. J., and Shader R. I. (1982) Liquid chromatographic determination of chlordiazepoxide and metabolites in plasma. *Pharmacology* **24**, 261–266.

Dixon R. and Martin D. (1981) Tricyclic antidepressants: A simplified approach for the routine clinical monitoring of parent drug and metabolites in plasma using HPLC. *Res. Commun. Chem. Path. Pharmacol.* **33**, 537–545.

Dorey R. C., Landa B. L., and Narasimhachari N. (1983) Analysis of *cis*-thiothixene in plasma by HPLC. *Clin. Chem.* **29**, 1210.

Edelbroek P. M., de Haas E. J. M., and de Wolff F. A. (1982) Liquid-chromatographic determination of amitriptyline and its metabolites in serum, with adsorption onto glass minimized. *Clin. Chem.* **28**, 2143–2148.

Edinboro L. E. and Backer R. C. (1985) Preliminary report on the application of a high performance liquid chromatographic method for alprazolam in postmortem blood specimens. *J. Anal. Toxicol.* **9**, 207–208.

Emanuelsson B. and Moore R. G. (1978) Quantitation of zimelidine and norzimelidine in plasma using high-performance liquid chromatography. *J. Chromatogr. Biomed. Appl.* **146**, 113–119.

Faulkner R. D. and Lee C. (1983) Comparative assays for doxepin and desmethyldoxepin using high-performance liquid chromatography and high-performance thin-layer chromatography. *J. Pharm. Sci.* **72**, 1165–1167.

Fekete J., Del Castilho P., and Kraak J. C. (1981) Reversed-phase liquid chromatography for the separation of chlorpromazine, imipramine and some of their metabolites. *J. Chromatogr.* **204**, 319–327.

Foreman J. M., Griffiths W. C., Dextraze P. G., and Diamond I. (1980) Simultaneous assay of diazepam, chlordiazepoxide, N-desmethyldiazepam, N-desmethylchlordiazepoxide, and demoxepam in serum by high performance liquid chromatography. *Clin. Biochem.* **13**, 122–125.

Gerson B., Chan S., Bell F., and Pappalardo K. M. (1986) Trazodone: A simple, clean extraction and rapid quantification by high pressure liquid chromatography. *J. Psychiat. Res.* **20**, 69–76.

Gill R., Law B., and Gibbs J. P. (1986) High performance liquid chromatography systems for the separation of benzodiazepines and their metabolites. *J. Chromatogr.* **356**, 37–46.

Gillilan R. and Mason W. D. (1981) High-pressure liquid chromatographic determination of viloxazine in human plasma and urine. *J. Pharm. Sci.* **70**, 220–221.

Glassman A. H., Schildkraut J. J., Orsulak P. J., Cooper T. B., Kupfer D. J., Shader R. I., Davis J. M., Carroll B., Perel J. M., Klerman G. L., and Greenblatt J. (1985) Tricyclic antidepressants—blood level measurements and clinical outcome: An APA task force report. *Am. J. Psychiatry* **142**, 155–162.

Godbillon J. and Gauron S. (1981) Determination of clomipramine or imipramine and their non-demethylated metabolites in human blood or plasma by high-performance liquid chromatography. *J. Chromatogr.* **204**, 303–311.

Goldstein S. A. and Van Vunakis H. (1981) Determination of fluphenazine, related phenothiazine drugs and metabolites by combined high-performance liquid chromatography and radioimmunoassay. *J. Pharmacol. Exp. Ther.* **217**, 36–43.

Good T. J. and Andrews J. S. (1981) The use of bonded-phase extraction columns for rapid sample preparation of benzodiazepines and metabolites from serum for HPLC analysis. *J. Chromatogr. Sci.* **19**, 562–566.

Gottschalk L. A., Dinvovo E., Biener R., and Nandi B. R. (1978) Plasma concentrations of thioridazine metabolites and ECG determinations. *J. Pharm. Sci.* **67**, 155–157.

Grassi E., Passetti G. L., and Trebbi A. (1977) Quantitative determination of pinazepam and its metabolites in blood and urine by high-performance liquid chromatography. *J. Chromatogr.* **144**, 132–135.

Greizerstein H. B. and Wojtowicz C. (1977) Simultaneous determination of chlordiazepoxide and its N-demethyl metabolite in 50-μL blood samples by high pressure liquid chromatography. *Anal. Chem.* **49**, 2235–2236.

Gupta R. N. and Lew M. (1985) Determination of trazodone in human plasma by liquid chromatography with fluorescence detection. *J. Chromatogr. Biomed. Appl.* **342**, 442–446.

Gupta R. N., Bartolucci G., and Molnar G. (1981) Analysis of chlorpromazine in plasma: Effect of specimen storage. *Clin. Chim. Acta.* **109,** 351–354.

Hackett L. P. and Dusci L. J. (1979) The use of high-performance liquid chromatography in clinical toxicology. II. Tricyclic antidepressants. *Clin. Toxicol.* **15,** 55–61.

Hackman M. R. and Brooks M. A. (1981) Differential pulse amperometric detection of drugs in plasma using a dropping mercury electrode as a high-performance liquid chromatographic detector. *J. Chromatogr. Biomed. Appl.* **222,** 179–190.

Hale P. W. and Poklis A. (1984) Determination of diastereoisomeric pairs of thioridazine 5-sulfoxide by high-performance liquid chromatography. *J. Chromatogr. Biomed. Appl.* **336,** 452–455.

Hawes E. M., Hubbard J. W., Martin M., McKay G., Yeung P. K. F., and Midha K. K. (1986) Therapeutic monitoring of chlorpromazine. III. Minimal interconversion between chlorpromazine and metabolites in human blood. *Ther. Drug Monit.* **8,** 37–41.

Heazlewood R. L. and Lemass R. W. J. (1984) Simple high-performance liquid chromatographic assay for the routine monitoring of clonazepam in plasma. *J. Chromatogr. Biomed. Appl.* **336,** 229–233.

Heizmann P., Geschke R., and Zinapold K. (1984) Determination of bromazepam in plasma and of its main metabolites in urine by reversed-phase high-performance liquid chromatography. *J. Chromatogr. Biomed. Appl.* **310,** 129–137.

Hequet D., Jarry C., Rouquette C., and Brachet-Liermain A. (1985) Plasma assay of haloperidol by high performance liquid chromatography. *Ann. Biol. Clin.* **43,** 739–742.

Hirayama H. and Kasuya Y. (1983) High-performance liquid chromatographic determination of bromazepam in human plasma. *J. Chromatogr. Biomed. Appl.* **277,** 414–418.

Ho P. C., Triggs E. J., Heazlewood V., and Bourne D. W. A. (1983) Determination of nitrazepam and temazepam in plasma by high-performance liquid chromatography. *Ther. Drug Monit.* **5,** 303–307.

Hojabri H. and Glennon J. D. (1985) Determination of cianopramine in human plasma by high-performance liquid chromatography and gas-liquid chromatography with ultraviolet, fluorescence and electron-capture detection. *J. Chromatogr. Biomed. Appl.* **342,** 97–103.

Hubbard J. W., Cooper J. K., Hawes E. M., Jenden D. J., May P. R. A.; Martin M., McKay G., Van Putten T., and Midha K. K. (1985) Therapeutic Monitoring of Chlorpromazine. I. Pitfalls in Plasma Analysis. *Ther. Drug Monit.* **7,** 222–228.

Jatlow P. I., Miller R., and Swigar M. (1982) Measurement of haloperidol in human plasma using reversed-phase high-performance liquid chromatography. *J. Chromatogr. Biomed. Appl.* **227,** 233–238.

Jewesson P. J. and Remick R. A. (1986) Influence of blood collection tubes on tricyclic antidepressant concentration. *J. Clin. Psychopharmacol.* **6,** 51–52.

Johnson S. M., Chan C., Cheng S., Shimek J. L., Nygard G., and Wahba Khalil S. K. (1982) Isocratic high-performance liquid chromatographic method for the determination of tricyclic antidepressants and metabolites in plasma. *J. Pharm. Sci.* **71,** 1027–1030.

Johnson S. M., Nygard G., and Wahba Khalil S. K. (1984) Isocratic liquid chromatographic method for the determination of amoxapine and its metabolites. *J. Pharm. Sci.* **73,** 696–699.

Jozefczak C., Ktorza N., and Uzan A. (1982) High-performance liquid chromatographic determination of indalpine, a new nontricyclic antidepressant, in human plasma. *J. Chromatogr. Biomed. Appl.* **230,** 87–95.

Kabra P. M. and Nzekwe E. U. (1985) Liquid chromatographic analysis of clonazepam in human serum with solid-phase (Bond-Elut) extraction. *J. Chromatogr. Biomed. Appl.* **341,** 383–390.

Kabra P. M., Stevens G. L., and Marton L. J. (1978) High-pressure liquid chromatographic analysis of diazepam, oxazepam and N-desmethyldiazepam in human blood. *J. Chromatogr.* **150,** 355–360.

Kabra P. M., Mar N. A., and Marton L. J. (1981) Simultaneous liquid chromatographic analysis of amitriptyline, nortriptyline, imipramine, desipramine, doxepin, and nordoxepin. *Clin. Chim. Acta.* **iii,** 123–132.

Kamimura H., Sasaki H., Yokoi K., and Kawamura S. (1985) Determination of indeloxazine in plasma by liquid chromatography and gas chromatography-mass spectrometry. *J. Pharm. Sci.* **74,** 559–561.

Kelly H., Huggett A., and Dawling S. (1982) Liquid-chromatographic measurement of nitrazepam in plasma. *Clin. Chem.* **28,** 1478–1481.

Ketchum C., Robinson C. A., and Scott J. W. (1983) Analysis of amoxapine, 8-hydroxyamoxapine, and maprotiline by high-pressure liquid chromatography. *Ther. Drug Monit.* **5,** 309–312.

Kilts C. D., Patrick K. S., Breese G. R., and Mailman R. B. (1982) Simultaneous determination of thioridazine and its S-oxidized and N-demethylated metabolites using high performance liquid chromatography on radially compressed silica. *J. Chromatogr. Biomed. Appl.* **231,** 377–391.

Kilts C. D., Dew K. L., Ely T. D., and Mailman R. B. (1985) Quantification of R-(+)-7-chloro-8-hydroxy-1-phenyl-2,3,4,5-tetrahydro-1H-3-methyl-3-benzazepine in brain and blood by use of reversed-phase high-performance liquid chromatography with electrochemical detection. *J. Chromatogr. Biomed. Appl.* **342,** 452–457.

Kimball D. F. and Lampert A. A. (1982) Analysis of amoxapine and its metabolites in human plasma by high pressure liquid chromatography. *S.D. J. Med.* **35**, 31–33.

Kobayashi A., Sugita S., and Nakazawa K. (1985) Determination of amoxapine and its metabolites in human serum by high-performance liquid chromatography. *Neuropharmacol.* **12**, 1253–1256.

Kobayashi A., Sugita S., and Nakazawa K. (1984) High-performance liquid chromatographic determination of imipramine and desipramine in serum. *J. Chromatogr. Biomed. Appl.* **336**, 410–414.

Kogan M. J., Pierson D., and Verebey K. (1983) Quantitative determination of haloperidol in human plasma by high-performance liquid chromatography. *Ther. Drug Monit.* **5**, 485–489.

Korpi E. R., Phelps B. H., Granger H., Chang W.-H., Linnoila M., Meek J. L., and Wyatt R. J. (1983) Simultaneous determination of haloperidol and its reduced metabolite in serum and plasma by isocratic liquid chromatography with electrochemical detection. *Clin. Chem.* **29**, 624–628.

Koteel P., Mullins R. E., and Gadsden R. H. (1982) Sample preparation and liquid-chromatographic analysis for tricyclic antidepressants in serum. *Clin. Chem.* **28**, 462–466.

Kozu T. (1984) High-performance liquid chromatographic determination of nitrazepam and its metabolites in human urine. *J. Chromatogr. Biomed. Appl.* **310**, 213–218.

Kraak J. C. and Bijster P. (1977) Determination of amitriptyline and some of its metabolites in blood by high-pressure liquid chromatography. *J. Chromatogr. Biomed. Appl.* **143**, 499–512.

Kruger R., Mengel I., and Kuss H. J. (1984) Determination of benperidol in human plasma by high-performance liquid chromatography. *J. Chromatogr. Biomed. Appl.* **311**, 109–116.

Kuss H. J. and Feistenauer E. (1981) Quantitative high-performance liquid chromatographic assay for the determination of maprotiline and oxaprotiline in human plasma. *J. Chromatogr.* **204**, 349–353.

Lagerstrom P.-O., Marle I., and Persson B.-A. (1983) Solvent extraction of tricyclic amines from blood plasma and liquid chromatographic determination. *J. Chromatogr. Biomed. Appl.* **273**, 151–160.

Larsen N.-E., Hansen L. B., and Knudsen P. (1985) Quantitative determination of perphenazine and its dealkylated metabolite using high-performance liquid chromatography. *J. Chromatogr. Biomed. Appl.* **341**, 244–250.

Larsson M. and Forsman A. (1983) A high-performance liquid chromatographic method for the assay of perphenazine and its dealkylated metabolite in serum after therapeutic doses. *Ther. Drug Monit.* **5**, 225–228.

Leelavathi D. E., Dressler D. E., Soffer E. F., Yachetti S. D., and Knowles J. A. (1985) Determination of promethazine in human plasma by automated high-performance liquid chromatography with electrochemical detection and by gas chromatography-mass spectrometry. *J. Chromatogr. Biomed. Appl.* **339,** 105–115.

Lensmeyer G. L. and Evenson M. A. (1984) Stabilized analysis of antidepressant drugs by solvent-recycled liquid chromatography: Procedure and proposed resolution mechanisms for chromatography. *Clin. Chem.* **30,** 1774–1779.

Lensmeyer G. L., Rajani C., and Evenson M. A. (1982) Liquid-chromatographic procedure for simultaneous analysis for eight benzodiazepines in serum. *Clin. Chem.* **28,** 2274–2278.

Le Roux Y., Gaillot J., and Beider A. (1982) High-performance liquid chromatographic determination of pipotiazine in human plasma and urine. *J. Chromatogr. Biomed. Appl.* **230,** 401–408.

Lindberg R. L. P. (1985) Selective and sensitive high-performance liquid chromatographic assay for the metabolites of nomifensine in human plasma. *J. Chromatogr. Biomed. Appl.* **341,** 333–339.

Lindberg R. L. P., Salonen J. S., and Iisalo E. I. (1983) Determination of nomifensine in human serum: A comparison of high-performance liquid and gas-liquid chromatography. *J. Chromatogr. Biomed. Appl.* **276,** 85–92.

Linnoila M., Dorrity F., and Jobson K. (1978) Plasma and erythrocyte levels of tricyclic antidepressants in depressed patients. *Am. J. Psychiatry* **135,** 557–561.

Linnoila M., Insel T., Kilts C., Potter W. Z., and Murphy D. L. (1982) Plasma steady-state concentrations of hydroxylated metabolites of clomipramine. *Clin. Pharmacol. Ther.* **32,** 208–211.

MacKichan J. J., Jusko W. J., Duffner P. K., and Cohen M. E. (1979) Liquid-chromatographic assay of diazepam and its major metabolites in plasma. *Clin. Chem.* **25,** 856–859.

McBurney A. and George S. (1984) High-performance liquid chromatography of haloperidol in serum at the concentrations achieved during chronic therapy. *J. Chromatogr. Biomed. Appl.* **308,** 387–392.

McCormick S. R., Nielsen J., and Jatlow P. (1984) Quantification of alprazolam in serum or plasma by liquid chromatography. *Clin. Chem.* **30,** 1652–1655.

McCutcheon J. R. (1979) Reverse-phase HPLC determination of thioridazine and mesoridazine in whole blood. *J. Anal. Toxicol.* **3,** 105–107.

McIntyre I. M., Norman T. R., and Burrows G. D. (1981) *In vitro* stability of nomifensine in plasma. *Clin. Chem.* **27,** 203–204.

McKay G., Geddes J., Cowper M. J., and Gurnsey T. S. (1983) Recent advances in the analysis of phenothiazine drugs and their metabolites using high performance liquid chromatography. *Prog. Neuropsychopharmacol. Biol. Psychiatry* **7**, 703–707.

McKay G., Cooper J. K., Hawes E. M., Hubbard J. W., Martin M., and Midha K. K. (1985) Therapeutic monitoring of chlorpromazine. II. Pitfalls in whole blood analysis. *Ther. Drug Monit.* **7**, 472–477.

Mellstrom B. and Braithwaite R. (1978) Ion-pair liquid chromatography of amitriptyline and metabolites in plasma. *J. Chromatogr.* **157**, 379–385.

Mellstrom B. and Tybring G. (1977) Ion-pair liquid chromatography of steady-state plasma levels of chlorimipramine and demethylchlorimipramine. *J. Chromatogr. Biomed. Appl.* **143**, 597–605.

Midha K. K., Cooper J. K., McGilveray I. J., Butterfield A. G., and Hubbard J. W. (1981) High-performance liquid chromatographic assay for nanogram determination of chlorpromazine and its comparison with a radioimmunoassay. *J. Pharm. Sci.* **70**, 1043–1046.

Miller R. L. and Devane C. L. (1986a) Analysis of trazodone and m-chlorophenyl piperazine in plasma and brain tissue by high-performance liquid chromatography. *J. Chromatogr. Biomed. Appl.* **374**, 388–393.

Miller R. L. and Devane C. L. (1986b) Measurement of haloperidol and reduced haloperidol in human plasma using reversed-phase high-performance liquid chromatography. *J. Chromatogr. Biomed. Appl.* **374**, 405–408.

Miyao Y., Suzuki A., Noda K., and Noguchi H. (1983) A sensitive assay method for pimozide in human plasma by high-performance liquid chromatography with fluorescence detection. *J. Chromatogr. Biomed. Appl.* **275**, 443–449.

Miyazaki K., Arita T., Oka I., Koyama T., and Yamashita I. (1981) High-performance liquid chromatographic determination of haloperidol in plasma. *J. Chromatogr. Biomed. Appl.* **223**, 449–453.

Morselli P. L., Bianchetti G., Tedeschi G., and Braithwaite R. A. (1981) Haloperidol: Clinical Pharmacokinetics and Significance of Therapeutic Drug Monitoring, in *Therapeutic Drug Monitoring* (Richens A. and Marks A., eds.) Churchill Livingstone, Edinburgh.

Moyes R. B. and Moyes I. C. A. (1977) Metabolism and pharmacokinetics of clomipramine: Measurement of plasma antidepressant levels by high-performance liquid chromatography. *Postgrad. Med. J.* **53** (suppl. 4), 117–123.

Murakami K., Murakami K., and Ueno T. (1982) Simultaneous determination of chlorpromazine and levomepromazine in human plasma and urine by high-performance liquid chromatography using electrochemical detection. *J. Chromatogr. Biomed. Appl.* **227**, 103–112.

Naranjo C. A., Abel J. G., Sellers E. M., and Giles H. G. (1980) Unaltered diazepam plasma binding using indwelling heparinized cannulae for sampling. *Br. J. Clin. Pharmacol.* **9,** 103–105.

Narasimhachari N., Dorey R. C., Landa B. L., and Friedel R. O. (1984) Improved high-performance liquid chromatographic method for the quantitation of *cis*-thiothixene in plasma samples using *trans*-thiothixene as internal standard. *J. Chromatogr. Biomed. Appl.* **311,** 424–429.

Norman T. R. and Maguire K. P. (1985) Analysis of tricylic anti-depressant drugs in plasma and serum by chromatographic techniques. *J. Chromatogr. Biomed. Appl.* **340,** 173–197.

Orsulak P. J., Sink M., and Weed J. (1984) Blood-collection tubes for tricyclic antidepressant drugs: A reevaluation. *Ther. Drug Monit.* **6,** 444–448.

Oyehaug E., Ostensen E. T., and Salvesen B. (1982) Determination of the antidepressant agent citalopram and metabolites in plasma by liquid chromatography with fluorescence detection. *J. Chromatogr. Biomed. Appl.* **227,** 129–135.

Oyehaug E., Ostensen E. T., and Salvesen B. (1984) High-performance liquid chromatographic determination of citalopram and four of its metabolites in plasma and urine samples from psychiatric patients. *J. Chromatogr. Biomed. Appl.* **308,** 199–208.

Pacakova V., Stulik K., and Tomkova H. (1984) Determination of some tricyclic neuroleptics by reversed-phase high-performance liquid chromatography with ultraviolet and polarographic detection. *J. Chromatogr.* **298,** 309–318.

Parkinson D. (1985) Sensitive analysis of butyrophenone neuroleptics by high-performance liquid chromatography with ultraviolet detection at 254 nm. *J. Chromatogr. Biomed. Appl.* **341,** 465–472.

Peat M. A. and Kopjak L. (1979) The screening and quantitation of diazepam, flurazepam, chlordiazepoxide, and their metabolites in blood and plasma by electron-capture gas chromatography and high pressure liquid chromatography. *J. Forens. Sci.* **24,** 46–54.

Peat M. A., Finkle B. S., and Deyman M. E. (1979) High-pressure liquid chromatographic determination of chlordiazepoxide and its major metabolites in biological fluids. *J. Pharm. Sci.* **68,** 1467–1468.

Peng G. W. (1984) Assay of adinazolam in plasma by liquid chromatography. *J. Pharm. Sci.* **73,** 1173–1175.

Perchalski R. J. and Wilder B. J. (1978) Determination of benzodiazepine anticonvulsants in plasma by high-performance liquid chromatography. *Anal. Chem.* **50,** 554–557.

Petters I., Peng D.-R., and Rane A. (1984) Quantitation of clonazepam and its 7-amino and 7-acetamido metabolites in plasma by high-performance liquid chromatography. *J. Chromatogr. Biomed. Appl.* **306,** 241–248.

Proelss H. F., Lohmann H. J., and Miles D. G. (1978) High-performance liquid-chromatographic simultaneous determination of commonly used tricyclic antidepressants. *Clin. Chem.* **24,** 1948–1953.

Przyborowska M., Szumilo H., Misztal G., and Olajossy M. (1982) Determination of chlorpromazine in blood by high-performance liquid chromatography (HPLC) *Acta. Pol. Pharm.* **42,** 467–72.

Puglisi C. V. and de Silva J. A. F. (1981) Determination of the anxiolytic agent 8-chloro-6-(2-chlorophenyl)-4H-imidazo-[1,5-*a*][1,4]-benzodiazepine-3-carboxamide in whole blood, plasma or urine by high-performance liquid chromatography. *J. Chromatogr. Biomed. Appl.* **226,** 135–146.

Puglisi C. V., Pao J., Ferrara F. J., and de Silva J. A. F. (1985) Determination of midazolam (Versed®) and its metabolites in plasma by high-performance liquid chromatography. *J. Chromatogr. Biomed. Appl.* **344,** 199–209.

Quattrocchi F., Karnes H. T., Robinson J. D., and Hendeles L. (1983) Effect of serum separator blood collection tubes on drug concentrations. *Ther. Drug Monit.* **5,** 359–362.

Raisys V. A., Friel P. N., Graaff P. R., Opheim K. E., and Wilensky A. J. (1980) High-performance liquid chromatographic and gas-liquid chromatographic determination of diazepam and nordiazepam in plasma. *J. Chromatogr. Biomed. Appl.* **183,** 441–448.

Rao S. N., Dhar A. K., Kutt H., and Okamoto M. (1982) Determination of diazepam and its pharmacologically active metabolites in blood by Bond Elut column extraction and reversed-phase high-performance liquid chromatography. *J. Chromatogr. Biomed. Appl.* **231,** 341–348.

Ratnaraj N., Goldberg V. D., Elyas A., and Lascelles P. T. (1981) Determination of diazepam and its major metabolites using high-performance liquid chromatography. *Analyst* **106,** 1001–1004.

Ratnaraj N., Goldberg V., and Lascelles P. T. (1984) Determination of clobazam and desmethylclobazam in serum using high-performance liquid chromatography. *Analyst* **109,** 813–815.

Reece P. A., Zacest R., and Barrow C. G. (1979) Quantification of imipramine and desipramine in plasma by high-performance liquid chromatography and fluorescence detection. *J. Chromatogr. Biomed. Appl.* **163,** 310–314.

Root I. and Ohlson G. B. (1984) Trazodone overdose: Report of two cases. *J. Anal. Toxicol.* **8,** 91–94.

Rop P. P., Viala A., Durand A., and Conquy T. (1985) Determination of citalopram, amitriptyline and clomipramine in plasma by reversed-phase high-performance liquid chromatography. *J. Chromatogr. Biomed. Appl.* **338,** 171–178.

Rop P. P., Conquy T., Gouezo F., Viala A., and Grimaldi F. (1986) Determination of metapramine, imipramine, trimipramine and their major metabolites in plasma by reversed-phase column liquid chromatography. *J. Chromatogr. Biomed. Appl.* **375**, 339–347.

Routledge P. A., Kitchell B. B., Bjornsson T. D., Skinner T., Linnoila M., and Shand D. G. (1980) Diazepam and N-desmethyl diazepam redistribution after heparin. *Clin. Pharmacol. Ther.* **27**, 528–532.

Rovei V. and Sanjuan M. (1980) Simple and specific high performance liquid chromatographic method for the routine monitoring of clonazepam in plasma. *Ther. Drug Monit.* **2**, 283–287.

Sajgo M., Gesztesi A., and Sido T. (1981) Determination of tofisopam in serum by high-performance liquid chromatography. *J. Chromatogr. Biomed. Appl.* **222**, 303–307.

Salonen J. S. and Scheinin M. (1983) Determination of maprotiline and N-desmethylmaprotiline from biological fluids by HPLC. *J. Anal. Toxicol.* **7**, 175–177.

Sankey M. G., Holt J. E., and Kaye C. M. (1982) A simple and sensitive HPLC method for the assay of prochlorperazine in plasma. *Br. J. Clin. Pharmacol.* **13**, 578–580.

Shaw W., Long G., and McHan J. (1983) An HPLC method for analysis of clonazepam in serum. *J. Anal Toxicol.* **7**, 119–122.

Shibanoki S., Kubo T., and Ishikawa K. (1986) Chromatographic assay of 10-[3-(3-hydroxypyrrolidinyl)propyl]-2-trifluoromethyl phenothiazine using electrochemical detection. *J. Chromatogr. Biomed. Appl.* **377**, 436–440.

Silber B., Lo M., and Riegelman S. (1980) The influence of heparin administration on the plasma protein binding and disposition of propranolol. *Res. Commun. Chem. Pathol. Pharmacol.* **27**, 419–429.

Skellern G. G., Meier J., Knight B. I., and Whiting B. (1978) The application of HPLC to the determination of some 1,4-benzodiazepines and their metabolites in plasma. *Br. J. Clin. Pharmacol.* **5**, 483–487.

Skinner T., Gochnauer R., and Linnoila M. (1981) Liquid chromatographic method to measure thioridazine and its active metabolites in plasma. *Acta Pharmacol. Toxicol.* **48**, 223–226.

Smith D. J. (1981) The separation and determination of chlorpromazine and some of its related compounds by reversed-phase high performance liquid chromatography. *J. Chromatogr. Sci.* **19**, 65–71.

Smith G. A., Schulz P., Giacomini K. M., and Blaschke T. F. (1982) High-pressure liquid chromatographic determination of amitriptyline and its major metabolites in human whole blood. *J. Pharm. Sci.* **71**, 581–583.

Sonsalla P. K., Jennison T. A., and Finkle B. S. (1982a) Quantitative liquid-chromatographic technique for the simultaneous assay of tricylic antidepressant drugs in plasma or serum. *Clin. Chem.* **28,** 457–461.

Sonsalla P., Jennsion T., and Finkle B. (1982b) Importance of evaporation conditions and two internal standards for quantitation of amitriptyline and nortriptyline. *Clin. Chem.* **28,** 1401–1402.

Stoll A. L., Baldessarini R. J., Cohen B. M., and Finklestein S. P. (1984) Assay of plasma thioridazine and metabolites by high-performance liquid chromatography with amperometric detection. *J. Chromatogr. Biomed. Appl.* **307,** 457–463.

Streator J. T., Eichmeier L. S., and Caplis M. E. (1980) Determination of tricylic antidepressants in serum by high pressure liquid chromatography on a silica column. *J. Anal. Toxicol.* **4,** 58–62.

Strojny N. and de Silva J. A. F. (1985) Determination of diclofensine, an antidepressant agent, and its major metabolites in human plasma by high-performance liquid chromatography with fluorometric detection. *J. Chromatogr. Biomed. Appl.* **341,** 313–331.

Suckow R. F. (1983) A simultaneous determination of trazodone and its metabolite 1-m-chlorophenylpiperazine in plasma by liquid chromatography with electrochemical detection. *J. Liq. Chromatogr.* **6,** 2195–2208.

Suckow R. F. and Cooper T. B. (1981) Simultaneous determination of imipramine, desipramine, and their 2-hydroxy metabolites in plasma by ion-pair reversed-phase high-performance liquid chromatography with amperometric detection. *J. Pharm. Sci.* **70,** 257–261.

Suckow R. G. and Cooper T. B. (1982) Simultaneous determination of amitriptyline, nortriptyline and their respective isomeric 10-hydroxy metabolites in plasma by liquid chromatography. *J. Chromatogr. Biomed. Appl.* **230,** 391–400.

Suckow R. G. and Cooper T. B. (1984) Determination of trimipramine and metabolites in plasma by liquid chromatography with electrochemical detection. *J. Pharm. Sci.* **73,** 1745–1748.

Suckow R. F. and Cooper T. B. (1985) Determination of amoxapine and metabolites in plasma by liquid chromatography with electrochemical detection. *J. Chromatogr. Biomed. Appl.* **338,** 225–229.

Suckow R. F., Cooper T. B., Quitkin F. M., and Stewart J. W. (1982) Determination of mianserin and metabolites in plasma by liquid chromatography with electrochemical detection. *J. Pharm. Sci.* **71,** 889–892.

Sumirtapura Y. C., Aubert C., Coassolo P., and Cano J. P. (1982) Determination of 7-amino-flunitrazepam (Ro 20-1815) and 7-amino-desmethylflunitrazepam (Ro 5-4650) in plasma by high-performance liquid chromatography and fluorescence detection. *J. Chromatogr. Biomed. Appl.* **232,** 111–118.

Sutfin T. A. and Jusko W. J. (1979) High-performance liquid chromatographic assay for imipramine, desipramine, and their 2-hydroxylated metabolites. *J. Pharm. Sci.* **68**, 703–705.

Sutfin T. A., D'Ambrosio R., and Jusko W. J. (1984) Liquid-chromatographic determination of eight tri- and tetracyclic antidepressants and their major active metabolites. *Clin. Chem.* **30**, 471–474.

Tada K., Moroji T., Sekiguchi R., Motomura H., and Noguchi T. (1985) Liquid chromatographic assay of diazepam and its major metabolites in serum, and application to pharmacokinetic study of high doses of diazepam in schizophrenics. *Clin. Chem.* **31**, 1712–1715.

Tasset J. J. and Hassan F. M. (1982) Liquid-chromatographic determination of amoxapine and 8-hydroxyamoxapine in human serum. *Clin. Chem.* **28**, 2154–2157.

Taylor E. H., Sloniewsky D., and Gadsden R. H. (1984) Automated extraction and high-performance liquid chromatographic determination of serum clonazepam. *Ther. Drug Monit.* **6**, 474–477.

Taylor G. and Houston J. B. (1982) Simultaneous determination of promethazine and two of its circulating metabolites by high-performance liquid chromatography. *J. Chromatogr. Biomed. Appl.* **230**, 194–198.

Thoma J. J., Bondo P. B., and Kozak C. M. (1979) Tricyclic antidepressants in serum by a Clin-Elut column extraction and high pressure liquid chromatographic analysis. *Ther. Drug Monit.* **1**, 335–358.

Tjaden U. R., Lankelma J., Poppe H., and Muusze R. G. (1976) Anodic coulometric detection with a glassy carbon electrode in combination with reversed-phase high-performance liquid chromatography. *J. Chromatogr. Biomed. Appl.* **125**, 275–286.

Tjaden U. R., Meeles M. T. H. A., Thys C. P., and Van Der Kaay M. (1980) Determination of some benzodiazepines and metabolites in serum, urine and saliva by high-performance liquid chromatography. *J. Chromatogr. Biomed. Appl.* **181**, 227–241.

Tomasini J. L., Bun H., Coassolo P., and Aubert C. (1985) Determination of clobazam, N-desmethylclobazam and their hydroxy metabolites in plasma and urine by high-performance liquid chromatography. *J. Chromatogr. Biomed. Appl.* **343**, 369–377.

Vandermark F. L., Adams R. F., and Schmidt G. J. (1978) Liquid-chromatographic procedure for tricyclic drugs and their metabolites in plasma. *Clin. Chem.* **24**, 87–91.

Van Rooij H. H., Waterman R. L., and Kraak J. C. (1979) Dynamic cation-exchange systems for the separation of drugs derived from butyrophenone and diphenylpiperidine by high-performance liquid chromatography and applied in the determination of halopemide in plasma. *J. Chromatogr. Biomed. Appl.* **164**, 177–185.

Varian Instruments. Rapid sorbent extraction of tricyclic antidepressants from serum. *Sample Preparation Digest No. 2.* Sunnyvale, California.

Vasiliades J. and Sahawneh T. H. (1981) Determination of midazolam by high-performance liquid chromatography. *J. Chromatogr. Biomed. Appl.* **225**, 266–271.

Viala A., Hou N., Durand A., Ba B., Aaes-Jorgensen T., and Jorgensen A. (1983) Dosage du cis(z)-clopenthixol et de la fluphénazine dans le sang total et le plasma par chromatographie en phase liquide à haute performance avec étalonnage interne. *J. Pharm. Belg.* **38**, 299–303.

Violon C, Pessemier L., and Vercruysee A. (1982) High-performance liquid chromatography of benzophenone derivatives for the determination of benzodiazepines in clinical emergencies. *J. Chromatogr.* **236**, 157–168.

Visser T., Oostelbos M. C. J. M., and Toll P. J. M. M. (1984) Reliable routine method for the determination of antidepressant drugs in plasma by high-performance liquid chromatography. *J. Chromatogr. Biomed. Appl.* **309**, 81–93.

Vree T. B., Baars A. M., Hekster Y. A., and Van Der Kleijn E. (1979) Simultaneous determination of diazepam and its metabolites N-desmethyldiazepam, oxydiazepam and oxazepam in plasma and urine of man and dog by means of high-performance liquid chromatography. *J. Chromatogr. Biomed. Appl.* **162**, 605–614.

Vree T. B., Baars A. M., Booij L. H. D., and Driessen J. J. (1981) Simultaneous determination and pharmacokinetics of midazolam and its hydroxymetabolites in plasma and urine of man and dog by means of high-performance liquid chromatography. *Arzneimittelforsch.* **31**, 2215–2219.

Wallace J. E., Shimek E. L., and Harris S. C. (1981a) Determination of tricyclic antidepressants by high-performance liquid chromatography. *J. Anal. Toxicol.* **5**, 20–23.

Wallace J. E., Shimek E. L., Harris S. C., and Stavchansky S. (1981b) Determination of promethazine in serum by liquid chromatography. *Clin. Chem.* **27**, 253–255.

Wallace J. E., Shimek E. L., Stavchansky S., and Harris S. C. (1981c) Determination of promethazine and other phenothiazine compounds by liquid chromatography with electrochemical detection. *Anal. Chem.* **53**, 960–962.

Watson I. D. and Stewart M. J. (1977) Quantitative determination of amitriptyline and nortriptyline in plasma by high-performance liquid chromatography. *J. Chromatogr.* **132**, 155–159.

Weinfield R. E. and Miller K. F. (1981) Determination of the major urinary metabolite of flurazepam in man by high-performance liquid chromatography. *J. Chromatogr. Biomed. Appl.* **223**, 123–130.

Wells C. E., Juenge E. C., and Furman W. B. (1983) Simultaneous assay of thioridazine and its major metabolites in plasma at single dose levels with a novel report of two ring sulfoxides of thioridazine. *J. Pharm. Sci.* **72**, 622–625.

Westenberg H. G. M., Drenth B. F. H., de Zeeuw R. A., De Cuyper H., Van Praag H. M., and Korf J. (1977a) Determination of clomipramine and desmethylclomipramine in plasma by means of liquid chromatography. *J. Chromatogr.* **142**, 725–733.

Westenberg H. G. M., de Zeeuw R. A., de Cuyper H., Van Praag H. M., and Korf J. (1977b) Bioanalysis and pharmacokinetics of clomipramine and desmethylclomipramine in man by means of liquid chromatography. *Postgrad. Med. J.* **53** (suppl. 4), 124–130.

Westerlund D. and Erixson E. (1979) Reversed-phase chromatography of zimelidine and similar dibasic amines. I. Analysis in biological material. *J. Chromatogr.* **185**, 593–603.

Westerlund D., Nilsson L. B., and Jaksch Y. (1979) Straight-phase ion-pair chromatography of zimelidine and similar divalent amines. I. Bioanalysis. *J. Liq. Chromatogr.* **2**, 373–405.

Whall T. J. and Dokladalova J. (1979) High-performance liquid chromatographic determination of (Z)- and (E)-doxepin hydrochloride isomers. *J. Pharm. Sci.* **68**, 1454–1456.

Wills D. H. and Gelder M. G. (1980) Plasma fluphenazine levels by radioimmunoassay in schizophrenic patients treated with depot injections of fluphenazine decanoate. *Adv. Biochem. Psychopharmacol.* **24**, 599–602.

Wong A. S. (1983) An evaluation of HPLC for the screening and quantitation of benzodiazepines and acetaminophen in post mortem blood. *J. Anal. Toxicol.* **7**, 33–36.

Wong S. H. Y. and Waugh S. W. (1983) Determination of the antidepressants maprotiline and amoxapine, and their metabolites, in plasma by liquid chromatography. *Clin. Chem.* **29**, 314–318.

Wong S. H. Y., Waugh S. W., Draz M., and Jain N. (1984) Liquid-chromatographic determination of two antidepressants, trazodone and mianserin, in plasma. *Clin. Chem.* **30**, 230–233.

Wood M., Shand D. G., and Wood A. J. J. (1979) Altered drug binding due to the use of indwelling heparinized cannulas (heparin lock) for sampling. *Clin. Pharmacol. Ther.* **25**, 103–107.

Yang S. and Evenson M. A. (1983) Simultaneous liquid chromatographic determination of antidepressant drugs in human plasma. *Anal. Chem.* **55**, 994–998.

Yufu N., Itoh M., Notomi A., and Nakao H. (1984) Simultaneous measurement of various antidepressants in the plasma of depressed patients by high performance liquid chromatography. *Fol. Psychiat. Neurol. Jap.* **38**, 57–64.

Principles of Radioreceptor Assays

Larry Tune

1. Introduction

The scientific principles behind the radioreceptor assay (RRA) have been articulated in a number of excellent reviews (Enna, 1981, 1982), and they will be summarized here. In order to appreciate the strengths and limitations of RRAs, the basic principles of receptor binding technology must first be appreciated, since they provide the basis for our scientific understanding of the RRA (Hollenberg and Cuatrecasas, 1975; Yamamura et al., 1978).

First, one must demonstrate that ligand binding has taken place. This requires a sufficiently high signal-to-noise ratio for radioactively bound ligands so that one can reliably distinguish specific from nonspecific binding. The ratio between the amount of radioactivity bound to a specific binding side (specific binding), relative to the amount of radioactivily ligand that inevitably adheres to other membrane components (nonspecific binding) must be large enough so that these two types of binding can be distinguished. In most instances the number of receptors present in brain tissue is extremely small. As a result, this distinction has only been possible since the recent development of ligands with sufficiently high specific activity (greater than 1 mCi/mmol).

Once ligand binding has been observed, the criteria of saturability must be satisfied. One must demonstrate that the ligand saturates its receptors, since it is assumed that the number of targeted receptors in question is relatively small. By adding increasing concentrations of radioligand, one should find a point at which the ligand binding plateaus as a result of saturation. Demonstrating saturability is also important in distinguishing specific from nonspecific binding. Nonspecific binding increases linearly with increasing concentrations of ligand. Analysis of a typical saturation curve yields an estimate of the number of binding sites ("binding site concentrations," B_{max}), and the point at which binding plateaus. An analysis of the rate of saturation will give an estimate of the affinity of the receptor for this binding site.

The next condition to satisfy is that the ligand is specific for the targeted receptor. Most drugs bind to more than one receptor. Thus, distinguishing the particular receptor in a given tissue preparation is of considerable importance. To evaluate the specificity of a given compound, two things must be done. First, the conditions of the assay must be quite exact, since it is clear that small variations in the time of incubation, pH, temperature, type of buffer, type of tissue, and tissue concentration are all very important in characterizing a particular receptor. As one example, tritiated spiroperidol and haloperidol label dopamine-2 receptors in the corpus striatum. However, these same ligands bind to pricipally serotonin-2 receptors in the cortex. Thus, choice of tissue sample in this instance is of great importance. Once the exact conditions of the assay are established, binding experiments must be conducted using the radiolabeled ligand in competition with unlabeled compounds that are known to be agonists and antagonists of the receptor site. Once these principles of receptor binding assays have been satisfied, the important methodologic issues specific to individual RRAs must be appreciated. First, the tissue preparation may be different when comparing receptor and RRA studies utilizing the same radioactive ligand. Enna (1982) gives the example that to study the γ-aminobutyric acid (GABA) receptor in a receptor binding assay, crude synaptic membranes are frozen and used essentially fresh. For the RRA, however, these frozen pellets are homogenized in Tris-citrate buffer; this is followed by suspension and addition of Triton X-100, which functions as a detergent in this assay. Similar comparisons can be made for receptor assays and RRAs involving the dopamine receptor, benzodiazepine receptor, and muscarinic acetylcholine receptor assays (Enna, 1982). In all these assays, although one utilizes the same radioactive ligand, there are differences in tissue preparation and tissue requirements for the two assays (RRA and receptor binding assay). The assays are tailored to the particular ligand and situation.

Similar observations can be made for sample preparation. As Enna (1982) notes, one of the main advantages of RRAs is that little or no extraction is generally required prior to analysis. For example, in the measurement of cerebrospinal fluid (CSF), serum GABA, neuroleptics, and muscarinic acetylcholine antagonists, no sample preparation is required. One adds the CSF or serum directly into the assay system. This is not true for all ligands, since the opiate receptor does require homogenization in 0.1N HCL, centrifugation, lyophilization, resuspension in acid, and then neutralization with 0.1N KOH, in order to measure opiod peptides using radioreceptor methods (Simantov et al., 1977).

Although drugs and neurotransmitters can be detected in CSF and serum without elaborate sample preparation, this is not without problems since the addition of CSF and serum can profoundly affect binding characteristics within the assay. Mailman (1984) has eloquently demonstrated that serum can dramatically affect the standard curve for spiperone binding. Figures 1a, and 1b show the effect when serum is added to this assay. Because of daily fluctuations within the assays caused by serum as well as other factors, it is wise to calculate a standard curve for the displacement of radioligands in each experiment. One can determine the "behavior" of the assay on each given day. By investigating compounds over a two-log unit range around the dissociation constant (K_d), one obtains information that is useful for data analysis and for characterizing the assay. Studying an unknown sample in the range between 20 and 80% displacement of specific binding normally yields the most accurate results.

The sensitivity of RRAs is a function of the affinity of the receptor for the radioligand and the incubation volume (Enna, 1982). If receptor affinity were maximal, then the concentration of unlabeled antagonists necessary to define the lower limit of study (20% displacement) would be reduced, making it possible to detect smaller amounts of unlabeled compound. With regard to incubation volume, the amount of unlabeled compound required in a smaller incubation assay would be decreased compared to a large one. As noted above for assays of serum or CSF, the effect of larger amounts of serum and CSF on the assay binding characteristics becomes increasingly problematic.

The last issue concerns the specificity of RRAs. Two problems exist when looking at CSF, and particularly serum, using RRA methods. First, there are few radiolabeled compounds that adhere specifically to a single receptor. For example, as noted previously, spiroperidol (a compound that was generally thought in clinical circles to specifically label dopamine D-2 receptors) labels both D-2 and serotonin-2 receptors. Indeed, nowhere is this more apparent than in the case of tricyclic antidepressants, which adhere with great affinity to a number of receptors (Enna, 1982). The second issue concerns reproducability in comparison to other assays. Depending on the comparison assay, the relationship between serum

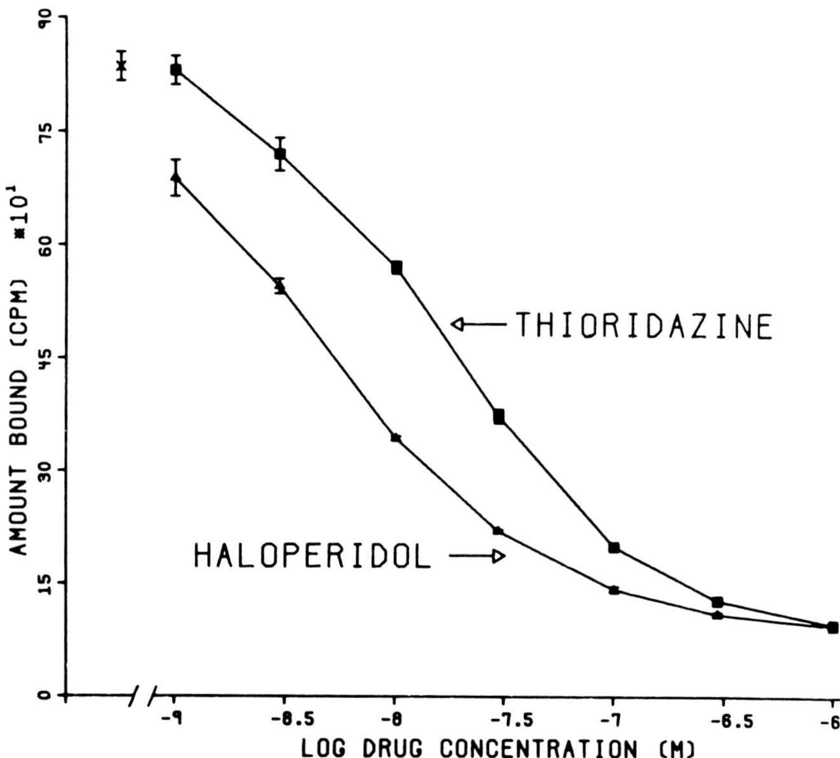

Fig. 1a. Displacement curves of ^{3}H-spiperone from rat striatal membranes by thioridazine and haloperidol with no serum present in the assay (Mailman et al., 1984).

neuroleptic levels done by RRA and other more standard methods can be quite high ($r = 0.83$) or very low ($r = 0.2$) (Ko, et al, 1985). Since receptor assays measure the interaction of drugs, their metabolites, and drug combinations that interact at a particular receptor, such comparisons may not be valid. If one were measuring the neuroleptic compound thioridazine, there is a pharmacologically active metabolite, mesoridazine, that is equally potent as a neuroleptic compound. If this drug were compared to, for example, a gas chromatography (GC) method, the dopamine receptor assay might show levels that are artifically high simply because it is measuring more than thioridazine.

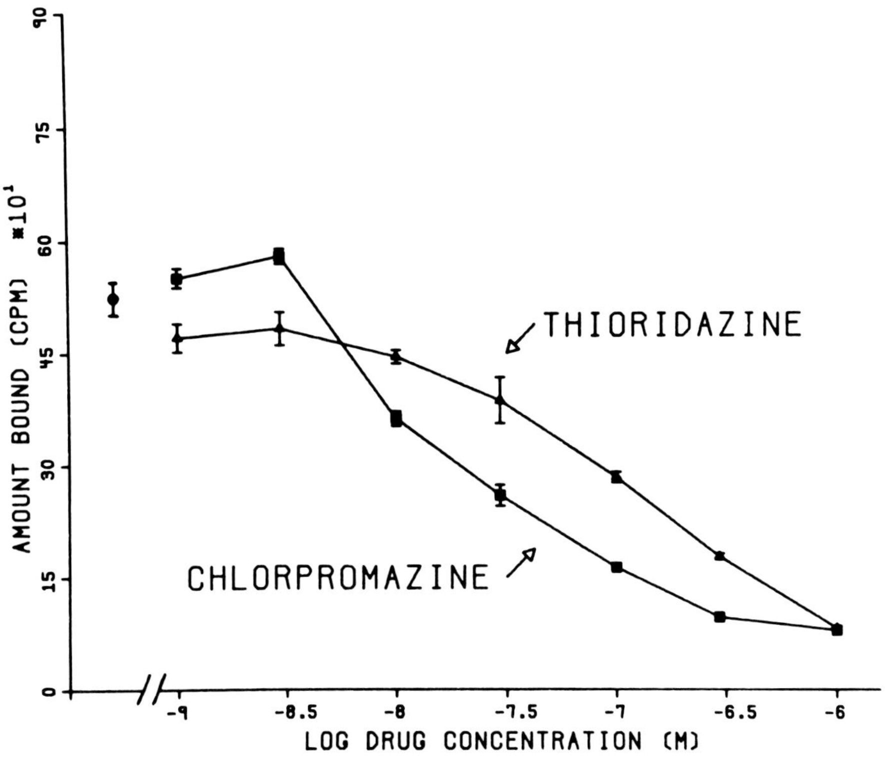

Fig. 1b. Displacement of ^{3}H-spiperone from rat striatal membranes by thioridazine or chlorpromazine in the presence of 50 μL serum per 1 mL assay (Mailman et al., 1984).

2. Clinical Applications of Radioreceptor Assays: Focus on Antimuscarinic RRA

2.1. Site of Action of Anticholinergics

In recent years, the mechanisms whereby anticholinergic drugs reduce the acute extrapyramidal symptoms (EPS) caused by neuroleptics have been characterized by neuroanatomic and neurophysiologic studies (Calne et al., 1975). Current evidence suggests that the therapeutic effects of neuroleptics derive from their blockade of dopamine receptors, particularly the D-2 subtype

(Creese et al., 1976; Kebabian and Calne, 1979) within the striatum. These receptors mediate the inhibitory effects of dopamine on cholinergic interneurons. Blockade of the striatal D-2 receptors results in disinhibition of the cholinergic neurons, and the increased turnover and release of acetylcholine causes an excessive stimulation of postsynaptic striatal muscarinic cholinergic receptors (Racagni et al., 1976). Fortunately, the dopaminergic–cholinergic synaptic sequence appears to be unique to the extrapyramidal system, and therefore pharmacologic restitution of cholinergic tone by coadministration of muscarinic receptor antagonists does not disturb the therapeutic action of neuroleptics that presumably occurs at dopamine synapses in the corticolimbic projections. Although this description of striatal cholinergic–dopaminergic neuronal synaptic relationships appears operationally valid, it does not adequately reflect the emerging evidence of the topographic organization of the striatal connections or the second-order interactions between striatal neurons and dopaminergic afferents.

Recent studies have uncovered major cholinergic projections from the magnocellular nuclei in basal forebrain, including the nucleus basalis of Meynert, diagnonal band of Broca, and medial septal nucleus that innervates the cerebral cortex and hippocampal formation (Johnston et al., 1981). Evidence is emerging that this cholinergic system appears to play a critical role in higher cognitive functions. Destruction of the hippocampal cholinergic pathway in experimental animals profoundly impairs recent memory (Olton and Feustle, 1981). Also, the basal forebrain cholinergic pathways selectively degenerate in Alzheimer's dementia, a disorder characterized by a profound deterioration of higher cognitive functions in humans (Whitehouse et al., 1982). Since anticholinergic drugs impair recent memory (Drachman, 1971) and, at high doses, cause more global disruptions of cognitive functions (Longo, 1966), interference with neurotransmission at the corticohippocampal cholinergic synapses is the likely mechanism of this effect of anticholinergics.

The muscarinic receptors, the predominant population of receptors mediating the central effects of acetylcholine, have been well characterized by means of ligand binding techniques (Birdsall and Hulme, 1976). These receptors exhibit relatively uniform characteristics with regard to antagonist interactions throughout major brain regions (Birdsall et al., 1980). Displacement studies with agonists reveal a subpopulation of sites uniformly labeled by antagonists that can be resolved into low, high, and "super" high-affinity compounds (Birdsall et al., 1978). The most commonly used antagonist ligand to label the musarinic receptors is ^{3}H-quinuclidinyl benzilate (QNB); this ligand exhibits a subnanomolar affinity for the receptors, with an extremely slow rate of dissociation that results in a very favorable ratio of specific to nonspecific binding. The compelling correlation between the affinity of muscarinic antagonists for the ^{3}H-QNB-labeled sites and their ability to antagonize the physiologic effects of cholinergic agonists has established that this ligand is an excellent probe for the antagonist conformation of central muscarinic receptors (Yamamura and Snyder, 1974). Furthermore, cross-species studies indicate a high degree of conservancy of muscarinic receptor characteristics as revealed by the specific binding of ^{3}H-QNB.

2.2. Characterization of the Radioreceptor Assay for Anticholinergics

Since several RRAs for psychotropic medications have been developed to detect these drugs on the basis of their interactions with specific receptor sites (Ferkany and Enna, 1982), we felt that the muscarinic receptor labeled with ^{3}H-QNB might serve as an effective probe for detecting drugs on the basis of their anticholinergic properties. Preliminary experiments indicated that the addition of 200 µL of drug-free human serum to the 2-mL buffer mix used for measuring the specific binding of ^{3}H-QNB to suspensions of membranes prepared from the rat forebrain caused a $29 \pm 2\%$ inhibition of total binding and a $30 \pm 5\%$ inhibition of nonspecific binding of ^{3}H-QNB (0.3 nM) to the receptor sites (Tune and Coyle, 1981). Thus, the ratio of total to nonspecific binding was approximately 8:1. Displacement curves generated by the addition of increasing concentrations of drugs known to have varying potency as anticholinergics revealed comparable K_i values in the absence or in the presence of 200 µL of drug-free human serum. Notably, the assay detected the classical muscarinic antagonists such as benztropine, as well as the anticholinergic properties of drugs with other primary effects, such as diphenhydramine, amitriptyline, and thioridazine. In order to have a norm against which individual displacement values for serum could be compared, inhibition curves with atropine in the presence of drug-free human serum were run for each assay. Thus, values for serum samples could be expressed in terms of the equivalent amount of atropine

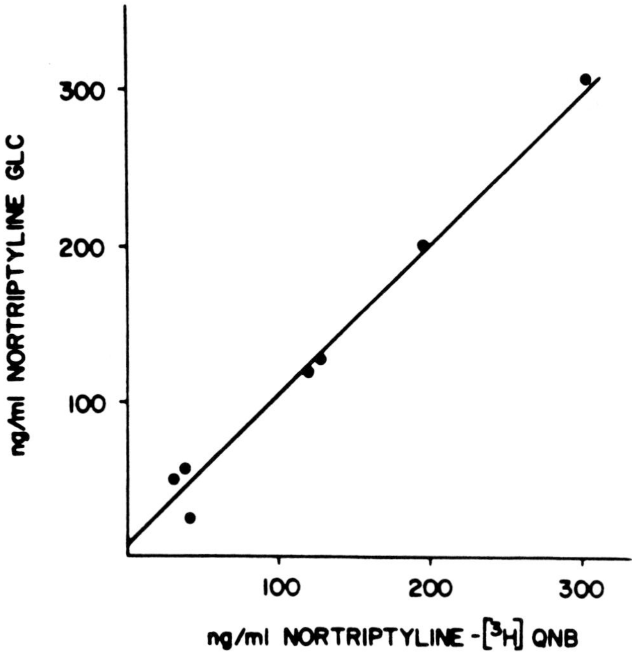

Fig. 2. Correlation between serum nortriptyline levels measured by gas chromatography (GLC) and by the radioreceptor assay. Nortriptyline was extracted from serum obtained from seven patients.

that produced comparable inhibition of the specific finding of ^{3}H-QNB.

To validate the assay, serum levels of nortriptyline in patients receiving this drug as an antidepressant were measured both by the muscarinic receptor assay and by GC (Tune and Coyle, 1981). Since the gas chromatographic method measures total serum levels of nortriptyline, the nortriptyline was extracted from serum in heptane-isoamyl alcohol (99:1) and then back-extracted in 0.1N HCl for its detection of the basis of its muscarinic blocking properties (Fig. 2). Although the serum levels of nortriptyline ranged over 10-fold in seven patients (25–305 μg/L), there was an excellent correlation between the values obtained with the radioreceptor assay and those determined by GC ($r = 0.99$; $p < 0.001$).

Since whole serum has been used in the assay, only anti-cholinergic drugs dissociated from serum proteins are free to compete with ^{3}H-QNB at the muscarinic receptors. Because the serum is diluted 10-fold with the assay buffer, the equilibrium between free and protein-bound drug is shifted, resulting in a constant increase in the unbound drug. To assess the degree of protein binding of benztropine, we examined the relationship between the values measured with whole serum added and the levels obtained in split samples, in which the serum proteins were first precipitated by treatment with perchloric acid. The values for whole serum, as a percent of the total amount of anticholinergic in the perchloric extract of the serum, varied from 1.4 to 10% in samples from 34 patients receiving benztropine. Nevertheless, the two values correlated significantly ($p < 0.05$), and an average of 95% of the benztropine was bound to serum proteins under the conditions of the assay. Since free drug represents that available to interact with receptors in vivo, we felt that the measurements of whole serum gave a better insight into the physiologically active component of the anticholinergic drugs.

2.3. Acute Extrapyramidal Symptoms and Serum Neuroleptic and Anticholinergic Levels

In a cross-sectional study, the relationship between serum levels of anticholinergics and the total daily dose of benztropine was examined in 49 patients who were receiving this drug in addition to neuroleptics. No correlation was observed between the serum levels of the anticholinergics and total daily doses of benztropine. On the contrary, serum anticholinergic activity varied on the average of 10-fold from the lowest to the highest level in patients receiving 2, 4, and 6 mg/d of benzotropine (Fig. 3). In six patients, who were followed serially with increasing oral doses of benztropine, a markedly nonlinear relationship between daily dose and serum anticholinergic levels was observed. In most cases, 2-mg increments in oral dose were associated with severalfold increases in the serum level of anticholinergic activity.

A cohort of 109 patients, who were receiving at least 400 mg/d of chlorpromazine equivalents (Davis, 1976), were evaluated for the presence and severity of EPS with the DiMascio scale, and serum levels of neuroleptics were determined by the RRA of Creese and Syder (1977). In this cross-sectional study, the patients' drug doses were determined by their treating physician; and the majority of these patients was receiving haloperidol as the neuroleptic. Notably, a poor and nonsignificant correlation between serum levels of neuroleptics and the severity of EPS was observed ($r = 0.029$; $n = 109$; $p < 0.1$). This finding is consistent

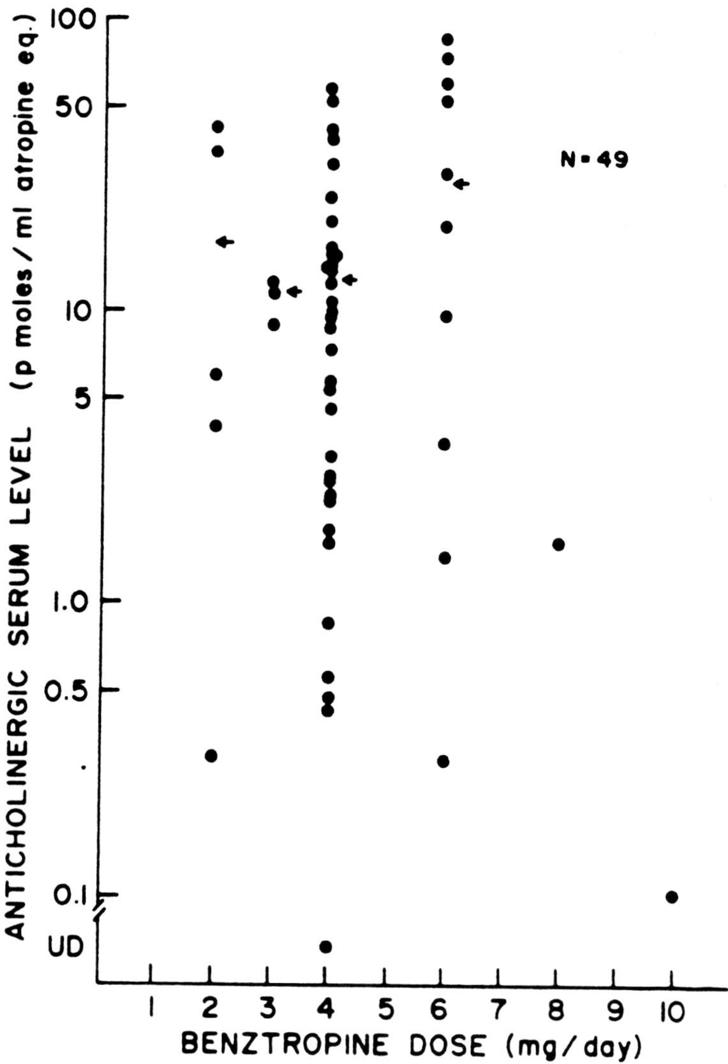

Fig. 3. Relationship between total daily dose of benztropine and serum levels of free anticholinergics. Serum samples were obtained from 49 in-patients who were receiving a fixed dose of benztropine for at least 4 d prior to phlebotomy. Results for the radioreceptor assay are expressed in terms of the amount of atropine (picomoles) that produced the same amount of inhibition. Arrows indicate the mean level for each daily dose.

with clinical observations that the development of EPS is a poor predictor of therapeutic doses of neuroleptics (Alpert et al., 1978), and that individuals vary in vulnerability to the EPS-inducing effects of neuroleptics (Chase et al., 1970).

For 76 of these patients, whose serum levels of anticholinergics had been measured in the same samples, there was a highly significant inverse relationship between the severity of EPS and the serum levels of anticholinergics (Fig. 4) detected by the RRA ($r = 0.441$; $p < 0.005$). Of patients with serum levels of less than 10 pmol/mL of atropine equivalents, 44 experienced EPS with a DiMascio score greater than or equal to 2, whereas only 9% of patients with serum levels in excess of 10 nmol/L of atropine equivalents experienced significant EPS ($x^z = 8.9$; $n = 76$; $p < 0.005$). The value, 10 nmol/L atropine equivalents, is approximately five times the K_i of atropine for the muscarinic receptor in the assay, and thus reflects a serum level associated with substantial blockade of muscarinic receptors. To rule out an artifactual skew in the serum neuroleptic levels that might account for the low incidence of EPS in patients with higher serum anticholinergic levels, the relationship between serum neuroleptic and serum anticholinergic levels was plotted. However, no significant correlation was observed in the whole population ($r = 0.16$) or in those patients experiencing EPS ($r = 0.21$) or free of EPS ($r = 0.25$).

Since the mechanisms of action of anticholinergics in reversing EPS involve correcting the neuroleptic-induced cholinergic dysfunction in the striatum (Calne et al., 1975), we were curious as to whether the ratio of anticholinergic activity to dopamine receptor-blocking activity might play a critical role in control of EPS. In other words, we wondered whether EPS resulting from low serum neuroleptic levels might be corrected by low anticholinergic levels, whereas high serum neuroleptic levels would require high anticholinergic levels. However, plotting the relationship between EPS and the ratio of serum anticholinergics to serum neuroleptics, both measured by the RRA techniques, did not reveal a significant inverse correlation ($r = 0.257$, $p < 0.05$; $n = 76$). Nevertheless, patients with serum anticholinergic/neuroleptic ratios of 0.5 or greater had a significantly lower incidence of EPS as compared to those with a lower ratios ($x_2 = 7.5$; $p < 0.02$). Thus, an "ideal ratio" between anticholinergics and neuroleptics does not appear to be involved in EPS control, although an excess of anticholinergic is associated with a more favorable response.

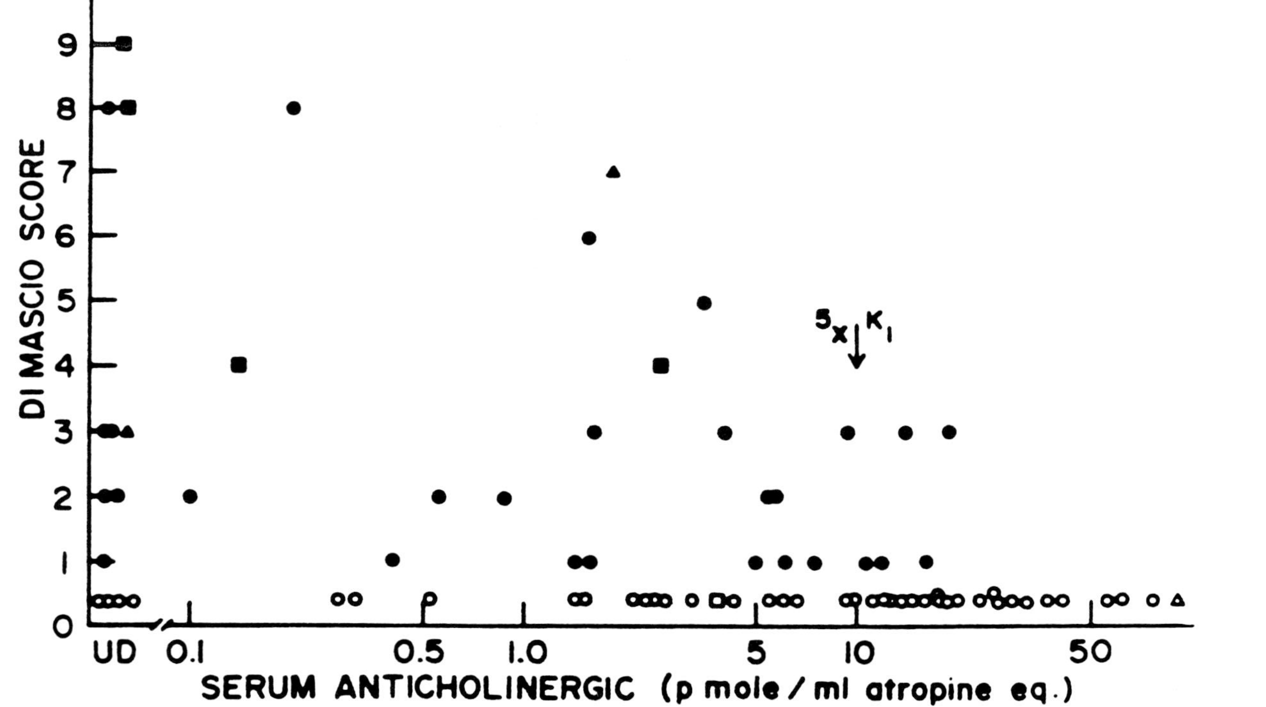

Fig. 4. Correlation between acute extrapyramidal side effects and serum anticholinergic levels. Seventy-six patients, who were receiving both neuroleptics (at least 400 mg/d of chlorpromazine or its equivalent) and anticholinergic medication for at least 4 d of constant dosage, were rated for EPS with the DiMascio scale, and a blood sample was obtained. Results for the RRA are present in terms of the amounts of atropine (picomoles) that produce the same degree of inhibition of specific binding of ^{3}H-QNB. Open symbols are EPS-positive patients. UD, undetectable serum levels. ●, benztropine; ■, trihexyphenidyl; ▲, diphenhydramine. 5 K indicates the concentration of atropine equivalent to five times the K_1 for atropine. $r = -0.83$; $n = 24$; $p < 0.005$.

2.4. Cognitive Effects of Anticholinergic Drugs

As discussed above, the anticholinergic drugs not only interfere with cholinergic neurotransmission in the striatum, but also with the function of the cholinergic projections to the cerebral cortex and hippocampal formation, which are thought to play an important role in cognitive functions. Accordingly, we wondered whether the serum levels of anticholinergics associated with reduction in EPS might also result in impairments in recent memory. A cohort of 24 patients, all of whom were receiving neuroleptics and satisfied Research Diagnostic Criteria (Spitzer et al., 1975) for schizophrenia, were examined with regard to the severity of schizophrenic symptoms as measured by the Mini-Present State Exam of Wing et al. (1974), verbal intelligence quotient (verbal IQ) as measured by the Wechsler Adult Inventory Scale (WAIS), and performance on a recent memory task (Tune et al., 1982). Serum levels of neuroleptics and anticholinergics were measured by the RRA techniques. All of the patients were in remission; none of the patients were disoriented as to person, place, or time. No significant correlation was observed between the serum levels of anticholinergics or neuroleptics with regard to the severity of schizophrenic symptoms. However, a significant inverse correlation between serum levels of anticholinergics and performance on the recent memory test was demonstrated ($r = 0.51$; $p < 0.01$; $n = 24$) (Table 1). Notably, the serum levels of anticholinergics measured in these patients were within the range associated with control of EPS; thus, it appears that serum levels of anticholinergics effective in reducing EPS may also cause subtle but significant impairments in higher cognitive functions in the absence of frank delirium.

These findings have received additional support from a prospective study of postoperative delirium in patients undergoing cardiac surgery (Tune et al., 1981). In this study, patients were examined for cognitive function by the Mini-Mental Status Exam (Folstein et al., 1975), and were rated for delirium by clinical assessment. After surgery, blood was obtained at the time of clinical evaluation for measurement of serum anticholinergic levels. When all the samples taken throughout the course of the study were analyzed, 14 out of 16 samples obtained from patients who were rated as clinically delirious had serum anticholinergic levels of greater than 7.5 nmol/L of atropine equivalents, whereas only 5 of 33 samples obtained from cognitively intact patients had serum

Table 1
Effects of Anticholinergics on Recent Memory Function in Schizophrenic Patients[a]

Parameter	Mean ± SEM	Correlation coefficients		
		Atropine equivalents	Chlorpromazine equivalents	Recent memory
Mini PSE	4.7 ± 0.8	0.05	0.07	−0.29
Atropine equivalents (nmol/L)	12.0 ± 2.5	—	0.12	−0.51[b]
Chlorpromazine equiv-alents (μg/L)	32.0 ± 10	—	—	0.20
Recent memory	7.2 ± 0.2	—	—	—

[a]The correlation coefficients for the various parameters measured in 24 patients are presented. Recent memory refers to the average number of correct responses for 15 trials (Tune et al., 1982).
[b] $p < 0.01$.

levels of greater than this ($x^2 = 23.8; p < 0.001$). Comparison of the change in Mini-Mental State Examination from before surgery with serum anticholinergic levels revealed a highly significant inverse correlation (Fig. 5). Thus, as serum levels of anticholinergic drugs increased, there was a progressive impairment in higher cognitive functions as assessed by the difference in scores before and after surgery. Although multiple factors contribute to postoperative delirium in patients undergoing cardiac surgery, these results suggest that drugs with anticholinergic effects may play a significant contributory role.

2.5. Discussion of Anticholinergic RRA

In the example provided, the assessment of the role of serum concentrations of anticholinergics in treating EPS is complex. For instance, some neuroleptic drugs have significant anticholinergic effects of their own and will tend to reduce the incidence of EPS associated with their use (Tune and Coyle, 1980, 1981). A number of anticholinergic drugs with widely varying chemical structures and properties are in clinical use, making comparisons among drugs difficult. Last, metabolites of both neuroleptic and anticholinergic drugs may have different degrees of anticholinergic activity and contribute to the reduction of EPS. The RRA method described here takes these issues into account to a large extent, since it measures interaction at the muscarinic receptor, the physiologic site of action of these drugs, rather than measuring specific drug compounds.

Using RRA methodology, a highly significant inverse relationship is found between the presence of EPS and serum anticholinergics. Of particular interest is the fact that considerable variation in serum anticholinergic drug concentrations is observed for a given dose; a plot of the relationship between total daily dose and serum anticholinergic concentrations showed a lack of correlation. It is also noteworthy that the average serum concentration at four different daily doses, serum concentrations that varied up to 100-fold, fell below the serum anticholinergic concentration associated with a low incidence of EPS. Finally, when we looked at the effect of the serum neuroleptic concentration on EPS, it was clear that there was no significant relationship between serum neuroleptic concentration (measured with RRA) and the presence and severity of EPS.

These findings have theoretical implications. Despite the wealth of evidence supporting a sequential rather than a parallel

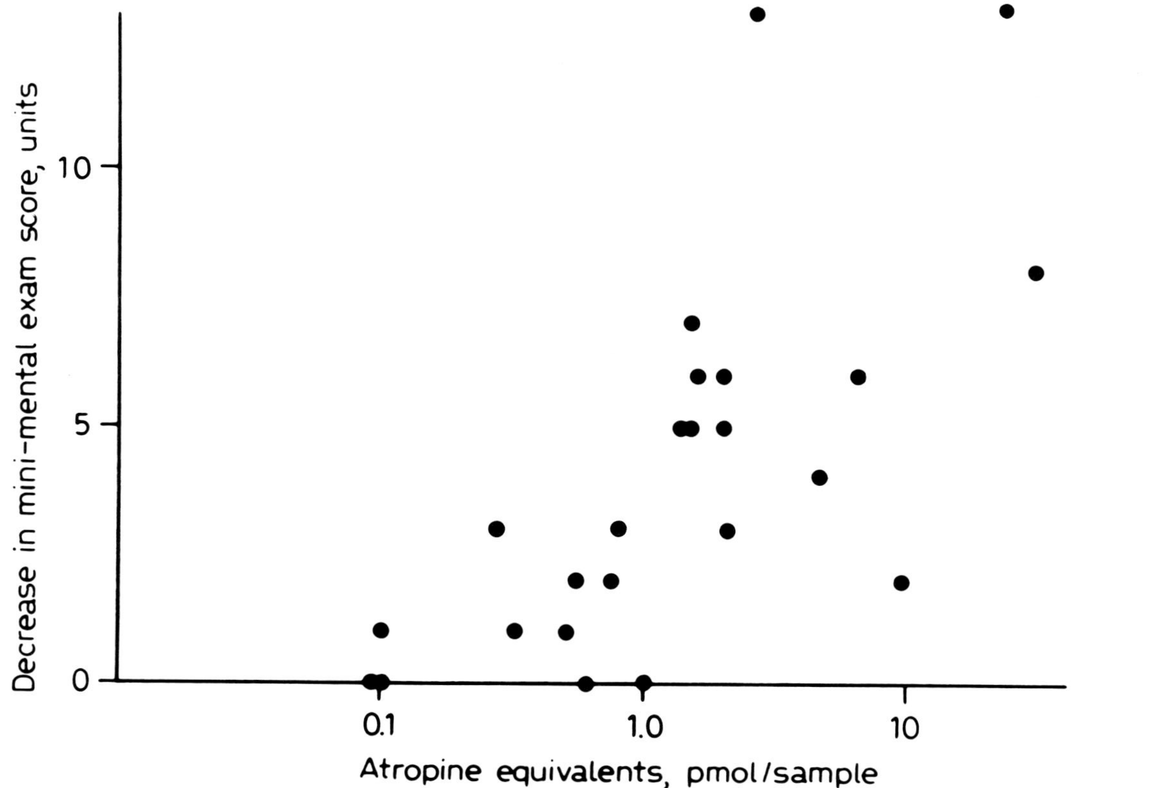

Fig. 5. Results of postcardiotomy study. Change in clinical state in relation to serum anticholinergic levels in 18 patients who underwent 24 postcardiotomy assessments. Clinical state is expressed as the change in Mini-Mental State Exam score between precardiotomy and postcardiotomy ratings. The correlation coefficient was $r = -0.83$; $n = 24$; $p < 0.001$.

synaptic relationship between these two neuronal systems in the striatum (Racagni et al., 1976), the generally accepted view of the mechanism of action of anticholinergics in treating EPS is that of restoring the "balance" between cholinergic and dopaminergic function in the striatum. It was possible for us to investigate the ratio of serum concentrations of drugs affecting these two systems—neuroleptics and anticholinergics—to evaluate this relationship in the treatment of EPS. Our clinical data support a sequential relationship. The clearest relationship between treatment of EPS and drug is the relationship between serum anticholinergic concentration and clinical response.

The role of anticholinergics in the generation of confusional states is clearer. From earlier literature it is known that anticholinergics are involved in the production of defects in memory and attention in normal volunteers (Drachman, 1971). Our data extend these observations to a variety of patient groups. The utility of RRA techniques in this assessment is underscored when one investigates the patients with postcardiotomy delirium. In no instance did patients receive a "toxic" dose of single anticholinergic compound. However, when the collective effects of compounds known occasionally to produce delirium are taken into account, it seems reasonable that the "net effect" of large numbers of anticholinergic compounds is appreciated and is significantly related to confusional states.

A subtle but nonetheless persistent effect has been demonstrated with anticholinergics in chronic schizophrenic patients. Defects of recent memory have been correlated with rising serum anticholinergic drug concentration. This is significant because few guidelines exist with which to manage patients on standard anticholinergic drugs. Often the assumption is that when patients complain of difficulties with their memory and attention, this is merely the cognitive decline that is clearly demonstrated to be associated with schizophrenia. Our findings would challenge this. Perhaps as a result of studies of this type, new guidelines can be generated so that maximum therapeutic response with minimized drug toxicity will result in an approved clinical state for the patient.

3. Other Possible Clinical Applications of Radioreceptor Assays: Antidepressants and Benzodiazepines

Although this chapter has focused on the neuroleptic and muscarinic acetylcholine radioreceptor assays, this methodology

Table 2
In Vivo Measurement of Nortriptyline in Patient Plasma (adapted from
Innes, et al., 1979)

Patient	Radioreceptor assay	GLC, ng/mL
1	300	305
2	195	200
3	132	135
4	120	120
5	42	25
6	33	57
7	30	50

has been used to measure benzodiazepine and antidepressant medications. Enna, 1982 has shown the ability of antidepressants to inhibit ligand binding at three separate sites: muscarinic acetylcholinergic, alpha-1 adrenergic, and serotonin-2 receptors. That antidepressants are able to potently inhibit multiple receptor binding ligands is of some interest, since it is possible to characterize the activity of these drugs in several neurotransmitter systems. To keep the issue clear, however, a simple application would be to develop one of these radioreceptor methodologies as a means of measuring tricyclic antidepressants. Indeed, Innis et al. (1979) have shown an excellent relationship between the anticholinergic activity of tricyclic antidepressants and existing methods for the determination of tricyclic antidepressant drug levels (Table 2).

Lastly, in those instances in which endogenous compounds are known to exist (e.g., opiate-like substances to include enkephalins) or in which receptor sites have been discovered that are not clearly identified with known neurotransmitters, use of radioreceptor assay methodology may be useful for detecting interactions of other compounds with the receptor sites or for the detection of endogenous substances. A review of the study of opioid peptides clearly demonstrates the potential fullness of this type of research.

References

Alpert M., Diamond F., Weisenfreund J., Talepros E., and Friedhoff A. J. (1978) The neuroleptic hypothesis: A study of the co-variation of extrapyramidal and drug effects. *Br. J. Psychiat.* **133,** 169–175.

Ayd F. J. (1961) A survey of drug-induced extrapyramidal reactions. *J. Am. Med. Assoc.* **175,** 1054–1060.

Birdsall N. J. M. and Hulme E. C. (1976) Biochemical studies on muscarinic acetylcholine receptors. *J. Neurochem.* **27,** 7–16.

Birdsall N. J. M., Burgen A. S. V., and Hulme E. C. (1978) The binding of agonists to brain muscarinic receptors. *Mol. Pharmacol.* **14,** 723–736.

Birdsall N. J. M., Hulme E. C., and Burgen D. S. V. (1980) The character of the muscarinic receptors in different regions of the rat brain. *Proc. Roy. Soc. Lond.* (Biol.) **207,** 1–12.

Calne D., Chase T. W., and Barbeau A. (1975) *Dopaminergic Mechanisms* Raven, New York.

Chase T. N., Schnur J. A., and Gordon E. K. (1970) Cerebrospinal fluid monoamine catabolites in drug-induced extrapyramidal disorders. *Neuropharmacology* **9,** 265–268.

Creese I. and Snyder S. H. (1977) A simple and sensitive radio-receptor assay for antischizophrenic drugs in blood. *Nature* **270,** 261–263.

Creese I., Burt D. R., and Snyder S. H. (1976) Dopamine receptor binding predicts clinical and pharmacological potencies of antiparkinsonian drugs. *Science* **192,** 481–483.

Davis J. M. (1976) Comparative doses and costs of antipsychotic medications. *Arch. Gen. Psychiat.* **33,** 858–863.

DiMascio A. (1971) Towards a more rational use of antiparkinsonian drugs in psychiatry. *Drug Ther.* **1,** 23–37.

Donlon P. T. and Stenson R. L. (1976) Neuroleptic-induced extrapyramidal symptoms. *Dis. Nerv. Syst.* **37,** 629–635.

Drachman D. A. (1971) Memory and Cholinergic Systems, in *Neurotransmitter Function* (W. S. Fields ed.), Yearbook, New York.

Eadie M. J. and Tyrer J. H. (1974) *Anticonvulsant Therapy: Pharmacological Basis and Practice,* Churchill Livingstone, London.

Enna S. J. (1981) Radioreceptor Assays, in *Physico-chemical Methodologies in Psychiatric Research* (I. Hanin and S. H. Koslow, eds.) Raven, New York.

Enna S. J. (1982) Radioreceptor Assays for Neurotransmitters and Drugs, in *Handbook of Psychopharmacology* vol. 15 (Iversen L. L., Iversen S. D., Snyder S. H., eds.) Plenum, New York.

Ferkany J. W. and Enna S. J. (1982) Radioreceptor Assays, in *Analysis of Biogenic Amines* (G. B. Baker and R. T. Coutts, eds.) Elsevier/North-Holland, Amsterdam.

Folstein M., Folstein S., and McHugh P. R. (1975) Mini-Mental State. A practical method for grading the cognitive state of patients for the clinician. *J. Psychiatr. Res.* **12,** 289–298.

Hollenberg M. D. and Cuatrecasas P. (1975) Biochemical Identification of Membrane Receptors Principles and Techniques, in *Handbook of Psychopharmacology* vol. 1 (Iversen L. L., Iversen S. D. Snyder S. H., eds.) Plenum, New York.

Innes, R. B., Tune, L., Rock, R., DePaulo, J. R., U'Prichard, D., and Snyder, S. H. (1979). Tricyclic antidepressant Radioreceptor Assay. *Eur J. Pharmacol.* 58:473–77.

Johnston M. V., McKinney M., and Coyle J. T. (1981) Neocortical cholinergic innervation: A description of extrinsic and intrinsic components in the rat. *Exp. Brain Res.* **43,** 159–172.

Kebabian J. W. and Calne D. B. (1979) Multiple receptors for dopamine. *Nature* **277,** 93–96.

Ko G., Korpi E. R., and Linnoila M. (1985) On clinical relevance and methods of quantification of plasma concentrations of neuroleptics. *J. Clin. Psychopharm.* **5,** 253–262.

Long V. G. (1966) Behavioral and electroencephalographic effects of atropine and related compounds. *Pharmacol. Rev.* **18,** 965–989.

Mailman R. B., Peise J. P., Crofton K. M., Petitto J., DeHaven D. L., Kilts C. D., and Lewis M. H. (1984) Thioridazine and the neuroleptic radioreceptor assay. *Biol. Psychiatr.* **19,** 833–47.

Olton D. S. and Feustle W. A. (1981) Hippocampal function required for nonspatial working memory. *Exp. Brain Res.* **41,** 380–389.

Racagni G., Cheney D. L., Trabuchi M., and Costa E. (1976) In vivo actions of clozapine and haloperidol on the turnover rate of acetylcholine in rat striatum. *J. Pharmacol. Exp. Ther.* **96,** 323–332.

Richelson E. and Divinitz-Romero S. (1977) Blockade by psychotropic drugs of the muscarinic acetylcholine receptor in cultured nerve cells. *Biol. Psychiatr.* **12,** 771–782.

Rish S. C., Hicey L. Y., and Janowsky D. S. (1979) Plasma levels of tricyclic antidepressants and clinical efficacy. Review of the literature. *J. Clin. Psychiat.* **40,** 4–116.

Simantov R., Childers S., and Snyder S. (1977) Opioid peptides: Differentiation by radioimmunoassay and radioreceptor assay. *Brain Res.* **135,** 358–367.

Spitzer R. L., Endicott J., and Robins E. (1975) *Research Diagnostic Criteria* 2nd Ed., New York State Psychiatric Institute, New York, Biometrics Research.

Tune L. E. and Coyle J. T. (1980) Serum levels of anticholinergic drugs in treatment of acute extrapyramidal side effects. *Arch. Gen. Psychiat.* **37,** 293–297.

Tune L. E. and Coyle J. T. (1981) Acute extrapyramidal side effects: Serum levels of neuroleptics and anticholinergics. *Psychopharmacology* **75,** 9–15.

Tune L. E., Creese I., Coyle J. T., Pearlson G., and Snyder S. H. (1979) Low neuroleptic serum levels in patients receiving fluphenazine decanoate. *Am. J. Psychiatry* **137,** 80–82.

Tune L. E., Holland A., Folstein M. F., Damlouji N. F., Gardner T. J., and Coyle J. T. (1981) Association of post-operative delirium with raised serum levels of anticholinergic drugs. *Lancet* **ii,** 651–653.

Tune L. E., Strauss M. E., Lew M. F., Breitlinger E., and Coyle J. T. (1982) Serum levels of anticholinergic drugs and impaired recent memory in chronic schizophrenic patients. *Am. J. Psychiat.* **139,** 1460–1462.

Van Putten T. (1974) Why do schizophrenic patients refuse to take their drug? *Arch. Gen. Psychiat.* **31,** 67–72.

Whitehouse P. J., Price D. L., Struble R. G., Clark A. W., Coyle J. T., and DeLong M. (1982) Alzheimer's disease and senile dementia: Loss of neurons in the basal forebrain. *Science* **215,** 1237–1239.

Wing J. K., Cooper J. E., and Sartorius N. (1974) *Measurement and Classification of Psychiatric Symptoms,* Cambridge University Press, Cambridge.

Yamamura H. I. and Snyder S. H. (1974) Muscarinic cholinergic receptor binding in the longitudinal muscle of guinea pig ileum with [^{3}-H]-quinuclidinyl benzilate. *Mol. Pharmacol.* **10,** 861–867.

Yamamura H. I., Enna S. J., and Kuhar M. J., eds. (1978) *Neurotransmitter Receptor Binding* Raven, New York.

Application of Immunoassay Techniques in Psychopharmacology

Dennis E. Schmidt and Michael H. Ebert

1. Introduction

The advent of polyclonal immunoglobin-based technology, originated by Yalow and Berson in 1959, for qualitative or quantitative determination of specific substances in biological tissue has revolutionized the analytical aspects of biomedical research. The refinement of this technology by the application of monoclonal antibody (Mab) techniques that were pioneered by Kohler and Milstein (1975) increased the already numerous applications of the methodology. Consequently, the published information relevant to the relationship between immunology and neurobiology is of immense proportions. A comprehensive review of even a specialized area is difficult, and new methods are evolving so rapidly that any review is doomed to early obsolescence.

Immunoglobins are glycoproteins produced by specific plasma cells in all higher animals in response to the presence of a foreign substance, or antigen (At). The antigen may have one, several, or many determinants (epitopes), each of which will stimulate one or more β-lymphocytes to produce specific immunoglobins that will "recognize" that particular combination of epitopes that initiated the response. Each stimulated lymphocyte, in turn, has the capacity to differentiate into a family of daughter plasma cells, each of which will secrete only the specific immunoglobin that will react only with the exact antigenic determinant(s) that stimulated the original lymphocyte. Administration of a multideterminant At (i.e., containing one or more epitopes per molecule of At) will therefore evoke the release of a heterogeneous or polyclonal group of antibodies (Abs) into the plasma of the challenged animal. Subsequently, Kohler and Milstein (1975) were able to develop a method to fuse actively secreting "challenged" lymphocytes with a suitable myeloma cell line to produce clones that exhibited both the parent lymphocytes' property of specific Ab production and the immortal character of myeloma cell lines.

Selective isolation and cultivation of these clones followed by assessment of the Abs they secrete can be used to produce unlimited amounts of a single specific Mab that will recognize a single specific set of At determinants. The potential application of Mabs, which can be designed specifically to "recognize" the endogenous or exogenous substance(s) of choice, appears to be limited only by the imagination of the investigator and the pace of technological innovations that put these ideas into action. The use of Abs in neurobiology can be divided into three broad, though not mutually exclusive, areas; analytical, diagnostic, and therapeutic. A large number of reviews (not an inclusive list) have described recent advances in techniques for the production and isolation of both murine and human polyclonal antibodies (Pabs)* and Mabs (Nakamura, 1983; Schonherr and Houwink, 1984; Abrams et al., 1983; Goding, 1980; Roome and Reading, 1984; Marullo et al., 1985; Kozbar, and Roder, 1983). These techniques have been applied in clinical medicine in such diverse fields as the diagnosis and treatment of cancer; in organ transplantation; in the use of pharmacological agents directed toward specific surface antigens on malignant cells or against specific pathogenic organisms; in nuclear medicine as tissue- or organ-specific imaging agents or as carriers of radionuclides or cytotoxic drugs that are able to be directed to specific in vivo sites; and in other diverse areas of genetics and cell biology (Zalcberg, 1985; Krakauer, 1985; Haauman et al., 1985; Keenan et al., 1985; Dick, 1985; Carrasquillo et al., 1984; Larson et al., 1984; Mach et al., 1981; Engelberg and Eisenstein, 1984; Vitetta et al., 1983; Valentino et al., 1985). The following discussion will focus mainly on a portion of the third application, i.e., the use of antibody-based techniques to develop quantitative immunoassays (IAs) for small molecular weight (mw) endogenous and exogenous compounds. The rationale for designing new methods suitable to specific applications in the field of neurobiology, and especially neuropharmacology, will be discussed.

2. General Immunoassay Principles

The importance of sensitive and specific IA procedures as analytical methods that can be applied to the quantification of

---

*Polyclonal antibodies formed in response to an antigen are most frequently used by employing aliquots of the diluted blood plasma, or antisera. Therefore, although polyclonal antibodies are not usually isolated *per se,* the terms polyclonal antibodies and polyclonal antisera will be used interchangeably in this review.

psychotropic drugs, neurotransmitters, and their metabolites, is increasing rapidly. By exploiting the high affinity and remarkable specificity of Abs, it is usually possible to develop an IA method that will be able to utilize readily available laboratory equipment and procedures and measure very small quantities of a target analyte (10^{-9} to 10^{-11} mol or less) contained in commonly encountered tissue or fluid samples. Moreover, because of the unique specificity possessed by a carefully designed Ab, most biological samples can be analyzed directly, without prior extraction or other processing. As a result of the development of nonradioactive markers that can assess the degree of the At–Ab complex formation and the progress that has been made in developing stable "dry" chemistry procedures, many of the IAs are moving rapidly from the laboratory to the clinic. IAs have traditionally been developed empirically and have been based on the assumptions that both the Ab and the At are univalent (i.e., contain only a single binding site) and obey the laws of mass action. Willey (1985), however, has recently presented convincing evidence that the kinetic mechanisms that determine the interaction of Abs and Ats are divided into two distinct classes. In one case, all of the Ab–At complexes are formed by monovalent noninteractive combinations of one At at a single Ab site. Although the Abs that are developed for use in haptenic IAs are frequently capable of polyvalent interactions, IA assays for low mw compounds are in this class because the free At molecules, by virtue of their small physical size, can combine with only one site on any Ab at any one time. All monovalent, noninteractive Ab–At reactions obey the laws of mass action, and standard kinetic analysis can be applied successfully. There is, however, a second group of IAs in which interactive combinations between multiple, multivalent Ats and/or Abs predominate; i.e., there is more than one Ab and/or one At in each molecule of Ab–At complex that forms. Moreover, many of the intermediate complexes that form are largely resistant to dissociative rearrangement, and prolonged equilibration times are required (as long as 30 h) before the final thermodynamically favored multivalent Ab–At complex is formed. Therefore multivalent Ab–At complex formation violates required assumptions and invalidates the use of standard kinetic analysis for these assays.

The combination of a monovalent Ab with a monovalent At is a simple bimolecular combination. The primary determinant of whether interactions between polyclonal Abs and Ats will be simple or complex is the molecular size of the At (Willey, 1985). The drugs or other target Ats discussed in this review have low molecular weights, and their molecular size is therefore very small relative to that of the Ab. These small Ats are virtually engulfed and surrounded by the specific Ab binding site (Chard, 1982); therefore, even though the Ab (and frequently the At) contains polyvalent binding sites, the small At is not capable of binding simultaneously more than a single Ab molecule. This is true for both Mabs and for the multiple-determinant Abs found in polyclonal antisera. When employing polyclonal antisera, individual Ats may bind to different sites on a given Ab and/or to an entirely different Ab type; but each individual Ab–At reaction constitutes a noninteractive combination of one At with one Ab site. Atassi (1978) and Willey (1985) independently determined that the minimum molecular size required in order for an At to be able to simultaneously bind more than one Ab is approximately 1 kdalton, well above the size of small polypeptides (<10 amino acids), neurotransmitters, and virtually all nonpolymeric drug molecules and their metabolites.

Classical Scatchard analysis is a simple and straightforward kinetic method that will determine whether an At–Ab reaction is simple or complex. Noninteractive monovalent Ab–At reactions yield linear Scatchard plots; and kinetic analysis of the data will determine the dissociation constant, or affinity, of either a polyclonal antisera or a Mab for a specific target At. In the case of polyclonal antisera, the dissociation constant resulting from Scatchard analysis will be determined primarily by the Ab binding site(s) that has the highest affinity for the antigen (Willey, 1985). The affinity of the Ab binding will determine the absolute sensitivity limit of the resulting assay, the range of At concentrations that can be measured, and the appropriate concentration of Abs to be used in the assay.

Conversely, Scatchard analysis of complex Ab–At reactions will yield curvilinear plots. Methods for the kinetic analysis of nonlinear Scatchard plots are available, but in order to be valid they require that the binding that is being analyzed (in this case multivalent Ab–At binding) is occurring in an independent, noninteractive fashion. In this context, this assumption requires that the binding of the At at any site, on any Ab, does not exert any negative or positive cooperative effects on the subsequent binding of that At–Ab complex to another unbound Ab or At or to another identical or dissimilar Ab–At complex. However, Willey (1985) has clearly demonstrated that multivalent Ab–At binding results in a variety of definable Ab–At complexes that can exhibit significant positive and negative cooperativity. Although standard nonlinear Scatchard analysis of cooperative multivalent binding data will still yield affinity constants for the Ab being used, the failure of the reactions to fulfill the assumptions required to validate such analysis make the resulting dissociation constants meaningless. These values will not represent the affinity of the Ab for the At, and they are not a valid or useful parameter for establishing appropriate Ab concentrations, for determining the range of the assay, or for estimating equilibrium times. In practice, polyvalent interactive Ab–At reactions are characterized by very long equilibrium times. Moreover, the system undergoes gradual changes in the average binding cooperativity as the percentage of various forms of the At–Ab complex (polymers, oligomers, circular dimers) change over time. This, in turn, results in constant changes in the calculated affinity value as the actual affinity and rate of dissociation of various different complexes change over time, until the final equilibrium is obtained.

Scatchard analysis is therefore an essential step when establishing a new IA assay, in order to determine the affinity constant of an Ab for the target At and to make certain that an assay will be based on monovalent Ab–At reactions (i.e., a linear Scatchard plot). Rather than performing a large number of empirical experiments to establish experimental conditions, the resulting affinity constant can be used to directly estimate the sensitivity of the assay and to establish the concentrations of Ab and At that are appropriate. Kinetic analysis of the affinity constant is also required for each new batch of polyclonal antisera or Mab to reestablish that the affinity of the Ab remains appropriate for the assay conditions. When curvilinear Scatchard plots are obtained from an Ab preparation, it is suggested that the early work on multivalent binding by Crothers and Metzger (1972) and the more recent discussions of complex IAs (Willey, 1985) be reviewed before proceeding with assay development.

As previously stated, IAs for all low mw target Ats ($< 1k$-dalton) will be based on monovalent Ab–At reactions and therefore follow mass action kinetics. At equilibrium, a portion of the free Ab will be combining monovalently with free At, with the rate constant k_1, to form the Ab–At complex.

$$\text{At} + \text{Ab} \underset{k_2}{\overset{k_1}{\rightleftharpoons}} \text{At–Ab} \tag{1}$$

Simultaneously, some of the Ab–At complex will be dissociating, with the rate constant k_2, to reform free At and Ab. Although all IAs are based on this reaction, they differ greatly in their intended purpose and in the methods that have been devised to measure the extent of At–At reaction. For example, in addition to using a specific Ab to quantitate a target At, it is also possible to develop an Ab that can be used to detect and/or quantitate endogenous Ab that has been formed in vivo in response to a naturally occurring At. Similarly, it is possible to generate Abs that will have a high affinity for the surface structure of specific cells. An Ab that is specific for a surface antigen of a malignant cell line could be used for tumor imaging or to carry a toxic drug specifically to that tissue. The nonanalytical use of Abs is of increasing importance in diverse areas of diagnosis and therapy. We will review some of the procedures used to develop Abs that will recognize a specific low mw target At and describe how these Abs can be used to quantitate substances such as endogenous neurotransmitters, psychotropic drugs, and/or their metabolites.

2.1. Production of a Suitable Antibody

As discussed above, most psychopharmacological drugs, neurotransmitters, and their metabolites and some of the biologically active peptides are capable of reacting with only a single Ab site. Such compounds are also below molecular weight 600, however, which is approximately the minimum molecular size required to be recognized by a normally active immune system. Therefore, drugs and other similar-sized molecules are nonantigenic or at best elicit only a very weak antigenic response when administered in unmodified form. In order to develop Abs to these compounds, they must first be linked to a much larger macromolecule (bovine serum, albumin, thyroglobulin, hemocyanin etc.) in order to elicit an immunogenic response in a suitable experimental animal. The combination of the high mw substance and the target drug or other target molecule is called a hapten. There is a large body of literature concerning techniques for the synthesis of appropriate haptens and the associated methods for inducing and isolating the resulting Abs. For examples of recent reviews in hapten development, see Zalcberg, 1985; Schonherr and Howink, 1984; James et al., 1984; Nakamura, 1983; Oellerich, 1980, 1984; Hawes et al., 1983; Marullo et al., 1985; Sherman-Gold et al., 1983; DeBlas et al., 1985; Flurkey et al., 1985; Kosfeld et al., 1985; Knoll and Wisser, 1984; Guesdon et al., 1986; Tiefenauer and Andres, 1984.

Some general principles, however, should be kept in mind when developing and/or using IAs. A primary concern in developing a hapten-generated Ab that will be suitable for analytical use is that the resulting Ab must also be able to recognize the At when it is not coupled to the haptenic macromolecule. The ability of the Ab to recognize the unconjugated At molecule and its ability to distinguish between the free At, its metabolites, and other closely related compounds will be the major determinant of the specificity of the assay. Moreover, the affinity of the Ab for the free At will also be a major determinant of the minimum sensitivity that will be possible. It is more efficient to expend effort in developing a haptenic conjugate that will generate a suitably specific Ab than to attempt to separate or screen multiple antisera for Abs in hope of finding one with acceptable specificity. When choosing a hapten, it is also important to consider the circumstances in which the IA is intended to be used. For example, in pharmacokinetic studies it is essential that the Mab or polyclonal antisera be able to distinguish between the parent drug and closely related metabolites. In this case, production of Abs with absolute specificity for the parent drug must be the primary criterion in evaluating the suitability of the hapten. Conversely, in forensic work the ability to detect the metabolites of the target drug at extended postadministration times, when the level of the unmetabolized drug might be very low, would be desirable. In this case an extensive effort to develop an Ab that is specific for only the parent drug would be counterproductive because it would ignore the metabolites that would indicate recent drug use.

In synthesizing haptenic protein conjugates, the functional group in the target At molecule that is involved in the covalent linkage to the hapten macromolecule will not be recognized in its unconjugated form by the subsequent Abs. This effect is especially important when the target molecule contains only one or two functional groups or if it does not have at least some type of unique tertiary spatial configuration. For these reasons, it has been difficult to develop satisfactory Abs directed against some of the small mw endogenous neurotransmitter compounds [such as serotonin, dopamine, norepinephrine (Knoll and Wisser, 1984; Flurkey et al., 1985; Manz et al., 1985)], whereas others [for example histamine (Guesdon et al., 1986)] are less troublesome.

There are many instances in which the presence of the hapten-At linkage has an adverse effect on the specificity of the resulting Ab. Most do not appear in the literature, however, because they represent problems that were solved in the course of successful development of an Ab with appropriate specificity. A single example will serve to illustrate the concepts involved. The primary metabolite of morphine in humans is the glucuronide formed on the 3-hydroxy group. A morphine hapten can be made by oxidation of the phenolic 3-hydroxy to form 3-*O*-carboxymethyl morphine, which is then coupled to a protein amino group by the carbodiimide reaction (Spector and Parker, 1970; Spector, 1971). This hapten elicits Abs that cannot differentiate between morphine, morphine glucuronides, or heroin (3-acetyl-morphine). These Abs, therefore, cannot be used to measure morphine levels without prior separation steps (Sandouk et al., 1984; Grabinski et al., 1983; Edwards et al., 1986). IAs that use these Abs are very useful, however, for detecting illicit narcotic use because the glucuroride metabolites have relatively long half-lives and can detect narcotic use in humans at long post-use times. Alternatively, azo-morphine can be synthesized (which leaves the ring hydroxy group unchanged) and coupled to keyhole limpet hemocyanin as the hapten (Gross et al., 1984) or can be conjugated to the haptenic protein via the 6-hydroxy group in the ring (Moore et al., 1984). The resulting Ab to these haptens is able to readily differentiate heroin and morphine from each other and from their metabolites.

Despite the fact that they contain sufficient functional groups to permit hapten synthesis and still maintain specific molecular features, attempts to develop Pabs and/or Mabs to the common endogenous monoamine transmitters, especially serotonin (5-hydroxytryptomine, 5-HT), dopamine (DA), and norepinephrine (NE), have been unexpectedly difficult. A primary reason for this difficulty appears to be the fact that mammalian systems contain widespread enzyme systems that are capable of rapidly inactivating free monoamine neurotransmitters. If the haptenic protein conjugate is designed so that essential structural elements of the free transmitter are preserved, the haptenic transmitter conjugate is also recognizable by the neurotransmitter's metabolic enzymes (Knoll and Wisser, 1984). This results in rapid in vivo metabolism of the transmitter portion of the haptenic complex. The Abs that develop in response to the hapten have high affinity for the metabolized rather than for the unmetabolized neurotransmitter. This has been especially true for catacholamines. Haptens containing monoamine neurotransmitter metabolites do not encounter this difficulty. Development of IAs for the endogenous neurotransmitter metabolites has been successful, and the number of available assays is increasing. There are as yet unpublished reports that novel organic chemistry techniques and/or modification of in vivo metabolism during hapten exposure have enabled Abs to be successfully developed for measuring free neurotransmitters.

A problem encountered in developing an Ab for 5-HT illustrates the "arm" or "bridge" effects that are frequently encountered when linking low mw molecules to large antigenic molecules. Chemical conjugations that use the dimethylpimylimidate or N-carboxymethyl derivatives of 5-HT and DA to link them to carrier proteins result in Abs that have a high affinity for the chemical coupling "arm," but have low affinity for the free transmitters (Flurkey et al., 1985). Similar "arm" effects are also frequently encountered when synthesizing haptens for drugs or other nonendogenous compounds. For certain specific applications, an Ab that has high affinity for the chemical "arm", however, can be very advantageous. For example, 5-HT that is coupled to a haptenic protein via the formaldehyde reaction (Mannich reaction) and endogenous 5-HT in brain tissue that has been fixed in formaldehyde are attached to a protein by the identical chemical "arm." Antibodies with a high affinity for formaldehyde-coupled 5-HT are therefore very useful for investigating regional and subcellular immunocytological localization of 5-HT following formaldehyde fixation of brain tissue (Steinbusch et al., 1982; Milstein et al., 1983).

Another novel exploitation of the "arm" effect, and one that appears to have other potential applications, has recently been reported by Sato and Yamamoto (1983). In attempting to develop an IA for a β-adrenergic blocking agent, the authors had considerable difficulty in isolating antisera with adequate specificity for the free drug. They could, however, readily obtain antisera that had high affinity and specificity for the succinylate drug/protein "bridge" used in the hapten. To overcome this difficulty, the authors partially purified their drug-containing tissue samples from their experimental animals and chemically "succinylated" the fraction to convert any drug to a drug succinyl-protein "bridge." The resulting conjugate was recognized by the Ab with high affinity and specificity, and an acceptable IA procedure that utilized the "bridge-specific" antisera was developed. The authors also reported that the conjugation procedure increased the sensitivity of the assay 100-fold and significantly reduced cross-reactivity with other closely related drugs and metabolites when compared to the assay that used antisera for the free drug. Because extra steps are required in such "bridge-dependent" assays, they may be less desirable than developing antisera specific for a free drug. This innovative approach should be considered, however, when antisera or Abs that do not exhibit "bridge" effects are difficult or impossible to obtain.

In addition to actual chemical alteration, less dramatic molecular modifications that affect the three-dimensional configuration of a drug molecule also play a significant role in determining Ab specificity. It has been demonstrated that Abs that can readily distinguish between the stereometric configurations of a compound containing a chiral carbon can be isolated. For example, Abs with significantly differential affinity for the *d* and *l* versions of ephedrine (Midha et al., 1983b) or amphetamine (Faraj, 1976) can be used to analyze the individual isomers of these drugs without prior separation. The converse is also true. Some IA assays used for measuring amphetamine (Mason et al., 1983) and methamphetamine (Tamaki et al., 1983; Aoki and Kuroiwa, 1983) are not able to distinguish between isomers. If for example a patient had received an isomeric mixture of one of these compounds, these IAs would measure only total drug, and levels of the active (*d* isomer) and inactive (*l* isomer) drug could not be measured without prior separation. Faraj's (1976) studies also suggest that unanticipated changes in the stereochemical configurations of molecules that are not considered to be isomers could easily complicate hapten formation. For example, an energetically unfavorable spatial configuration that does not occur when a nonrigid molecule is in solution can be imposed when the molecule is covalently coupled to the carrier protein. The Ab generated in response to this hapten will therefore recognize the "forced" steric configuration of the molecule. Because it is thermodynamically unfavorable, the molecule does not freely assume the "forced" configuration in the test sample, and as a result the specificity and affinity of the Ab for the configuration of the At as it occurs in the assay system may be significantly reduced.

Another example of innovative hapten design has been the intentional development of an Ab that is specific for a metabolite but will not recognize the parent drug. Such a strategy has made it possible to study the pharmacology and pharmacokinetics of the sulfoxide metabolite of chlorpromazine (Yeung, 1983) and trifluoperazine (Aravagiri, 1984) without having to separate them from the parent drug.

Deliberate development of an Ab that cross-reacts with closely related compounds can also be exploited. Dixon and coworkers (1984) developed an Ab that had high affinity for naloxone and also for a new, chemically related opioid antagonist, nalmefene. Because of the dual affinity, the resulting radioimmunoassay (RIA) for quantitation of nalmefene was able to utilize commercially available radiolabeled naltrexone as the labeled At. Labeled nalmefene was not available, and an Ab that would not cross-react with naltrexone would not have resulted in an acceptable IA for nalmefene.

Finally, the intentional development and isolation of an Ab that exhibts broad cross-reactivity among multiple members of a chemically related class of drugs that are usually prescribed individually can be used to develop a single IA method that can be used to assay for any of the individual drugs in the class. For example, a single broad-specificity Ab can be used to assay for any individual member of a large group of benzodiazepine tranquilizers (DeBlas et al., 1985; Sherman-Gold et al., 1983), tricyclic antidepressants (Marullo et al., 1985; Hubbard et al., 1978; Scoggins et al., 1980), and phenothiazines (Hawes et al., 1983).

The number of drug molecules attached per molecule of the haptenic protein has also been found to be an important consideration in developing suitable Abs. It is frequently stated in the literature that haptens with high or maximal drug:carrier ratios are more desirable or that a minimum of 10–20 drug molecules per molecule of protein is required in order to produce antisera of acceptably high titer. In practice, however, suitable Abs have been produced using a hapten with a drug:protein ratio as low as 2 [although a longer than usual time was required to elicit the desired response (Beiser et al., 1976)]. Conversly, it has been reported that there is a target "window" for the substitution ratio and that very low (<25 drug molecules/protein) or excessively high levels of substitution (>100–200 of drug molecules/protein) will result in significantly diminished immunogenicity (Robinson et al., 1975). Alternatively, Kato et al. (1984) recently reported that the best immunogenic response to a hapten consisting of the drug methotrexate attached to keyhole limpet hemocyanin (KLC) was obtained at the highest possible level of substitution, which was reported as 430:1. It also has been suggested that the use of a highly substituted carrier protein, such as KLC, as opposed to the more traditional hapten proteins such as bovine serum albumin, increases the efficiency of Ab production (Kato et al., 1984).

Using a completely different approach, it has been reported that increasing the distance between the drug and the carrier protein by introduction of a "spacer" molecule, such as a succinate group, will be able to diminish "arm" effects and result in an antiserum that is more specific for the free At (Jeffcoate and Searle,

1972). Although such an effect was clearly demonstrated in their study, the ultimate usefulness of "spacer molecules" in a wide range of other applications is questionable. Careful consideration of attachment chemistry, the structural configuration of the target At, and the functional group on the At molecule that will be used as the attachment point is more likely to result in useful Abs than is undertaking more drastic chemical modification of the At molecule. The use of methods other than increasing "spacer" distance to develop an acceptable hapten is illustrated by the successful development of an IA for melatonin by Tiefenauer and Andres (1984).

If all measures fail and a suitable functional group for hapten synthesis simply cannot be identified in a target At molecule, it may be worthwhile to attempt more significant chemical modification of the molecule to obtain a suitably reactive site. A successful example of this was the synthesis of t-3'-succinylmethylnicotine by Langone et al. (1973). Despite repeated attempts, it was not possible to obtain an Ab that had acceptable affinity for unconjugated nicotine, even though various haptens were attempted. When the nicotine analog (t-3'-succinylmethylnicotine) was coupled to a haptenic protein, however, an Ab was raised that had high affinity and acceptable specifity for free nicotine.

As suggested by these specific examples and confirmed in the remainder of the extensive literature that describes the preparation of many different drug haptens, broad "rules" that accurately describe Ab development do not exist. Rather, the affinity and specificity of Abs that result from the immunogenicity of a specific drug:protein hapten can vary greatly in each specific case. The success or failure of each specific hapten depends upon such multiple factors as the nature of protein carrier used, the ratio of drug:protein, the specific chemistry involved, the chemical or stereochemical characteristics of the target At, and probably most significantly, the response of the individual animal of the chosen species that was immunized. When a hapten for an IA of a new compound must be developed, those procedures that have been used successfully with chemically related molecules should be attempted first.

Once the chosen hapten has been synthesized, and purified if necessary, the immunization of subjects from the chosen species can be carried out according to a number of widely accepted, though largely empirical, methods. Immunization procedures that are suitable for development of Mabs have recently been reviewed in considerable detail by Schonherr and Houwink (1984). Some general characteristics are common to all of the immunization methods. When the goal is to produce a polyclonal antisera that will be used directly in an IA, it is very important that the hapten preparation be as free from any other antigenic contaminants as possible. Such contamination will produce antigenic responses to the contaminating immunogens, and this will invariably reduce the specificity, titer, and affinity of the resulting antisera for the target At. Conversely, however, if the goal is production of a Mab, extensive purification of the hapten is unnecessary. The screening procedures that are used to isolate the desired Mab eliminate all of the unwanted or nonspecific Abs; and if the design of the hapten is adequate, an Ab with the desired specificity and affinity can be identified. In fact, using a high-purity hapten may be a disadvantage when attempting to develop a Mab. The broader the range of epitopes that stimulate the immune response, the more likely it is that an Ab with the desired affinity and specificity for a unique combination of determinants in the unconjugated At will be generated. One of the greatest advantages of Mab development rests on the fact that even a very minor Ab, if it has the most desirable characteristics, can be used to produce an unlimited supply of that same specific Mab.

A common method to ensure that a suitable immune response will occur is to administer the immunogen in a substance called an adjuvant. The effect of an adjuvant is to increase or maximally stimulate the immune system to respond to an immunological challenge. Common and widely used methods for preparation and use of suitable adjuvants have been described by Herbert (1973). As a general rule, the development of suitable polyclonal antisera requires multiple immunizations. The affinity of the antiserum will increase with time or with longer intervals between repeated challenge. It is usually not possible to harvest acceptable polyclonal antisera without making a minimum of 2–3 injections over 2–4 mo. For production of Mabs, several closely spaced immunizations and harvesting at much shorter intervals (2–3 wk) is commonly employed. The overall strength of the immune response is less important than providing the maximal number of different Abs to select from. Even when developing Mabs, it is usually wise to immunize a minimum of 2–3 animals, because neither the specificity nor the intensity of the immune response of individual animals is reproducible or predictable.

A detailed immunochemical characterization of lymphocyte populations and the complex synthetic and regulatory biochemical processes that govern Ab production is beyond the scope of this review, but comprehensive reviews of these areas appear regularly. Briefly, although all lymphocytes appear homogenous under a light microscope, they are in fact heterogenous, and each lymphocyte type produces a distinct class of immunoglobin. It has been estimated that an individual has the potential to produce 10^7–10^8 different Abs. The different immunoglobins (Igs) are classified as IgG, IgM, IgE, IgA, or IgD, depending on their specific molecular structure. The predominant Abs that develop in response to most haptens, or are secreted from hapten-generated hybridomas used to produce Mabs, are almost exclusively IgG antibodies, though some IgM Abs do occur. The detailed biochemical structure of immunoglobin classes has been studied extensively, and it is known that IgG Abs contain two long (or heavy) and two short (or light) peptide chains that are connected to each other by specific disulfide bonds. Each Ab has a constant region (Fc) that is a characteristic of the specific mammalian species in which it is formed. The identical "Fc fragment" therefore is contained in all IgG Abs from that species. Each Ab also contains a variable region (Fab) that is unique to the Ab that is produced by each individual lymphocyte or its clone. It is the Fab "fragment" that is responsible for the specificity of the binding of that Ab with the specific epitope(s) on the At molecule. This "lock-and-key" design is responsible for the remarkable specifity and high affinity of Ab–At binding.

Two sets of genes, V genes and C genes, direct the synthesis of variable and constant portions of these peptide chains, respectively. The V and C peptide chains are synthesized independently on separate genes, and the Ab is then "assembled" by the lymphocyte from these individual polypeptide fragments. The Fc fragment of an Ab is itself antigenic when administered to any other species than the one from which it was obtained. Conversely, Fab fragments are not antigenic. This property has been innovatively exploited via elegant genetic engineering techniques to produce chimeric Abs. Chimeric Abs have many unique and powerful applications that are not possible with naturally occurring Abs.

As a normal part of the immunological response, each stimulated β-lymphocyte generates a family of daughter cells, each of which secretes Abs that are genetically identical to those formed by the parent lymphocyte. The serum of a subject that has been immunized with a multideterminant hapten will contain a large number of genetically distinct Abs, each of which is secreted by the clone arising from a single β-lymphocyte. Such serum is said to contain Pabs or to be a polyclonal antiserum. To circumvent the finite time that β-lymphocytes will produce Abs following their isolation in vitro, Kohler and Milstein (1975) described their development of a method for fusing an immunogenically stimulated β-lymphocyte with an immortal myeloma cell line. The result was a continuously growing hybridoma cell line that secreted large amounts of the Ab that was identical to that produced by each original β-lymphocyte. By isolating and growing separately a number of single hybridoma cells and screening the resulting Abs to determine their specific characteristics, it was possible to identify and obtain a limitless amount of a genetically distinct Mab possessing a unique set of desired characteristics. The original hybridoma methods employed murine cell lines, but methods for producing human Mabs have now been developed. Descriptions of various methods that are used for cell fusion, hybridoma growth, Mab screening, and isolation appear frequently in the literature and are periodically reviewed. Recent reviews include those of Schonherr and Houwink (1984), Nakamura (1983), James et al. (1984), and Roome and Reading (1984). The new advances that appear regularly in genetic engineering demonstrate that the future is not limited by what can be accomplished by using naturally occurring Abs. For example, the combination of human C genes with nonhuman V genes results in production of a chimeric Ab that contains essential characteristics derived from each host. By this means, the specificity and affinity characteristics of an Ab that would be impossible or dangerous to develop in humans can be elicited in a nonhuman species. The genes that produce the relevant Fab fragments can be combined with a human Fc gene. The chimeric Ab that will result could be used to "deliver" the desired Fab-determined characteristics to a human subject without triggering the immunogenic reaction that would normally result from the Fc portion of a foreign Ab. Reports of current progress in formation and application of chimeric Abs are becoming increasingly common (Oi and Morrison, 1986; Roome and Reading, 1984; Boulianne, 1984; Morrison et al., 1984; Takada et al., 1985). It has also been possible to physically separate Abs into their separate Fab and Fc fragments, and the resulting Fab fragments retain the At affinity and specificity of the parent Ab. Because they are not antigenic, Fab fragments have many potential uses in diagnosis and therapy of human diseases.

Continuing progress in genetics may alter significantly these traditional approaches to development of specific Abs. Gene modification and synthesis techniques suggest that the development of altered or completely synthetic C or V genes will be possible. Such genes could be used to code directly the formation of an Ab with predetermined specificity or affinity. Progress is limited because the knowledge of how specific peptide sequences in an Ab confer specificity and affinity characteristics is not yet available. The most important implications for synthesis or alteration of V genes appear to lie in the potential to develop Ab specificities for small mw compounds, especially for in vivo structures that are difficult or impossible to generate via traditional in vivo immunogenic response.

From this overview of general IA considerations it should be obvious that the identification of a suitable hapten and subsequent preparation of acceptable polyclonal sera or identification and isolation of Mabs is a lengthy and frequently complicated process. The number of commercially available antisera and/or Mabs for drugs and endogenous compounds that are important in neurobiology continues to increase dramatically. Many of them have been specifically designed for quantitative analysis. In many instances, the Abs are supplied only in kits that provide complete sets of reagents and predesigned assay methods for specific drugs or endogenous compounds. Unfortunately, if one requires the Ab for an atypical use, or intends to significantly alter the method, the pure Ab frequently cannot be obtained alone. Some companies have begun to offer technical services that will custom-develop a specific antiserum or Mab that can be tailored to the individual needs of the investigator. The preceding discussion was designed to serve as an introduction to some of the basic theories and ideas that are useful in development of a new IA.

2.2. Polyclonal Antibodies vs Monoclonal Antibodies

When developing a quantitative IA, especially for a small mw compound that requires a hapten, the question arises as to which will produce a more acceptable method—a Mab, a partially purified Pab fraction, or simply diluted polyclonal antisera? As previously discussed, compounds with a molecular size below 1 kdalton always react monovalently with Abs and, therefore, obtaining first-order kinetics need not be a factor in choosing the Ab preparation. The availability of limitless quantities of Mabs that do not vary significantly in specificity or affinity from batch to batch confers an overwhelming advantage on Mabs for most applications. There are several factors, however, among them the high cost of some Mabs, that under certain circumstances may make the use of a Pab desirable. The important limitations of Pabs include (1) the fact that specificity, affinity, and titer of a polyclonal antiserum will vary from animal to animal, from batch to batch, and even over time from the same animal; and (2) the fact that the sudden loss of the immunized doner animal is always a possibility. Improper techniques in propagating hybridoma clones can also lead to considerable "drift" in Mab affinity and or specifity. Moreover, mutations that alter Mab production or the loss of cell lines can also occur. Appropriate reculture and storage of the original hybridoma are essential to ensure Mab integrity. When these are carried out properly, an unlimited supply of an identical Mab can be expected.

It was originally thought that a Mab could not have an affinity equal to that of a high-affinity polyclonal antiserum. This misconception apparently arose because only a few of the thousands of lymphocytes responsible for a typical immune response produced an Ab with high affinity, and initial isolation and screening methods were not adequate to reliably isolate and identify the Abs produced by those few cells. Improved screening techniques have identified Mabs that have affinities equal to or better than affinities of a polyclonal antiserum (10^{-9} to 10^{-11}). If an important consideration is the need to develop an IA as rapidly as possible, however, the quickest method will be to use polyclonal antisera. The affinity and specificity of a Mab depend on the ability of a single Fab region to differentiate the specific set of determinants on the At. A typical immune response will normally produce only a few cells that have the required combination of affinity and specificity. Considerable time is frequently required to select and culture the appropriate cell line and characterize the resulting Mab. In contrast, the specificity of a polyclonal antiserum, in effect, results from the summing of the specificities of all the different Abs for the different determinants of the At, whereas the affinity depends primarily on the affinity of the Abs with the highest affinity for the At. Therefore a polyclonal antiserum of high specificity and affinity can be obtained much more quickly and polyclonal-based methods still represent a valid and powerful IA system.

2.3. Heterogeneous vs Homogeneous Immunoassay (IA) Techniques

A heterogeneous IA refers to a method in which the Ab–At complex must be separated from the unbound Ab or At prior to determining the extent of the Ab–At interaction. Homogeneous IAs, conversely, make use of some unique characteristic that can be conferred on either the Ab or the At so that a measurable property will be altered when they are combined in the Ab–At complex. Because heterogeneous assays require a separation step, they are more complicated and time-consuming than homogeneous assay methods. Numerous procedures have been developed to make the separation step more convenient. A variety of solid-phase separators (starch, charcoal, plastic beads) to which the Ab will readily be adsorbed, or to which it can be attached, have been developed. The separation obtained in heterogeneous assays frequently permits more precise or sensitive measurements than are possible with homogeneous assays. Regardless of whether one employs a heterogeneous or a homogeneous IA, a general note of caution is in order. An IA is almost always developed and validated for measurement of a particular substance in a given tissue preparation or other medium. Kubasik and Sine (1979) tested 23 different commercial radioimmunoassay kits designed to measure a variety of different drugs. When these methods were applied to otherwise identical serum and plasma samples, there were significant differences in the results of these tests. McKay et al. (1985) have reported that significant differences in the apparent blood levels of chlorpromazine, as measured by a radioimmunoassay (RIA) procedure, will vary significantly depending on the preparation of the blood sample for analysis. If an IA is to be used under any other set of circumstances than the ones used to validate the method, the assay must be revalidated under the new circumstances.

The traditional RIA, which is a heterogeneous assay, was widely adopted before homogeneous IA techniques were developed. The advent of nonisotopic IA markers, however (e.g., enzymes, luminescence, fluorescence), have permitted ingenious homogeneous IA schemes to be developed. Homogeneous assays reduce the number of precise measurements and transfers that are required per assay, and this usually improves the precision of an analytical assay. Except when an RIA has some unique advantage, it appears that development of a homogeneous IA would be the method of choice. The specific techniques that have been developed for separation of free and bound fractions in heterogeneous assays and for utilizing the markers that identify the extent of Ab–At formation in homogeneous assays are discussed in detail in the sections that follow.

2.4. Antibody-Limited (Ab-Limited) vs Antibody-Excess (Ab-Excess) Immunoassay Techniques

Originally, RIAs were based on the competition between the target At in the sample and a precise amount of the identical radiolabeled At for a limited number of Ab binding sites. In such assays, the number of Ab binding sites must be in excess of the At present, i.e., the assay is Ab-limited. In virtually all Ab-limited IAs, the correct Ab concentration is that which will bind approximately 50% of the labeled At at equilibrium. Although excellent results can be achieved with Ab-limited assays, they have a number of disadvantages. A potential disadvantage, which will vary depending on the experience and competence of the user, is the relatively large number of precise volumetric measurements that are required (the dilution and addition of the Ab, the preparation and amount of target sample added, and the dilution and addition of the labeled At). Multiple sequential measurements decrease the average precision that is possible because of the summing effect of any pipeting errors that occur.

In order for an Ab-limited IA to be valid, the steady-state equilibrium that is established between the limited Ab binding sites and the labeled and unlabeled At must be maintained. Great care must be exercised to ensure that the separation method does not disturb the steady-state equilibrium. Because the Ab–At reaction is readily reversible, and the Ab site is in an unsaturated state, small changes in equilibrium can have a significant effect on the precision of the assay, especially near the upper or lower limits of the assay (i.e., when the difference between the amount of labeled and unlabeled At is greatest). Another frequently encountered disadvantage of an Ab-limited assay is its limited range. Because this type of assay depends on competitive binding between the labeled and unlabeled At, it is valid only when the Ab site is unsaturated, and ideally the labeled/unlabeled At ratio should be less than 10:1. Moreover the absolute sensitivity and the range (usually about three-fold) of the assay are predetermined factors imposed on the assay by the dissociation constant of the Ab and cannot be changed by altering other aspects of the assay. For example, if the dissociation constant of an Ab is $3 \times 10^{-7} M$, the sensitivity and range in which the assay will be maximally accurate is confined to measuring concentrations of the At between 1.5 and $4.5 \times 10^{7} M$.

In Ab-excess IA, the Ab is present in excess of that needed to bind all of the At. For this type of assay to be valid, the Ab must combine with 100% of the At. In accordance with the principles of mass action, the 10- to 100-fold excess of one of the reactants (Ab) in an equilibrium reaction will shift the reaction toward incorporation of all of the other reactant (At) into the product (Ab–At complex). In a heterogeneous Ab-excess IA, this factor reduces significantly the likelihood that the separation step will influence the precision of the assay by disturbing the equilibrium of the reaction. Moreover, the presence of excess Ab expands greatly the range of At concentrations that can be measured by an Ab with a given affinity. In this system, the range is determined by the upper and lower limit of sensitivity of the system used for measuring the amount of Ab–At complex formed. The major disadvantages of an Ab-excess assay had been that it required a relatively pure Ab preparation to be used and that it consumed relatively large amounts of Ab. Ab-excess assays were therefore of limited value until Mab techniques appeared. The easy availability of large amounts of Mab has made Ab-excess assays the method of choice when developing a new IA. The methods that are available for quantitating the amount of Ab–At complex formed by an Ab-excess assays are discussed in detail in the sections that describe the specific IA methods.

3. Markers Used to Detect the Extent of Antibody–Antigen (Ab–At) Formation

An increasing number of methods continue to be developed for quantitating the extent of Ab–At formation in IA methods. We have attempted to review the most important and innovative of these methods, but as with the larger field of immunology, methods are evolving so rapidly that many aspects of this review will rapidly become obsolete.

The first use of Abs for a quantitative assay, developed by Yalow and Bernson (1959), was a limited Ab-heterogeneous RIA that depended on the competition of radiolabeled insulin and unlabeled insulin from the biological sample for a limited number of insulin-specific Ab sites. Free and bound fractions were separated. The amount of labeled insulin in each fraction allowed the authors to quantitate the amount of insulin that was present in the original biological sample. RIAs are still among the most sensitive IAs possible because labeled Ats of very high specific activity can be prepared. Other methods that can be used to quantitate the extent of Ab–At complex formation were originally developed primarily to circumvent some of the disadvantages of RIAs (e.g., use of radioactivity, absence of a labeled At) or to be able to take advantage of less expensive instrumentation. They were not developed in order to increase the sensitivity of IAs. A major disadvantage of many of the nonisotopic IAs has been that their sensitivities are generally far below those of RIAs.

Previous statements in the literature have indicated that the affinity of the Ab is the primary factor one must consider in determining the sensitivity of an IA. In fact, the absolute sensitivity of any IA method actually depends on a combination of the affinity of the Ab and the sensitivity of the system that has been chosen to measure the number of "marker" molecules or atoms that have been attached to either the Ab or the At. The affinity of the Ab for the At determines how many molecules of the Ab–At complex will be formed by the concentrations of Ab and At under the conditions that exist in the incubation medium. The Ab affinity is important because it must be in the correct range so that a significant portion of the target At will be converted to the At–Ab complex at the At concentration that is being measured. The final sensitivity of the IA is then determined by the ability to measure the minimum number of Ab–At complexes that is possible. RIAs and many types of flourescent and luminescent IAs accomplish this task directly by simply measuring the number of Ab- or At-linked marker molecules that are in the Ab–At complex. The presence of an Ab or At that has high specific activity indicates that there are a relatively large number of radioactive atoms per molecule of labeled substance that will undergo radioactive disintegration during the time period normally used to measure the radioactivity (<10 min). Because modern scintillation counters can measure virtually 100% of the radioactive disintegrations that occur, it is possible to detect as little as 10^9 to 10^{12} mol of tritum-labeled or 10^{-11} to 10^{-15} mol of a ^{125}I-labeled Ab–At complex. This molar sensitivity is lower than the molar affinity constant of the lowest affinity Abs, so the minimum sensitivity of an RIA will therefore be determined primarily by the affinity of the Ab. Stated another way, the measuring system is capable of measuring fewer Ab–At complexes than the Ab will form at the lowest concentration of the target At that can be measured by an Ab of a given affinity. It should be noted, however, that the sensitivity of a RIA will always increase or decrease as a function of the specific activity of the labeled Ab or At, regardless of the affinity of the Ab.

There are large differences in the efficiency of various luminescent, fluorescent, or colorimetric marker molecules used in IAs. Even when employing very efficient molecules, however, the measurement systems used in these IAs require that the "output" of many marker molecules must be summed in order to generate a measurable signal. The minimum amount of a nonisotopic labeled Ab–At complex that can be measured by these systems is in the range of 10^{-5} to 10^{-8} mol. It is not unusual to obtain an Ab with a molar affinity below this value, so the minimum sensitivity of direct nonisotopic IAs will be determined primarily by the measurement system, rather than by the affinity of the Ab. This means that direct nonisotopic methods cannot take advantage of the ability of high affinity Abs to complex with very low concentrations of a target At because they will not produce enough At–Ab complexes for the measurement system to detect. In order to circumvent this inherent lack of sensitivity, nonisotopic IA methods have been developed that employ various multiple-step systems that are able to "enhance" or "multiply" the number of detectable marker molecules that will be generated as a function of the formation of the Ab–At complex. A detailed discussion of these methods is presented in the following sections.

3.1. Radioimmunoassay (RIA)

In a classic paper, Yalow and Berson (1959) were the first to exploit the mammalian immune response to develop a valid quantitative assay for a biological substance on a crude tissue preparation. The authors generated an antiserum containing a high titer of Abs that bound specifically to insulin, and synthesized a radiolabeled insulin analog that had a high affinity for the Ab. RIAs measure the degree to which an unknown amount of the target At that is present in a test sample is able to inhibit the binding of a known amount of the radiolabeled At to a limited number of the specific Ab binding sites. The target At is quantified by comparing the inhibition of binding of the labeled At produced by the unknown sample to a standard curve. The standard curve is made by measuring the inhibition of binding of an identical amount of the radiolabeled At to the same number of Ab sites that is produced by known, graded amounts of pure target At standard. The technique has now been adapted for analysis of a variety of classes of important psychotropic drugs, and a survey of these methods are given in Table 1.

RIAs continue to have a number of unique advantages over other types of IAs, but also involve disadvantages that have stimulated development of other IA systems. As discussed above, the primary advantages of RIAs continue to be their comparative ease of development, their simplicity of use, and their high sensitivity. It is relatively simple to obtain a high-affinity polyclonal antiserum for the many target Ats, and if it is technically possible to label the At to a high specific activity, a sensitive RIA can be developed (10^{-9} to 10^{-12} mol is not uncommon). Moreover, a high degree of precision is possible because of the low level of background "noise" that is inherent in RIAs. There are virtually no naturally occurring radioactive materials in biological systems, so that if adequate techniques are developed to completely separate bound and free labeled At from the Ab–At complex, the radioactivity contributed by background radiation will be essentially zero. This is in direct contrast to many enzyme or fluorescent IAs, which suffer from high background or blank values against which the separated free and/or bound samples must be measured.

A significant disadvantage of RIA is the requirement for the target At to be in sufficiently pure form that it can be radiolabeled. In assays for drugs or common low mw endogenous compounds, the pure At is normally readily available. In many therapeutic or diagnostic applications, the target At frequently cannot be obtained in chemically pure form without extraordinary difficulty, and RIA methods are therefore not practical. The ability to insert adequate levels of tritium or radioiodine into the At molecule also varies greatly, depending on the chemical characteristics of the individual target At molecule. Another frequently encountered difficulty is that the addition of the radiolabeled atoms to the At molecule significantly reduces its affinity for the Ab binding site. The small tritium atom rarely alters At affinity significantly. It may be difficult to attain an adequate specific activity, however, especially in low mw compounds that may accommodate only a limited number of tritium atoms. The short-half life of ^{125}I and ^{131}I permits synthesis of Ats with very high specific activity. The significantly larger size of the iodine atom increases the possibility, however, that its presence will adversely affect the affinity of the labeled At, especially in low mw Ats or when more than one iodine atom is inserted per molecule of At. Many reviews discuss the problems and procedures used in labeling drug Ats for use in RIA (e.g., Ly and Hegesippe, 1984).

Table 1
References to Some Radioimmunoassays for Determination
of Psychotropic Drugs

Drug	Sample type	Reference
Alfentanil and sufentanil	Plasma	Michiels et al. (1983)
Alfentanil and sufentanil	Serum	Schuttler and White (1984)
Benzodiazepines (review article)	Plasma	Maguire and Fulton (1980)
Bromperidol	Plasma	Tischio et al. (1984)
Brotizolam	Plasma	Bechtel and Weber (1985)
Bupropion	Plasma	Butz et al. (1983)
Cannabinoids	Urine	Cais et al. (1983)
Chlordiazepoxide	Blood, saliva	Dixon et al. (1979)
Chlordiazepoxide	Plasma	Dixon et al. (1975)
Chlorpromazine	Plasma	Hubbard et al. (1978a)
Chlorpromazine	Plasma	Hubbard et al. (1978)
Chlorpromazine	Plasma and brain (rat)	Kawashima et al. (1975)
	Plasma	Midha et al. (1979)
Chlorpromazine sulfoxide	Plasma	Yeung et al. (1983)
Diazepam	Plasma	Bourne et al. (1978)
Diazepam	Blood, saliva	Dixon and Crews (1978)
Diazepam	Serum	Gelbke et al. (1977)
Diazepam and desmethyldiazepam	Blood	Peskar and Spector (1973)
Doxepin and desmethyldoxepin	Plasma	Midha and Charette (1980a)
d- and l-Ephedrine	Plasma	Midha et al. (1983b)
Flupenthixol	Serum	Jorgenson (1978)
Fluphenazine	Plasma	Midha et al. (1980c)
Fluphenazine	Plasma	Goldstein and Vunakis (1981)
Fluphenazine	Plasma	Wiles and Franklin (1978)
Haloperidol	Serum	Terauchi et al. (1984)
Haloperidol (reduced)	Plasma and urine	Browning et al. (1985)
7-Hydroxychlorpromazine	Plasma	Yeung et al. (1985)
7-Hydroxytrifluorperazine	Plasma	Aravagiri et al. (1985)

Imipramine and desipramine	Plasma	Midha et al. (1980b)
Methadone	Plasma and urine	Robinson and Smith (1983)
Methimazole	Plasma	Cooper et al. (1984)
5-Methoxytryptophol	Plasma and pineal gland	Kennaway (1983)
Morphine	Urine	Litman et al. (1983)
Morphine	Plasma and urine	Edwards et al. (1986)
Morphine	Plasma	Moore et al. (1984)
Nalmefene	Plasma	Dixon et al. (1984)
Naloxone	Plasma	Hahn et al. (1983)
Nortriptyline	Plasma	Midha et al. (1978)
Perphenazine	Plasma	Midha et al. (1981b)
Phencyclidine	Saliva	McCarron et al. (1984)
Phencyclidine and metabolites	Plasma and urine	Lindgren and Holmstedt (1983)
Phenothiazines	Plasma and urine	Robinson and Smith (1985)
Phenothiazines and metabolites	Plasma	Hawes et al. (1983)
Prazepam and metabolites	Plasma	Kohler-Schmidt and Bohn (1983)
Prochlorperazine	Plasma	Midha et al., (1983a)
Sultopride and sulpiride	Rat brain	Mizuchi et al. (1983)
Tetrahydrocannabinol and metabolites	Urine	Jones et al. (1984)
Tetrahydrocannabinol and metabolites	Human body fluid	Law et al. (1984)
Tricyclic anti-depressants	Plasma	Hubbard (1978b)
Tricyclic anti-depressants	Body fluids	Mason et al. (1984)
Tricyclic anti-depressants	Plasma and urine	Robinson and Smith (1985)
Trifluoperazine	Plasma	Midha et al. (1981a)
Trifluoperazine	Tablets	Midha et al. (1984)
Trifluoperazine	Plasma	Midha et al. (1983c)
Trifluoperazine and fluphenazine	Plasma	Hawes et al. (1983)
Trifluoperazine and fluphenazine	Plasma	Hawes et al. (1984)
Trifluoperazine sulfoxide	Plasma	Aravagiri et al. (1984)
Trimeprazine	Plasma	McKay et al. (1984)
Triprolidine	Plasma and urine	Findlay et al. (1984)

The major disadvantage of RIAs is the difficulty inherent in the use of radioactivity; this is significantly greater with radioiodine than with tritum. Expensive, dedicated equipment is required to measure radioactivity. Although such equipment is standard in research laboratories and large medical centers, it is far less common in smaller clinics. The use of radioactive compounds, especially those containing radioiodine, requires special training to ensure safe handling and disposal.

Another disadvantage inherent in all RIAs is that they are heterogeneous assays, i.e., they require the Ab–At complex to be separated from free At (or Ab) prior to measuring radioactivity. A number of innovative solid-phase separation methods have been developed, however, that make this separation more accurate and less cumbersome. A common separation method uses second Ab precipitation. For example, if the primary Ab is a rabbit IgG (or in the case of a chimeric Ab, Fc fragments from a rabbit), an anti-rabbit IgG Ab can be obtained from a goat by immunization with an appropriate rabbit IgG. The resulting anti-rabbit IgG goat anti-sera can be used to rapidly precipitate rabbit IgG Abs after they have equilibrated with a labeled and unlabeled target At. To increase the speed and convenience of second Ab separation, a variety of methods have been developed in which the second Ab is attached to solid supports such as easily separated plastic or latex beads or the walls of a small tube. An absolute requirement of any separation method is that the steady-state equilibration of At–Ab binding must not be altered significantly during the separation process.

To circumvent the problems encountered when attempting to radiolabel low mw Ats, the label is sometimes attached to the Ab, rather than the At. Because Abs are high mw glycoproteins, they contain a number of reactive sites and normally can be radiolabeled to comparatively high specific activities, even when using tritium. Because of their larger molecular size, this can usually be accomplished without affecting their affinity for the At. RIAs that utilize radiolabeled Abs are normally capable of attaining very high sensitivity. The obvious drawback of Ab labeling is that large amounts of highly purified Ab must be available. Prior to the advent of Mab techniques, this was an overriding handicap because the isolation of purified Abs from polyclonal antisera is a long and cumbersome procedure. The current availability of large quantities of pure Mab suggest that RIAs that utilize labeled Abs could be used to develop RIAs with high sensitivity for many compounds. Progress and problems in RIA technology have been reviewed recently by Ahuja and Pillai (1983), Dwenger (1984), and Willey (1985).

A recently developed RIA, based on scintillation proximity radioimmunoassay (SPRIA), appears to make a homogeneous RIA possible (Udenfriend et al., 1985). The new method makes innovative use of the low-energy Auger elections that are emitted by [125]I. Additional modification of the Udenfriend (1985) procedure should allow one to use a high specific activity ^{125}I-labeled protein to quantitate low mw target Ats that could not themselves be successfully labeled with iodine. The published method uses the technique of scintillation proximity that was first described in principle by Hart and Greenwald (1975). In that report the light produced by a tritium-activated fluorophor was measured by scintillation counting when brought into appropriate proximity by an Ab–At reaction. Udenfriend et al. (1985) concluded that the low-energy [125]I Auger electrons could also be used for SPRIA. They developed an ^{125}I-based homogeneous SPRIA that could quantitate morphine, tissue enkephalin, or thyroxin. The Ab was attached to fluorophor containing polyvinyl beads, and the target At (thyroxine, morphine, or enkephalin) was labeled with [125]I. Because the path length of the [125]I Auger elections is quite short, the fluorophor on the Ab-coated beads was not activated until it was brought into close proximaty to the ^{125}I by the specific Ab–At reaction. When the proper amount of fluorophor-Ab beads is incubated with ^{125}I iodine-labeled At and unlabeled target At from the biological sample, the level of fluorescence decreases as a function of the amount of target At in the biological sample.

The current method requires that the At of interest be iodine-labeled. Therefore their method is subject to the same limitations as conventional iodine-dependent RIA methods. The authors also briefly described a converse application of the method described above in which the target At (in this case a protein), rather than the Ab, was attached to the fluorophor beads, and the specific Ab was labeled with [125]I. When appropriate amounts of the At-coupled beads, ^{125}I-labeled Ab, and target At from the biological sample were then incubated, the competition between the fluor-coupled At and the unlabeled target At for the ^{125}I-labeled Ab determined the amount of proximity-initiated fluorescence.

This alternative method appears to have great promise because it appears that Ab-labeled SPRIA may be adaptable to a variety of low mw biological compounds and drugs. A SPIRA method that uses an ^{125}I-labeled Ab has the high sensitivity that results from using a high specific activity ^{125}I-labeled ligand and the speed and accuracy of a homogeneous IA, but not the limitations that accompany the labeling of low mw Ats with iodine. There do not appear to be any major technical limitations in developing the proposed procedures. It has already been demonstrated that most Abs can be iodinated to an adequate specific activity without appreciable loss of specifity or affinity for the target At. Similarly, the chemistry that is required to couple a drug or other low mw biological target Ats to fluorophor beads is identical to methods that are required for coupling the Ats to their haptenic proteins.

Regardless of whether an At- or Ab-labeled SPRIA procedure was used, the authors found that the relatively dilute concentrations of fluor-containing beads and the ^{125}I activator that are required, because of the limited range of [125]I Auger elections, resulted in background fluorescence that was very low. As a result, after a suitable equilibration time, the incubation mixture could immediately be counted without prior separation of the beads and the unbound ^{125}I-containing reagent. It appears that an Ab-labeled ^{125}I-based SPRIA method has significant potential as an improved RIA method for therapeutic drug monitoring and could find wide application as a quantitative assay for a number of low mw endogenous substances of interest to investigators in many areas of neurobiology.

3.2. Enzyme Immunoassay (EIA)

The concept of coupling the activation of an enzyme as the marker for the extent of Ab–At complex formation is a rapidly growing area of IA research. When sensitivity is adequate, EIAs are frequently replacing more traditional analytical methods (including RIAs) for measuring many biological substances and drugs. The majority of commercial IAs are enzyme-based assays, and their development continues to be an important commercial undertaking.

The rationale for developing EIAs is based on their ability to combine the power of two important biological phenomena: the high specificity and affinity of an Ab for its target At and the amplification that can be achieved by using the At–Ab immunoreaction to initiate specific chemical reactions that are catalyzed by certain enzymes. As discussed previously, in traditional RIAs and fluorescent IAs, each Ab–At complex usually contains only a single marker molecule (or at best two or three), and there is no amplification between formation of the marker-containing Ab–At complex and measurement of the units of marker that are present. In an EIA, however, each molecule of Ab–At complex that forms is able to catalyze the formation of multiple molecules of the enzyme product. Theoretically, under ideal conditions, a single Ab–At complex could be made to catalyze the formation of a virtually unlimited amount of enzyme product. A great deal of the amplification that results from the accumulation of the enzyme product is lost, however, because the conventional methods that are used to quantitate enzyme products are, on a molar basis, less sensitive than other methods used to detect the markers used to indicate Ab–At formation (e.g., radioactivity, fluorescence). As a result, virtually all currently available EIAs are less sensitive than comparable RIAs or other types of IA methods. Commercially available EIAs have limited use in neurobiological research and in therapeutic monitoring of psychopharmacological drugs because their absolute sensitivity is inadequate. When sensitivity is adequate, EIAs are usually advantageous because they require relatively inexpensive reagents, use common laboratory equipment, and can be performed easily and quickly.

A major difficulty encountered in many EIA methods is lack of specificity and a high level of background "noise." To some extent this is inherent in most EIA systems because of the functional way in which the activation of the enzyme is linked to Ab–At formation. The enzyme response in an EIA assay will be activated when any interfering cross-reacting At reacts with the Ab and the resulting enzyme activity is not a "graded" response. The enzyme is fully activated by At–Ab complex formation regardless of the affinity of the interfering At for the Ab site. Therefore all nonspecific or low-affinity reactions that occur between any cross-reacting non-target At and the enzyme-coupled Ab during the incubation period are equally magnified by the enzyme system. The net result is that reduced specificity or high background readings occur to a significant degree in many EIA methods.

Another factor to be considered in developing an EIA is that synthesis of the At–enzyme conjugate (At–E) can have a variety of influences on subsequent formation of the At–E–Ab complex. As discussed previously, all of the drugs or low mw target Ats being reviewed must be attached to high mw haptenic molecules in order to be immunogenic. Even though the resulting Abs are selected on the basis of their specificity and affinity for the unconjugated At, Abs that simultaneously have significant affinity for some determinants contained in the At-haptenic protein "bridge" are common. EIAs require that the At must be attached to the marker or marker-generating molecule, in this case an enzyme, to form the At–E complex that competes with free target At for Ab binding sites. The chemistry that is used to attach the At to the haptenic molecule and to the enzyme can be chosen deliberately to be either identical or different. If the chemistry is identical, the Ab will "see" the same chemical link in the At–E complex that occurred in its generating hapten. It is therefore conceivable that the Ab will have greater affinity for the At–E complex than for the free At. When this occurs it will alter significantly the kinetics of the system. If the At–E conjugate affinity for the Ab is higher than that of the free At, it will compete more effectively for Ab sites, and a higher concentration of free At will be required to displace it. Alternatively, the Ab–E–Ab complex may dissociate more slowly than the Ab–At complex, and this would affect the time needed to attain steady state. In any case it is always essential that the kinetics of the At–E (or other At complex) reaction with the Ab be thoroughly investigated prior to establishing the final procedures. This applies equally to polyclonal antisera or when screening for a suitable Mab.

An impressive number of different EIA techniques and modifications that use the principle of enzyme amplification have been developed, and this diversity makes an organized review somewhat difficult. This diversity has also led to considerable confusion in the various terms and acronyms that have been coined by various authors to describe different EIA types. We will survey the important principles that underlie the most prevalent forms of EIAs. Recent reviews of EIA procedures can be consulted for additional information (Oellerich, 1980, 1984; Stanley et al., 1985; Ngo, 1983; Ngo and Lenhoff, 1980; Ishikawa et al., 1983; Price, 1984). A survey of recent EIA methods for analysis of a variety of psychotropic drugs is given in Table 2.

3.2.1. Enzyme-Multiplied Immunoassay and Enzyme-Linked Immunosorbant Assay

EIAs can be divided into heterogeneous and homogeneous types. Systems in which the activity of the enzyme is not affected directly by formation of the Ab–At complex require a separation prior to measuring enzyme activity. The most prevalent heterogeneous EIAs are simple competitive assays identical in rationale to the original RIA of Yalow and Berson (1959). Pure At is chemically coupled to the chosen purified enzyme to form an At–E complex.

Table 2
Reference to Some Enzyme Immunoassays for Determination
of Psychotropic Drugs

Drug	Sample type	Reference
Acetominophen	Serum	Hepler et al. (1984a)
Antiepileptic drugs	Serum	Hayashi et al. (1983)
Barbiturates	Serum	Watson et al. (1983)
Barbiturates	Serum	Drost et al. (1984)
Barbiturates	Serum	Pape et al. (1983a,b)
Benzodiazepines	Serum	Drost et al. (1984)
Benzodiazepines	Serum	Manchon et al. (1985)
Caffeine	Plasma	Zysset et al. (1984)
Caffeine	Serum	Miceli et al. (1984)
Cannabinoids	Urine	Irving et al. (1984)
Cannabinoids	Urine	Schwartz et al. (1985)
Methamphetamine	Plasma	Tamaki et al. (1983)
Morphine	Urine	Littman et al. (1983)
Opiates	Plasma	Hausmann et al. (1983)
Opiates	Body fluids	Hepler et al. (1984b)
Pentobarbital	Serum	Karnes et al. (1984)
Pentobarbital	Plasma	Pape et al. (1983b)
Pentobarbital	Plasma	Wellington and Rutkowski (1982)
Pentobarbital	Serum	Sarandis et al. (1984)
Phencyclidine	Serum	Walberg et al. (1983)
Phenytoin	Serum	Kulpmann et al. (1984)
Salicylic acid	Serum	Fraser (1983)
Tricyclic antidepressants	Serum or plasma	Sonsalla et al. (1984)

Predetermined amounts of the At–E conjugate (analagous to radio-labeled At in RIA) compete with unconjugated target At from the test sample for a limited number of Ab binding sites. Following equilibration, the bound fraction (At–Ab plus Ab–At–E) must be separated from the remaining unbound At–E activity. Following separation, the tenzyme activity in the free and/or bound fraction is measured. The activity in the bound fraction will decrease in direct proportion to the amount of unlabeled target At present in the incubation mix. If the assay is one in which the primary Ab is in soluble form, any of the classical Ab separation techniques can be employed, although second or double Ab precipitation is most commonly employed. This type of EIA typically is called an enzyme-multiplied immunoassay technique (EMIT). If the primary Ab has been preattached to some type of solid-phase separator prior to incubation, the assay is most often termed an enzyme-linked immunosorbent assay (ELISA). Though ELISAs are common in many areas of neuropharmacology, they are not commonly employed for quantitative analysis of low mw Ats. Finally, if the second Ab, rather than the primary Ab, is affixed to a solid phase as the means for isolating the primary Ab, the method is commonly referred to as a double-antibody solid-phase (DASP) technique.

In determining whether to conjugate the Ab or At to the enzyme when developing hetergeneous EIAs, or in selecting the method to be used to separate the free and bound fractions, several factors should be considered. The chemical conjugations require that the enzyme and either the At or Ab must be available in purified form. Consequently, EIA methods that will entail synthesis of Ab–E conjugate are ideally suited when a Mab is available. Pabs are acceptable when the marker enzyme will be conjugated to the At. Second Ab precipitation is a very efficient method for separating the free and bound fractions in EIA in either case. It is possible, however, to attach either the At or the Ab (whichever is not conjugated to the enzyme) to a solid support to aid in separation of the free and bound fractions. Regardless of the separation method employed, enzyme activity is initiated by adding the substrate and any other factor required for enzyme activity following separation of the fractions.

A large number of different heterogeneous and homogeneous "sandwich" EIAs have appeared in the literature. All "sandwich" assays, as well as a significant number of the more complex types of EIAs, depend on the ability of the target At to simultaneously bind to two different Abs that recognize two different, completely independent epitopes on the At molecule. Because they contain a relatively small number of functional groups (compared to proteins and other high mw compounds), virtually all psychopharmacological drugs and other low mw biological molecules (including neurotransmitters and their metabolites) are not able to simultaneously bind to more than one Ab. These factors make them unsuitable for analysis by sandwich EIAs or other assays that require two-site binding by the At, and these assays are not reviewed.

Although heterogeneous EIAs described above have proven to be satisfactory for measuring a number of relevant biological substances, greater emphasis in the commercial arena has been placed on development of homogeneous EIAs. They represent a growing system of versatile assays that are especially well suited to clinical settings.

The most common type of homogeneous EIAs are also termed EMIT assays. In these assays the activity of the marker enzyme is either inhibited or enhanced directly as a consequence of formation of the Ab–At complex. Because enzyme activity is regulated in proportion to the amount of target At present in the incubation, enzyme activity can be determined immediately after equilibration without prior separation. This is accomplished by designing the At–E conjugate so that the activity of the enzyme is either significantly increased or decreased, compared to its activity in the unbound state, when the At–E complex is bound in the Ab–At complex. As a result, the level of enzyme product formed per unit of time will vary as a function of the amount of unconjugated At from the test sample that is present in the incubation media. If enzyme activity is inhibited by binding of the At–E complex to the Ab, the level of product formation will be directly proportional to the target Ab concentration. Alternatively, if formation of the At–E–Ab complex stimulates enzyme activity, product formation will vary inversely in relation to the level of target At in the incubation. In practice, most homogeneous EIAs require an incubation period during which equilibrium binding is reached prior to the addition of the enzyme substrate. This procedure prevents significant amounts of the enzyme product to be formed prior to regulation of enzyme activity by formation of the At–E–Ab complex. In some systems, however, it has been reported that At–E–Ab equilibrium is so rapid that the substrate can be included in the original incubation and equilibration, and measurement of the rate of product formation can occur simultaneously. Homogeneous IAs require that the addition of substrate and other reagents does not alter steady-state Ab–At binding.

The methods that are used to measure the rate of enzyme activity vary greatly. The sensitivity of enzyme product measurement, rather than Ab affinity, is usually the limiting factor in determining the sensitivity of an EIA. Many of the colorimetric and spectrophotometric methods that traditionally have been used to measure enzyme products in biochemistry are relatively insensitive (10^{-5} to 10^{-8} mol) and, therefore, the potential gains in sensitivity that result from enzyme amplification of the rate of Ab–At formation are largely canceled. The advantages of EIA over competing IA methods currently depend primarily on factors other than sensitivity, such as simplicity, stability of reagents, and low cost of detection instruments. Newer, more sensitive, methods for measuring enzyme activity are continually being developed, however (e.g., chemi- or bioluminescence, electrochemical detection), and the sensitivity that is ultimately possible by using EIAs is beginning to be more fully exploited.

Similar to the more complex heterogeneous EIAs, most of the multistep homogeneous EIA methods depend on the ability of a multiepitope target At to bind simultaneously to multiple Ab sites, a property not possessed by the small, haptenic molecules whose analysis is being reviewed here.

3.2.2. Enzyme Modulated Immunoassay

Another common type of EIA is termed enzyme modulated immunoassay (EMIA). In this context, an enzyme modulator is defined as any substance that will suppress or enhance the catalytic activity of the indicator enzyme. Inhibitory EMIA modulators are classical competitive, or noncompetitive, inhibitors of enzyme activity. Facilitory modulators, which are less commonly used in EMIA, are usually biochemical cofactors that are required by some enzymes in order to catalyze their characteristic biochemical reactions. A covalent At–modulator (At–M) conjugate is synthesized from the purified At and modulator, and the ability of this conjugate to perform its dual function forms the basis for EMIAs. The At portion, which must maintain its high affinity for the Ab binding site, competes in the usual fashion with free At from the test sample for a limited number of Ab binding sites. The At–M conjugate must also maintain its ability to modulate the indicator enzyme when in the unbound state. When bound in the Ab–At–M complex, its ability to affect the indicator enzyme must either be eliminated or greatly reduced. When these criteria are met, the degree to which activity of the indicator is modulated will be proportional to the amount of unconjugated At in the incubation mixture. If the At–M complex is an enzyme inhibitor, activity will increase as the level of free At increases, and conversely, if the complex is facilitatory, enzyme activity will increase as the level of free At increases. Because the level of enzymatic activity is maximal at these points, EMIA assays that employ an inhibitory modulator will be maximally sensitive when At levels are near the lower range of the assay, whereas assays employing an excitatory modulator will be maximally sensitive when the At concentration is near the highest range of the assay. When developing an EMIA procedure, comparison of the concentration of At that is to be determined with the affinity of the specific Ab for the target At can be used to determine whether a inhibitory or facilitory modulator will result in a more sensitive EMIA assay.

An acceptable enzyme modulator must have certain properties. The modulator must be capable of significantly altering the activity of the target enzyme at low modulator concentrations (10^{-6} M or below); it must be absent in the biological test samples; it must be able to modulate enzyme activity after it has been covalently linked to the At; and finally, the ability to alter enzyme activity must be abolished, or very significantly attenuated, when the At–M complex is bound to the Ab.

A recent development that can potentially improve EMIA technology significantly is the use of specific anti-enzyme Abs (AEnAbs) as inhibitory EMIA modulators, as first reported by Ngo and Lenhoff (1980). This is accomplished by immunizing a test animal with the purified EMIA indicator enzyme in order to develop an Ab, preferably a Mab, that will complex with high affinity to the enzyme. The use of AEnAbs has several very significant advantages in EMIA. First, they are potent specific, irreversible inhibitors of their target enzymes. Second, because the enzymes used in EIA are large immunogenic proteins, development of specific monoclonal AEnAbs is relatively easy. They can be conjugated with a target At using well-established chemical methods. Mabs directed against some of the enzymes that are used as indicators in EIAs are available commercially. AEnAbs have also made it possible to consider as EMIA indicators certain enzymes that otherwise are very suitable, but for which inhibitory modulators were previously unavailable. Finally, they have proven to be well suited to EMIA because the ability of the AEnAb–At to inhibit the indicator enzyme is completely inactivated by formation of the AEnAb–At–Ab complex. The use of AEnAbs increases greatly the potential range of applications of inhibitory EMIA techniques in many areas of neurobiology, and further development of such methods appears desirable.

3.2.3. Enzyme-Amplified Immunoassay

As discussed previously, despite the potential of EIA methods for attaining high sensitivity, a major disadvantage of most currently available EIA procedures is their relative lack of sensitivity, which results from the relatively low sensitivity of the methods that are presently used to measure enzyme activity. A recent advance in EIA methodology, enzyme-amplified immunoassay (EAIA), promises to increase significantly the sensitivity of these assays (Stanley et al., 1985). The rationale can be applied to any standard EMIT assay that uses alkaline phosphatase (AP) as the target enzyme. In a standard AP-EMIT assay, the enzyme product, NAD, is monitored spectrophotometrically as it is formed by dephosphorylation of NADP by AP activity. Alternatively, in an EAIA assay the NAD formed by AP activity is not measured directly, but rather functions as the essential cofactor that is required to initiate the activity of a second enzyme system that is present in the system. In this case the second enzyme system consists of an enzyme-catalyzed oxidation-reduction cycle that has an absolute requirement for NAD cofactor in order to function and yields an intensely colored formazon dye as one of its products. As outlined previously, EMIT assays are designed so that the activity of the enzyme, in this case AP, generates a reaction product (NAD) in direct proportion to the level of target At in the incubation. Each At molecule thereby catalyzes formation of multiple NAD molecules, and this constitutes the normal level of enzyme enhancement obtained using AP-EMIT. In the EAIA system, however, each NAD molecule now acts as an activator of the re-dox system that results in the formation of multiple molecules of the intensely colored dye molecule for each molecule of NAD formed by the first enzyme. Thus the second enzyme system in effect is able to "reamplify" the NAD-generating EMIT reaction in direct proportion to the original target At concentration. The authors (Stanley et al., 1985) were able to quantitate accurately 1×10^{-5} IU/L of TSH, which was 70 times more sensitive than a conventional ELISA. This sensitivity compared favorably to that attainable by RIA assays for TSH. The dye-producing re-dox system used appears to be adaptable to any current EIA that employs AP or other NAD-producing enzyme system.

Because of the specific requirements that must be met before an enzyme can be used is an EIA method, the number of enzymes that have been used is limited when compared to the total number of enzymes that have been adequately characterized and purified. First, the enzyme must have a high turnover rate, because this characteristic is a major factor in determining ultimate sensitivity of the assay, i.e., the higher the turnover rate of the enzyme, the more molecules of product that will be formed by a given amount of activated enzyme in a given time. Ideally, a relatively simple method that can measure the enzyme's activity with high sensitivity should be available. This requirement is not absolute. If an otherwise desirable enzyme is identified, a suitably sensitive assay could be developed, but obviously this would involve additional time. In addition to these properties, which determine the sensitivity of the EIA method, the enzyme must also be available in large quantities in a highly purified form at affordable cost and must be absent in the biological samples being tested. There is also an extensive literature describing the chemical methods used to attach enzymes to Ats, Abs, or solid supports. A comprehensive and detailed survey of these methods was recently published by Ishikawa et al. (1983).

3.3. Luminescence Immunoassay

The adaptation of the phenomenon of luminescence as a marker for IAs has significantly increased the level of sensitivity that is currently attainable by nonisotopic IAs. These assays (LIAs), which consist of an otherwise typical IA method, employ as the marker the light that is emitted by certain naturally occurring (bioluminescent) or synthetic organic (chemiluminescent) molecules under specific circumstances. Although there are specific differences in the chemistry of bio- and chemiluminescent compounds, for the purpose of this review they will be considered equivalent and will be referred to by the generic term luminescence. When a luminescent molecule is oxidized (hydrogen peroxide or NAD are usually the oxidant in LIA), the luminescent molecule is raised to an excited chemical state. The energy that accompanies its return to a ground state is emitted in the form of photons. The intensity of the emitted light can be measured with great sensitivity using relatively simple and inexpensive instruments. Depending on the nature of the assay, the light intensity will depend either on the amount of luminescent marker present or on the amount of oxidizing agent added to the system to catalyze light formation. In either case, LIAs are designed so that the light intensity generated by the bound fraction will be proportionate to the concentration of target At in the test sample. By exploiting these unique properties, a number of LIAs have attained sensitivities that approach those of RIAs (Seitz, 1984; Takayasu, 1985; Wannlund et al., 1980).

Improvements in both the precision and the sensitivity of LIAs resulted from development of the modern solid-state photomultiplier circuit. Analogous to enzyme amplification, a photomultiplier system can take a given amount of light and linearly amplify the resulting signal by a factor of 10^5 to 10^7. The factors that limit the theoretical detection limits of luminescence (Seitz, 1981) and its potential for application to clinically relevent assays (Whitehead et al., 1979) have been reviewed. A second factor that permits attainment of high sensitivity is the absence of extraneous light sources involved in measuring luminescence that contribute to background "noise." The only significant instrumental interference arises from spontaneous "dark" counts that are inherent in all photomultipliers. This source of interference can be reduced in modern instruments by using two or more photomultipliers. In a multicell system, the light emitted by the luminescent reaction will simultaneously activate both multipliers, whereas the "dark" events are random and are statistically unlikely to coincide in both detectors. Instruments employing multiple photomultiplier circuits are significantly more expensive than single channel "luminometers," however. The low cost and high sensitivity of the simple single-channel machines are, in fact, attractive features of LIA assays, and these instruments are adequate for all but the most sensitive applications of LIA.

As with other IA methods, LIAs can be designed in alternative forms. In one method the luminescent molecule, analagous to the radioactive molecule in RIA, is attached to the Ab or At. Following an appropriate heterogeneous IA procedure, the number of luminescent molecules, and therefore the light that will be emitted by the free and/or bound fraction, will be proportional to the target At concentration.

There are two major types of luminescent labels that can be used in this manner. In one category are molecules (e.g., luminal) that will be irreversibly oxidized and thereby consumed during the light-producing reaction. In this case light production, which is proportional to the amount of luminescent label present, will not be continuous and will decrease rapidly as the luminescent molecules are oxidized. Therefore the light-detection system must be sensitive enough to detect the level of light produced immediately after the oxidizing reaction is initiated. A consumable luminescent label is hypothetically similar to a radioisotope in which both the half-life *and* the initiation of decay are controllable by the investigator. The sensitivity of the luminescent marker–Ab conjugate, analagous to radioactive specific activity, can be increased by adding more luminescent molecules to the Ab molecule. Luminescent efficiency is equivalent to the radioactive half-life, and the light-gathering efficiency of the photomultiplier is analogous to the radioisotope counting efficiency.

The second category of luminescent labels acts as a catalyst for an energy transfer and will emit light after accepting energy from another energy donor molecule in the system. In this case the luminescent label is not consumed, and the rate of light production will depend both the number of luminescent molecules present and on the amount of the energy-providing consumable reactants that are added. Because the rate of emission from the luminescent marker can be increased by adding an increased energy supply, the light-detection system can have a lower absolute sensitivity. This method of increasing sensitivity, however, invokes serious handicaps. In order to maintain acceptable precision, it is essential that light production remain exactly proportional to the target At concentration. In order to attain this precision, the amount of added reactants and all other factors that affect the luminescence-inducing conditions, such as temperature, pH, and ionic concentration, must be rigorously controlled. The potential for increasing sensitivity is further compromised by the fact that this method of light production also results in significant levels of background luminescence that further decrease the degree of precision and accuracy of light production. Therefore any significant increase in sensitivity that might be attained by using nonconsumable luminescent compounds has not been exploited to date.

The LIA procedures discussed above represent the direct method, whereby the rate of light generation is determined by the number and efficiency of the luminescent molecules contained in the free or bound At fraction. Direct LIAs therefore are limited by the "luminosity" that can be built into an At–luminescent molecule conjugate. A more promising means to exploit the potential sensitivity and precision that is possible with luminescence methods is to use a product formed by a marker enzyme in an EIA system as the energy source to catalyze a luminescent reaction. Such a system would, in effect, be able to attain a double amplification of the At-generated marker signal. The EIA system represents the first amplification because multiple molecules of the enzyme product are generated by each activated enzyme molecule. The product is then used as the oxidative energy source to drive a luminescent reaction, resulting in light production that will be directly proportional to the target At concentration. Sensitivity is increased because the light that is produced is amplified a second time by the photomultiplier system. Takayasu et al. (1985) have recently published a combined enzyme-luminescent assay that used a β-D-galactosidase-EMIT assay that generated glucose as one of its products. The glucose served as the source of the substrate for a glucose oxidase system that generated H_2O_2 as a product. The H_2O_2 was then used as the energy source for a highly efficient peroxy-oxalate-H_2O_2-luminescent dye system. The ultimate result of the combined assay system was the activation of light emission by multiple molecules of a highly efficient luminescent compound for each molecule of target At in the test sample. The assay system was 10–50 times more sensitive than the comparable RIA for measuring phenytoin. Similar results have been obtained by modifying EIAs for 17-α-hydroxy progesterone (Arakawa et al., 1982) and thyrodine (Arakawa et al., 1985). Similarly Wannlund et al. (1980) were able to quantitatively measure subpicomolar levels of methotrexate using an LIA assay. Seitz (1984) has investigated a number of chemi- and bioluminescent systems and has estimated that enzyme-coupled luminescent assays are ultimately capable of measuring attamole (10^{-18} mol) concentrations of luminescent compounds. Luminescent methods that are able to measure femtogram levels of peroxidase already exist (Puget et al., 1977), and coupling these methods to peroxidase-containing EIAs should result in assays with exceptional sensitivities. Naturally occurring bioluminescence systems similar to the bacterial system used by Wannlund et al. (1982) and Wannlund and DeLuca (1983) have been isolated and purified, and they are among the most efficient light-emitting reactions known (Seitz, 1984).

The technical constraints that apply to EIAs also apply, in general, to LIA technology. The luminescent marker must be available in purified form so that it can be attached to an Ab or At. Care must be taken so that covalent attachment of the luminescent molecule to the Ab or At does not significantly diminish the luminescence efficiency or significantly reduce the affinity or specificity of the AT–luminiscent conjugate. The methodological details employed in a variety of specific types of luminescent assays have been evaluated and compared (DeLuca and McElroy, 1981; Kricka and Thorpe, 1981; Kricka and Carter, 1982; Wood et al., 1984; Burkot et al., 1985). LIA techniques do not appear to be exceptionally difficult to develop, they do not require expensive and/or specialized equipment, and there is no unusual hazard involved. Therefore, there do not appear to be significant technical obstacles that would prevent sensitive enzyme-coupled LIA techniques from being developed for a variety of drugs and other low mw endogeneous compounds that are of interest in psychopharmacology.

3.4. Electrochemical Immunoassay

As discussed previously, the sensitivity of most EIAs is limited by the minimum sensitivity of the method used to measure the product of the enzyme, rather that on the affinity of the specific Ab. Because the assay methods for many enzymes are not adequately sensitive, EIAs have been only marginally useful in biological problems that require high sensitivity. In the continuing search to identify methods that will increase the sensitivity of the relevant enzyme assays, the sensitive detection limits (10^{-13} to 10^{-16} mol) of modern electrochemical detectors (ECDs) have recently been employed as a method to measure EIA-generated enzyme products (Hechemy and Anaker, 1983). Some of these efforts have resulted in successful IA methods, and the evolving field has recently been reviewed by Heineman and Halsall (1985). These assays can also be divided into two classes. In one type, an "electroactive" molecule is used as a direct marker for the Ab or At. An "electroactive" molecule (i.e., a molecule that can be detected with high sensitivity by an ECD) is chemically conjugated to an Ab or At. Subsequently, an ECD is then used to measure the level of the electrochemical label that is found in the separated free and/or bound fraction. The alternative approach, which results in a more sensitive assay system, is to use a marker enzyme in an EIA system that will yield an "electroactive" enzyme product that can be measured with high sensitivity using ECD.

The latter method has resulted in EIA-ECD assays with sensitivities similar to those of EAIAs (i.e., 10^{-8} to 10^{-10} mol; Heineman and Halsall, 1985). Further improvements in sensitivity may be possible by increasing the electroactive "specific activity" of the marker molecule by attaching more than one electroactive molecule to each molecule of Ab or At or to use marker compounds that can be oxidized (or reduced) at more than one site in each molecule. A significant number of enzymes produce products that are "electroactive." Therefore a potential advantage of coupling ECD/EIA methods is that it might significantly expand the category of enzymes that could be considered as marker enzymes in ECD-EIA coupled systems. A number of enzymes have very high turnover rates and are otherwise desirable for use as an EIA marker enzyme (for example cholinesterase), but have not been used because of the lack of a suitably sensitive assay for their product(s). Using an ECD to measure the products of such enzymes with precision and sensitivity may allow some useful new enzymes to be used as EIA markers. For example, because of its very high turnover, cholinesterase would be capable of a very high level of amplification in an EIA system. By incorporating choline oxidase in a hypothetical cholinesterease-EIA-ECD system, the choline produced would be converted to an equivalent amount of H_2O_2, which in turn could be measured with high sensitivity by an ECD. Mathematical consideration of the turnover rate of cholinesterease suggests that if a cholinesterease-EIA/choline oxidase/H_2O_2/ECD system were developed, levels of target At as low as 10^{-14} mol could be measured accurately. Such a method is speculative, but illustrates a potential application of a combined EID-ECD technique.

3.5. Fluorescent Immunoassay (FIA)

Another major system used in IAs employs fluorescence labeling of Abs or Ats. Because these methods also depend on quantitative measurement of light, FIAs are able to exploit the same theoretical and technical advancements previously described for light monitoring in luminescent immunoassays. Because of the rapid rate at which FIA methods are evolving, descriptions of the details of specific FIAs would soon be of limited value. Alternatively, a brief overview of the general principles and of the innovative new FIA techniques is presented in order to acquaint the reader with the primary advantages or limitations of FIA. This background can then serve as the foundation for the literature search, which will be required for the reader to identify or develop an FIA method specifically suited to the needs of the intended analysis.

In certain molecular structures the energy that is acquired when a specific wavelength or photon of light is absorbed causes the molecule to be raised to an excited electronic state in which a number of the outer electrons are promoted to a higher energy level. If the energy that is released when the electrons return to their ground state is also emitted in the form of a photon, which is always of longer wavelength than the exciting energy, the molecule is said to fluoresce. In order to be suitable for use in IAs, a fluorescent compound must meet several minimum criteria. Fluorescent molecules are excited only by a specific wavelength of light. Therefore there must be a sufficiently large difference between the wavelengths of the excitation and the emission energy so that the latter can be measured without interference from the former. Second, because of technical aspects of light generation and measurement, the excitation and emission energies of the fluorophor must be in or near the visible regions of the spectrum in order to be practical for use in FIAs. Finally, the fluorescent efficiency of molecule must be high. Although the theoretical and technical properties that determine the fluorescent efficiency that a compound will exhibit under any given set of circumstances is beyond the scope of this discussion, the term is reasonably self-explanatory. Efficiency simply represents the amount of energy that is emitted in relation to the amount of energy absorbed. A variety of distinct structural classes of fluorescent compounds has been characterized and is widely used for markers in various fluorescent technologies, including immunoassay. Probably the most critical step that is unique to development of an FIA is developing a suitable method for covalently attaching the chosen fluorescent molecule to an Ab or At molecule without disrupting either the specificity or affinity of the immune molecule or disrupting the fluorescent characteristics of the marker molecule (*see* Gadow et al., 1984; Hemmilae and Loevgren, 1982; Hicks, 1984; Smith et al., 1981, for some recent review of methods for synthesis and use of fluorescent immunoassay reagents).

Similar to other light-dependent IAs, the ability of fluorometry to attain high sensitivity ($10^{-14}M$) depends on the high sensitivity of modern solid state photomultiplier systems that are able to measure minute amounts of light while generating very low levels background noise. In fluorescence, however, the rate and stability of the light that is emitted is completely dependent on the intensity and stability of the excitation energy. It is technically difficult and very costly to develop energy sources capable of maintaining absolutely constant outputs, especially when operating at high sensitivity. To overcome this potentially limiting factor, the ratio-recording fluorometer was developed. In these instruments, the intensity of the emission energy is continually compared to, and then is expressed as a ratio of, the excitation intensity. As a result, variability in the emission energy that would otherwise result from small, virtually unavoidable variations in excitation intensity are continually compensated for and no longer contribute to variability.

The application of FIA methods to biological samples is subject to the same limitations described previously for other light-dependent IA methods; for example, light scattering, absorbtion of excitation or emission energy by other biological constituents in the incubation media, contamination by naturally occurring fluorescent substances, and so on. FIAs cannot be employed for analysis of biological samples that contain significant levels of several commonly occurring endogenous fluorescent compounds unless sample preparation steps are employed to reduce or eliminate these interfering substances. The reader is referred to previous sections for additional details concerning the advantages and limitations of light-dependent IA methods. In addition to these general considerations, FIA methods are specifically subject to the phenomena of quenching. Quenching occurs when diverse factors such as temperature, pH, and autoxidation of the fluorophor by constituents in the incubation, cause a reduction in the fluorescence efficiency of the fluorophor–Ab or –At complex. Each fluorophor–Ab or –At complex has unique quenching characteristics that must be determined empirically so that optimal emission conditions can be established from the incubation mixture.

An expanding number of different FIA methods continue to be developed and a detailed listing of specific FIA methods suitable for measuring specific substances would rapidly become obsolete. A number of recently reported flourescent assays are listed in Table 3. In general, most of the previously described techniques that are used to separate the free and bound fractions in other types of heterogeneous IAs are also applicable to heterogeneous FIA methods. Following separation, the concentration of Ab- or At-bound fluorescent label in the free or bound fraction is then determined by measuring the intensity of the emitted light and is used to determine concentration of the target Ab or At in the incubation mixture.

As in other areas of IA, the greatest effort continues to be devoted to development of new and innovative homogeneous FIA methods or to increase the sensitivity of FIAs. A number of commercial homogeneous FIAs pioneered by Abbott Laboratories are based on polarization FIA, a technique that was first introduced by Dandliker et al. (1973) and Spencer et al. (1973). When the light source used to excite the fluorphor is polarized, the degree of polarization of the emitted light depends on the amount of random Brownian movement experienced by the absorbing molecule in the time span between absorption and emission. Brownian motion is

Table 3
References to Some Fluorescence Immunoassays for Psychotropic Drugs

Drug	Sample type	Reference
Barbiturates	Urine	Colbert et al. (1984)
Caffeine	Plasma	Pearson et al. (1984)
Pentobarbital	Serum	Li et al. (1984)
Phenobarbital	Serum	Dean et al. (1983)
Phenytoin	Serum	Dean et al. (1983)
Phenytoin	Serum	Kurtz et al. (1983)
Theophylline	Serum	Kurtz et al. (1983)

inversely related to molecular size. When a fluorophor is attached to a small molecule, a relatively high level of motion takes place that, in turn, reduces the degree of polarization of emitted light. If the photomultiplier in the fluorometer is designed to measure only light of a specific polarization, the amount of light that is detected when a small unbound fluorophor is excited therefore will be relatively small. The reagents in a polarization FIA are designed so that the small fluorescent-containing molecule will become bound to a large molecular weight molecule or to a solid support as a result of the specific Ab/At reaction. The bound fluorophor exhibits significantly decreased Brownian movement when it is attached to the large molecule. The reduced motion produces a specific incremental increase in the amount of polarized light that is emitted and detected by the photomultiplier, which is directly proportional to the extent of the specific Ab/At reaction. A major drawback of polarization methods has been that a relatively expensive and dedicated polarization fluorometer is required to run these assays. Extensive commercial development of polarization FIA methods that are applicable to quantitative analysis of numerous drugs and other biological substances, however, many of which are relevant to psychopharmacology, has resulted in significantly increased availability and steadily declining cost of the required instrumentation.

Another novel approach to FIA, introduced commercially by the Syva Corporation, involves the use of a quenching substance that will absorb the emitted photons and reduce the intensity of the fluorescent signal only when spatially located in proximity of the fluorophor. In the free, or unbound, state, only a small fraction of the randomly oriented quenching molecules are close enough to act as a quenching agent, and the emission intensity is proportional to the total amount of fluorophor present. The fluorophor and quench-containing reagents are designed so that the Ab/At reaction orients the quenching substance in a position to efficiently absorb the photons emitted by the fluorophor. The net result is that the light intensity emitted by the equilibrated incubation mixture will decrease in direct proportion to the amount of target Ab or At present in the incubation mixture. Because this type of assay depends on measuring a graded reduction in light intensity, rather than comparing light emission against a dark or zero background control, these assays have comparatively poor minimum sensitivities. The inverse of this system has also been investigated. In this case, the quenching agent is able to significantly inhibit fluorescence in the unbound state in the reaction mixture. The immunoreactive agents are designed so that the formation of specific Ab/At complex disrupts the ability of the quenching agent to absorb the emitted photons, and the emission intensity therefore increases, rather than decreases, in proportion to the target Ab or At concentration. This method also appears to be of limited sensitivity, however, because the concentration of quenching agent needed to significantly inhibit emission intensity in the unbound state is relatively high, and therefore a relatively high concentration of the target Ab or At is also required in order to produce a significant increase in emission intensity.

Since the primary limitation of many FIAs is lack of adequate sensitivity, it would appear that a simple and direct way to increase sensitivity would be the attachment of multiple molecules of a fluorophor to a single Ab or At molecule. The resulting increase in the "specific activity" of the indicator molecule would significantly enhance the light intensity resulting from each individual Ab/At reaction. "Concentration quenching," however, which is defined as the significant decrease in emission intensity that occurs when multiple molecules of a fluorophor are in close proximity of each other, is characteristic of most fluorophors and has effectively limited this approach. The use of Abs that can simultaneously bind to multiple fluorophor-labeled At molecules is also technically feasible, but as discussed previously, the complex kinetics exhibited by multivalent Ab/At reactions are not well suited to quantitative IAs. It has been observed recently that certain classes of fluorophors, for example umbelliferone-type analogs, might be less subject to concentration quenching. If this proves to be true, high "specific activity" umbelliferone-labeled Ab or At molecules might be used to increase significantly the sensitivity FIAs.

Recent technical advances in instrumentation have enabled the phenomena of "photochemical bleaching" to be explored as a means to improve the sensitivity of FIAs. In conventional fluorometry, the solution of fluorescent molecules is excited with a fixed wavelength of light of constant intensity. Within microseconds the rate at which the fluorophors are being excited to higher electronic states is balanced by the rate of deactivation, or fluorescence, and a steady state is achieved. The intensity of the observed fluorescence depends both on the number of fluorophors present and on the intensity of the excitation energy. Within reason, therefore, the use of high excitation energy allows as little as 10–12 mol of many fluorophors to be measured with conventional equipment. Each fluorescent compound, however, has a threshold of excitation intensity above which the molecule will be permanently altered and, after an initial release of photons, it ceases to fluoresce. This phenomena is called "photochemical bleaching," and the bleaching time of a substance has been found to be inversely proportional to its fluorescence halflife (i.e., the time span between excitation and emission). Although the destruction of the fluorophor would normally be a serious disadvantage, photochemical bleaching can be exploited in several ways. Hirschfield (1976a,b) demonstrated that when a high "specific activity" multiple fluorophor labeled marker molecule, which would normally be of limited use because it exhibited marked concentration quenching, was excited with a pulsed laser of sufficient energy to produce complete photochemical bleaching, the molecule emitted the maximal theoretical yield of photons possible, i.e., concentration quenching was eliminated. Although the intensity of the emission produced by pulsed bleaching of a multiple labeled fluorophor is very intense, however, the time span of the emission is short and the technical aspects of monitoring the signal are critically important. A closely related and more practical application of photochemical bleaching, first reported by Hirshfield in 1979, is currently the subject of increasing research interest. In this system time-delayed light measurement corresponding to time delays in post-bleaching light emission that result from differences in fluorescent lifetime of different fluorescence molecules in the incubation mixture are employed as a method to reduce or eliminate interfering background fluorescence.

The principle of time-delayed measurements of emission energy that can exploit differences in fluorescent half-life have also been applied to conventional fluorometry. Most of the biological substances that normally interfere with FIAs by contributing to the background fluorescent signal have significantly shorter fluorescent lifetimes than that of the labeled fluorophors. Therefore, rather than using constant excitation energy, the excitation of the incubation mixture is accomplished by using a precisely timed series of intense, but very short, pulses of energy that are followed by a specific predetermined time delay before the emission intensity is monitored. The delay is precisely timed to permit the rapidly emitted light from the nonspecific fluorescence substances to pass without being measured and to coincide with the maximal emission from the labeled fluorophor. In addition to reducing interference, the high intensity and reproductivity of the excitation laser results in significantly improved minimum sensitivity to be attained by FIAs. The recent advent of reliable, moderately priced pulsed visible wavelength lasers and precisely timed, solid-state electronic circuitry will undoubtedly lead to continued rapid development of this promising technique, though dedicated instruments will always be required.

4. Conclusion

In summary, it is clear from the current pace of development that immunochemistry-based techniques will play a major role in the quantitative analysis of endogenous and exogenous substances relevant to psychopharmacology and neurochemistry. It is difficult to conceive of other analytical techniques that can compete with the broad, yet exquisitely specific, mammalian immune response. Moreover, it appears that continuing advances in protein chemistry and molecular genetics will allow the manipulation of Abs or the immune system itself, in ways that will allow even further exploitation of the specificity and sensitivity of IAs. The handicaps resulting from the low sensitivity displayed by earlier nonradioactive IAs are quickly receding before powerful and innovative techniques such as enzyme amplification, enzyme channeling assays, and advances in luminescent and fluorescent techniques. Moreover, coupling immunoenzyme amplification with the sensitive methods developed for measuring fluorescent- or luminescent enzyme-generated products permits the advantages of the two methods to be effectively combined. IAs that generate fluorescent products have taken nonradioactive IA methods to sensitivities approaching 10^{-16} mol. These limits exceed those of all but the most sensitive radioimmunoassays and would have been considered impossible not many years ago. Nor have advances been confined to increasing the specificity or the sensitivity of esoteric research assays. By virtue of their relative speed and simplicity, immunoassay methods are ideally suited to the clinical environment, and one no longer needs to be an immunologist or analytical expert to successfully use immunoassay methods. For example, progress in dry chemistry and Ab and At immobilization techniques have led to the commercial development of impressively straightforward "mix and measure" prepackaged IAs. The advent of simple IAs that could be used on an outpatient basis for purposes such as therapeutic drug monitoring do not seem far beyond the horizon. A simple 10-min enzyme amplification test strip for qualitative determination of the presence or absence of morphine in human urine has already been reported (Littman et al., 1983).

References

Abrams P. G., Knost J. A., Clarke G., Wilburn S., Oldham R. K., and Foon K. A. (1983) Determination of the optimal cell lines for development of human hybridomas. *J. Immunol.* **131**, 1201–1204.

Ahuja M. M. S. and Pillai N. K. (1983) Radioimmunoassay: Technical aspects and clinical applications (1). *J. Assoc. Physicians India* **31**, 453–456.

Aoki K. and Kuroiwa Y. (1983) Enzyme immunoassay for methamphetamine. *J. Pharmacobiodyn.* **6**, 33–38.

Arakawa H., Maeda M., and Tsuji A. (1982) Chemiluminescence enzyme immunoassay of 17-alpha-hydroxyprogesterone using glucose oxidase and bis(2,4,6-trichlorophenyl)-oxalate-fluorescent dye system. *Chem. Pharm. Bull.* **30**, 3036–3040.

Arakawa H., Maeda M., and Tsuji A. (1985) Chemiluminescence enzyme immunoassay for thyroxin with use of glucose oxidase and a bis(2,4,6-trichlorophenyl)-oxalate-fluorescent dye system. *Clin. Chem.* **31**, 430–434.

Aravagiri M., Hawes E. M., and Midha K. K. (1984) Radioimmunoassay for the sulfoxide metabolite of trifluoperazine and its application to a kinetic study in humans. *J. Pharm. Sci.* **73**, 1383–1387.

Aravagiri M., Hawes E. M., and Midha K. K. (1985) Radioimmunoassay for the 7-hydroxy metabolite of trifuorperazine and its application to a kinetic study in human volunteers. *J. Pharm. Sci.* **11,** 1196–1202.

Atassi M. Z. (1978) Precise determination of the entire antigenic structure of lysozyme: Molecular features of protein antigenic structures and potential of "surface-stimulation" synthesis in a powerful new concept for protein binding sites. *Immunochemistry* **15,** 909–936.

Bechtel W. D. and Weber K. H. (1985) Brotizolam radioimmunoassay: Development, evaluation and application to human plasma samples. *J. Pharm. Sci.* **74,** 1265–1269.

Beiser S. M., Butler V. P., and Erlanger B. F. (1976) *Textbook of Immunopathology* vol. 1, 2nd Edn., Grune and Stratton, Inc., Orlando, FL.

Boulianne G. L., Hozumi N., and Shulman M. J. (1984) Production of functional chimeric mouse/human antibody. *Nature* **312,** 643–464.

Bourne R. C., Robinson J. D., and Teale J. D. (1978) A simple radioimmunoassay for plasma diazepam and its application to single dose studies in man. *Br. J. Clin. Pharmacol.* **63,** 371P.

Browning J. L., Harrington C. A., and Davis C. M. (1985) Quantification of reduced haloperidol by radioimmunoassay. *J. Immunoassay* **6,** 45–66.

Burkot T. R., Wirtz R. A., and Lyon J. (1985) Use of fluorodinitrobenzene to identify monoclonal antibodies which are suitable for conjugation to periodate-oxidized horseradish peroxidase. *J. Immunol. Meth.* **84,** 25–31.

Butz R. F., Smith P. G., Schroeder D. H., and Findlay J. W. (1983) Radioimmunoassay for bupropion in human plasma: Comparison of tritiated and iodinated radioligands. *Clin. Chem.* **29,** 462–465.

Cais M., Dani S., and Shimoni M. (1983) A novel non-centrifugation radioimmunoassay for cannabinoids. *Arch. Toxicol.* **6,** 105–113.

Carrasquillo J. A., Krohn K. A., and Beaumier P. (1984) Diagnosis and therapy of solid tumors with radiolabelled antibodies. *Cancer Treat. Rep.* **68,** 317–328.

Chard T. (1982) *An Introduction to Radioimmunoassay and Related Techniques* Elsevier Biomedical, Amsterdam.

Colbert D. L., Smith D. S., Landon J., and Sidki A. M. (1984) Single-reagent polarization fluoroimmunoassay for barbiturates in urine. *Clin. Chem.* **30,** 1765–1769.

Cooper D. S., Bode H. H., Nath B., Saxe V., Maloof F., and Ridgway E. C. (1984) Methimazole pharmacology in man: Studies using a newly developed radioimmunoassay for methimazole. *J. Clin. Endocrinol. Metab.* **58,** 473–479.

Crothers D. M. and Metzger H. (1972) The influence of polyvalency on the binding properties of antibodies. *Immunochemistry* **9,** 341–357.

Dandliker W. B., Kelly R. J., Dandliker J., Farquhar J., and Levin J. (1973) Fluorescence polarization immunoassay. Theory and experimental method. *Immunochemistry* **10**, 219–227.

Dean K. J., Thompson S. G., Burd J. F., and Buckler R. T. (1983) Simultaneous determination of phenytoin and phenobarbital in serum or plasma by substrate-labelled fluorescent immunoassay. *Clin. Chem.* **29**, 1051–1056.

DeBlas A. L., Sangameswaran L., Haney S. A., Pank D., Abraham C. J., and Rayner C. A. (1985) Monoclonal antibodies to benzodiazepines. *J Neurochem.* **45**, 1748–1753.

DeLuca M. A. and McElroy W. D. (1981) Proceedings of the Symposium on Bioluminescence and Chemiluminescence, in *Basic Chemistry and Analytical Applications* (DeLuca M. A. and McElroy W. D., eds.) Academic, New York.

Dick H. M. (1985) Monoclonal antibodies in clinical medicine. *Br. Med. J.* **291**, 762–764.

Dixon W. R. and Crews T. (1978) Diazepam: Determination in micro samples of blood, plasma and saliva by radioimmunoassay. *J. Anal. Toxicol.* **2**, 210–213.

Dixon R., Hsiao J., Taaffe W., Hahn E., and Tuttle R. (1984) Nalmefene: Radioimmunoassay for a new opioid antagonist. *J. Pharm. Sci.* **73**, 1645–1646.

Dixon W. R., Earley J., and Postma E. (1975) Radioimmunoassay of chlordiazepoxide in plasma. *J. Pharm. Sci.* **64**, 937–939.

Dixon W. R., Lucek R., Earley J., and Perry C. (1979) Chlordiazepoxide: A new, more sensitive and specific radioimmunoassay. *J. Pharm. Sci.* **68**, 261.

Drost R. H., Plomp T. A. A., and Maes R. A. A. (1984) Applicability of the EMIT-st serum and urine assays for screening of barbiturates and benzodiazepines in serum. *Arch. Formacol. Tocicol.* **10**, 184–190.

Dwenger A. (1984) Radioimmunoassay: An overview. *J. Clin. Chem. Clin. Biochem.* **22**, 883–894.

Edwards D. J., Popouski Z., Baumann T. J., and Biurns B. A. (1986) Specific ^{125}I radioimmunoassay for morphine. *Clin. Chem.* **32**, 157–158.

Engleberg N. C. and Eisenstein B. I. (1984) The impact of new cloning techniques on the diagnosis and treatment of infectious diseases. *N. Eng. J. Med.* **311**, 892–901.

Faraj B. A., Israili Z. H., Knight N. E., Smissman E. E., and Pazdernik D. J. (1976) Specificity of an antibody directed against D-amphetamine: Studies with rigid and nonrigid analogs. *J. Med. Chem.* **19**, 2–25.

Flurkey K., Bolger M. B., and Linthicum D. S. (1985) Preparation and characterization of antisera and monoclonal antibodies to serotonergic and dopaminergic ligands. *J. Neuroimmunol.* **8,** 115–127.

Fraser A. D. (1983) Clinical evaluation of the EMIT salicylic acid assay. *Ther. Drug. Mon.* **5,** 331–334.

Gadow A., Fricke H., Strasburger C. J., and Wood W. G. (1984) Synthesis and evaluation of luminescent tracers and hapten-protein conjugates for use in luminescence immunoassays. *J. Clin. Chem. Clin. Biochem.* **22,** 337–347.

Gelbke H. P., Schlicht H. J., and Schmidt G. G. (1977) Radioimmunological screening and gas chromatographic identification of diazepam in blood and serum. *Arch. Toxicol.* **38,** 295–305.

Goding J. W. (1980) Antibody production by hybridoma. *J. Immunol. Meth.* **39,** 285–295.

Goldstein S. A. and Vunakis H. V. (1981) Determination of fluphenazine related phenothiazine drugs and metabolites by combined high-performance liquid chromatography and radioimmunoassay. *J. Pharmacol. Exp. Ther.* **217,** 36–43.

Grabinski P. Y., Kaiko R. F., Walsh T. D., Foley K. M., and Houde R. W. (1983) Morphine radioimmunoassay specificity before and after extraction of plasma and cerebrospinal fluid. *J. Pharm. Sci.* **72,** 27–30.

Gross S. J., Grant J. D., Wing R., Schuster R., Lomax P., and Campbell D. H. (1984) Critical antigenic determinants for production of antibodies to distinguish morphine, heroin, codeine and dextromethorphan. *Immunochemistry* **11,** 453–456.

Guesdon J. L., Chevrier D., Mazie J-C., David B., and Aurameas S. (1986) Monoclonal anti-histamine antibody: Preparation, characterization and application to enzyme immunoassay of histamine. *J. Immunol. Meth.* **87,** 69–78.

Haauman J. J., Deen C., Krose C. J. M., Zulstra J. J., Coolen J., and Radu J. (1985) Monoclonal antibodies in cytology, a jungle of pitfalls. *Immunol. Today* **5,** 56–58.

Hahn E. F., Lahita R., Kreek M. J., Duma C., and Inturrisi C. E. (1983) Naloxone radioimmunoassay: An improved antiserum. *J. Pharm. Pharmacol.* **35,** 833–836.

Hart H. E. and Greenwald E. B. (1979) Scintillation Proximity Assay (SPA)—A new method of immunoassay. Direct and inhibition mode with human albumin and rabbit human anti-albumin. *Mol. Immunol.* **16,** 265–267.

Hausmann E., Kohl B., vonBoehmer H., and Wellhoner H. H. (1983) False-positive EMIT indication of opiates and methadone in doxylamine intoxication. *J. Clin. Chem. Clin. Biochem.* **21,** 599–600.

Hawes E. M., Aravagari M., Dulos A., Rauw G. A., and Stonkus M. D. (1983) Radioimmunoassays for phenothiazine drugs and their major metabolites in plasma. *Prog. Neuropsychopharmacol. Biol. Psychiat.* **7,** 709–714.

Hawes E. M., Shetty H. U., Cooper J. K., Rauw G., McKay G., and Midha K. K. (1984) Radioimmunoassay for psychotropic drugs. III. Synthesis and properties of haptens for trifluoperazine and fluphenazine. *J. Pharm. Sci.* **73,** 247–250.

Hayashi S., Kurooka S., Arisue K., Kohda K., and Hayashi C. (1983) Semi-automated continuous-flow enzyme immunoassay for anti-epileptic drugs in serum. *Clin. Chem.* **29,** 1790–1792.

Hechemy K. E. and Anaker R. L. (1983) Coating of polymeric surfaces for immunoassay by forced absorption technique. *J. Immunoassay* **4,** 147–157.

Heineman W. R. and Halsall H. B. (1985) Strategies for electrochemical immunoassay. *Anal. Chem.* **57,** 1321–1331.

Hepler B., Weber J., Sutheimer C., and Sunshine I. (1984a) Homogeneous enzyme immunoassay of acetaminophen in serum. *Am. J. Clin. Pathol.* **81,** 602–610.

Hepler B., Wutheimer C., and Sunshine I. (1984b) Combined enzyme immunoassay-LCEC method for the identification, confirmation and quantitation of opiates in biological fluids. *J. Anal. Toxicol.* **8,** 78–90.

Herbert W. J. (1973) Passive haemagglutinatim with special reference to the tanned cell technique, in *Handbook of Experimental Immunology* 2nd Edn. (Weir D. M., ed.) Blackwell Scientific Publications, Oxford.

Hicks J. M. (1984) Fluorescence immunoassay. *Human Pathol.* **15,** 112–116.

Hirschfeld T. (1976a) Optical microscopic observation of single small molecules. *Appl. Opt.* **15,** 2965–2966.

Hirschfeld T. (1976b) Quantum efficiency independence of the time integrated emission from a fluorescent molecule. *Appl. Opt.* **15,** 3135–3139.

Hirschfeld T. (1979) Fluorescence background discrimination by prebleaching. *J. Histochem. Cytochem.* **27,** 96–101.

Hubbard J. W., Midha K. K., McGilveray I. J., and Cooper J. K. (1978a) Radioimmunoassay for psychotropic drugs. I. Synthesis and properties of haptens for chlorpromazine. *J. Pharm. Sci.* **67,** 1563–1571.

Hubbard J. W., Midha K. K., Cooper J. K., and Charette C. (1978b) Radioimmunoassay for psychotropic drugs. II. Synthesis and properties of haptens for tricyclic antidepressants. *J. Pharm. Sci.* **67,** 1571–1578.

Irving J., Leeb B., Foltz R. C., Cook C. E., Bursey J. T., and Willette P. E. (1984) Evaluation of immunoassays for cannabinoids in urine. *J. Anal. Toxicol.* **8,** 192–196.

Ishikawa E., Imagawa M., Hashida S., Yoshitake S., Hamaguchi Y., and Ueng T. (1983) Enzyme-labeling of antibodies and their fragments for enzyme immunoassay and immunohistochemical staining. *J. Immunoassay* **4**, 209–327.

James K., Boyd J. E., Micklem L. R., Ritchie A. W. S., Dawes J., and McClelland D. B. L. (1984) Monoclonal antibodies, their production and potential in clinical practice. *Scott Med. J.* **29**, 67–83.

Jeffcoate S. L. and Searle J. E. (1972) Preparation of a specific antiserum to estradiol-17 coupled to protein through the B-ring. *Steroids* **19**, 181–188.

Jones A. B., Elsohly H. N., and Elsohly M. A. (1984) Analysis of the major metabolite of delta-9-tetrahydrocannabinol in urine. *J. Anal. Toxicol.* **8**, 252–254.

Jorgenson A. (1978) A sensitive and specific radioimmunoassay for cis-(Z)-flupenthixol in human serum. *Life Sci.* **23**, 1533–1542.

Karnes H. T., Iafrate P., Gudat J. C., and Hendeles L. (1984) Evaluation of a rapid immunoassay for monitoring serum pentobarbitol concentrations. *Am. J. Hosp. Pharm.* **41**, 2642–2646.

Kato Y., Paterson A., and Lagone J. J. (1984) Monoclonal antibodies to the therapeutic agent methotrexate: Production, properties and comparison with polyclonal antibodies. *J. Immunol. Meth.* **67**, 321–326.

Kawashima K., Dixon R., and Spector S. (1975) Development of radioimmunoassay for chlorpromazine. *Eur. J. Pharmacol.* **32**, 195–202.

Keenan A. M., Harbert J. C., and Larson S. M. (1985) Monoclonal antibodies in nuclear medicine. *J. Nucl. Med.* **26**, 531–537.

Kennaway D. J. (1983) Radioimmunoassay of 5-methoxy tryptophol in sheep plasma and pineal glands. *Life Sci.* **32**, 2461–2469.

Knoll E. and Wisser H. (1984) Problems in the development of radioimmunoassay of catecholamines. *J. Clin. Chem. Clin. Biochem.* **22**, 741–749.

Kohler G. and Milstein C. (1975) Continuous cultures of fused cells secreting antibody of predefined specificity. *Nature* **256**, 495–497.

Kohler-Schmidt H. and Bohn G. (1983) Radioimmunoassays for determination of prazepam and its metabolites. *Forensic Sci. Intl.* **22**, 243–248.

Kosfeld B. M. H., Harbauer G., Grill H. J., and Pollow K. (1985) Radioimmunoassay of human serotonin. *J. Clin. Chem. Clin. Biochem.* **23**, 657–662.

Kozbar D. and Roder J. C. (1983) The production of monoclonal antibodies from human lymphocytes. *Immunol. Today* **4**, 72–79.

Krakauer H. (1985) Clinical applications of monoclonal antibodies. *Eur. J. Clin. Microbiol.* **4**, 1–9.

Kricka L. J. and Carter T. J. N. (1982) *Clinical and Biochemical Luminescence* (Kircka L. J. and Carter T. J. N., eds.) Marcel Dekker, New York.

Kricka L. J. and Thorpe G. H. G. (1981) Luminescent immunoassay. *Ligand Rev.* **3,** 17–24.

Kubasik N. P. and Sine H. E. (1979) A further comparison of radioassay results for serum and plasma. *Clin. Chem.* **25,** 135–136.

Kulpmann W. R., Gey S., Beneking M., Kohl B., and Oellerich M. (1984) Determination of total and free phenytoin in serum by non-isotopic immunoassays and gas chromatography. *J. Clin. Chem. Clin. Biochem.* **22,** 773–779.

Langone J. J., Gjika H. B., and VanVanakis H. (1973) Nicotine and its metabolites: Radioimmunoassay for nicotine and cotinine. *Biochemistry* **212,** 5025–5030.

Larson S. M., Carrasquillo J. A., and Reynolds S. C. (1984) Radioimmunodetection and radioimmunotherapy. *Cancer Invest.* **2,** 263–381.

Law B., Mason P. A., Moffat A. C., and King L. J. (1984) A novel ^{125}I radioimmunoassay for the analysis of beta-9-tetraphydrocannabinol and its metabolites in human body fluids. *J. Anal. Toxicol.* **8,** 14–18.

Li P. K., Lee J. T., and Schreiber R. M. (1984) Rapid quantification of pentobarbital in serum by fluorescence polarization immunoassay. *Clin. Chem.* **30,** 307–308.

Lindgren J. E. and Holmstedt B. (1983) Guide to the analysis of phencyclidine and its metabolites in biological material. *Arch. Toxicol.* **6,** 61–73.

Litman D. J., Lee R. H., Jeong H. J., Tom H. K., Stiso S. N., Sizto N. C., and Ullman E. F. (1983) An internally referenced test strip immunoassay for morphine. *Clin. Chem.* **29,** 1598–1603.

Ly B. and Hegesippe M. (1984) Labelling of drug antigens. *Int. J. Nucl. Med. Biol.* **11,** 79–83.

Mach J. P., Buchegger F., and Forni M. (1981) Use of radiolabelled monoclonal anticarcinoembryonic antigen for the detection and localization of diverse cancers by external photoscanning. *Immunol. Today* **2,** 239–249.

Maguire K. P. and Fulton A (1980) The Measurement of Anti-Anxiety Drugs in Plasma, in *Handbook of Studies on Anxiety* (Burrows G. D. and Davies B., eds.) Elsevier/North Holland Biomedical, Amsterdam.

Manchon M., Verdier M. F., Pallud P., Vialla A., Beseme F., and bienvenu J. (1985) Evaluation of EMIT-TOX enzyme immunoassay for the analysis of benzodiazepines in serum: Usefulness and limitations in an emergency laboratory. *J. Anal. Toxicol.* **9,** 209–212.

Manz B., Kosfeld H., Harbauer G., Grill H. J., and Pollow K. (1985) Radioimmunoassay of human serum serotonin. *J. Clin. Chem. Clin. Biochem.* **23,** 657–662.

Marullo S., Hoebeke J., Guillet J-G., and Strosbeg A. D. (1985) Structural analysis of the epitope recognized by a monoclonal antibody directed against tricyclic antidepressants. *J. Immunol.* **135,** 471–477.

Mason P. A., Bal T. S., Law B., and Moffat A. C. (1983) Development and evaluation of a radioimmunoassay for the detection of amphetamine and related compounds in biological fluids. *Analyst* **108,** 603–607.

Mason P. A., Rowan K. M., Law B., Moffat A. C., Kitner E. A., and King L. A. (1984) Development and evaluation of a radioimmunoassay for analysis of body fluids to determine the presence of tricyclic antidepressant drugs. *Analyst* **109,** 1213–1215.

McCarron M. M., Walberg C. B., Soares J. R., Gross S. J., and Basett R. C. (1984) Detection of phencyclidine usage by radio immunoassay of saliva. *J. Anal. Toxicol.* **8,** 197–201.

McKay G., Cooper J. K., Hawes E. M., Hubbard J. W., Martin M., and Midha K. K. (1985) Therapeutic monitoring of chlorpromazine. II. Pitfalls in whole blood analysis. *Ther. Drug Monit.* **1,** 372–477.

McKay G., Rauw G. A. J., Stonkus M. D., Dulos R. A., Gedir R. G., Hawes E. M., and Midha K. K. (1984) Radioimmunoassay for trimeprazine in human plasma. *J. Pharmacol. Meth.* **12,** 203–211.

Miceli J. N., Aravind M. K., and Ferrell W. J. (1984) Analysis of caffeine: Comparison of the manual enzyme multipled immunoassay (EMIT), automated EMIT and HPLC procedures. *Therap. Drug Mon.* **6,** 344–347.

Michiels M., Hendriks R., and Heykants J. (1983) Radioimmunoassay of the new opiate analgesics alfentanil and sufentanil. Preliminary pharmacokinetic profile in man. *J. Pharm. Pharmacol.* **35,** 86–93.

Midha K. K. and Charette C. (1980a) Radioimmunoassay for total doxepin and N-desmethyldoxepin in plasma. *Commun. Psychopharmacol.* **4,** 121–129.

Midha K. K., Charette C., Cooper J. K., and McGilveray I. J. (1980b) Comparison a new GLC-AFID method with a GLC-MS selected ion monitoring technique and a radioimmunoassay for the determination of plasma concentrations of imipramine and desipramine. *J. Anal. Toxicol.* **4,** 237–243.

Midha K. K., Cooper J. K., and Hubbard J. W. (1980c) Radioimmunoassay for fluphenazine in human plasma. *Commun. Psychopharmacol.* **4,** 107–114.

Midha K. K., Hawes E. M., Rauw G., McVittie J., McKay G., Cooper J. K., and Shetty H. U. (1983a) Radioimmunoassay for prochlorperazine in human plasma. *Ther. Drug. Monit.* **5,** 117–121.

Midha K. K., Hubbard J. W., Cooper J. K., and Mackonka C. (1983b) Stereospecific radioimmunoassays for 1-ephedrine and d-ephedrine in human plasma. *J. Pharm. Sci.* **72,** 736–739.

Midha K. K., Hawes E. M., Rauw G., McKay G., Cooper J. K. and Shetty H. U. (1983c) Development of radioimmunoassays for trifluoperazine and their application to metabolic studies of the drug. *J. Pharmacol. Meth.* **9,** 283–293.

Midha K. K., Hubbard J. W., Cooper J. K., Hawes E. M., Fournier S., and Yeung P. (1981a) Radioimmunoassay for trifluoperazine in human plasma. *Br. J. Clin. Pharmacol.* **12,** 189–193.

Midha K. K., Mackona C., Cooper J. K., Hubbard J. W., and Yeung P. K. F. (1981b) Radioimmunoassay for perphenazine in human plasma. *Br. J. Clin. Pharmacol.* **1,** 85–88.

Midha K. K., Korchinski E. D., Roscoe R. M. H., Hawes E. M., Cooper J. K., and McKay G. (1984) Relative bioavailability of a commercial trifluoperazine tablet formation using a radioimmunoassay technique. *J. Pharm. Sci.* **73,** 261–263.

Midha K. K., Loo J. C. K., Charette C., Rowe M. L., Hubbard J. W., and McGilversay I. J. (1978) Monitoring of therapuetic concentrations of psychotropic drugs in plasma by radioimmunoassays. *J. Anal. Toxicol.* **2,** 185–192.

Midha K. K., Loo J. C. K., Hubbard J. W., Rowe M. L., and McGilversay I. J. (1979) Radioimmunoassay for chlorpromazine in plasma. *Clin. Chem.* **25,** 166–168.

Milstein C., Wright B., and Cuello C. (1983) The discrepancy between the cross-reactivity of a monoclonal antibody to serotonin and its immunohistochemical specificity. *Mol. Immunol.* **20,** 113–123.

Mizuchi A., Kitagawa N., and Miyachi Y. (1983) Regional distribution of sultopride and sulpiride in rat brain measured by radioimmunoassay. *Psychopharmacology* **81,** 195–198.

Moore R. A., Baldwin D., Allen M. C., Watson P. J. Q., Bullingham R. E. S., and McQuay H. J. (1984) Sensitive and specific morphine radioimmunoassay with iodine label: Pharmacokinetics of morphine in man after intravenous administration. *Ann. Clin. Biochem.* **21,** 318–325.

Morrison S. L., Johnson M. J., Herzenberg L. A., and Oi V. T. (1984) Chimeric human antibody molecules: Mouse antigen-binding domains with human constant region domains. *Proc. Natl. Acad. Sci. USA* **81,** 6851–6855.

Nakamura R. M. (1983) Monoclonal antibodies: Methods and clinical laboratory applications. *Clin. Physiol. Biochem.* **1,** 160–172.

Ngo T. T. (1983) Enzyme modulator mediated immunoassay (EMMIA). *Int. J. Biochem.* **15,** 583–590.

Ngo T. T. and Lenhoff H. M. (1980) Enzyme modulators as tools for the development of homogeneous enzyme immunoassays. *FEBS Lett.* **116,** 285–288.

Oellerich M. (1980) Enzyme immunoassays in clinical chemistry: Present status and trends. *J. Clin. Chem. Clin. Biochem.* **18,** 197–208.

Oellerich M. (1984) Enzyme-immunoassay: A review. *J. Clin. Chem. Clin. Biochem.* **22,** 895–904.

Oi V. T. and Morrison S. L. (1986) Chimeric antibodies. *Biotechniques* **4,** 214–220.

Pape B. E., Cary P. L., and Clay L. C. (1983a) Reactivity of EMIT serum barbiturate reagents: Potential application for the quantitation of selected barbiturates. *Therap. Drug Mon.* **5,** 473–477.

Pape B. E., Cary P. L., Clay L. C., and Godolphin W. (1983b) Pentobarbital quantitation using EMIT barbiturate assay reagents: Application to monitoring of high-dose pentobarbital therapy. *Therp. Drug Mon.* **5,** 467–471.

Pearson S. and Smith J. M. (1984) Measurement of plasma caffeine concentrations by substrate labelled fluoroimmunoassay. *Ann. Clin. Biochem.* **21,** 208–212.

Peskar B. and Spector S. (1973) Quantitative determination of diazepam in blood by radioimmunoassay. *J. Pharmacol. Exp. Ther.* **186,** 167–172.

Price C. P. (1984) Analytical techniques for therapeutic drug monitoring. *Clin. Biochem.* **17,** 52–56.

Puget K., Michelson A. M., and Ayrameas S. (1977). Light emission techniques for the microestimation of femtogram levels of peroxidase. *Anal. Biochem.* **79,** 447–456.

Robinson K. and Smith R. N. (1983) Methadone radioimmunoassay: Two simple methods. *J. Pharm. Pharmacol.* **35,** 566–570.

Robinson K. and Smith R. N. (1985) Radioimmunoassay of tricyclic antidepressant and some phenothiazine drugs in forensic toxicology. *J. Immunoassay* **6,** 11–22.

Robinson J. D., Morris B. A., and Marks V. (1975) Development of a radioimmunoassay for etorpine. *Res. Commun. Chem. Path. Pharmacol.* **10,** 1–8.

Roome A. J. and Reading C. L. (1984) The use of Epstein-Barr virus transformation for the production of human monoclonal antibodies. *Exp. Biol.* **43,** 35–55.

Sandouk P., Scherrman J. M., and Bourdon R. (1984) Combined liquid-solid chromatography and radioimmunoassay for determination of morphine in human fluids. *J. Pharmacol. Meth.* **11,** 227–237.

Sarandis S., Pichon R., Miyada D., and Pirkle H. (1984) Quantitation of pentobarbital in serum by enzyme immunoassay. *J. Anal. Toxicol.* **8,** 59–60.

Sato S. and Yamamoto I. (1983) Enzyme immunoassays for β-adrenoreceptor blocking agent, befunolol and its main metabolite, M1. *J. Immunoassay* **4,** 351–371.

Schonherr O. T. and Houwink E. H. (1984) Antibody engineering, a strategy for the development of monoclonal antibodies. *Antonie Van Leeuwenhoek* **50,** 597–623.

Schuttler J. and White P. F. (1984) Optimization of the radioimmunoassays for measuring fentanyl and alfentanil in human serum. *Anesthesiology* **61**, 315–320.

Schwartz R. H., Hayden G. F., and Riddle M. (1985) Laboratory detection of marijuana use: Experience with a photometric immunoassay to measure urinary cannabinoids. *Am. J. Diseases Child.* **139**, 1093–1096.

Scoggins B. A., Maguire K. P., Norman T. R., and Burrows G. D. (1980) Measurement of tricyclic antidepressants. 1. A Review of methodology. *Clin. Chem.* **26**, 5–17.

Seitz W. R. (1981) Chemiluminescence and bioluminescence analysis: Fundamentals and biomedical applications. *CRC Crit. Rev. Anal. Chem.* **13**, 1–58.

Seitz W. R. (1984) Immunoassay labels based on chemiluminescence and bioluminescence. *Clin. Biochem.* **17**, 120–125.

Sherman-Gold R., Dundai Y., Fogelfeld L., and Fuchs S. (1983) Production of a high affinity antiserum to benzodiazepines. *J. Immunol.* **4**, 135–146.

Sidki A. M., Pourfarzaneh M., Rowell F. J., and Smith D. S. (1982) Direct determination of phenobarbital in serum or plasma by polarization fluoroimmunoassay. *Ther. Drug. Monit.* **4**, 397–403.

Smith D. S., Al-Hakiem M. H-H., and Landon J. (1981) A review of fluoroimmunoassay and immunofluorometric assay. *Ann. Clin. Biochem.* **18**, 253–274.

Sonsalla P. K., Bridges R. R., Jennison T. A., and Collins C. (1984) An evaluation of the EMIT-st assay for the detection of tricyclic antidepressant drugs in plasma or serum. *Clin. Toxicol.* **22**, 63–67.

Spector S. (1971) Quantitative determination of morphine in serum by radioimmunoassay. *J. Pharmacol. Exp. Ther.* **178**, 253–258.

Spector S. and Parker G. W. (1970) Morphine: Radioimmunoassay. *Science* **168**, 1347–1348.

Spencer R. D., Toledo F. B., Williams B. T., and Yoss N. L. (1973) Design, construction and two applications for an automated flow-cell polarization fluorometer with digital readout: Enzyme-inhibitor (atitrypsin) assay and antigen-antibody (insulin-insulin antiserum) assay. *Clin. Chem.* **19**, 838–844.

Stanley C. J., Johannsson A., and Self C. H. (1985) Enzyme amplification can enhance both the speed and the sensitivity of immunoassays. *J. Immunol. Meth.* **83**, 89–95.

Steinbusch H. W. M., Verhofstad A. A. J., and Joosten W. J. (1982) Antibodies to serotonin for neuroimmunocytochemical studies. *Histochem. Cytochem.* **30**, 756–759.

Takada S. T., Naito T., Hama K., Noma T., and Hanjo T. (1985) Construction of chimeric processed immunoglobin genes containing mouse variable and human constant region sequences. *Nature* **314**, 452–454.

Takayasu S., Maeda M., and Tsuji A. (1985) Chemiluminescent enzyme immunoassay using β-D-galactoidase as the label and the bis (2,4,6-trichlorophenyl) oxalatefluorescent dye system. *J. Immunol. Meth.* **83**, 317–325.

Tamaki Y., Fukuda M., Kishida T., and Takahashi N. (1983) Solid phase micro-ELISA for methamphetamine. *Jpn. J. Legal Med.* **37**, 417–420.

Terauchi Y., Watari S., Utsui Y., and Sekine Y. (1984) Sensitive determination of haloperidol in human serum by radioimmunoassay. *Radioisotopes* **33**, 376–379.

Tiefenauer L. X. and Andres R. Y. (1984) Prevention of bridge binding effects in haptenic immunoassay systems exemplified by an iodinated radioimmunoassay for melatonin. *J. Immunol. Meth.* **74**, 293–298.

Tischio J., Hetyei N., and Patrick J. (1984) Bromperidol radioimmunoassay: Human plasma levels. *J. Pharm. Sci.* **73**, 546–548.

Udenfriend S., Gerber L. D., Brink L., and Spector S. (1985) Scintillation proximity radioimmunoassay utilizing ^{125}I-labelled ligands. *Proc. Natl. Acad. Sci. USA* **82**, 8672–8676.

Valentino K. L., Winter J., and Reichardt L. F. (1985) Applications of monoclonal antibodies to neuroscience research. *Ann. Rev. Neurosci.* **8**, 199–232.

Vitetta E. S., Krolick K. A., Miyama-Inaba M., Cushley W., and Uhr J. W. (1983) Immunotoxins: A new approach to cancer therapy. *Science* **219**, 644–650.

Walberg C. B., McCarron M. M., and Schulze B. N. (1983) Quantitation of phencyclidine in serum by enzyme immunoassay: Results in 405 patients. *J. Anal. Toxicol.* **7**, 106–110.

Wannlund J. and DeLuca M. (1983) Bioluminescent immunoassays. *Meth. Immunol.* **92**, 426–432.

Wannlund J., Arari J., Levine L., and DeLuca M. (1980) A bioluminescent immunoassay for methotrexate at the subpicomole level. *Biochem. Biophys. Res. Comm.* **96**, 440–446.

Wannlund J., Egghart J., and DeLuca M. (1982) Bioluminescent Immunoassays: A Model System for Detection of Compounds at the Attamole Level, in *Serano Symposium* Raven, New York.

Watson A. T., Manno J. E., and Manno R. R. (1983) Quantitation of barbiturates by a modification of the EMIT-tox serum barbiturate assay. *J. Anal. Toxicol.* **7**, 257–261.

Wellington P. and Rutkowski R. B. (1982) Rapid estimation of plasma pentobarbital levels by an enzyme-immunoassay. *Ther. Drug Monit.* **4**, 319–324.

Whitehead T. P., Kricka L. J., Carter T. J. N., and Thorpe G. H. G. (1979) Analytical luminescence: Its potential in the clinical laboratory. *Clin. Chem.* **25**, 1531–1546.

Wiles D. H. and Franklin M. (1978) Radioimmunoassay for fluphenazine in human plasma. *Br. J. Clin. Pharmacol.* **5**, 265–268.

Willey K. P. (1985) Simple and complex antibody reactions in radioimmunoassay and the predictions of assay characteristics. *J. Immunol. Meth.* **84**, 343–358.

Wood W. G., Fricke H., Haritz J., Gadow A., Krausz H-S., Tode B., Strasburger C. J., and Scriba P. C. (1984) An evaluation of four different luminescence immunoassay systems: CELIA (chemiluminescent immunoassay), SPLAT (solid-phase antigen luminescence technique), ILMA (immunoluminometricassy) and ILSA (immunoluminometric labelled second antibody). *J. Clin. Chem. Clin. Biochem.* **22**, 349–356.

Yalow R. S. and Berson S. A. (1959) Assay of plasma insulin in human subjects by immunological methods. *Nature* **184**, 1648–1649.

Yeung P. K. F., Hubbard J. W., Cooper J. K., and Midha K. K. (1983) A study of the kinetics of chlorpromazine sulfoxide by a specific radioimmunoassay after a single oral dose of chlorpromazine in healthy volunteers. *J. Pharmacol. Exp. Ther.* **226**, 833–838.

Yeung P. K. F., McKay G., Ramshaw I. A., Hubbard J. W., and Midha K. K. (1985) A comparison of two radioimmunoassays for 7-hydroxychlorpromazine: rabbit polyclonal antibodies vs. mouse monoclonal antibodies. *J. Pharmacol. Exp. Ther.* **233**, 816–822.

Zalcberg J. R. (1985) Monoclonal antibodies to drugs: Novel diagnostic and therapeutic reagents. *Pharmacol. Ther.* **28**, 273–285.

Zysset T., Wahllander A., and Preisig R. (1984) Evaluation of caffeine plasma levels by an automated enzyme-immunoassay (EMIT) in comparison with a high-performance liquid chromatographic method. *Ther. Drug Mon.* **6**, 348–354.

Drug Screening for Benzodiazepines, Antidepressants, and Neuroleptics

Graham R. Jones and Peter P. Singer

1. Introduction

It might be said that "drug screening" is one of the most overused, misunderstood, and poorly performed tests in the armament of medical science. Ideally, a drug screen should be a single test that can accurately detect and identify all known drugs and toxins present in a small amount of blood or urine, using readily available technology and performed by staff who have only rudimentary analytical training.

In reality, however, a drug screen is composed of numerous tests, the collective effectiveness of which depends on several factors including: (1) equipment available, (2) precise methods used, their accuracy and sensitivity, (3) dosage, metabolism, and chemical structure of the drugs of interest, (4) type of specimens submitted, and (5), most important, the experience of the analytical personnel.

Because of these variables, a drug screen can *never* be truly comprehensive. There are always drugs in a pharmacological class that for reasons of unique structure or high potency are difficult to detect.

For example, the daily dose of trifluoperazine or fluphenazine may be less than 5 mg, whereas for chlorpromazine the daily dose can exceed 1000 mg. In fact, many of the phenothiazines are difficult to detect in blood because of their low dosage and extensive metabolism. Conversely, unique color reactions make phenothiazine metabolites relatively easy to detect in urine by thin-layer chromatography (TLC), even if the parent drug is absent because of extensive biotransformation. Even for a procedure like gas chromatography (GC), precise details like the solvent used, buffer pH, and choice of column phase will affect specificity and sensitivity. Matrix effects and interference from endogenous compounds are a constant problem.

Not least, the efficiency and scope of a drug screening system is *always* directly proportional to the experience of the people running it. With the possible exception of automated immunoassays, drug screening methods should always be run by personnel who perform those tests on a regular, ongoing basis and never by people who "rotate" through a laboratory area or who try to run drug screens as a minor or occasional part of their normal work. The reason is that any test, particularly those involving chromatography, is subject to interference by endogenous substances and exogenous contaminants (e.g., plasticizers, food additives). It requires considerable experience to recognize these "artifacts" and to judge what can be ignored and which GC peaks or TLC spots may be significant. The recognition of a wide variety of "common" [over-the-counter (OTC) and prescription] medications and of metabolite patterns is equally important.

Quite apart from the knowledge and expertise required to interpret chromatograms, chemists and technologists should know the *limitations* of their tests. In this regard, the purpose of the drug testing is very important in deciding which methods are suitable. For example, it is perfectly reasonable to request a drug screen to find out if a patient is taking a neuroleptic or benzodiazepine medication that has not been prescribed. If the drug is detected and properly identified, then the patient can be confronted and counseled. It is very difficult, however, to use a drug screen to assess whether a patient is regularly taking a medication that *has been* prescribed. The analyst must know whether that drug is detectable in a particular specimen (e.g., blood or urine) at the concentration expected from the dosage used. For practical purposes it is unlikely that haloperidol, for example, would be detected in any TLC- or GC-based drug screen, since therapeutic levels are in the low ng/mL range for blood. Haloperidol is also thermally labile. Similarly, even massive "therapeutic" doses of trifluoperazine may be difficult to detect because the drug is so rapidly and extensively metabolized.

It is also important to realize that there is almost never a predictable relationship between the concentration of a drug or its metabolites in urine and the corresponding blood level, or indeed a particular therapeutic effect. Another important factor is the terminal half-life of a drug. The simple detection of a drug does not mean it is being taken according to a physician's instructions, but simply that it has been taken at some point prior to the test. Several drugs, including some of the benzodiazepines and neuroleptics, have elimination half-lives that are measured in days. Consequently, some of these drugs or their metabolites can be detected in the urine for weeks after dosage has been stopped.

In an era of growing sophistication in laboratory techniques, there are literally hundreds of methods published as "comprehensive" drug screens or group-specific screens. Any "drug screen" method is a compromise, however, of the several factors discussed above. Even the *number* of specimens to be screened each day has an important bearing on the choice of method. For example, any GC method takes a minimum of 12–15 min chromatography time per sample, plus the initial work-up (i.e., extraction) time and is therefore unsuitable for mass screening unless a considerable amount of equipment is available. For example, only 60–80 samples can be screened using one GC instrument, assuming a mean time of 15 min per specimen extract, even if it is run 24 h/d with an auto-injector. On the other hand, automated immunoassay procedures are ideally suited for screening large numbers of specimens, although these methods may be less comprehensive and are often only group-specific.

It is the purpose of this chapter to outline the advantages and disadvantages of the various methods for screening and to indicate, when possible, those drugs that can be detected by each of these methods.

2. Drug Screening Methods

2.1. Immunoassays

Immunoassay techniques are gaining a good reputation for drug screening purposes. Generally, capital costs are low to moderate, although the cost per test can range from only a few cents for an in-house developed radioimmunoassay (RIA) method, to a $3–5 per sample (reagents only) for some of the more expensive commercial fluorescence polarization immunoassay (FPIA) tests.

The EMIT® (enzyme multiplied immunoassays) chemistries probably require the least expensive equipment with capital costs of only $3000–6000 for semi-automated techniques. FPIA methodology can require an expensive, automated analyzer ($40,000–50,000), although lease/loan terms can usually be negotiated with reagent contracts. Alternatively, RIA and radioreceptor assay (RRA) techniques require at least a gamma counter ($15,000–30,000) for ^{125}I-based assays and a scintillation counter ($30,000–40,000) for tritium-based assays.

Immunoassay sensitivity is usually good, however, and very little, if any, workup is required for the specimen, making the technique ideal for automation. A large number of samples can be processed by one person and with far less training than for most chromatographic methods.

The disadvantages of immunoassays are that they may cross-react with undesired drugs and are subject to nonspecific interferences and even deactivation of the enzyme or antibody. Furthermore immunoassays are far more sensitive to matrix changes than are the chromatography methods. Of all the techniques available, however, FPIA (e.g., Abbott TDx® system) is remarkably robust and can often be used on quite badly hemolyzed blood specimens without further treatment.

One other disadvantage of immunoassay screens is their inability to distinguish drugs within a class or to estimate the amount present without knowing the precise identity of the drugs. For example, whether a given assay response is caused by, say, an "overdose" or "toxic" quantity of a potent drug or a therapeutic level of a less potent drug cannot be determined. This is not the case with receptor assays (e.g., RRA), however, which usually give a response that is quantitatively proportional to the inherent pharmacologic activity (e.g., anxiolytic) of the sample, i.e., a product of molar amount and potency.

By their nature, immunoassays designed for one class of drugs give no information about another class. Therefore many assays may have to be run in order to cover a wide range of drugs. There will always be examples of drugs within a pharmacological class that are structurally different and therefore will cross react poorly, if at all.

Furthermore, depending upon the purpose of the drug screen, it is usually important to confirm a positive assay result by another method. For forensic purposes, this should normally be a *non*-immunoassay method and, preferably, mass spectrometry. Valid confirmations may also be made, however, by suitable GC, TLC, or high-performance liquid chromatography (HPLC) methods.

2.1.1. Enzyme Immunoassay

Homogenous enzyme immunoassays (EMIT®, Syva Diagnostics, Division of Syntex, PO Box 10058, Palo Alto, CA) have been commercially available for at least 15 years. For a detailed description of the technology and principles involved, readers are referred to the manufacturer's literature (Syva Diagnostics) and Maggio (1980).

Briefly, however, the assay is based on the rate of conversion of NAD to NADH by the enzyme glucose-6-phosphate dehydrogenase (G6PDH), measured spectrophotometrically at 340 nm. Antibody and NAD substrate are added to the specimen, whereupon antibody binds to any drug present. G6PDH-tagged drug is then added, and binds to any remaining antibody sites, therefore reducing the G6PDH activity. Thus the amount of unbound G6PDH is directly proportional to the amount of analyte drug present in the specimen and therefore proportional to the rate of conversion of NAD to NADH.

One of the major disadvantages of EMIT® systems is that they are prone to nonspecific interference from exogenous substrates (e.g., salt added to urine) or from nonideal specimens (e.g., hemolyzed plasma). Since changes in optical density are measured directly, even a small amount of hemoglobin in plasma can reduce the efficiency and sensitivity of the assay. By their nature, EMIT systems tend to be less sensitive than RIA or FPIA systems.

2.1.1.1. BENZODIAZEPINES.   The antibody in the EMIT-TOX® serum assay for benzodiazepines was raised toward nordiazepam, but cross-reacts well with diazepam and oxazepam in addition to many other benzodiazepines. Since many of the "classical" benzodiazepines form nordiazepam as a long-acting major metabolite (e.g., diazepam, chlordiazepoxide, clorazepate, ketazolam, prazepam, halazepam), their use can be effectively detected. The system is of limited value for most of the lower dose benzodiazepines, however, such as lorazepam, triazolam, and alprazolam, since plasma levels are too low after therapeutic doses. Even the detection of some of the intermediate-dose benzodiazepines that do not form nordiazepam, such as bromazepam and nitrazepam, is difficult except after moderately large doses.

The Syva EMIT® urine benzodiazepine screen is based on the detection of oxazepam, the major metabolite of nordiazepam and, consequently, of all benzodiazepines that form nordiazepam. Although many of the parent benzodiazepines cross-react well with the antibody in the urine assay, its effectiveness for detecting non-oxazepam-forming drugs is uncertain. Considering the extremely high percentage of urinary oxazepam, which exists as the glucuronide conjugate, it is likely that the conjugate cross-reacts in the assay to some extent. Similarly, it is possible, even likely, that many other benzodiazepine metabolites also cross-react, although very little has been published in this regard. Readers are referred to cross-reactivity studies published for both the EMIT® urine and serum benzodiazepine screening kits (Poklis, 1981; Pegon et al., 1982; Slightom et al., 1982; Manchon et al., 1985). Inconsistencies occur in the apparent cross-reactivities reported by different research groups, although it is quite possible that this may be caused as much by batch-to-batch changes in the assay as by different study conditions.

It is probably safe to assume that the EMIT® urine assays are far less likely to detect use of the low-dose drugs such as lorazepam and triazolam after therapeutic doses, however, even if their novel metabolites cross-react well. The cut-off for the urine-based screen is given by Syva as 0.7 μg/mL (based on oxazepam) and 0.3 μg/mL (based on diazepam) for the serum screen.

2.1.1.2. ANTIDEPRESSANTS.   Three separate EMIT® systems are available for tricyclic antidepressants (TCAs). Two kits are designed for the quantitative assay of amitriptyline and nortriptyline in plasma and of imipramine and desipramine in plasma. A third, slightly less sensitive kit is designed as a broader, semiquantitative screen for toxic levels of TCAs in plasma. Since most of the TCAs have similar chemical structures and a similar dosage range, most are easily detected by this method in plasma (e.g., amitriptyline, nortriptyline, imipramine, desipramine, doxepin, trimipramine, protriptyline). The hydroxy metabolites also cross-react, and so the serum assay may also be effective for urine, which may *not* contain high concentrations of the parent drug or the desalkyl metabolites. The EMIT® TCA quantitative assay methods for plasma include a procedure for removal of the hydroxy metabolites by solid-phase extraction, if this is desired.

More problematic is the fact that other nontricyclic nonantidepressant drugs, such as chlorpromazine and cyclobenzaprine (Schroeder et al., 1986), may cross-react. One other disadvantage is that the pharmacologically similar antidepressants such as maprotiline and amoxapine cross-react relatively poorly in the assay. None of the EMIT® screens are likely to detect any of the monoamine oxidase inhibitor (MAOI) antidepressants at either therapeutic or toxic levels.

2.1.2. Fluorescence Polarization Immunoassay (FPIA)

FPIA drug assays (e.g., Abbott Diagnostics TDx®) are quickly gaining acceptance as both specific assays and for drug-class screening. In principle, for screening kits, a suitable example of the drug class (e.g., nordiazepam for the benzodiazepines) is tagged with fluorescein, which absorbs blue light and emits green fluorescence about 4 ns later. Unbound drug is distinguished from bound drug by the loss of plane polarization. Since the sample is usually diluted 1:10 as part of the assay, and only light emitted at right angles to the source is measured, the TDx® assays and screens are relatively unaffected by even moderately severe hemolysis.

Currently, TDx® benzodiazepine screens are available for urine and plasma, in addition to a plasma-based assay for tricyclic antidepressants. A TDx® assay for neuroleptics is not available at this time.

2.1.2.1. BENZODIAZEPINES.   The TDx® urine kit was developed as a "class screen" for those benzodiazepines that are metabolized to nordiazepam and oxazepam. In fact, the antibody cross-reacts 50–150% with a wide range of 1,4-benzodiazepines (e.g., diazepam, lorazepam, bromazepam, nitrazepam, flurazepam, and temazepam) and the newer triazolobenzodiazepines (e.g., triazolam, alprazolam). There is some evidence that the antibody may also cross-react with many of the hydroxylated and possibly conjugated metabolites (Baselt, 1984). In fact, the major limiting factor with the class benzodiazepine immunoassays, such as the TDx® system, is the extremely wide dosage range and resulting blood and urine levels. For example, the dosage of triazolam is 0.125–0.5 mg, whereas for chlordiazepoxide the dosage is normally 10–25 mg. Accordingly, peak plasma levels of triazolam resulting from therapeutic doses are generally less than 20 μg/L, whereas chlordiazepoxide plasma levels up to at least 2000 μg/L are not uncommon. In other words, a class benzodiazepine screen will only indicate if the drug or its metabolite may be present, but not which one. Consequently, it is impossible to quantitatively interpret benzodiazepine screen results. One cannot distinguish, say, a moderately high level of triazolam from the "residue" of a diazepam dose ingested 2d or more previously.

The Abbott TDx® system is, in the opinion of the authors, inherently more sensitive toward the newer low-dose benzodiazepines than is the Syva EMIT® system. The cut-off for the urine screen is stated as 0.2 mg/L although it may be useful at levels as low as 0.05 mg/L if the laboratory is able to perform satisfactory confirmatory methods (e.g., GC or MS).

2.1.2.2. ANTIDEPRESSANTS.   A semiquantitative serum-based screen for the TCAs is available (TDx®, Abbott Diagnostics, Chicago, IL). The kit is standardized against imipramine, with a low calibrator of 75 ng/mL. The antibody cross reacts with all the major tricyclic antidepressants and their desalkyl metabolites with efficiencies of 50–110%. The Abbott TDx® tricyclic antidepressant assay will tolerate a remarkable degree of hemolysis without seriously affecting its usefulness.

2.1.3. Radioimmunoassay (RIA)

For RIA, as with any immunoassay, an antibody must be raised against the analyte. The more specific the antibody is for that drug, the more specific the assay. RIA procedures intended as screens are generally "class-specific," and those intended for accurate quantitation tend to be drug- or analyte-specific. To a large extent, the potential use of an RIA method may depend on whether a particular laboratory already has a radioisotope lisence and either a gamma counter or a liquid scintillation counter.

One significant disadvantage of RIA procedures is the requirement to separate bound from unbound drug as an integral part of the assay procedure. This may be accomplished by using dextran-coated charcoal, precipitation methods, florisil absorption, or antibody-coated tubes.

Since radioactivity may be measured at extremely low levels, however, RIA is inherently more sensitive than other immunoassay procedures. Commercial RIA tests are now available for the benzodiazepines, TCAs, and some of the neuroleptics.

2.1.3.1. BENZODIAZEPINES.   Recently, Roche Diagnostics (Nutley, NJ) has released an ^{125}I-based "class assay" for the detection of benzodiazepines in urine. Although the cut-off for the assay is stated as 100 µg/L, the sensitivity is 5 µg/L (95% confidence). The assay cross-reacts with a wide range of benzodiazepines, including oxazepam, diazepam, nordiazepam, triazolam, and alprazolam.

2.1.3.2. TRICYCLIC ANTIDEPRESSANTS.   Wien Laboratories (Succasunna, NJ) markets a kit for the measurement of TCAs in serum (Tri-Cy®) based on [^{3}H]-imipramine. It is more tedious to run than either the EMIT® or TDx® immunoassay TCA kits, but is intended primarily as a quantitative procedure. The assay cross-reacts reasonably well with both the parent drug and dealkylated metabolites, although, as with the other immunoassays, it may cross-react weakly with the hydroxy metabolites and some of the neuroleptic drugs.

2.1.3.3. NEUROLEPTICS.   Janssen Pharmaceuticals markets an RIA kit (HAL-RIA-200) for the determination of haloperidol in plasma, saliva, and urine. Apparently, haloperidol metabolites do not cross-react significantly. Although the assay is based on [^{3}H]-haloperidol, the other butyrophenones (bromperidol, droperidol, moperone, and trifluperidol) all have a high degree of cross-reactivity, making it useful as a screen. The sensitivity of the assay is stated as 0.5 μg/L.

Janssen also market a similar RIA kit for pimozide, a low-dose neuroleptic that is difficult to measure by GC or HPLC.

2.1.4. Radioreceptor Assay (RRA)

The RRA method is, potentially, an extremely powerful screening tool for drugs in blood or plasma. If the specific receptor for a drug can be isolated (not necessarily pure, but free of other receptors), it can be used in place of an antiserum preparation in an assay. The principle of RRA is similar to RIA, except for the fact that the receptor preparation is used in place of the antiserum.

Since the degree to which a drug binds to the receptor is proportional to the potency or pharmacological activity of that drug (within certain limitations), the RRA method is equally sensitive to a low-dose, strongly binding drug as to a high-dose, weakly binding drug. For example, triazolam at a level of 20 μg/L and nordiazepam at a level of 1000 μg/L might give a similar response since they are pharmacologically roughly equivalent. In this respect, a RRA method differs from a RIA method in that the latter gives no information about potency or pharmacological activity, unless the *identity* of the drug is known and no other cross-reacting material is present.

It is also clear that RRA methods may be of limited use for detecting drugs in urine, since pharmacologically inactive (and therefore receptor inactive) metabolites predominate and the parent drug is frequently absent.

2.1.4.1. BENZODIAZEPINES.   To the authors' knowledge, commercial RRA kits are not available for benzodiazepines or antidepressants, although several methods have been described in the literature (e.g., Aaltonen and Scheinin, 1982; Jockemsen et al., 1982; Lund, 1981). Crude benzodiazepine receptor suspensions are usually prepared from rat brain homogenates and are stable for many months, especially if freeze-dried. Both [^{3}H]-diazepam and [^{3}H]-flunitrazepam are commercially available (Amersham, Arlington Heights, IL).

2.1.4.2. NEUROLEPTICS.   Wellcome Diagnostics (Dartford, England) markets a commercial RRA kit for the neuroleptic drugs. A preparation of dopamine receptors on calf caudate membrane is used with [^{3}H]-spiperone as the radiotracer. The kit is designed to detect and measure a wide range of neuroleptics in serum. Serum levels are expressed as "neuroleptic units (NU)/L," with reference to a haloperidol standard (1 NU of dopamine receptor blocking activity is produced by 1 nmol of haloperidol). Sensitivity of the assay is 8.5 NU/L.

This method has many advantages in that it can detect drugs such as thiothixene, pimozide, haloperidol, and flupenthixol, which are difficult to detect and measure by chromatographic procedures. The method suffers from two main disadvantages, however. It measures "dopamine receptor binding affinity" rather than true "neuroleptic activity." For example, thioridazine has a high binding affinity and gives NU/L values higher than expected considering the pharmacological activity observed. Also, TCAs have a significant affinity for dopamine receptors and may give a "false positive" in the absence of neuroleptics. Other groups of drugs do not appear to interfere, however.

2.2. Chromatographic Methods

2.2.1. Thin-Layer Chromatography (TLC)

TLC has traditionally been one of the major techniques employed in drug screening (Baselt, 1980; Chamberlain, 1985). Capital and operating costs are low, and up to 20 specimens may be screened using one plate. Several plates may be run simultaneously to cope with a heavy workload. In general, commercial plates give more reproducible results than do user-made plates, unless the user is very experienced. A certain amount of skill and experience is required, not only in applying extracts to the TLC plates, but particularly for interpretation of the results.

Some companies such as Whatman publish application notes for the use of their plates to separate groups of drugs such as benzodiazepines or narcotics (e.g., Whatman TLC application note 156).

One particularly useful commercial system for drug screening is marketed under the name Toxi-Lab® (Analytical Systems, Division of Marion Scientific, Kansas, MO). The "plate" is made from compressed silicate fiber and is already impregnated with 26 drug standards distributed over four different channels. A specimen extract (e.g., of urine) is concentrated onto a silicate disk in a porcelain well or an aluminum cup and then "plugged" into the bottom of the plate. Although the concentration stage is not 100% efficient, it overcomes the problem of large diffuse spots, since the disk is of a finite size. The plate is developed in the normal way, and the drugs visualized by Mandelin's reagent, a water wash, UV light, and Dragendorff's reagent. The entire process of extraction through to visualization only takes 45–50 min for 1–10 specimens. The Toxi-Lab® system is particularly useful because it is marketed with a compendium of data characterizing over 100 drugs and their metabolites (R_F values, color at each stage), including photographs of representative chromatograms.

2.2.1.1. BENZODIAZEPINES.   TLC is generally a poor technique for the detection of benzodiazepines in blood or plasma. Even the relatively high-dose drugs like diazepam and oxazepam give only weak fluorescence spots on most systems, except at very high, toxic levels. The lower-dose benzodiazepines such as lorazepam, alprazolam, and triazolam are, for practical purposes, almost impossible to detect by TLC at therapeutic levels. There are some exceptions, such as nitrazepam and flunitrazepam, in which specific high-performance thin-layer chromatography (HPTLC) methods have been devised for quantitation (Haefelfinger, 1975, 1979).

All benzodiazepines are very extensively biotransformed and consequently difficult to detect by TLC without hydrolysis of the glucuronide metabolites. One of the few exceptions is flurazepam, which gives a characteristic metabolite pattern with the Toxi-A TLC system. In fact, the most widely used screening methods have involved acid hydrolysis of the benzodiazepine metabolites to the corresponding benzophenones (Roets and Hoogmartens, 1980; Peel and Perrigo, 1980; de Silva et al., 1976). These products have good TLC properties and usually fluoresce strongly under UV light. There are three disadvantages to the method, however. First, several benzodiazepines give rise to the same benzophenones; second, the triazolo benzodiazepines (e.g., triazolam, alprazolam) do not hydrolyze to form benzophenones and so are very difficult to detect; and, third, many other drugs and endogenous compounds in urine hydrolyze to give fluorescent products.

2.2.1.2. ANTIDEPRESSANTS.   Most of the antidepressants except the MAOIs are relatively easy to detect by TLC in urine either directly, or as the desalkyl or hydroxylated metabolites. The Toxi-Lab® system is particularly useful for this group of drugs since the R_F values and color reactions of the parent drugs and the major metabolites are characteristic. Most of the antidepressants, however, are relatively difficult to detect in blood or plasma by TLC methods at therapeutic levels (0.05–0.3 mg/L).

The only MAOI that can sometimes be detected by TLC is tranylcypromine, in urine, but not in blood.

2.2.1.3. NEUROLEPTICS.   The higher-dose phenothiazines are easy to detect by TLC in urine as the metabolites, although often the parent compounds are absent. Chlorpromazine, thioridazine, and methotrimeprazine, in particular, can give quite characteristic metabolite patterns. The relatively low-dose phenothiazines (e.g., trifluoperazine, fluphenazine, and perphenazine), however, are a much more difficult problem; at best they usually only give a very weak "pink phenothiazine" colored spot at a very low R_F value. The phenothiazines are rarely present at a high enough concentration in blood and plasma to be detected by routine TLC drug screening methods.

The butyrophenone neuroleptics (e.g., haloperidol) are extremely difficult to detect in either urine or blood by TLC.

2.2.2. High Performance Liquid Chromatography (HPLC)

The number of papers published on the use of HPLC for the quantitation of benzodiazepines, antidepressants, and neuroleptics in blood or plasma has increased enormously in the past few years. This has been largely because of improvements in HPLC pumps, the quality and reproducibility of HPLC columns, and the smaller size and higher quality of UV detector cells.

HPLC has been used relatively little for drug screening, however, especially for basic and low-dose neutral drugs. One of the major drawbacks of HPLC is that the retention time or retention volume of a drug does not give an analyst any more information than a retention time does on a GC system. Furthermore, the resolving power of HPLC for a mixture of many drugs with a single system is less than that of a mediocre capillary GC system. The inherent separating power of HPLC for only a few drugs is potentially much greater than for GC, however, because of the greater variety of "mobile phases" that may be used. Capillary HPLC columns are available, but are very prone to overloading and contamination with particulate material.

One further, related problem is the 20–30 min required to complete a programmed solvent gradient that covers a useful polarity range.

Diode array detectors have increased the power of HPLC as a screening tool, but UV spectra are often pH-dependent. Neither are they as characteristic as, say, a mass spectrum. One major advance that might make HPLC more attractive as a screening method is the improvement being made in the HPLC/mass spectrometry interface, such as the "thermospray" system (Vestec; Hewlett Packard).

Some HPLC methods, however, have been published for the detection of benzodiazepines (e.g., Suitheimer and Sunshine, 1982) and antidepressants (Visser et al., 1984; Beierle and Hubbard, 1983), as well as for general drug screens (Kabra et al., 1981).

2.2.3. Gas Chromatography (GC)

GC is generally regarded as the most useful, "comprehensive" screening technique for a wide range of drugs and toxins if both sensitivity and resolving power are considered. It is beyond the scope of this chapter to cover all the options available for a GC system, but injectors, column types, and detectors should be discussed briefly.

2.2.3.1. COLUMNS. For general drug screening, fused silica capillary columns are infinitely preferable over packed columns. Capillary columns have a far greater resolving capability and are usually more inert. Even injectors designed to take packed columns can be used with the "new" so-called Megabore® (0.5 or 0.75 mm id) fused silica columns by using relatively simple and inexpensive adaptor kits including the use of a makeup gas at the detector (Bogusz et al., 1986; Franke et al., 1986). Such systems will vastly improve the resolution obtainable by the best-packed columns. At low carrier gas flows (e.g., 2 mL/min), the resolution of a 0.5-mm id megabore column approaches that of a 0.32-mm column ("capillary mode"), whereas at high flow rates it may be used for rapid quantitations with a decrease in resolution ("packed mode").

Capillary columns of 0.32-mm id seem to give the best compromise of resolution and loading capacity, although they may require a split or splitless capillary injection port. A wide range of drugs can be separated within 15 min on a 12-m column; these drugs range from the amphetamines to strychnine (Fig. 1) and include all of the benzodiazepines, antidepressants, and most of the neuroleptics. Several papers describing drug screening methods using capillary GC have been described recently (e.g., Taylor et al., 1986; Lora-Tamayo et al., 1986).

In fact, a 10–12-m capillary column is sufficient for virtually all screening purposes, using a fast temperature program (e.g., 80°C for 1 min, 20°C/min rise, and the final temperature held at 300°C for 3 min, a total of 15 min). If it is necessary to double the resolution, the column length must be increased to 40–48-m, and the run increased to an unacceptable time of 60 min.

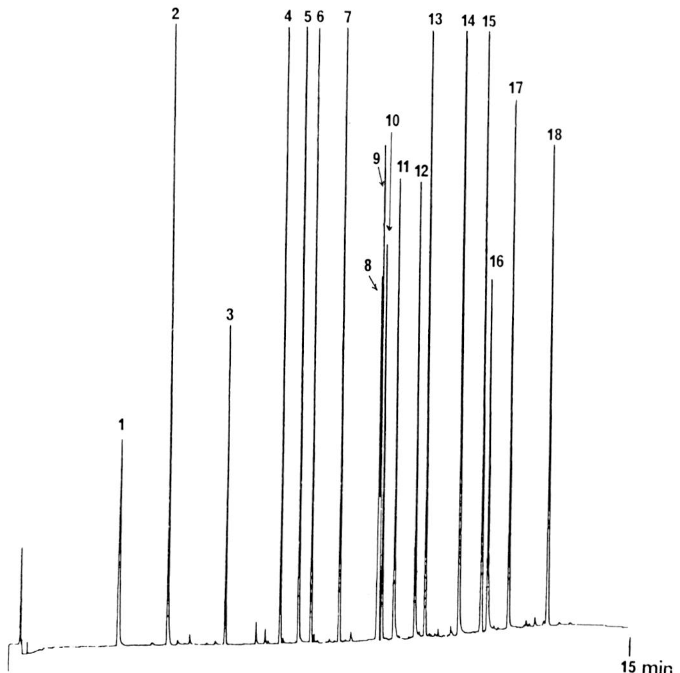

Fig. 1. GC separation of several drugs on a 12-m, 0.32-mm id fused silica column coated with 5% phenylmethylsilicone using a temperature programmed run and NP detection. The following drugs were extracted from an aqueous standard: methamphetamine (1), nicotine (2), methyprylon (3), meperidine (4), caffeine (5), lidocaine (6), chlorpheniramine (7), propoxyphene (8), amitriptyline (9), nortriptyline (10), pentazocine (11), codeine (12), diazepam (13), amoxapine (14), flurazepam (15), quinidine (16), haloperidol (17), strychnine (18).

Columns of 0.25 mm id are available, and these give higher resolution, but are prone to overloading, causing misshapen peaks and altered retention times (Ettre, 1979).

The number of phases required for capillary chromatography does not need to be very large, since their resolving power is inherently much greater than that of packed columns. A crosslinked phenylmethylsilicone column is a good general, slightly polar system.

2.2.3.2. Detectors.   The most common and reliable, cheapest, and easiest detector to use is the flame ionization detector (FID). The FID is relatively nonspecific, however, responding to virtually all organic compounds that chromatograph, including not only drugs, but plasticizers, lipids, and other endogenous compounds.

The best general detector for drug screening is the nitrogen-specific detector [nitrogen-phosphorus (NPD) or thermionic-specific detector (TSD)]. The NPD has excellent sensitivity to most organic molecules containing nitrogen and phosphorus, and has a very low response to compounds lacking those elements. Hence, nitrogen-containing drugs, including the benzodiazepines, anti-depressants, and neuroleptics, give a good response, whereas lipids, hydrocarbons, and most plasticizers are not detected, thus reducing the chance of misidentifying endogenous compounds or contaminants as drugs. The enhanced sensitivity and selectivity of the NPD over the FID enables drugs to be routinely screened for in blood or plasma at levels at least down to 0.1 mg/L, and often lower, from a 1 mL specimen. The absolute sensitivity of a capillary/NPD system would normally be at least two or three orders of magnitude lower.

The electron-capture detector (ECD) can also be considered selective, since it responds to compounds that absorb an electron emitted from the nickel-63 foil; that is, molecules that have electronegative atoms or moieties—especially halogens. The ECD is therefore particularly suitable for the detection of benzodiazepines, being much more sensitive and selective than the NPD for this purpose. Very few antidepressants give an ECD response (clomipramine and amoxapine are exceptions), although many neuroleptics contain one or more halogens and may be detected. Lipids and hydrocarbons are not ECD-sensitive, although some plasticizers and solvent impurities will interfere.

In an effort to increase the power of GC systems for detection and identification, two additional approaches have been taken. Many authors have described the technique of splitting the injection onto two GC columns of different polarities and using the dual retention characteristics for identification (Fretthold et al., 1986; Watts and Simonick, 1986). Often in the past, however, the two columns used were only slightly different in polarity (e.g., methyl-silicone versus 5% phenylmethylsilicone and DB1 versus DB5) and did not change the relative retention times of most compounds sufficiently to give a valid discrimination. A second approach is to split the effluent from one column onto two different detectors (e.g., NPD and FID or NPD and ECD). The NPD/ECD split has been used in the authors' laboratory for about 2 years with good results. The NPD chromatogram generally is good for most nitrogen-containing drugs, whereas the ECD chromatogram gives enhanced response for halogen-containing neuroleptics and especially for the benzodiazepines.

The detection of benzodiazepines, antidepressants, and neuroleptics is summarized below.

2.2.3.3. BENZODIAZEPINES.   Blood, plasma or serum are the specimens of choice for the detection of benzodiazepines by GC. Urine, conversely, is a poor choice since all benzodiazepines are extensively metabolized and only excreted unchanged in trace amounts. Furthermore, the metabolites are mostly present as glucuronide conjugates, which require hydrolysis before they can be chromatographed (Schutz, 1982).

With a few exceptions, most of the benzodiazepines can be chromatographed unchanged on either packed or capillary systems. All of the benzodiazepines are lipid-soluble and easily extracted from biological fluids with any one of a number of solvents. The main problem lies with the wide range of potency of the benzodiazepines and, consequently, of therapeutic blood levels, which may cover two or more orders of magnitude (e.g., triazolam vs. chlordiazepoxide). Thus, it is difficult to screen for all benzodiazepines using a single GC system. Until the mid-1980s, the vast majority of benzodiazepine use was of diazepam, chlordiazepoxide, clorazepate, prazepam, or other nordiazepam precursors. So, benzodiazepine use was easily detectable using GC by monitoring for nordiazepam. Packed-column systems were used, which also chromatograph flurazepam and the active metabolite, *N*-desalkylflurazepam, although therapeutic levels are 2–10 times lower than for diazepam. The best-packed system we have used is a 1.2-m glass column packed with GP 3% SP-2250DB on 100/120 Supelcoport (Supelco 1-1983) (Fig. 2). Bromazepam is detectable at therapeutic levels on the same system, although it tails slightly; nitrazepam also chromatographs, but usually with a moderate amount of tailing. Alprazolam, triazolam, and some of the newest annelated benzodiazepines (e.g., brotizolam, estazolam) are more problematic because of their low therapeutic levels, although they do chromatograph fairly well on the same 3% SP-2250 DB system, but with a longer retention time.

The 3-hydroxybenzodiazepines, however, such as oxazepam, temazepam, lorazepam, and lormetazepam all decompose partially or completely on most GC systems and require derivatization prior to chromatography (Joyce et al., 1984). Trimethylsilyl (TMS) derivatives are most commonly formed, although there is some

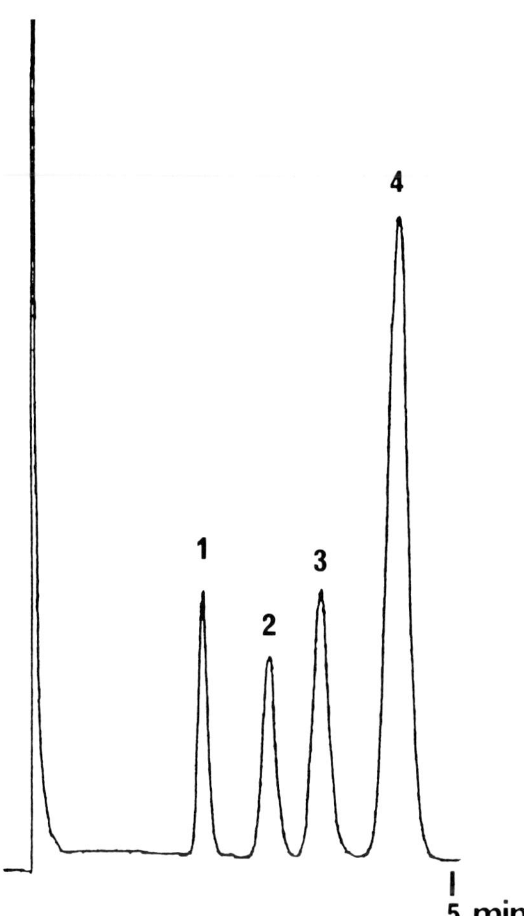

Fig. 2. Gas chromatography separation of benzodiazepines on a 1.2-m 2-mm id column packed with GP 3% SP-2250DB on 100/120 mesh Supelcoport, oven temperature 240°C, and EC detection. The drugs were extracted from whole blood using flurazepam (4) as internal standard: diazepam (1), desalkylflurazepam (2), and nordiazepam (3).

evidence that *t*-butyldimethylsilyl derivatives [formed with *N*-methyl-*N*-(*t*-butyldimethylsilyl)-trifluoracetamide; MTBSTFA] may be more stable (Stanworth, 1984).

Chlordiazepoxide always decomposes partially and sometimes totally on most, if not all, GC systems. HPLC is the method of choice for measuring chlordiazepoxide and its metabolites.

Most of the benzodiazepines will tail to some degree on capillary column systems, and this is presumably because of active sites or pyrolyzed biological residues in the injection port. Some workers have employed direct injection onto a short precolumn (e.g., Grob, 1984; Baktir and Bircher, 1985) of uncoated 0.5-mm id fused silica butt-connected to the main analytical column (e.g., 0.32-mm id fused silica) to overcome the problem.

The preferred detector for benzodiazepines is the ECD. NP detectors are also satisfactory for the higher dose "classical" benzodiazepines (e.g., diazepam, nordiazepam), but they usually lack sufficient sensitivity for the newer triazolo benzodiazepines such as triazolam and alprazolam.

2.2.3.4. ANTIDEPRESSANTS.   All of the tricyclic, tetracyclic, and related antidepressants can be chromatographed well on a good deactivated phenylmethylsilicone-coated packed column such as 3% SP-2250DB on 100/120 Supelcoport (Supelco 1-1983 clinical packing), or on virtually any methylsilicone- or phenylmethylsilicone-fused silica column (e.g., DB-1, DB-5, SE-54). The preferred detector is the NPD because of good selectivity and sensitivity. Most of the published quantitative GC methods for TCAs in blood or plasma may be adapted for screening purposes.

Urine, however, may be a better specimen for screening purposes. Although the TCAs are extensively metabolized, sufficient of the parent compound, or, in the case of a tertiary amine, the desalkyl metabolite, is excreted to allow detection in urine by GC/NPD. Furthermore, unconjugated hydroxylated tertiary or secondary amines are also excreted, giving a fairly characteristic metabolite pattern. To see this pattern, however, the urine should be extracted after making it only relatively weakly basic (pH < 10).

Of the MAOIs, only tranylcypromine can readily be detected by GC. Therapeutic blood or plasma levels of tranylcypromine are generally <0.1 mg/L. Therefore, urine is a much better specimen to screen, since concentrations are an order of magnitude or more higher. Most GC systems that will detect amphetamines will detect tranylcypromine—preferably using an NP detector. The hydrazine MAOIs are extremely difficult to screen for by GC, although quantitative methods have been described.

2.2.3.5. NEUROLEPTICS.   Gas chromatography is not a good comprehensive method for detecting neuroleptic use. There are three main problems. First, daily doses of neuroleptics, including the phenothiazines, range over at least two, and sometimes three, orders of magnitude. Thioridazine or mesoridazine doses can range up to 800 mg or more daily and even higher in exceptional cases, whereas fluphenazine may be given in doses of only 12.5 mg/mo. Second, some of the higher molecular weight phenothiazines such as perphenazine, piperacetazine, and thiopropazate are difficult to detect with conventional drug screening systems because of their very long retention time and extensive metabolism. Third, some neuroleptics such as pimozide and haloperidol have very poor chromatographic characteristics—pimozide because of its inherent high polarity, and haloperidol because it is thermolabile.

Nonetheless, chlorpromazine, thioridazine, mesoridazine (as its metabolite thioridazine), and methotrimeprazine are reasonably easy to detect in both blood or plasma and urine providing the dosage is not too small. The presence of metabolites of all these compounds in urine can aid in detection, if properly characterized.

Again, the RIA and RRA procedures referred to earlier would serve as much more sensitive and reliable screening methods for the presence of neuroleptic drugs.

2.2.4. Mass Spectrometry (MS)

In clinical and forensic toxicology, mass spectrometry is usually used in combination with gas chromatography (GC/MS). The introduction of the Mass Selective Detectors® (MSD; Hewlett Packard, Palo Alto, CA) and the Ion Trap® (Finnigan Corporation, San Jose, CA) instruments have brought the minimum cost for a GC/MS system down from about $250,000 to as little as $75,000. Nonetheless, GC/MS is still usually regarded as a confirmation technique and is still too expensive for most laboratories to use for routine purposes.

It is beyond the scope of this chapter to describe the use or principles of GC/MS in any detail (for an introductory text, *see* Watson, 1985). Most laboratories use electron-impact (EI) ionization almost exclusively, although chemical ionization (CI) can be extremely useful for some confirmations or molecular weight determination.

Laboratories considering the purchase of a full-size GC/MS system or an "MSD" should bear in mind that an experienced operator is required to both run the system and interpret mass spectra. A good GC/MS data system used in conjunction with a good MS drug library will indicate the most likely match, if the authentic MS of that drug, drug metabolite, or toxin is in the library. It is up to the operator to determine if that computer match is correct!

Two screening methods for antidepressants and neuroleptics by GC/MS have been published (Maurer and Pfleger, 1984a,b), and a GC-MS approach to the screening of butyrophenones (Hattori et al., 1986) has also been reported.

3. Conclusions

Drug screening is performed for a variety of reasons, some good, some bad. Regardless of those specific reasons, it is important that testing be performed by experienced personnel who know the limitations and pitfalls of the methods and equipment they use. Partial drug screening is becoming easier to perform with the growing number of automated immunoassays available. Indeed, immunoassays are a valuable and powerful supplement to the armament of chromatography tests. Unfortunately, this improvement in technology also makes it easier for less experienced or less vigilant laboratories to report erroneous or incomplete results, where chromatographic confirmatory methods should also be used, whether or not that includes mass spectrometry. It is important that all positive results be confirmed by an independent method that is based on different principles than the original screen, if there is any possibility those results could be used to the detriment of the person being tested or if they will influence a course of medical treatment. With the pharmaceutical industry producing a constantly changing spectrum of drugs, it is important that all laboratories engaged in drug testing learn to scientifically apply a multi-method approach and not rely solely on immunoassays.

References

Aaltonen L. and Scheinin M. (1982) Application of radioreceptor assay of benzodiazepines for toxicology. *Acta Pharmacol. Toxicol.* **50,** 206–212.

Baktir G. and Bircher J. (1985) Capillary gas-liquid chromatographic determination of the benzodiazepine triazolam in plasma using a retention gap. *J. Chrom.* **339,** 192–197.

Baselt R. C. (1980) *Analytical Procedures for Therapeutic Drug Monitoring and Emergency Toxicology* Biomedical Publications, Foster City, California.

Baselt R. C. (1984) Drug Screening Interpretation, in *Advances in Analytical Toxicology* vol. 1 (Baselt, R. C., ed.) Biomedical Publications, California.

Beierle F. A. and Hubbard R. W. (1983) Liquid chromatographic separation of antidepressant drugs. I. Tricyclics. *Ther. Drug Monit.* **5,** 279–292.

Bogusz M., Bialka J., Gierz J., and Klys M. (1986) Use of short, wide-bore capillary columns in GC toxicological screening. *J. Anal. Toxicol.* **10,** 135–138.

Chamberlain J. (1985) *Analysis of Drugs in Biological Fluids* CRC, Florida.

de Silva J. A. F., Bekersky I., Puglisi C. V., Brooks M. A., and Weinfeld R. E. (1976) Determination of 1,4-benzodiazepines and diazepin-2-ones in blood by electron-capture gas-liquid chromatography. *Anal. Chem.* **48,** 10–19.

Ettre L. S. (1979) *Introduction to Open Tubular Columns* Perkin-Elmer, Connecticut.

Franke J. P., de Zeeuw R. A., and Wijsbeek J. (1986) Potentials of wide-bore fused silica capillary columns for substance identification by means of retention indices. *J. Anal. Toxicol.* **10,** 132–134.

Fretthold D., Jones P., Sebrosky G., and Sunshine I. (1986) Testing for basic drugs in biological fluids by solvent extraction and dual capillary GC/NPD. *J. Anal. Toxicol.* **10,** 10–14.

Grob K. (1984) Effect of "dirt" injected on-column in capillary chromatography: Analysis of the sterile fraction of oils as an example. *J. Chromatogr.* **287,** 1–4.

Haefelfinger P. (1975) Determination of nanogram amounts of primary aromatic amines and nitro compounds in blood and plasma. *J. Chromatogr.* **111,** 323–329.

Haefelfinger P. (1979) Determination of the 7-amino metabolites of the 7-nitro-benzodiazepines in human plasma by thin-layer chromatography. *J. High Res. Chrom. Chrom. Comm.* **1,** 39–42.

Hattori H., Suzuki O., and Brandenberger H. (1986) Positive- and negative-ion mass spectrometry of butyrophenones. *J. Chromatogr.* **382,** 135–145.

Jochemsen R., Horback G. J. M. J., and Breimer D. D. (1982) Assay of nitrazepam and triazolam in plasma by a radioreceptor technique and comparison with a gas chromatographic method. *Res. Commun. Chem. Pathol. Pharmacol.* **35,** 259–273.

Joyce J. R., Bal T. S., Ardey R. E., Stevens H. M., and Moffat A. C. (1984) The decomposition of benzodiazepines during analysis by capillary gas chromatography/mass spectrometry. *Biomed. Mass Spectrom.* **11,** 284–289.

Kabra P. M., Stafford B. E., and Marton L. J. (1981) Rapid method for screening toxic drugs in serum with liquid chromatography. *J. Anal. Toxicol.* **5,** 177–182.

Lora-Tamayo C., Rams M. A., and Chacon J. M. R. (1986) Gas chromatographic data for 187 nitrogen- or phosphorus-containing drugs and metabolites of toxicological interest analysed on methylsilicone capillary columns. *J. Chromatogr.* **374,** 73–85.

Lund J. (1981) Radioreceptor assay for benzodiazepines in biological fluids using a new dry and stable receptor preparation. *Scan. J. Clin. Lab. Invest.* **41,** 275–281.

Maggio E. T. (1980) *Enzyme-Immunoassay.* CRC Press, Inc., Boca Raton, Florida.

Manchon M., Verdier M. F., Pallud P., Vialle A., Beseme F., and Bienvenu J. (1985) Evaluation of EMIT-TOX enzyme immunoassay for the analysis of benzodiazepines in serum: usefulness and limitations in an emergency laboratory. *J. Anal. Toxicol.* **9,** 209–212.

Maurer H. and Pfleger K. (1984a) Screening procedure for detection of phenothiazine and analogous neuroleptics and their metabolites in urine using a computerized gas chromatographic-mass spectrometric technique. *J. Chromatogr.* **306,** 125–145.

Maurer H. and Pfleger K. (1984b) Screening procedure for detection of antidepressants and their metabolites in urine using a computerized gas chromatographic-mass spectrometric technique. *J. Chromatogr.* **305,** 309–323.

Peel H. W. and Perrigo B. J. (1980) Toxicological analysis of benzodiazepine-type compounds in post-mortem blood by gas chromatography. *J. Anal. Toxicol.* **4,** 105–113.

Pegon Y., Pourcher E., and Vallon J. J. (1982) Evaluation of the EMIT TOX enzyme immunoassay for toxicological analysis of benzodiazepines in serum. *J. Anal. Toxicol.* **6,** 1–3.

Poklis A. (1981) An evaluation of EMIT-dau benzodiazepine metabolite assay for urine drug screening. *J. Anal. Toxicol.* **5,** 174–176.

Roets E. and Hoogmartens J. (1980) Thin-layer chromatography of the acid hydrolysis products of nineteen benzodiazepine derivatives. *J. Chromatogr.* **194,** 262–269.

Schroeder T. J., Tasset J. J., Otten E. J., and Hedges J. R. (1986) Evaluation of Syva EMIT® toxicological serum tricyclic antidepressant assay. *J. Anal. Toxicol.* **10,** 221–224.

Schutz H. (1982) *Benzodiazepines. A Handbook* Springer-Verlag, New York.

Slightom E. L., Cagle J. C., McCurdy H. H., and Castagna F. (1982) Direct and indirect homogenous enzyme immunoassay of benzodiazepines in biological fluids and tissues. *J. Anal. Toxicol.* **6,** 22–25.

Stanworth D. (1984) The Analysis of Benzodiazepines in Blood Using Bonded Phase Capillary Gas Chromatography With Electron Capture Detection, in *Proceedings of the 21st International Meeting of the International Association of Forensic Toxicologists* (TIAFT) (Dunnett N. and Kimber K. J., eds.) TIAFT, Newmarket, UK.

Suitheimer C. and Sunshine I. (1982) Benzodiazepines by Reverse-Phase HPLC, in *Methodology for Analytical Toxicology* vol. 2 (Sunshine I. and Jatlow P. I., eds.) CRC, Florida.

Taylor R. W., Greutink C., and Jain N. (1986) Identification of underivatized basic drugs in urine by capillary column gas chromatography. *J. Anal. Toxicol.* **10,** 205–208.

Visser T., Oostelbos M. C. J. M., and Toll P. J. M. M. (1984) Reliable routine method for the determination of antidepressant drugs in plasma by high-performance liquid chromatography. *J. Chromatogr.* **309,** 81–93.

Watson J. T. (1985) *Introduction to Mass Spectrometry* Raven, New York.

Watts V. W. and Simonick T. F. (1986) Screening of basic drugs in biological samples using dual column capillary chromatography and nitrogen-phosphorus detectors. *J. Anal. Toxicol.* **10,** 198–204.

In Vitro and Ex Vivo Neurochemical Screening Procedures for Antidepressants, Neuroleptics, and Benzodiazepines

Glen B. Baker and Andrew J. Greenshaw

1. Introduction

Our knowledge of the biochemical actions of antidepressants, neuroleptics, and anxiolytics at the cellular and molecular level has led to the development of a number of rather rapid and inexpensive in vitro and ex vivo screening procedures that have proven useful in identifying new drugs with psychotropic properties. The biochemical actions most frequently tested include effects on uptake of radiolabeled biogenic amines by nerve-ending preparations, effects on monoamine oxidase (MAO) activity, and actions on binding of radioligands to binding sites prepared from brain tissue. In the following review, each of these, along with effects on adenylate cyclase, will be considered, and the descriptions will include examples of specific applications and a discussion of how these tests fit into the overall strategies for testing potential psychotropic drugs.

2. Effects of Drugs on Uptake of Radiolabeled Biogenic Amines into Nerve-Ending Preparations

The principal means of terminating the actions of biogenic amines such as dopamine (DA), noradrenaline (NA), and 5-hydroxytryptamine (5-HT, serotonin) after they have acted on postsynaptic receptors is by an active reuptake process into the presynaptic nerve terminal (Iversen, 1971). This is of particular interest to researchers in biological psychiatry since it has been known for some time that tricyclic antidepressants have a relatively potent blocking action on this active reuptake (Carlsson et al., 1969a,b; Shaskan and Snyder, 1970; Lidbrink et al., 1971; Ross et al., 1972), with this effect being particularly strong on NA and 5-HT. A detailed discussion of the characteristics of this active reuptake mechanism is beyond the scope of this chapter, but a number of detailed reports and reviews are available (Snyder et al., 1970; Iversen, 1971; Horn, 1978; Paton, 1976, 1980; Koe, 1976; Randrup and Braestrup, 1977; Ross, 1980; Baker and Dyck, 1985).

Basically, the in vitro and ex vivo incubation procedures for studying the effects of drugs on uptake of biogenic amines involve incubating radiolabeled NA, 5-HT, or DA with the nerve terminal-containing preparation in a physiological salt solution [the drug under consideration is added to the solution (in vitro) or has been injected into the animal from which the tissue has been taken (ex vivo)]. The tissue is separated from the solution at the end of the incubation period by centrifugation or filtration, and the amount of radioactivity in the pellet or on the filter is counted. Often the incubation is done in the presence of an MAO inhibitor to prevent the metabolic breakdown of the biogenic amine of interest. Even if this is the case, preliminary experiments should be conducted to determine that most of the radioactivity remaining is in the form of the unmetabolized amines. This usually involves separation of the amine from its metabolites using chromatographic procedures after extraction of the filter or pellet. Examples of procedures used are those by Jonsson et al. (1969), Sachs and Johnson (1972), Hamberger and Tuck (1973), Raiteri et al. (1975), de Belleroche et al. (1976), Dyck and Boulton (1980), and Urwyler and von Wartburg (1980).

2.1. Tissue Preparations Used for Conducting Uptake Experiments

A search through the literature in this area soon makes it obvious that a variety of tissue preparations has been used in vitro and ex vivo to study the effects of drugs on uptake of biogenic amines. Preparations used include slices, prisms, minces, homogenates (with or without previous removal of nuclear debris), P_2 synaptosomal fractions [also called crude mitochondrial fractions and containing myelin fragments, synaptosomes (nerve terminals), and mitochondria], and purified synaptosomes. Slices, prisms, minces, and P_2 fractions can be prepared more rapidly than purified synaptosomes, but obviously they contain several other components that may affect the uptake of the biogenic amines. Synaptosomes do represent a concentrated sample of nerve endings, but may undergo damage during preparation (e.g., centrifuging through hyperosmotic solutions) and/or during washing and collections (Wheeler, 1978). Brain slices are relatively easy to prepare and have the advantage of more closely representing the normal physiological environment, but swelling during incubation and the presence of a large extracellular space may introduce problems for uptake experiments. Comprehensive articles discussing the preparation and use of brain slices (Dingledine, 1984; Lipton, 1985) and synaptosomes (Whittaker, 1984; Rodriquez de Lores Arnaiz and Pellegrino de Iraldi, 1985) are available and are recommended reading for anyone venturing into this area of research. Factors to be considered with regard to the separation of tissue from incubation medium have also been discussed (Raiteri and Levi, 1973; Wheeler, 1978; Baker and Dyck, 1985).

In Table 1, we have compiled a list of references that deal with various methodological aspects of uptake studies. Table 2 consists of selected examples of studies in which the effects of antidepressant drugs on biogenic amine uptake were investigated in vitro or ex vivo. Researchers planning to use this technique for screening their compounds are recommended to check the appropriate references to get details such as method of preparation of tissue, composition of incubation medium, time of incubation, and separation of tissue and medium. Table 3 is a comprehensive comparison of the relative effects of several antidepressant drugs on biogenic amine uptake (Hyttel, 1982). Other useful references also containing such comprehensive comparative tables are those by Koe (1976), Richelson (1984), Fuller and Wong (1985), and Baldessarini (1985).

2.2. Other Factors to Consider when Performing In Vitro and Ex Vivo Uptake Experiments

2.2.1. In Vitro vs Ex Vivo Studies

Uptake can also be studied by administering the radiolabeled amines or their labeled precursor amino acids, dissecting out the brain tissue and performing autoradiography, or extracting the labeled amines from the tissue and counting the amount of radioactivity associated with the amines. The density of the autoradiographic profile or the amount of label in the dissected tissue can then be compared in vehicle-treated animals and those pretreated with the drug of interest. Alternatively, brains can be removed from control or drug-treated animals, appropriate tissue preparations (slices, synaptosomes, and so on) made, and incubations carried out with the radiolabeled biogenic amines; tissue

Table 1

Papers Dealing with Some Methodological Aspects of
Uptake Experiments[a]

Reference	Factors investigated
Shaskan and Snyder, 1970	Kinetics of 5-HT accumulation into brain slices
Snyder et al., 1970	Review of uptake and subcellular distribution of catecholamines and 5-HT in brain
Dorris and Shore, 1971	Uptake and storage mechanisms in corpus striatum
Iversen, 1971	Review of role of transmitter uptake mechanisms
Ferris et al., 1972	Isomers of amphetamines, deoxypipradrol, and methylphenidate and inhibition of uptake of triated catecholamines into cerebral cortex slices, synaptosomal preparations from cerebral cortex, hypothalamus, and striatum, and into adrenergic nerves of rabbit aorta
Kuhar et al., 1972	Synaptosomal uptake of 5-HT in discrete brain regions following midbrain raphe lesions
White and Paton, 1972	Effects of external Na^+ and K^+ on NA uptake by synaptosomes
Harris and Baldessarini, 1973	Effects of cations on uptake of $[^3H]$-DA by homogenates of corpus striatum
Raiteri and Levi, 1973	Studies on effects of sudden cooling on radiolabeled amines taken up by synaptosomes
Heikkila et al., 1975	Studies on the distinction between uptake inhibition and release of $[^3H]$-DA in brain slices
Raiteri et al., 1975	Distinction between release and inhibition of reuptake in synaptosomes
Baumann and Maitre, 1976	Studies to determine if inhibition of DA uptake by drugs is a misinterpretation of in vitro experiments

330

deBellroche et al., 1976	Automated separation and analysis of DA, its amino acid precursors and metabolites: application to measurement of specific radioactivities of DA in striatal synaptosomes
Levi and Raiteri, 1976	Review of synaptosomal transport processes
Paton, 1976	A book containing chapters by several authors on mechanisms of neuronal and extraneuronal transport of catecholamines
White, 1976	Synaptic vesicles and rapid binding of newly accumulated NA within synaptosomes
Carruba et al., 1977b	Study of mazindol and amphetamines as inhibitors of uptake of, and releasers of, $[^3H]$-DA in striatal slices
Moisset, 1977	Factors contributing to modulation of uptake of NA by cortical synaptosomes
Choi and Abramson, 1978	Effects of pH changes and charge characteristics on synaptosomal uptake of NA
Horn, 1978	Review of characteristics of neuronal DA uptake
Raiteri et al., 1978	Review of uptake and release of DA in synaptosomes
Stahl and Meltzer, 1978	5-HT transport by human platelets
Wheeler, 1978	Discussion of problems with uptake studies in synaptosomes
Hunt et al., 1979	Use of synaptosomes and field-stimulated brain slices to differentiate DA uptake inhibitors and releasing agents
Ross, 1979	Effects of reserpine on the actions of compounds on accumulation of 5-HT and NA in homogenates of hypothalamus
Tuomisto and Tuomisto, 1979	Comparison of $[^3H]$-DA intake into crude, partially purified, and purified synaptosomal preparations from striatum

(continued)

Table 1 *(continued)*

Reference	Factors investigated
Logan and O'Donovan, 1980	Na^+, K^+-ATPase, and uptake of NA by synaptosomes
Paton, 1980	Review of neuronal transport of catecholamines
Ross, 1980	Review of neuronal transport of 5-HT
Ross and Ask, 1980	Study on structural requirements for uptake into serotoninergic neurons
Urwyler and von Wartburg, 1980	Uptake and metabolism of catecholamines in synaptosomes: contribution of MAO
West and Fillenz, 1980	Storage and release of NA in synaptosomes prepared from hypothalamus
Thompson et al., 1981	Accumulation and release of $[^3H]$-DA in striatal slices from young, mature, and senescent rats
Schoemaker and Nickolson, 1982	Interaction between influx and efflux of DA in striatal synaptosomes
Minchin et al., 1983	Electroconvulsive shock and uptake and release of NA and 5-HT in brain slices
Wood and Wyllie, 1983	Critical assessment of uptake of NA in synaptosomal preparations
Dingledine, 1984	Book with several articles on brain slices
Whittaker, 1984	Review article on the preparation and use of synaptosomes
Baker and Dyck, 1985	Review article on in vitro neuronal transport
Lipton, 1985	Uses and abuses of brain slices
Rodriquez de Lores Arnaiz and Pellegrino de Iraldi, 1985	Preparation of subcellular fractions, including synaptosomes

[a]Portions of this table were taken directly from Baker and Dyck, 1985, with permission.

Table 2

Selected Examples of Studies on the Effects of Antidepressants
on Biogenic Amine Uptake[a]

Reference	Drugs studied	Brain tissue preparation used
Hendley and Snyder, 1968	MAO inhibitors	Prisms
Carlsson et al., 1969a,b	Tricyclics	Effect of drugs on reduction of brain amine levels by depleting agents
Ross et al., 1972	Chloroimipramine	Slices
Tuomisto and Walaszek, 1974	Tranylcypromine	Platelets
Ross and Renyi, 1975a,b	Tricyclics	Synaptosomes and slices
Koe, 1976	Tricyclics and novel antidepressants	Synaptosomes
Pylatuk and McNeill, 1976	Variety of drugs	Whole brain homogenates
Raiteri et al., 1976	Mianserin and imipramine	Synaptosomes
Carruba et al., 1977a	Mazindol and chloroimipramine	Synaptosomes
Friedman et al., 1977	Various antidepressants	Synaptosomes
Gentil et al., 1977	Tranylcypromine	Platelets
Goodlet et al., 1977	Mianserin	Slices, synaptosomes
Randrup and Braestrup, 1977	Novel antidepressants and tricyclics	Synaptosomes
Thomas and Jones, 1977	Clomipramine and desmethylclomipramine	Slices
Yu and Smith, 1977	Desmethylimipramine	Slices
Baker et al., 1978	MAO inhibitors	Prisms
Hyttel, 1978	Variety of antidepressants	Synaptosomes
Martin et al., 1978	Viloxazine	Prisms
Maxwell and White, 1978	Tricyclics and MAO inhibitors	Various
Ross, 1979	Tricyclics and novel antidepressants	Synaptosomes
Waldmeier et al., 1979	Tricyclics and novel antidepressants	Synaptosomes, thrombocytes
Dyck and Boulton, 1980	MAO inhibitors	Homogenates
Lai et al., 1980	Clorgyline and l-deprenyl	Synaptosomes

(continued)

333

Table 2 (continued)

Reference	Drugs studied	Brain tissue preparation used
Lassen et al., 1980	Femoxatine, paroxetine and chloroimipramine	Blockade of p-chloroamphetamine-induced hypermotility and depletion of brain 5-HT
Maitre et al., 1980	Variety of antidepressants	Platelets, synaptosomes
Riblet and Taylor, 1981	Trazodone, clomipramine, desipramine, imipramine, and fluoxetine	Platelets, synaptosomes
Azzaro and Demarest, 1982	MAO inhibitors	Synaptosomes
Eckhardt et al., 1982	Tricyclics	Synaptosomes
Hall et al., 1982	Zimelidine	Synaptosomes
Hyttel, 1982	Tricyclics and novel antidepressants	Synaptosomes
Quinaux et al., 1982	Various antidepressants	
Sparatore et al., 1982	Rigid analogs of imipramine and amitriptyline	Synaptosomes
Waldmeier, 1982	Variety of antidepressant drugs	Reversal of catecholamine depletion induced by α-methyl-m-tyramine
Richelson, 1984	Tricyclics and novel antidepressants	Synaptosomes
Baker and Dyck, 1985	Review article containing a table of effects of various drugs on biogenic amine uptake	Various types
Baldessarini, 1985	Variety of antidepressant drugs	Synaptosomes
Fuller and Wong, 1985	Tricyclics and novel antidepressants	Antagonism of drug-induced depletion of amines
Lemberger et al., 1985	Fluoxetine	Platelets

[a]Modified from Baker and Dyck, 1985, with permission.

334

and medium are separated as described previously. These types of experiments in which the drugs are administered in vivo and the tissue is subsequently removed and analyzed are often called in vivo studies. More strictly, they should be termed ex vivo and will be referred to as such in the remainder of this chapter (including in the discussion on receptors). Autoradiographic procedures will not be discussed. In general, the in vitro studies are more rapid. As with many in vitro preparations, however, they may be criticized for representing a nonphysiological preparation. The ex vivo preparations also have some disadvantages, however. Considerable metabolism of antidepressant drugs can occur, and it is very difficult to determine in such a preparation the contribution that active metabolites are making to the overall inhibition of uptake. For example, tricyclic antidepressants can undergo extensive demethylation (in the case of tertiary amine compounds) and hydroxylation, and both the demethylated and hydroxylated metabolites are known to have rather potent effects on uptake of NA and/or 5-HT (*see* Baldessarini, 1985; Rudorfer and Potter, 1985, for reviews). In the in vitro situation, however, each of the metabolites can be tested separately to determine their strength relative to the parent compound.

Another criticism of the in vitro preparation is that the concentrations of drug that are being studied may not be pharmacologically significant, i.e., they may be much higher than those present in the brain and blood after administration of clinically relevant doses. This criticism can be circumvented by performing brain and blood measurements of levels of the drug and metabolite concentrations after administration of the antidepressant drugs and using appropriate molar concentrations in parallel in vitro preparations. There is, in fact, now considerable information available on brain levels (from animal experiments) and plasma levels (from human studies) of antidepressants, and many laboratories doing neurochemical research have the analytical facilities to do such measurements themselves.

2.3. Further Considerations about Uptake Screening Procedures

There are presently on the market a number of effective antidepressant drugs that are not particularly good inhibitors of NA or 5-HT uptake. These include some of the MAO inhibitors and the so-called novel antidepressants (*see* Feighner, 1981; Leonard, 1984; Coccaro and Siever, 1985; Damlouji et al., 1985; Hollister, 1986, for

Table 3
Uptake of [³H]-Amine into Rat Brain Synaptosomes[a]

IC$_{50}$, nM	5-HT	NA	DA	NA/5-HT	DA/5-HT	DA/NA
Citalopram-Lu 10-171	1.8	8800	41000	4889	22778	4.7
Desmethylcitalopram-Lu 11-109	7.4	780	26000	105	3514	33
Didesmethylcitalo-pram-Lu 11-161	24	1500	12000	63	500	8.0
Citalopram-N-oxide-Lu 11-305	56	3200	>100000	57	>1786	>31
Indalpine-LM 5008	2.4	2400	1300	1000	542	0.54
CGP 6085 A	0.90	560	1800	622	2000	3.2
Alaproclate	89	32000	11000	360	124	0.34
Paroxetine-FG 7051	0.31	88	5900	284	19032	67
Fluoxetine-Lilly 110140	6.9	380	5000	55	725	13
Zimelidine-H 102/09	59	3200	27000	54	458	8.4
Ro 11-2465	0.87	46	17000	53	19540	370
Femoxetine-FG 4963	8.3	410	1400	49	169	3.4
Norzimelidine	9.6	260	5600	27	583	22
Trazodone	580	11000	19000	19	33	1.7
Chlorimipramine	1.5	24	4300	16	2867	179
YM-08054-1	64	60	6800	0.94	106	113
Cocaine	260	220	310	0.85	1.2	1.4
Trimipramine	2100	1300	6600	0.62	3.1	5.1
Amitriptyline	40	24	5400	0.60	135	225
Imipramine	35	20	18000	0.57	514	900
Butriptyline	2100	1100	5500	0.52	2.6	5.0
Iprindole	3200	1200	14000	0.38	4.4	12
Opipramol	2200	700	3000	0.32	1.4	4.3

N-Methyl protriptyline	200	62	3900	0.31	20	63
Fluotracen	210	31	2300	0.15	11	74
Doxepin	280	40	13000	0.14	46	325
Lilly 5182	51	5.4	9.3	0.11	0.13	1.7
Bupropion	19000	1500	600	0.079	0.032	0.40
Melitracen	710	54	8200	0.076	12	152
Mazindol	30	0.65	18	0.022	0.60	28
Mianserin	1200	23	40000	0.019	33	1739
Viloxazine	15000	280	56000	0.019	3.7	200
Nortriptyline	590	7.7	3600	0.013	6.1	468
Dibenzepine	3200	41	>100000	0.013	>31	>2439
Benztropine	13000	150	110	0.012	0.0085	0.73
Desmethylchlorimipra-mine	41	0.46	2200	0.011	54	4783
Nomifensine	830	6.6	48	0.0080	0.058	7.3
Protriptyline	260	1.3	3300	0.0050	13	2538
Desipramine	210	0.97	9100	0.0046	43	9381
Maprotiline	3000	8.4	10000	0.0028	3.3	1190
Talopram-Lu 3-010	1400	2.9	44000	0.0021	31	15172
Litracen	650	0.75	8000	0.0012	12	10667
Talsupram-Lu 5-003	850	0.79	9300	0.00093	11	11772
Hydroxymaprotiline-C49-802 B-Ba	8000	1.1	21000	0.00014	2.5	19091
K_m	49	19	105			
s	10	10	12.5			
x	1.20	1.53	1.12			

[a]The following structures of rat brain were used for uptake. 5-HT, whole brain minus cerebelleum, pons and medulla oblongata; NA, occipital plus temporal cortex; DA, corpus striatum. The last line shows the K_m value (nM), substrate concentration, s (nM), and $x = 1 + s/K_m$ in the Cheng and Prusoff equation $K_i = (IC_{50})/(1 + s/K_m) = (IC_{50}/x)$ for competive inhibitors (Cheng and Prusoff, 1973). All results are the mean of at least two determinations each with five concentrations of test compounds in triplicate. (Reproduced from Hytell, 1982 with permission from the author and Pergamon Journals Ltd.)

reviews). The former drugs would be picked up by testing for inhibition of MAO activity, as discussed in a subsequent section of this chapter. There are a number of novel antidepressants (e.g., mianserin, viloxazine, and iprindole) that do not have marked effects on uptake of NA or 5-HT, and such drugs would be missed by both the uptake and MAO screening tests. These latter drugs do, however, alter the sensitivity of β-adrenergic receptors after long-term administration and share this property with the tricyclic antidepressants and MAO inhibitors; this forms the basis for two other types of screening procedures that will be discussed in this chapter, i.e., radioligand binding studies and measurement of adenylate cyclase activity.

3. Inhibition of MAO

MAO inhibitors have been used in psychiatry for many years and still form an important part of the psychiatrist's armamentarium of antidepressant drugs (for reviews on their use, *see* Pare, 1985; Klein et al., 1980; Murphy et al., 1984, 1985). Despite problems with regard to side effects and interactions with other drugs and certain food products, these drugs seem to have undergone a resurgence in recent years (White and Simpson, 1981). Interest was stimulated by the finding that there were at least two forms of MAO, denoted A and B (Johnston, 1968; Knoll and Magyar, 1972; Yang and Neff, 1974), and that deprenyl, a specific inhibitor of type B, did not cause the "cheese effect" (interaction with tyramine-containing food) (Knoll and Magyar, 1972). Some examples of important articles and books that contain discussions of various aspects of MAO and its inhibition are Fowler and Callingham (1978), Fuller (1978), Singer et al. (1979), Youdim and Paykel (1981), Gorkin (1983), Tipton et al. (1984), Youdim and Finberg (1985), Callingham (1986), Glover and Sandler (1986), and Yu (1986).

Various procedures are available for assaying MAO activity, and these are based on one or more of the following aspects of the enzymic reaction: consumption of oxygen, disappearance of amine substrates, or formation of the following: ammonium, aldehydes, hydrogen peroxide, or acid or alcohol metabolites (*see* Yu, 1986, for review). Measurement techniques include manometry, oxygen electrodes, spectrophometry, fluorometry, colorimetry, high-pressure liquid chromatography (HPLC) with various detectors, gas chromatography, and radioenzymatic procedures (Yu, 1986).

Probably one of the most commonly used procedures over the past 20 years has been the radioenzymatic procedure of Wurtman and Axelrod (1963). The procedure, which has been adapted by several researchers, involves incubating the source of enzyme with ^{14}C-labeled amine substrate, acidifying the incubation mixture to stop the reaction, and extracting the radiolabeled products of the reaction with an organic solvent. The amount of radioactivity in the extract is then counted using a liquid scintillation counter. By utilizing [^{14}C]-5-HT or [^{14}C]-β-phenylethylamine as substrates, activity of MAO-A and MAO-B can be determined, respectively. A comprehensive description of the experimental details is given by Yu (1986). The method is rapid and sensitive and can be readily learned by technical staff; the only obvious disadvantage, in common with any assay using radiochemicals, is the necessity to dispose of the radioactivity at the end of the experiments.

Table 4 (from Yu, 1986) provides a list of some procedures that have been used for analysis of MAO activity. These procedures have been utilized primarily for in vitro and ex vivo assays of MAO. In the former, the drug of interest and the radiolabeled amine are added to the enzyme source (e.g., brain tissue) in a buffer solution in a test tube. In the latter, the animals are treated with the drug or vehicle, the brains are removed at a given time interval after drug administration, and the radiolabeled amine is added to the brain homogenate. The disadvantage of the ex vivo procedure is that in the case of reversible, competitive inhibitors the drug is diluted in concentration and the full effect of the drug on inhibition of MAO is not apparent ex vivo (Tipton et al., 1983; Fowler and Ross, 1984; Waldmeier et al., 1984). To overcome this problem, "protection experiments" have been utilized to assess the reversibility and potency of MAO inhibitors in vivo (Green and Elliot, 1980); here the prevention of action of an irreversible inhibitor, by virtue of the competetive occupancy of enzyme binding sites by a potentially reversible compound, is used as a measure of the potency and reversibility of an inhibitor (Turkish et al., 1987). Such an approach is, however, quite complex (Green, 1984), and the need for other in vivo measurements for establishing the functional reversibility of an MAO inhibitor has been emphasized (Green, 1984; Waldmeier et al., 1984). In this regard behavioral tests have been used as indirect indices of MAO activity in vivo. Such tests include potentiation of the effects of L-5-hydroxytryptophan, L-DOPA, and β-phenylethylamine (Maitre et al., 1976; Ortmann et al., 1980, 1984). Turkish et al. (1987) recently reported a behavioral test

Table 4
Assay of Monoamine Oxidase

Principle	Method	Reference
Consumption of oxygen	Manometry	Creasey, 1956
	Oxygen electrode	Tipton and Dawson, 1968
Disappearance of amine substrates	Direct UV measurement (Kynuramine)	Weissbach et al., 1960
	Ninhydrin reagent (spectrophotometry)	Udenfriend et al., 1955
	2,4,6-Trinitrobenzene-1-sulfonic acid (TNBS) reagent (spectrophotometry)	Obata et al., 1971
	HPLC (electrochemical)	Yu et al., 1982
Formation of ammonium	Nessler's color reaction (spectrophotometry)	Cotzias and Greenough, 1959; Guha and Krishamurti, 1965
	Indolephenol color reaction (spectrophotometry)	Nagatsu and Yagi, 1966
	Coupled with glutamate dehydrogenase system (spectrophotometry)	Yamada et al., 1972
	Fluorometry	Harada and Nagatsu, 1973
	Ammonium sensitive	Meyerson et al., 1978; Nimmo and Tipton, 1979
Formation of aldehydes	Direct UV measurement with various amine derivatives as substrates	Tabor et al., 1954; Deitrich et al., 1962; McEwen and Cohen, 1963; Zeller et al., 1965; Houslay and Tipton, 1974

	p-Dimethylaminobenz-alamine as substrate (colorimetry)	Deitrich and Erwin, 1969
	2,4-Dinitrophenylhy-drazine reagent (colorimetry)	Green and Haughton, 1961
	Formation of 4-hy-droxyquinoline from kynuramine (fluorometry)	Kraml, 1965
	Formation of 4,5-dihydroxyquinoline from 5-hydroxyky-nuramine (fluorometry)	Takahashi and Taka-hara, 1968
	Coupled with aldehyde dehydrogenase system (spectrophotometry)	Houslay and Tipton, 1973b; Anderson et al., 1976
	Fluorometry	Udenfriend, 1962
	Bioluminesence	Tenne et al., 1985
	HPLC (spectrophotometry)	Nissinen, 1984
	HPLC (fluorophotometry)	Nohta et al., 1983
	HPLC (electrochemical detection)	Yu et al., 1986
Formation of H_2O_2	Amperometry	Mason and Olson, 1970
	Homovanillic acid as fluorescent acceptor (fluorometry)	Snyder and Hendley, 1968; Guilbault et al., 1968; Tipton, 1969
	Leuco-2',7'-dichloro-fluorescein as fluorescent acceptor (fluorometry)	Köchli and von Wartburg, 1978
	Seopolin as fluorescent acceptor (fluorometry)	Harris and Cooper, 1982

(continued)

Table 4 *(continued)*

Principle	Method	Reference
Formation of acid or alcohol metabolites	Direct fluorometry of indoleacetic acid or 5-hydroxyindoleacetic acid	Lovenberg et al., 1964; Hidaka et al., 1967
	Radioenzymatic (solvent extraction)	Wurtman and Axelrod, 1963; Otsuka and Kobayashi, 1964; Tipton and Youdim, 1976; Fowler and Oreland, 1980
	Radioenzymatic (ion-exchange chromatography)	Robinson et al., 1968; Southgate and Collins, 1969; Goridis and Neff, 1971
	Radioenzymatic (liquid chromatography)	Jain et al., 1973; Wu and Dyck, 1975
	Gas chromatography	Uji et al., 1980
	HPLC	Kobes et al., 1980; Kawai and Nagatsu, 1982; Yu et al., 1986

[a]Reproduced from Yu, 1986, with permission.

involving potentiation of the effects of an acute injection of β-phenylethylamine. Only computer-based behavioral measurements were used, and the test was used to assess the time course of type B inhibition after administration of (–)-deprenyl and MD240928; the results indicated a significant underestimation of MD240928-induced inhibition of MAO using ex vivo measures.

4. Neurotransmitter Receptor Binding Studies In Vitro and Ex Vivo

As mentioned above, novel antidepressants are now available that do not have the tricyclic structure, do not inhibit MAO, and, in some cases (iprindole, mianserin, viloxazine, bupropion), are not particularly strong inhibitors of NA or 5-HT. The existence of these drugs, as well as the well-known delay seen before antidepressant drugs begin to produce clinical improvement in depressed patients, led researchers to study more closely changes in receptor sensitivity after antidepressant administration. Such studies have been greatly accelerated by the development of in vitro receptor binding studies, which have also been applied extensively to studies on neuroleptics and anxiolytics. The basic principles of such binding have been admirably described in a number of excellent articles. Some examples of reviews and books in this area are: Yamamura et al. (1978); Williams and Lefkowitz (1978); Seeman (1980); Enna (1980); Burt (1980); Bylund (1980); Yamamura and Enna (1981); Eckelman (1982); Iversen et al. (1983); Grigoriadis and Seeman (1984); Boulton et al. (1986).

Before proceeding further, we must mention that an in vitro "binding site" cannot automatically be equated with a "receptor." The binding must result in vivo in a physiological or biochemical response at relevant doses of the radioligand involved in order for the site to be considered a receptor. This aspect has been discussed at length in several articles (e.g., Ahlquist, 1979; Ariens, 1979; Burt, 1980; Creese, 1978; Sulser, 1982, 1986; Ariens and Simonis, 1983; Laduron, 1984; Leysen, 1984; May and Minneman, 1986; Peroutka, 1986), and workers intending to embark on research in this area would do well to familiarize themselves with the binding site/receptor controversy.

4.1. Antidepressants and Neurotransmitter Receptors

4.1.1. Chronic Effects of Antidepressants

Vetulani and Sulser (1975) demonstrated that long-term administration of antidepressants caused a reduction in the activity of a NA-stimulated adenylate cyclase in rat brain tissue, and this observation stimulated considerable research into the chronic effects of antidepressants on receptor binding. It has now been demonstrated that chronic administration of tricyclics, MAO inhibitors, novel antidepressants, and electroconvulsive treatment produces a reduction in the number of β-adrenergic receptors in rat brain (Banerjee et al., 1977; Wolfe et al., 1978; Sulser, 1982; Sugrue, 1983; Koe and Vinick, 1984; Maj et al., 1984; Murphy et al., 1984; Snyder and Peroutka, 1984; McNeal and Cimbolic, 1986; Okada et al., 1986; Scott and Crews, 1986). Peroutka and Snyder (1980b) conducted experiments in which groups of rats were administered, for 21 d, tricyclics, MAO inhibitors, or novel antidepressants (iprindole, fluoxetine). All of these drugs except fluoxetine decreased the number of β receptors and 5-HT$_2$ receptors. Of the antidepressants, all except desipramine produced a greater reduction in 5-HT$_2$ receptor number than in β receptors. This long-term drug administration had no effect on muscarinic, α-adrenergic (as measured by [³H]-WB-4101 binding; see later discussion), or dopamine receptors. Scatchard analyses indicated that these antidepressants were decreasing numbers of β and 5-HT$_2$ receptors without causing an alteration in affinity. The down-regulation of 5-HT$_2$ receptors by various types of antidepressants has also been demonstrated by other researchers (Kellar et al., 1981, 1985; Blackshear and Sanders-Bush, 1982; Scott and Crews, 1986).

With regard to changes in α$_1$ receptors, it should be noted that whereas several groups of researchers have observed no changes in α$_1$ binding after chronic administration of antidepressants when [³H]-WB-4101 or [³H]-dihydroergocryptine were used as ligands (*see* Maj et al., 1984, for review), Campbell and McKernan (1982) and Maj et al. (1983) did observe an increase in binding (B_{max}) to α$_1$-adrenoreceptors after chronic administration of imipramine and using [³H]-prazosin as ligand. It has been suggested (Doxey et al., 1981; Massingham, 1981) that [³H]-WB-4101 may not be as specific for α$_1$-adrenoreceptors as originally thought.

Considerable research has also been done on the α$_2$-adrenergic site after chronic administration of antidepressants, but this area of investigation is controversial, with increases, decreases, and no change in binding having been reported (Johnson et al., 1980; Cohen et al., 1982; Pilc and Vetulani, 1982; Sethy et al., 1983; Reisine et al., 1982; Maj et al., 1984, for review). In this regard, it is of interest that Pilc and Enna (1986) recently examined the effect of antidepressants on the yohimbine-sensitive component of the cAMP response to adrenoreceptor agonists and obtained results suggesting that antidepressants decrease the responsiveness of the rat brain α$_2$-adrenergic system by shifting the receptor to the antagonist-preferring state.

4.1.2. Other Aspects of the Contribution of Receptor Blockade to the Actions and Side Effects of Antidepressants

4.1.2.1. HISTAMINE RECEPTORS.   Snyder and Peroutka (1984) have also speculated on the possible role of histamine blockade in the actions of antidepressants. It is known that tricyclics and the novel antidepressants iprindole and mianserin are relatively strong inhibitors of the histamine H$_2$ adenylate cyclase (Green and Maayani, 1977; Kanof and Greengard, 1978), and that certain tricyclics are very potent at competing for H_1 receptor binding sites (as labeled with [³H]-mepyramine) (Hill et al., 1977; Tran et al., 1978). In general, the antidepressants tested were about 50–100 times more potent in competing at H_1 than at H_2 receptors, and as much as 1000 times more potent than their effects on amine uptake (Snyder and Peroutka, 1984). There are, however, some discrepancies between histamine receptor blockade by these drugs and clinical potency. For example, doxepin and amitriptyline, despite being 90 and 15 times more potent, respectively, than protriptyline at H_1 receptors, have similar clinical potencies to protriptyline (Snyder and Peroutka, 1984).

In spite of the uncertainty about the role of histamine receptors in antidepressant actions, it is certainly of interest that several antidepressants block H_1 receptors at concentrations similar to those of clinically useful antihistamines (drugs known in many cases to cause sedation) and also block H_2 receptors, which are known to mediate cardiac effects of histamine. Thus the sedative and/or hypotensive side effects of these antidepressants may be mediated, at least in part, through histaminergic mechanisms (Snyder and Peroutka, 1984; Paul et al., 1985).

4.1.2.2. CHOLINERGIC RECEPTORS.   It is well known that tricyclic antidepressants cause a variety of anticholinergic side effects (dry mouth, urinary retention, excessive perspiration, and so on). Experiments on the effects of antidepressants on displacing the muscarinic ligand [³H]-quinuclidinyl benzilate ([³H]-QNB) from rat whole brain homogenates show that the potencies of this action agree well with the known anticholinergic side effects of the antidepressants (Snyder and Yamamura, 1977). Thus the MAO inhibitors were much weaker than the tricyclics, and of the tricyclics tested, amitriptyline had the strongest effect on displacing the [³H]-QNB, and desipramine, the weakest, in accord with their known anticholinergic effects in the clinical situation.

Janowsky and coworkers (*see* Risch and Janowsky, 1984, for review) have proposed that depression is the result of an adrenergic-cholinergic imbalance, with the symptoms being the result of a cholinergic predominance. The receptor findings on the acute anticholinergic effects of the tricyclics may give some support to this hypothesis, but it should be remembered that, as stated earlier in this review, long-term administration of antidepressants had no effect on muscarinic receptor number or affinity.

4.1.2.3. α-ADRENERGIC RECEPTORS.   Of the tricyclic antidepressants, the tertiary amine compounds are particularly effective in treating patients with psychomotor agitation and also elicit a relatively high incidence of sedative or hypotensive side effects. On the other hand, the secondary amine tricyclics induce psychomotor activation (and hence are more effective in retarded depression) and result in a lower incidence of sedative and hypotensive side effects (Klein et al., 1980; Snyder and Peroutka, 1984). This may reflect a difference in potency at blocking NA or 5-HT reuptake, but it has also been proposed that blockade of α-adrenergic receptors may contribute to these properties of the antidepressants (U'Prichard et al., 1978). Indeed, in a series of tricyclics it has been found (U'Prichard et al., 1978) that the most potent displacers of the α-adrenergicreceptor ligand [^{3}H]-WB-4101 were the drugs with the greatest sedative and hypotensive properties, whereas the weakest displacers were the ones with the greatest ability to cause psychomotor activation. This work has now been extended to other antidepressants, with similar findings (Richelson, 1983, 1984; Baldessarini, 1985), and cumulative results (taken from Baldessarini, 1985) are shown in Table 5.

4.2. Neuroleptics and Neurotransmitter Receptors

4.2.1. Dopamine Receptors and Clinical Potency

The dopamine hypothesis of schizophrenia (Iversen, 1975) and the discovery of multiple dopamine (DA) receptors have been very important factors in the development of in vitro screening procedures for neuroleptic drugs. Researchers showed that there was a very good correlation between clinical potencies of neuroleptics and their abilities to displace [^{3}H]-butyrophenones from caudate binding sites in vitro (Creese et al., 1976; Seeman et al., 1976; Peroutka and Snyder, 1980a; Seeman, 1980; Creese, 1985). It was also demonstrated that there was no such correlation with binding to 5-HT$_2$ receptors, histamine H$_1$ receptors (Peroutka and Snyder, 1980a), or muscarinic cholinergic receptors (Creese, 1983). Also of interest is the fact that the potencies of neuroleptics in competing for [^{3}H]-haloperidol binding correlate highly with their potencies in blocking apomorphine- and amphetamine-induced stereotypy and in preventing apomorphine-induced emesis (Creese et al., 1976, 1983; Creese, 1985).

In vivo (more correctly, ex vivo; *see* section 2.2.1 of this chapter) binding studies have also been conducted with radiolabeled dopamine antagonists and have been useful in measuring the relative bioavailability of the drugs under study. In such experiments, rats are injected with the radioligand just before or after being injected with the drug of interest. At specified time intervals, the animals are killed, the brains removed, and brain regions dissected out. Depending on the researcher, the radioactivity in the regions is counted directly or homogenates are prepared and filtered and the radioactivity on the resultant particulate fraction is counted. Kuhar (1982) and Williams and Wood (1986) have discussed the advantages and disadvantages of ex vivo versus in vitro studies. Laduron et al. (1978) found that the ex vivo approach using [^{3}H]-spiperone as the radioligand was a useful means of determining the accessability of various DA antagonists to CNS receptors. Kohler et al. (1979, 1981) and Bischoff et al. (1982) used this approach to show that substituted benzamides possess a selective affinity for DA receptors in the septum and hippocampus, and LeFur et al. (1980) reported that such binding of [^{3}H]-spiperone in the frontal cortex involved mainly 5-HT$_2$ receptors, not dopaminergic receptors.

4.2.2. Neurotransmitter Receptors and Side Effects of Neuroleptics

4.2.2.1. α-RECEPTORS.   Peroutka et al. (1977) compared a number of neuroleptics with regard to the ratios of their potencies as α-adrenoreceptor antagonists to their potencies as DA receptor antagonists (K_i, [^{3}H]-WB4101 binding/K_i, [^{3}H]-haloperidol binding) and found that those neuroleptics with the greatest tendency to cause orthostatic hypotension and sedation had the lowest ratios. There was a significant inverse correlation between the ratio and this side effect throughout the range of neuroleptics studied (Snyder et al., 1978).

4.2.2.2. HISTAMINE RECEPTORS.   Like tricyclics and some novel antidepressants, neuroleptics are blockers of histamine H$_1$ and H$_2$ receptors (Tran et al., 1978; Snyder and Peroutka, 1984). This effect is particularly strong with regard to H$_2$ receptors (Green and Mayani, 1977; Kanof and Greengard, 1978). It is possible that these effects may contribute to the sedation seen with some neuroleptics.

Table 6 contains a summary of the effects of a variety of neuroleptics on several types of receptors (Peroutka and Snyder, 1980a).

4.2.2.3. EXTRAPYRAMIDAL SIDE EFFECTS: DOPAMINE RECEPTORS AND MUSCARINIC CHOLINERGIC RECEPTORS.   Parkinson-like side effects of rigidity and akinesia are observed quite frequently with some neuroleptics. This is thought to be caused by blockade of DA receptors in the striatum, which results in an imbalance between the dopaminergic and cholinergic systems. Since anticholinergic

Table 5
Potency of Antidepressants and Other Agents for Amine Receptors (IC$_{50}$, nM)[a]

Agent	Cholinergic ACh$_m$	Adrenergic			Dopami-nergic (D$_2$)	Serotonin		Histamine	
		α_1	α_2	β		5-HT$_1$	5-HT$_2$	H$_1$	H$_2$
Typical tricyclic-type									
Amitriptyline	34	26	765	6,800	290	1,480	13	0.1	54
Butriptyline	36	560	4,800	—	—	—	—	—	—
Protriptyline	70	200	6,700	3,100	360	—	640	48	420
Trimipramine	92	34	1,050	—	—	42	—	0.1	—
Doxepin	138	24	1,000	7,100	380	720	250	0.03	160
Imipramine	186	74	3,200	38,000	610	5,000	245	16	153
Clomipramine	268	46	3,200	—	—	—	—	—	50
Nortriptyline	550	66	2,100	15,000	800	920	40	17	400
Maprotiline	590	90	9,100	—	—	—	—	—	—
Amoxapine	1,000	50	2,600	—	100	—	—	—	—
Desipramine	1,200	136	7,000	4,200	980	9,500	540	250	320

Atypical									
Nomifensine	250,000	900	4,600	—	>10,000	1,400	—	—	—
Mianserin	2,400	60	53	4,400	2,200	500	—	3	67
Iprindole	6,000	6,000	8,300	21,000	6,300	>10,000	1,900	100	200
Fluoxetine	8,000	8,000	13,000	10,000	6,600	>10,000	1,300	—	—
Bupropion	47,600	50,000	>100,000	—	—	—	—	—	—
Trazodone	320,000	70	1,000	>10,000	3,000	1,700	111	460	50,000
Others									
Atropine	2	—	—	—	—	—	—	—	—
Chlorpromazine	5,700	4	> 10,000	25	3,500	15	28	41	300

[a]Data are summarized from a review and references therein by Baldessarini, *Biomedical Aspects of Depression and Its Treatment* (American Psychiatric Press, Inc., Washington, DC, 1983); for muscarinic and adrenergic receptors from E. Richelson, Mayo Clinic (*Mayo Clinic Proc.* **58**, 40, 1983; and personal communication, 1984); and for D-2 dopaminergic receptors from B. Cohen, Harvard Medical School (personal communication, 1983).

Competition for binding of radiolabeled ligands included WB-4101 or prazosin (α_1), clonidine or rauwolscine (α_2), dihydroxyalprenolol (β), spiroperidol (D-2), serotonin (5-HT$_1$), spiroperidol (in cerebral cortex), or ketanserin (5-HT$_2$), mepyramine (or stimulation of cyclic-GMP synthesis) (H$_1$), stimulation of cyclic-AMP synthesis (H$_2$); most were evaluated with membrane preparations of rat (or human) brain tissue; muscarinic receptors (binding of quinuclidinyl benzilate, QNB) were evaluated in guinea pig ileum and rat brain tissue and the results pooled.

Note that none of these affinities correlates significantly with clinical potency, even within the tricyclic group of agents (the range of potency is limited, ranging from about 30–300 mg/d typical doses).

Amitriptyline and its congeners are strongly anticholinergic, while desipramine, amoxapine, and nomifensine are the least anticholinergic of the tricyclic-type and comparable agents. Amitriptyline, doxepin, and trimipramine are relatively strongly antiadrenergic and antihistaminic, as is chlorpromazine (which may correspond with sedative and autonomic side effects). Most of the atypical agents are only weakly antimuscarinic. (This table was reprinted from Baldessarini, 1985, pp. with permission from the authors and Harvard University Press.)

Table 6

Average Clinical Daily Dose for Neuroleptics and In Vitro Potency at Neurotransmitter Receptors[a]

| | | Inhibition of $[^3H]$-ligand binding K_1, nM | | | |
| | | $[^3H]$-Spiroperidol | | $[^3H]$-WB- | $[^3H]$- |
Neuroleptic	Average clinical dose, mg/d	At dopamine receptors (caudate)	At serotonin receptors (cortex)	4101 at α-adrenergic receptors	Mepyramine at histamine H_1 receptors
Phenothiazines					
Fluphenazine	15	3.7	25	13	58
Trifluoperazine	25	4.4	24	67	135
Chlorpromazine	600	25	19	4.3	28
Thioridazine	625	63	16	7.1	25
Promazine	660	280	130	9.4	25
Butyrophenones					
Spiroperidol	1.5	0.68	0.59	14	550
Benperidol	1.6	1.4	3.7	7.1	200

Clofluperol	2.4	4.2	42	56	300
Trifluoperidol	3	2.7	6.8	8.4	2,100
Bromperidol	4	3.7	26	100	700
Droperidol	8	3.0	4.6	1.4	2,500
Haloperidol	12	4.4	45	14	2,600
Moperone	22	9.3	52	22	22,000
Fluanisone	100	16	36	1.8	150
Pipamperone	450	360	6.1	100	450
Others					
Fluspirilene	2	2.6	5.9	240	1,700
α-Flupenthixol	4	3.0	13	8.2	39
Pimozide	4	3.6	25	20	1,100
(+)-Butaclamol	6	3.8	2.6	20	470
cis-Thiothixene	30	4.5	36	11	37
Penfluridol	30	7.8	150	280	3,400
Clozapine	725	380	29	17	20

[a]Reprinted from Peroutka and Snyder, 1980a, with permission from the authors and the American Psychiatric Association.

drugs are effective in antagonizing these extrapyramidal side effects, it was proposed that those neuroleptics that were effective in blocking muscarinic receptors would demonstrate the lowest incidence of extrapyramidal side effects. Using [^{3}H]-QNB and striatal membranes, this turned out to be the case (Miller and Hilley, 1974; Snyder et al., 1974). Some drugs, such as pimozide, were relatively weak at displacing [^{3}H]-QNB, but still display a relatively low incidence of extrapyramidal side effects; Laduron and Leysen (1979) suggested that this seeming paradox may be related to the relatively slow onset of action of these drugs. Overall, the screening of potentially useful antipsychotic drugs for muscarinic receptor activity appears to be a useful predictor for the propensity to induce extrapyramidal side effects in the clinical situation (Creese, 1985).

4.3. Benzodiazepine Receptors

The discovery of specific binding sites for benzodiazepines in brain (Mohler and Okado, 1977; Squires and Braestrup, 1977) has done much to increase our knowledge of the mechanisms of action of these frequently prescribed drugs. There is a high correlation between the affinity of numerous benzodiazepines for these binding sites and their potency in in vivo animal screening procedures such as the cat muscle relaxant test (predictive of anxiolytic action) (Mohler and Okada, 1977) and inhibition of pentylenetetrazole-induced convulsions (Braestrup and Nielsen, 1982).

In a recent investigation, Miller et al. (1987) used a tracer radioligand technique employing ^{3}H-Ro 15-1788 to measure in vivo receptor occupancy in mice; these authors studied the correlation between receptor occupancy and the pharmocodynamic actions of clonazepam and lorazepam to determine the degree of receptor occupancy that corresponded to the ED$_{50}$ values for the effects of these drugs in blocking pentylenetetrazole seizures and inducing rotarod ataxia.

At least three subclasses of benzodiazepine receptors have been identified in mammalian tissues by using radioligand binding studies, and it is possible that at least one of these may mediate the anxiolytic actions of the benzodiazepines, whereas the others may be involved in the mediation of sedative, ataxic, and anti-convulsant properties (Braestrup and Nielsen, 1982; Williams, 1984; Martin, 1986).

Batteries of receptor binding studies have not only allowed comparison of relative affinities of benzodiazepines at the binding site, but have permitted the distinction between partial agonists, agonists, agonist/antagonists, and inverse agonists for this receptor (Wood et al., 1984a) and have resulted in the identification of nine classes of nonbenzodiazepine anxiolytic agents (Williams, 1984). Ex vivo binding experiments with [^{3}H]-flunitrazepam as the radioligand have also been utilized to demonstrate bioavailability of benzodiazepine agonists, antagonists, and partial agonists (Chang and Snyder, 1978; Yokoyama et al., 1982; Peterson et al., 1984). Further, this technique has also been employed to evaluate drugs that enhance [^{3}H]-flunitrazepam binding and has resulted in division of such drugs into those that enhance binding only ex vivo and those that enhance both ex vivo and in vitro (Skolnick et al., 1980; Mennini et al., 1982; Saano, 1982; Koe, 1983; Oakley and Jones, 1983; Jeevanjee et al., 1984; Wood et al., 1984b).

Several detailed descriptions of the utilization of receptor binding in drug screening for benzodiazepine-like drugs and neuroleptics are available (Creese, 1978; Williams, 1983, 1984; Williams and U'Pritchard, 1984; Williams and Wood, 1986).

5. Adenylate Cyclase

In vitro and ex vivo studies on the effects of drugs on neurotransmitter-sensitive adenylate cyclase have also done much to enhance our knowledge about the modes of action of psychotropic drugs. Some examples have already been mentioned in this chapter. The work of Vetulani and Sulser (1975) showed that long-term administration of antidepressants resulted in a decreased activity of NA-sensitive adenylate cyclase (thought to represent β-adrenoceptors). This has led to a major rethinking of the biogenic amine hypotheses of affective disorders (Sulser, 1982) and has resulted in numerous experiments on the effects of chronic administration of antidepressants on binding of radioligands on neurotransmitter binding sites.

There has been much work done on identifying 5-HT receptor subtypes, and work thus far suggests that 5-HT$_1$ receptors are associated with adenylate cyclase (Peroutka, 1976; Enjalbert et al., 1978; Fillion et al., 1979; Peroutka et al., 1979).

Histamine H$_2$ receptors are linked to adenylate cyclase, and measurement of histamine-sensitive adenylate cyclase has been used to monitor such receptors. As has been discussed above, tricyclic antidepressants and novel antidepressants are reasonably potent inhibitors of the H$_2$ adenylate cyclase in brain homogenates, and the neuroleptic phenothiazines are very potent indeed (Green and Maayani, 1977; Kanof and Greengard, 1978).

Studies on DA-sensitive adenylate cyclase have played an important role in research on the nature of DA receptors. Although the existence of up to five types of DA receptor subtypes had been suggested (Costall and Naylor, 1981; Seeman, 1982), there now seems to be a concensus that there are only two types, with high- and low-affinity versions of each, as suggested on the basis of interactions with adenylate cyclase (*see* Kebabian and Calne, 1979, for reviews: Stoof and Kebabian, 1984; Grigoriadis and Seeman, 1984; Mishra, 1986). Thus the DA_1 receptor is coupled to adenylate cyclase, whereas the DA_2 receptor is not or inhibits adenylate cyclase. For details of the methodology of adenylate cyclase assays, the reader is referred to recent reviews by Mishra (1986) and Sulakhe et al. (1986).

6. Future Directions

Recent research indicates that with the development of new, more specific radioligands, investigations of multiple receptors for neurotransmitters and drugs will continue. It is hoped that this will further advance our knowledge of the functions of these subtypes, allowing us to design drugs with more specific clinical actions and fewer side effects. It seems likely that rapid, relatively inexpensive in vitro and ex vivo screening procedures will play an important role in the selection of such drugs.

There has been a great deal of interest lately in binding sites for [3H]-antidepressants, with emphasis being on [3H]-imipramine (Raisman et al., 1979; Rehavi et al., 1980; Paul et al., 1984; Hrdina, 1986; Schoemaker and Langer, this volume). The [3H]-imipramine binding site, which is present in brain and platelets, appears to be closely associated with, but not identical to, the 5-HT uptake site. As shown in Fig. 1 (Paul et al., 1984), there is a high correlation between the potencies of various antidepressants in inhibiting both the binding of [3H]-imipramine and the uptake of [3H]-5-HT in human platelets (Paul et al., 1981). Laduron et al. (1982) have reported differences, however, in the subcellular distribution of 5-HT uptake sites and [3H]-imipramine binding sites. In a recent report, Hrdina (1987) studied the effects of chronic (21-d) administration of nortriptyline, fluoxetine, iprindole, phenelzine, and maprotiline on high- and low-affinity [3H]-imipramine binding

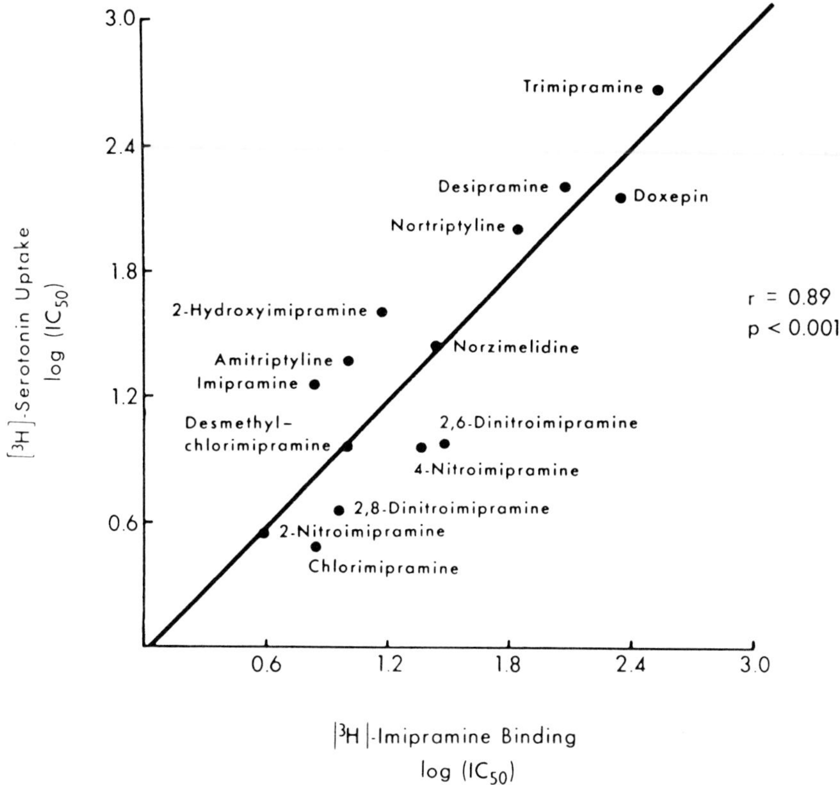

Fig. 1. Correlation between the potencies of various antidepressants in inhibiting the binding of ^{3}H-imipramine and the uptake of serotonin in human platelets. Values represent the logarithm of drug concentrations (nmol/L) necessary to inhibit 50% of specific drug binding or uptake (from Paul et al., 1984, with permission of the authors and the publisher).

sites. All of these drugs except iprindole significantly decreased the affinity, but not the proportion, of high-affinity sites for [^{3}H]-imipramine. Changes in [^{3}H]-imipramine's affinity for its binding sites after chronic administration were found by Hrdina to not be associated with the effects of the drugs on 5-HT levels or uptake. Thus, although studies on [^{3}H]-imipramine binding have yielded interesting data and suggest that this technique may have relevance for understanding the etiology of depression, further details on this binding must be obtained before it becomes a routine screening procedure for antidepressants.

The mention of the use of human platelets in the previous paragraph brings up another important issue, and that is the increased use of cells from the periphery in humans for future screening of drugs. Already platelets (source of MAO, [3H]-imipramine binding sites, [3H]-5-HT uptake sites, and α_2-adrenergic receptors), lymphocytes (source of β-adrenoreceptors), and fibroblasts (source of muscarninic and β-adrenoreceptors) are being used to study long-term effects of antidepressants and/or the etiology of affective disorders (e.g., Lingjaerde, 1984; Abel et al., 1985; Piletz and Halaris, 1985; Mann et al., 1985; Kafka et al., 1986; Honegger et al., 1986; Wood et al., 1986; Asberg et al., 1986). Such models will, at least to some degree, overcome criticisms of performing screening tests in animal models and applying the findings to the human situation.

Another area that may receive increased interest in the near future is the interaction of antidepressants with GABA binding sites. Lloyd et al. (1985) recently reported that long-term administration of several tricyclics, novel antidepressants, an MAO inhibitor, and electroshock up-regulated GABA-B binding in rat frontal cortex; the GABA-B site is bicuculline-insensitive and not associated with a chloride ionophore, but with a calcium channel (Lloyd, 1986). Thus GABA-B binding site measurement may be added to the repertoire of radioligand bindings tests already used as screening devices for new drugs.

Before closing, it should be mentioned that in vivo electrophysiological techniques have also been utilized extensively for investigating the effects of long-term administration of antidepressants on sensitivity of adrenergic and serotoninergic receptors. This aspect has not been covered in this volume, but the reader can obtain an overview from the review written by Blier and de Montigny (1985).

7. Summary

A variety of in vitro and ex vivo techniques for studying psychotropic drugs has been discussed. These have included uptake studies, measurement of MAO activity, receptor binding studies, and measurement of adenylate cyclase. Examples of applications of these techniques, when appropriate, to screening for potential antidepressants, neuroleptics, and anxiolytics have been discussed.

Although these screening procedures are for the most part inexpensive and convenient to perform, it should be remembered that other tests (behavioral, electrophysiological, pharmacokinetic, and so on) will eventually have to be carried to more clearly understand the overall action of the drug in question. Future screening procedures may be expanded to routinely include studies on binding sites for radiolabeled antidepressants and for GABA, as well as the use of peripheral models (e.g., platelets, lymphocytes, fibroblasts) in humans.

Acknowledgments

The authors thank the Provincial Mental Health Advisory Council (PMHAC) and the Alberta Heritage Foundation for Medical Research (AHFMR) for funding. The secretarial assistance of Miss Heather Schmidt is gratefully acknowledged.

References

Abel M. S., Clody D. E., Wennogle L. P., and Meyuerson L. R. (1985) Effect of chronic desmethylimipramine or electroconvulsive shock on selected brain and platelet neurotransmitter recognition sites. *Biochem. Pharmacol.* **34,** 679–683.

Ahlquist R. P. (1979) Adrenoceptors. *Trends Pharmacol. Sci.* **1,** 16–17.

Anderson R. A., Myerson L. R., and Tabakoff B. (1976) Characteristics of enzyme forming 3-methoxy-4-hydroxyphenylethylene-glycol (MOPEG) in brain. *Neurochem. Res.* **1,** 525–540.

Ariens E. J. (1979) Receptors: From fiction to fact. *Trends Pharmacol. Sci.* **1,** 11–15.

Ariens E. J. and Simonis A. M. (1983) Physiological and pharmacological aspects of adrenergic receptor classification. *Biochem. Pharmacol.* **32,** 1539–1545.

Asberg M. A., Eriksson B., Martensson B., Traskman-Bendz L., and Wagner A. (1986) Therapeutic effects of serotonin uptake inhibitors in depression. *J. Clin. Psychiatry* **46,** 23–35.

Azzaro A. J. and Demarest K. T. (1982) Inhibitory effects of type A and type B monoamine oxidase inhibitors on synaptosomal accumulation of [³H]-dopamine: A reflection of antidepressant potency. *Biochem. Pharmacol.* **31,** 2195–2197.

Baker G. B. and Dyck L. E. (1985) Neuronal Transport of Amines In Vitro, in *Neuromethods* vol. 2 *Amines and Their Metabolites* (Boulton A. A., Baker G. B., and Baker J. M., eds.) Humana, Clifton, New Jersey.

Baker G. B., McKim H. R., Calverley D. G., and Dewhurst W. G. (1978) Effects of the monoamine oxidase inhibitors tranylcypromine, phenelzine and pheniprazine on the uptake of catecholamines in slices from rat brain regions. *Proc. Eur. Soc. Neurochem.* **1**, 536.

Baldessarini R. J. (1985) *Chemotherapy in Psychiatry: Principles and Practice* Harvard University Press, Cambridge, Massachusetts.

Banerjee S. P., Kung L. S., Riggi S. J., and Chanda S. K. (1977) Development of beta-adrenergic subsensitivity by antidepressants. *Nature* **268**, 455–456.

Baumann P. A. and Maitre L. (1976) Is drug inhibition of dopamine uptake a misinterpretation of *in vitro* experiments? *Nature* **264**, 789–790.

Bischoff S., Bittiger H., Delini-Stula A., and Ortmann R. (1982) Septohippocampal system: Target for substituted benzamides? *Eur. J. Pharmacol.* **79**, 225–232.

Blackshear M. A. and Sanders-Bush E. (1982) Serotonin receptor sensitivity after acute and chronic treatment with mianserin. *J. Pharmacol. Exp. Ther.* **221**, 303–308.

Blier P., and de Montigny C. (1985) Neurobiological Basis of Antidepressant Treatments, in *Pharmacotherapy of Affective Disorders: Theory and Practice* (Dewhurst W. G. and Baker G. B., eds.) Croom and Helm, London.

Boulton A. A., Baker G. B., and Hrdina P. D., eds. (1986) *Neuromethods* vol. 4 *Receptor Binding* Humana, Clifton, New Jersey.

Braestrup C. and Nielsen A. (1982) Anxiety. *Lancet* **2**, 1030–1034.

Burt D. R. (1980) Basic Receptor Methods. II. Problems of Interpretation in Binding Studies, in *Receptor Binding Techniques* (Bylund D. B., ed.) Society for Neuroscience, Bethesda, Maryland.

Bylund D. B. (1980) Analysis of Receptor Binding Data, in *Receptor Binding Techniques* (Bylund D. B., ed.) Society for Neuroscience, Bethesda, Maryland.

Callingham B. A. (1986) Some aspects of monoamine oxidase pharmacology. *Cell Biochem. Func.* **4**, 99–108.

Campbell I. C. and McKernan R. M. (1982) Central and peripheral changes in α-adrenoreceptors in the rat after chronic tricyclic antidepressants. *Br. J. Pharmacol.* **75**, 100P.

Carlsson A., Corrodi H., Fuxe K., and Hokfelt T. (1969a) Effects of some antidepressant drugs on the depletion of intraneuronal brain catecholamine stores caused by 4,α-dimethyl-*meta*-tyramine. *Eur. J. Pharmacol.* **5**, 367–373.

Carlsson A., Corrodi H., Fuxe K., and Hokfelt T. (1969b) Effects of some antidepressant drugs on the depletion of intraneuronal brain 5-hydroxytryptamine stores caused by 4-methyl-α-ethyl-*meta*-tyramine. *Eur. J. Pharmacol.* **5**, 357–366.

Carruba M. O., Picotti G. B., Zambotti F., and Mantegazza P. (1977a) Effects of mazindol, fenfluramine and chlorimipramine on the 5-hydroxytryptamine uptake and storage mechanisms in rat brain: Similarities and differences. *Naunyn Schmiedbergs Arch. Pharmacol.* **300,** 227–232.

Carruba M. O., Picotti G. B., Zambotti F., and Mantegazza P. (1977b) Mazindol and amphetamine as inhibitors of the uptake and releasers of ^{3}H-DA by rat striatal synaptosomes. *Naunyn Schmiedebergs Arch. Pharmacol.* **298,** 15–22.

Chang R. S. L. and Snyder S. H. (1978) Benzodiazepine receptors: Labelling in intact animals with [^{3}H]flunitrazepam. *Eur. J. Pharmacol.* **48,** 213–218.

Cheng Y.-C. and Prusoff W. H. (1973) Relationship between the inhibition constant (K_1) and the concentration of inhibitor which causes 50 per cent inhibition (I_{50}) of an enzymatic reaction. *Biochem. Pharmacol.* **22,** 3099–3108.

Choi M. U. and Abramson M. B. (1978) Effects of pH changes and charge characteristics in the uptake or norepinephrine by synaptosomes of rat brain. *Biochim. Biophys. Acta* **540,** 337–345.

Coccaro E. F. and Siever L. J. (1985) Second generation antidepressants: A comparative review. *J. Clin. Pharmacol.* **25,** 241–260.

Cohen R. M., Ebstein R. P., Daly J. W., and Murphy D. L. (1982) Chronic effects of a monoamine oxidase-inhibiting antidepressant: Decreases in functional α-adrenergic autoreceptors precede the decrease in norepinephrine-stimulated cyclic adenosine 3':5'-monophosphate systems in rat brain. *J. Neurosci.* **2,** 1588–1595.

Costall B. and Naylor R. J. (1981) The hypothesis of different dopamine receptor mechanisms. *Life Sci.* **28,** 215–224.

Cotzias G. C. and Greenough J. J. (1959) Quantitative estimation of amine oxidase. *Nature* **183,** 1732–1733.

Creasey N. W. (1956) Factors which interfere in the monometric assay of monoamine oxidase. *Biochem. J.* **64,** 178–183.

Creese I. (1978) Receptor Binding as a Primary Drug Screening Device, in *Neurotransmitter Receptor Binding* (Yamamura H. I., Enna S. J., and Kuhar M. J., eds.) Raven, New York.

Creese I. (1983) Receptor Interactions of Neuroleptics, in *Neuroleptics: Neurochemical, Behavioral and Clinical Perspectives.* (Coyle J. T. and Enna S. J., eds.) Raven, New York.

Creese I. (1985) Dopamine and Antipsychotic Medications, in *Psychiatry Update, American Psychiatric Association, Annual Review* vol. 4. (Hales R. E. and Frances, A. J., eds.) American Psychiatric Press, Washington, DC.

Creese I., Burt D. R., and Snyder S. H. (1976) Dopamine receptor binding predicts clinical and pharmacological potencies of antischizophrenic drugs. *Science* **192,** 481–483.

Creese I., Sibley D. R., Hamblin M. W., and Leff S. E. (1983) The classification of dopamine receptors: Relationship to radioligand binding. *Ann. Rev. Neurosci.* **6,** 43–71.

Damlouji N. F., Feighner J. P., and Rosenthal M. H. (1985) Recent Advances in Antidepressants, in *Pharmacotherapy of Affective Disorders, Theory and Practice* (Dewhurst W. G. and Baker G. B., eds.) Croom Helm, London.

deBelleroche J., Dykes C. R., and Thomas A. J. (1976) The automated separation and analysis of dopamine, its amino acid precursors, and metabolites and the application of the method to the measurement of specific radioactivities of DA in striatal synaptosomes. *Anal. Biochem.* **71,** 193–203.

Deitrich R. A. and Erwin V. G. (1969) A convenient spectrophotometric assay for monoamine oxidase. *Anal. Biochem.* **30,** 394–402.

Deitrich R. A., Hellerman L., and Wein J. (1962) Diphosphopyridine nucleotide-linked aldehyde dehydrogenase. I. Specificity and sigma-rho function. *J. Biol. Chem.* **237,** 560–564.

Dingledine R., ed. (1984) *Brain Slices* Plenum, New York.

Dorris R. L. and Shore P. A. (1971) Amine uptake and storage mechanisms in the corpus striatum of rat and rabbit. *J. Pharmacol. Exp. Ther.* **179,** 15–19.

Doxey J. C., Hewlett D. R., and Roach A. G. (1981) Assessment of the α-adrenoreceptor selectivity of WB 4101: A comparison with prazosin and phentolamine. *Br. J. Pharmacol.* **61,** 262P–263P.

Dyck L. E. and Boulton A. A. (1980) The effect of reserpine and various monoamine oxidase inhibitors on the uptake and release of triated *meta*-tyramine, *para*-tyramine and dopamine in rat striatal slices. *Res. Commun. Psychol. Psychiat. Behav.* **5,** 61–78.

Eckelman W. C., ed. (1982) *Receptor-Binding Radiotracers* vol. 1 CRC, Boca Raton, Florida.

Eckhardt S. B., Maxwell R. A., and Ferris R. M. (1982) A structure–activity study of the transport sites for the hypothalamic and striatal catecholamine uptake systems. *Mol. Pharmacol.* **21,** 374–379.

Enjalbert A., Hamon M., Bourgoin S., and Bockaert J. (1978) Postsynaptic serotonin-sensitive adenylate cyclase in the central nervous system. *Mol. Pharmacol.* **14,** 11–23.

Enna S. J. (1980) Basic Receptor Methods, in *Receptor Binding Techniques* (Bylund D. B., ed.) Society for Neuroscience, Bethesda, Maryland.

Feighner J. P. (1981) Clinical efficacy of the new antidepressants. *J. Clin. Psycopharmacol.* **1** (suppl.), 235–265.

Ferris R. M., Tang F. L. M., and Maxwell R. A. (1972) A comparison of the capacities of isomers of amphetamine, deoxypipradrol, and methylphenidate to inhibit the uptake of tritiated catecholamines into rat cerebral cortex slices, synaptosomal preparations of rat cerebral cortex, hypothalamus, and striatum and into adrenergic nerves of rabbit aorta. *J. Pharmacol. Exp. Ther.* **181,** 407–416.

Fillion G., Rousselle J. C., Beaudoin D., Pradelles P., Goiny M., Dray F., and Jacob J. (1979) Serotonin-sensitive adenylate cyclase in horse brain synaptosomal membranes. *Life Sci.* **24,** 1813–1822.

Fowler C. J. and Callingham B. A. (1978) Monoamine oxidase A and B: A useful concept? *Biochem. Pharmacol.* **27,** 97–101.

Fowler C. J. and Oreland L. (1980) The nature of the substrate selective interaction between rat liver mitochondrial monoamine oxidase and oxygen. *Biochem. Pharmacol.* **29,** 2225–2233.

Fowler C. J. and Ross S. B. (1984) Selective inhibitors of monoamine oxidase A and B—biochemical and pharmacological properties. *Med. Res. Rev.* **4,** 323–358.

Friedman E., Fung F., and Gershon S. (1977) Antidepressant drugs and DA uptake in different brain regions. *Eur. J. Pharmacol.* **42,** 47–51.

Fuller R. W. (1978) Selectivity among monoamine oxidase inhibitors and its possible importance for development of antidepressant drugs. *Prog. Neuropsychopharmacol.* **2,** 303–311.

Fuller R. W. (1986) Pharmacologic modification of serotonergic function: Drugs for the study and treatment of psychiatric and other disorders. *J. Clin. Psychiatr.* **47,** 4–8.

Fuller R. W. and Wong D. T. (1985) Effects of antidepressants on uptake and receptor systems in the brain. *Prog. Neuro-psychopharmacol. Biol. Psychiat.* **9,** 485–490.

Gentil V., Alevizos B., and Lader M. (1977) Effects of single doses of tranylcypromine on platelet MAO and amine uptake in normal subjects. *Biochem. Pharmacol.* **27,** 1197–1201.

Glover V. and Sandler M. (1986) Clinical chemistry of monoamine oxidase. *Cell Biochem. Func.* **4,** 89–97.

Goodlet I., Mireyless S. E., and Sugrue M. F. (1977) Effects of mianserin, a new antidepressant on the *in vivo* uptake of monoamines. *Br. J. Pharmacol.* **61,** 307–313.

Goridis C. and Neff N. H. (1971) Evidence for a specific monoamine oxidase associated with sympathetic nerves. *Neuropharmacology* **10,** 557–564.

Gorkin V. Z. (1983) *Amine Oxidases in Clinical Research* Pergamon, Oxford.

Green A. L. (1984) Assessment of the Potency of Reversible MAO Inhibitors *In Vivo,* in *Monoamine Oxidase and Disease: Prospects for Therapy with Reversible Inhibitors* (Tipton K. F., Dostert P., and Strolin-Benedetti M., eds.) Academic, London.

Green A. L. and Elliot A. S. (1980) A new approach to the assessment of the potency of reversible monoamine oxidase inhibitors *in vivo,* and its application to (+)amphetamine, *p*-methyoxyamphetamine and harmaline. *Biochem. Pharmacol.* **29**, 2781–2789.

Green A. L. and Haughton T. M. (1961) A colorimetric method for the estimation of monoamine oxidase. *Biochem. J.* **78**, 172–175.

Green J. P. and Maayani S. (1977) Tricyclic antidepressant drugs block histamine H_2 receptor in brain. *Nature* **269**, 164–165.

Grigoriadis D. and Seeman P. (1984) The dopamine/neuroleptic receptor. *Can. J. Neurol. Sci.* **11**, 108–113.

Guha S. R. and Krishamurti C. R. (1965) Purification and solubilization of monoamine oxidase of rat liver mitochondria. *Biochem. Biophys. Res. Commun.* **18**, 350–354.

Guilbault G. G., Brignac P. J., Jr., and Juneau M. (1968) New substrates for the fluorometric determination of oxidative enzymes. *Anal. Chem.* **40**, 1256–1263.

Hall H., Ross S. B., and Ogren S. O. (1982) Effects of Zimelidine on Various Transmitter Systems in the Brain, in *Advances in Biochemical Psychopharmacology* vol. 31 (Costa F. and Racagni G., eds.) Raven, New York.

Hamberger B. and Tuck J. R. (1973) Effect of tricyclic antidepressants on the uptake of noradrenaline and 5-hydroxytryptamine by rat brain slices incubated in buffer or human plasma. *Eur. J. Clin. Pharmacol.* **5**, 1–7.

Harada M. and Nagatsu T. (1973) A sensitive fluorometric assay for monoamine oxidase activity. *Anal. Biochem.* **56**, 283–288.

Harris E. J. and Baldessarini R. J. (1973) The uptake of [^{3}H]-dopamine by homogenates of rat corpus striatum: Effects of cations. *Life Sci.* **13**, 303–312.

Harris E. J. and Cooper M. B. A. (1982) Monoamine oxidase activity of mitochondria prepared from rat liver and rat heart. *J. Neurochem.* **38**, 1068–1071.

Heikkila R. E., Orlansky H., and Cohen G. (1975) Studies on the distinction between uptake inhibition and release of [^{3}H]-dopamine in rat brain tissue slices. *Biochem. Pharmacol.* **24**, 847–852.

Hendley E. D. and Snyder S. H. (1968) Relationship between the action of monoamine oxidase inhibitors on the noradrenaline uptake system and their antidepressant efficacy. *Nature* **220**, 1330–1331.

Hidaka H., Nagatzu T., and Yagi K. (1967) Micro-determination of monoamine oxidase using serotonin as substrate. *J. Biochem.* (Tokyo), **62**, 621–623.

Hill S. J., Young J. M., and Marrian D. H. (1977) Specific binding of ^{3}H-mepyramine to histamine H_1 receptors in intestinal smooth muscle. *Nature* **270**, 361–363.

Hollister L. E. (1986) Current antidepressants. *Ann. Rev. Pharmacol. Toxicol.* **26**, 23–37.

Honegger U. E., Disler B., and Wiesmann U. N. (1986) Chronic exposure of human cells in culture to the tricyclic antidepressant desipramine reduces the number of beta-adrenoceptors. *Biochem. Pharmacol.* **35,** 1899–1902.

Horn A. S. (1978) Characteristics of Neuronal Dopamine Uptake, in *Advances in Biochemical Psychopharmacology* vol. 19 (Roberts P. J., Woodruff G. N., and Iversen L. L., eds.) Raven, New York.

Houslay M. D. and Tipton K. F. (1974) A kinetic evaluation of monoamine oxidase activity in rat live mitochondrial outer membranes. *Biochem. J.* **139,** 645–652.

Housley M. D. and Tipton K. F. (1973) The nature of the electrophoretically separable multiple forms of rat liver monoamine oxidase. *Biochem. J.* **135,** 173–186.

Hrdina P. D. (1986) Binding Sites for Antidepressants, in *Neuromethods* vol. 4 *Receptor Binding* (Boulton A. A., Baker G. B., and Hrdina P. D., eds.) Humana, Clifton, New Jersey.

Hrdina P. D. (1987) Regulation of high- and low-affinity ^{3}H-imipramine recognition sites in rat brain by chronic treatment with antidepressants. *Eur. J. Pharmacol.* (in press).

Hunt P., Raynaud J.-P., Leven M., and Schacht U. (1979) Dopamine uptake inhibitors and releasing agents differentiated by the use of synaptosomes and field-stimulated brain slices in vitro. *Biochem. Pharmacol.* **28,** 2011–2016.

Hyttel J. (1978) Inhibition of [^{3}H]-dopamine accumulation in rat striatal synaptosomes by psychotropic drugs. *Biochem. Pharmacol.* **27,** 1063–1068.

Hyttel J. (1982) Citalopram—pharmacological profile of a specific serotonin uptake inhibitor with antidepressant activity. *Prog. NeuroPsychopharmacol. Biol. Psychiat.* **6,** 278–291.

Iversen L. L. (1971) Role of transmitter uptake mechanisms in synaptic neurotransmission. *Br. J. Pharmacol.* **41,** 571–591.

Iversen L. L. (1975) Dopamine receptors in the brain. *Science* **188,** 1084–1089.

Iversen L. L., Iversen S. D., and Snyder S. H., eds. (1983) *Handbook of Psychopharmacology* vol. 18 *Biochemical Studies on CNS Receptors* Plenum, New York.

Jain M., Sands F., and von Korff R. W. (1973) Monoamine oxidase activity measurements using radioactive substrates. *Anal. Biochem.* **52,** 542–554.

Jeevanjee F., Johnson A. M., Loudon J. M., and Nicholass J. M. (1984) Enhancement of [^{3}H]-flunitrazepam binding by mianserin *in vivo.* *Neurosci. Lett.* **46,** 305–309.

Johnson R. W., Reisine T., Spotnitz L., Wiech N., Ursillo R., and Yamamura H. I. (1980) Effects of desipramine and yohimbine on α- and β-adrenoreceptor sensitivity. *Eur. J. Pharmacol.* **67**, 123–127.

Johnston J. P. (1968) Some observations upon a new inhibitor of monoamine oxiase in brain tissue. *Biochem. Pharmacol.* **17**, 1285–1297.

Jonsson G., Hamberger B., Malmfors T., and Sachs C. (1969) Uptake and accumulation of ^{3}H-noradrenaline in adrenergic nerves of rat iris. Effect of reserpine, monoamine oxidase and tyrosine hydroxylase inhibition. *Eur. J. Pharmacol.* **8**, 58–72.

Kafka M. S., Nurnberger J. I., Siever L., Targum S., Uhde T. W., and Gershon E. S. (1986) Alpha$_2$-adrenergic receptor function in patients with unipolar and bipolar affective disorders. *J. Affective Disord.* **10**, 163–169.

Kanof P. D. and Greengard P. (1978) Brain histamine receptors as targets for antidepressant drugs. *Nature* **272**, 329–333.

Kawai S. and Nagatsu T. (1982) Gas Chromatographic and High-Performance Liquid Chromatographic Determination of Monoamine Oxidase Activity Using Mixed Substrates, in *Monoamine Oxidase: Basic and Clinical Frontiers* (Kamijo K., Usdin E., and Nagatzu T., eds.) Excerpta Medica, Amsterdam.

Kebabian J. W. and Calne D. B. (1979) Multiple receptors for dopamine. *Nature* **277**, 93–96.

Kellar K. J., Cascio C. S., Butler J. A., and Kurtze R. N. (1981) Differential effects of electroconvulsive shock and antidepressant drugs on serotonin-2-receptors in rat brain. *Eur. J. Pharmacol.* **69**, 515–518.

Kellar K. J., Stockmeier C. A., and Gomez J. M. (1985) Regulation of serotonin neurotransmission by antidepressant drugs and electroconvulsive shock: The fall and rise of serotonin receptors. *Acta Pharmacol. Toxicol.* **56** (Suppl. 1), 138–145.

Klein D. F., Gittelman R., Quitkin F., and Ritkin A. (1980) *Diagnosis and Treatment of Psychiatric Disorders: Adults and Children* Williams & Wilkins, Baltimore, Maryland.

Knoll J. and Magyar K. (1972) Some Puzzling Pharmacological Effects of Monoamine Oxidase Inhibitors, in *Advances in Biochemical Psychopharmacology* vol. 5 *Monoamine Oxidases—New Vistas* (Costa E. and Sandler M., eds.) Raven, New York.

Kobes R. D., Wyatt R. J., and Neckers L. M. (1980) A sensitive rapid fluorometric assay for monoamine oxidase utilizing high-pressure liquid chromatography. *J. Pharmacol. Meth.* **3**, 305–310.

Köchli H. and von Wartburg J. P. (1978) A sensitive photometric assay for monoamine oxidase. *Anal. Biochem.* **84**, 127–135.

Koe K. (1983) Enhancement of benzodiazepine binding by progabide (SL 76002) and SL 75102. *Drug Dev. Res.* **3**, 421–432.

Koe B. K. (1976) Molecular geometry of inhibitors of the uptake of catecholamines and serotonin in synaptosomal preparations of rat brain. *J. Pharmacol. Exp. Ther.* **199,** 649–661.

Koe B. K. and Vinick F. J. (1984) Adaptive changes in central nervous system receptor systems. *Ann. Rep. Med. Chem.* **19,** 41–50.

Kohler C., Ogren S. V., Haglund L., and Angeby T. (1979) Regional displacement by sulpiride of [^{3}H]spiperone binding in vivo. Biochemical and behavioral evidence for a preferential action on limbic and nigral dopamine receptors. *Neurosci. Lett.* **13,** 51–56.

Kohler C., Haglund L., Ogren S. O., and Angeby T. (1981) Regional blockade by neuroleptic drugs of in vivo ^{3}H-spiperone binding in the rat brain. Relation to blockade of apomorphine induced hyperactivity and stereotypes. *J. Neural Transm.* **52,** 163–173.

Kraml M. (1965) A rapid microfluorometric determination of monoamine oxidase. *Biochem. Pharmacol.* **14,** 1683–1685.

Kuhar M. J. (1982) Localizing Drug and Neurotransmitter-Labelled Tracers, in *Receptor-Binding Radiotracers* vol. 1 (Eckelman W. C., ed.) CRC, Boca Raton, Florida.

Kuhar M. J., Aghajanian G. K., and Roth R. H. (1972) Tryptophan hydroxylase activity and synaptosomal uptake of serotonin in discrete brain regions after midbrain raphe lesions: Correlations with serotonin levels and histochemical fluorescence. *Brain Res.* **44,** 165–176.

Laduron P. M. (1984) Criteria for receptor sites in binding studies. *Biochem. Pharmacol.* **33,** 833–839.

Laduron P. and Leysen J. E. (1979) Is the low incidence of extrapyramidal side-effects of antipsychotics associated with antimuscarinic properties? *J. Pharmacol.* **30,** 120–122.

Laduron P. M., Janssen P. F. M., and Leysen J. E. (1978) Spiperone: A ligand of choice for neuroleptic receptors. 2. Regional distribution and in vivo displacement of neuroleptic drugs. *Biochem. Pharmacol.* **27,** 317–321.

Laduron P. M., Robbyns M., and Schotte, A. (1982) (^{3}H)-Desipramine and (^{3}H)-imipramine binding are not associated with noradrenaline and serotonin uptake in the brain. *Eur. J. Pharmacol.* **78,** 491–493.

Lai J. C., Leung T. K., Guest S. F., Lim L., and Davison A. N. (1980) The monoamine oxidase inhibitors and L-deprenyl also affect the uptake of dopamine, noradrenaline and serotonin by rat brain synaptosomal preparations. *Biochem. Pharmacol.* **29,** 2763–2767.

Lassen J. B., Lund J., and Sondergaard I. (1980) Central and peripheral 5-HT uptake in rats treated chronically with femoxetine, paroxetine and chlorimipramine. *Psychopharmacol.* **68,** 229–233.

LeFur G., Guilloux F., and Uzan A. (1980) In vivo blockade of dopaminergic receptors from different rat brain regions by classical and atypical neuroleptics. *Biochem. Pharmacol.* **29,** 267–270.

Lemberger L., Bergstrom R. F., Wolen R. L., Farid N. A., Enas G. G., and Aronoff G. R. (1985) Fluoxetine: Clinical pharmacology and physiologic disposition. *J. Clin. Psychiatry* **46,** 14–19.

Leonard B. E. (1984) Pharmacology of new antidepressants. *Prog. Neuro-Psychopharmacol. & Biol. Psychiat.* **8,** 97–108.

Levi G. and Raiteri M. (1976) Synaptosomal transport processes. *Int. Rev. Neurobiol.* **19,** 51–74.

Leysen J. E. (1984) Receptors for Neuroleptic Drugs, in *Advances in Human Psychopharmacology* vol. 3 (Burrows G. D. and Werry J., eds.) JAI Press, Greenswich, Connecticut.

Lidbrink P., Jonsson G., and Fuxe K. (1971) The effect of imipramine-like drugs and antihistamine drugs on uptake mechanisms in central noradrenaline and 5-hydroxytryptamine neurons. *Neuropharmacology* **10,** 521–536.

Lingjaerde O. (1984) Platelet serotonin uptake inhibition as a basis for monitoring antidepressant drug treatment. *J. Clin. Psychopharmacol.* **4,** 76–81.

Lipton P. (1985) Brain Slices: Uses and Abuses, in *Neuromethods* vol. 1 *General Neurochemical Techniques* (Boulton A. A. and Baker G. B., eds.) Humana, Clifton, New Jersey.

Lloyd K. G. (1986) GABA Receptor Binding, in *Neuromethods* vol. 4 *Receptor Binding* (Boulton A. A., Baker G. B., and Hrdina P. D., eds.) Humana, Clifton, New Jersey.

Lloyd K. G., Thuret F., and Pilc A. (1985) Upregulation of γ-aminobutyric acid (GABA) B binding sites in rat frontal cortex: A common action of repeated administration of different classes of antidepressants and electroshock. *J. Pharmacol. Exp. Ther.* **235,** 191–199.

Logan J. G. and O'Donovan D. J. (1980) Noradrenaline uptake by synaptosomes and (Na^+-K^+) ATPase. *Biochem. Pharmacol.* **29,** 2105–2112.

Lovenberg W., Levine R. J., and Sjoerdsma A. (1964) A sensitive assay of monoamine oxidase activity: In vitro application to heart and sympathetic ganglia. *J. Pharmacol. Exp. Ther.* **135,** 7–10.

Maitre L., Delini-Stula A., and Waldmeier P. C. (1976) Relations Between the Degree of Monoamine Oxidase Inhibition and Some Psychopharmacological Responses to Monoamine Oxidase Inhibitors in Rats, in *CIBA Foundation Symposium #39* (New Ser.) *Monoamine Oxidase and its Inhibition* (Wolstenholme G. E. W. and Knight J., eds.) Elsevier, Excerpta Medica, North Holland, Amsterdam.

Maitre L., Moser P., Baumann P. A., and Waldmeier P. C. (1980) Amine uptake inhibitors: Criteria of selectivity. *Acta Psychiatr. Scand.* **280** (Suppl.), 97–110.

Maj J., Gorka Z., Melzacka M., Rawlow A., and Pilc A. (1983) Chronic treatment with imipramine: Further functional evidence for the enhanced noradrenergic transmission in flex or reflex activity. *Naunyn Schmiedebergs Arch. Pharmacol.* **322,** 256–260.

Maj J., Przegalinski E., and Mogilnicka E. (1984) Hypotheses concerning the mechanism of action of antidepressant drugs. *Rev. Physiol. Biochem. Pharmacol.* **100,** 1–74.

Mann J. J., Brown R. P., Halper J. P., Sweeney J. A., Kocsis J. H., Stokes P. E., and Bilezikian J. P. (1985) Reduced sensitivity of lymphocyte beta-adrenergic receptors in patients with endogenous depression and psychomotor agitation. *New Eng. J. Med.* **313,** 715–720.

Martin I. L. (1986) The Benzodiazepine Receptor, in *Neuromethods* vol. 4 *Receptor Binding* (Boulton A. A., Baker G. B., and Hrdina P. D., eds.) Humana, Clifton, New Jersey.

Martin I. L., Baker G. B., and Mitchell P. R. (1978) The effect of viloxazine hydrochloride on the transport of noradrenaline, dopamine, 5-hydroxytryptamine and γ-aminobutyric acid in rat brain tissue. *Neuropharmacology* **17,** 421–423.

Mason W. D. and Olson C. L. (1970) Differential amperometric measurement of monoamine oxidase activity at tubular carbon electrode. *Anal. Chem.* **42,** 488–494.

Massingham R., Dubocovich M. L., Shepperson N. B., and Langer S. Z. (1981) In vivo selectivity of prazosin but not of WB 4101 for postsynaptic alpha-1 adrenoreceptors. *J. Pharmacol. Exp. Ther.* **217,** 467–474.

Maxwell R. A. and White H. L. (1978). Tricyclic and Monoamine Oxidase Inhibitor Antidepressants: Structure–Activity Relationships, in *Handbook of Psychopharmacology* vol. 14 (Iversen L. L., Iversen S. D., and Snyder S. H., eds.) Plenum, New York.

May J. M. and Minneman K. P. (1986) Adrenergic Receptors, in *Neuromethods* vol. 4 *Receptor Binding* (Boulton A. A., Baker G. B., and G. B. Hrdina P. D., eds.) Humana, Clifton, New Jersey.

McEwen C. M. and Cohen J. D. (1963) An amine oxidase in normal human serum. *J. Lab. Clin. Med.* **62,** 766–776.

McNeal E. T. and Cimbolic P. (1986) Antidepressants and biochemical theories of depression. *Psychol. Bull.* **99,** 361–374.

Mennini T., Abbiati A., Caccia S., Cotecchia S., Gomez A., and Garattini S. (1982) Brain levels of tofixopam in the rat and relationship with benzodiazepine receptors. *Naunyn Schmiedebergs Arch. Pharmacol.* **321,** 112–115.

Meyerson L. R., McMurtrey K. D., and Davis V. E. (1978) A rapid and sensitive potentiometric assay for monoamine oxidase using an ammonium-sensitive electrode. *Anal. Biochem.* **86,** 287–297.

Miller L. G., Greenblatt D. J., Paul S. M., and Shader R. I. (1987) Benzodiazepine receptor occupancy in vivo: Correlation with brain concentrations and pharmacodynamic actions. *J. Pharmacol. Exp. Ther.* **240,** 516–522.

Miller R. J. and Hiley C. R. (1974) Anti-muscarinic properties of neuroleptics and drug-induced Parkinsonism. *Nature* **248,** 596–597.

Minchin M. C., Williams J., Bowdler J. M., and Green A. R. (1983) Effect of electroconvulsive shock on the uptake and release of noradrenaline and 5-hydroxytryptamine in rat brain slices. *J. Neurochem.* **40,** 765–768.

Mishra R. K. (1986) Central Nervous System Dopamine Receptors, in *Neuromethods* vol. 4 *Receptor Binding* (A. A. Boulton, G. B. Baker, and G. B. Hrdina, eds.) Humana, Clifton, New Jersey.

Mohler H. and Okado T. (1977) Benzodiazepine receptor: Demonstration in the central nervous system. *Science* **198,** 849–851.

Moisset B. (1977) Factors contributing to the modulation of norepinephrine uptake by synaptosomes from mouse brain cortex. *Brain Res.* **121,** 113–120.

Murphy D. L., Garrick N. A., Aulakh C. S., and Cohen R. M. (1984) New contributions from basic science to understanding the effects of monoamine oxidase inhibiting antidepressants. *J. Clin. Psychiatry* **45,** 37–43.

Murphy D. L., Sunderland T., Campbell I., and Cohen R. M. (1985) Monoamine Oxidase Inhibitors as Antidepressants, in *Pharmacotherapy of Affective Disorders: Theory and Practice* (Dewhurst W. G. and Baker G. B., eds.) Croom Helm, London.

Nagatsu T. and Yagi K. (1966) A simple assay of monoamine oxidase and D-amino acid oxidase by measuring ammonia. *J. Biochem.* (Tokyo) **60,** 219–221.

Nimmo G. A. and Tipton K. F. (1979) The distribution of soluble and membrane bound forms of glutaminase in pig brain. *J. Neurochem.* **33,** 1083–1094.

Nissinen E. (1984) Determination of monoamine oxidase B activity by high-performance liquid chromatography. *J. Chromatogr.* **309,** 156–159.

Nohta H., Zaitzu K., Tsuruta Y., and Okura Y. (1983) Assay for monoamine oxidase A and B by high-performance liquid chromatography with fluorescence detection. *J. Chromatogr.* **280,** 343–349.

Oakley N. R. and Jones B. J. (1983) Buspirone enhances [^{3}H]-flunitrazepam binding in vivo. *Eur. J. Pharmacol.* **87,** 499–500.

Obata F., Ushiwata A., and Nakamura Y. (1971) Spectrophotometric assay of monoamine oxidase using 2,4,6-trinitrobenzene-2-sulfonic acid. *J. Biochem.* (Tokyo) **69,** 349–354.

Okada F., Tokumitsu Y., and Ui M. (1986) Desensitization of β-adrenergic receptor-coupled adenylate cyclase in cerebral cortex after *in vivo* treatment of rats with desipramine. *J. Neurochem.* **47,** 454–459.

Ortmann R., Schaub M., Felner A., Lauber J., Christen P., and Waldmeier P. C. (1984) Phenylethylamine-induced stereotypies in the rats: A behavioural test system for assessment of MAO-B inhibitors. *Psychopharmacology* **84,** 22–27.

Ortmann R., Waldmeier P. C., Radeke E., Felner A., and DeliniStula A. (1980) The effects of 5-HT uptake and MAO inhibitors on L-5-hydroxytryptophan-induced excitation in rats. *Naunyn Schmiedebergs Arch. Pharmacol.* **311,** 185–192.

Otsuka S. and Kobayashi Y. (1964) A radioisotopic assay for monoamine oxidase determinations in human plasma. *Biochem. Pharmacol.* **13,** 995–1006.

Pare, C. M. B. (1985) The present status of monoamine oxidase inhibitors. *Br. J. Psychiat.* **146,** 576–584.

Paton D. M., ed. (1976) *The Mechanisms of Neuronal and Extraneuronal Transport of Catecholamines* Raven, New York.

Paton D. M. (1980) Neuronal transport of noradrenaline and dopamine. *Pharmacology* **21,** 85–92.

Paul S. M., Janowsky A., Skolnick P. (1985) Monoaminergic Neurotransmitters and Antidepressant Drugs, in *Psychiatry Update: American Psychiatric Association Annual Review* vol. 4 (Hales R. E. and Frances A. J., eds.) American Psychiatric Press, Washington, DC.

Paul S. M., Rehavi M., Skolnick P., Ballenger J. C., and Goodwin F. K. (1981) Depressed patients have decreased binding of tritiated imipramine to platelet serotonin "transporter." *Arch. Gen. Psychiatry* **38,** 1315–1317.

Paul S. M., Rehavi M., Skolnick P., and Goodwin F. H. (1984) High Affinity Binding of Antidepressants to Biogenic Amine Transport Sites in Human Brain and Platelet: Studies in Depression, in *Neurobiology of Mood Disorders* (Post R. M. and Ballenger J. C., eds.) Williams & Wilkins, Baltimore, Maryland

Peroutka S. J. (1986) Serotonin Receptors, in *Neuromethods* vol. 4 *Receptor Binding* (Boulton A. A., Baker G. B., and Hrdina P. D., eds.) Humana, Clifton, New Jersey.

Peroutka S. J. and Snyder S. H. (1980a) Relationship of neuroleptic drug effects at brain dopamine, serotonin, alpha-adrenergic and histamine receptors to clinical potency. *Am. J. Psychiatry* **137,** 1518–1522.

Peroutka S. J. and Snyder S. H. (1980b) Long term antidepressant treatment decreases spiroperidol-labeled serotonin receptor binding. *Science* **210,** 88–90.

Peroutka S. J., Lebovitz R. M., and Snyder S. H. (1979) Serotonin receptor binding sites affected differentially by guanine nucleotides. *Mol. Pharmacol.* **16,** 700–708.

Peroutka S. J., U'Prichard D. C., Greenberg D. A., and Snyder S. H. (1977) Neuroleptic drug interactions with norepinephrine alpha-receptor binding sites in rat brain. *Neuropharmacology* **16,** 549–556.

Peterson E. N., Jensen L. H., Honore T., Braestrup C., Kehr W., Stephens D. N., Wachtel H., Seidelman D., and Schmiechen K. L. (1984) ZK91296, a partial agonist at benzodiazepine receptors. *Psychopharmacology* **83,** 240–248.

Pilc A. and Enna S. J. (1986) Antidepressant administration has a differential effect on rat brain α_2-adrenoceptor sensitivity to agonists and antagonists. *Eur. J. Pharmacol.* **132,** 277–282.

Pilc A. and Vetulani J. (1982) Attenuation by chronic imipramine treatment of (^{3}H)-clonidine binding to cortical membranes and clonidine-induced hypothermia: Influence of central chemosympathectomy. *Brain Res.* **238,** 499–504.

Piletz J. E. and Halaris A. (1985) Yohimbine and clonidine recognize different states of adrenergic receptors on human platelets: reconsideration of studies on platelet alpha-2-receptors in depressive illness. *Psychopharmacol. Bull.* **21,** 573–578.

Pylatuk K. L. and McNeill J. H. (1976) The effects of certain drugs on the uptake and release of ^{3}H-noradrenaline in rat whole brain homogenates. *Can. J. Physiol. Pharmacol.* **54,** 457–468.

Quinaux N., Scuvee-Moreau J., and Dresse A. (1982) Inhibition of *in vitro* and *ex vivo* uptake of noradrenaline and 5-hydroxytryptamine by five antidepressants; correlation with reduction of spontaneous firing rate of central monoaminergic neurones. *Naunyn Schmiedebergs Arch. Pharmacol.* **319,** 66–70.

Raisman R., Briley M., and Langer S. Z. (1979) Specific tricyclic antidepressant binding sites in rat brain. *Nature* **281,** 148–150.

Raiteri M. and Levi G. (1973) Depletion of synaptosomal neurotransmitter pool by sudden cooling. *Nature New Biol.* **243,** 180–182.

Raiteri M., Angelini F., and Bertollini A. (1976) Comparative study of the effects of mianserin, a tetracyclic antidepressant, and of imipramine on uptake and release of neurotransmitters in synaptosomes. *J. Pharm. Pharmacol.* **28,** 483–488.

Raiteri M., Bertollini A., Angelini F., and Levi G. (1975) *d*-Amphetamine as a releaser or reuptake inhibitor of biogenic amines in synaptosomes. *Eur. J. Pharmacol.* **34,** 189–195.

Raiteri M., Cerrito F., Cervoni A. M., del Carmine R., Ribera M. T., and Levi G. (1978) Studies on Dopamine Uptake and Release in Synaptosomes, in *Advances in Biochemical Psychopharmacology* vol. 19

(Roberts P. J., Woodruff G. N., and Iversen L. L., eds.) Raven, New York.

Randrup A. and Braestrup C. (1977) Uptake inhibition of biogenic amines by newer antidepressant drugs: Relevance to the dopamine hypothesis of depression. *Psychopharmacology* **53,** 309–314.

Rehavi M., Paul S. M., Skolnick P., and Goodwin F. K. (1980) Demonstration of specific high-affinity binding sites for ^{3}H-imipramine in human brain. *Life Sci.* **26,** 2273–2279.

Reisine T., Johnson R., Wiech N., Ursillo R., and Yamamura H. I. (1982) Rapid Desensitization of Central β-Receptors and Upregulation of α$_2$-Receptor Following Antidepressant Treatment, in *Typical and Atypical Antidepressants: Molecular Mechanisms* (Costa E. and Racagni G., eds.) Raven, New York.

Riblet L. A. and Taylor D. P. (1981) Pharmacology and neurochemistry of trazodone. *J. Clin. Psychopharmacol.* **1** (Suppl), 175–225.

Richelson E. (1983) Are receptor studies useful for clinical studies? *J. Clin. Psychiatry* **44,** 4–9.

Richelson E. (1984) The newer antidepressants: Structures, pharmacokinetics, pharmacodynamics and proposed mechanisms of action. *Psychopharmacol. Bull.* **20,** 213–223.

Risch S. C. and Janowsky D. S. (1984) Cholinergic-Adrenergic Balance in Affective Illness, in *Neurobiology of Mood Disorders* (Post R. M. and Ballenger J. C., eds.) Williams & Wilkins, Baltimore, Maryland.

Robinson D. S., Lovenberg W., Keiser H., and Sjoerdsma A. (1968). Effects of drugs on human blood platelet and plasma amine oxidase activity in vitro and in vivo. *Biochem. Pharmacol.* **17,** 109–119.

Rodriquez de Lores Arnaiz G. and Pellegrino de Iraldi A. (1985) Subcellular Fractionation, in *Neuromethods* vol. 1 *General Neurochemical Techniques* (Boulton A. A. and Baker G. B., eds.) Humana, Clifton, New Jersey.

Ross S. B. (1979) Interactions between reserpine and various compounds on the accumulation of [^{14}C] 5-hydroxytryptamine and [^{3}H]noradrenaline in homogenates from rat hypothalamus. *Biochem. Pharmacol.* **28,** 1085–1088.

Ross S. B. (1980) Neuronal transport of 5-hydroxytryptamine. *Pharmacology* **21,** 123–131.

Ross S. B. and Ask A.-L. (1980) Structural requirements for uptake into serotoninergic neurones. *Acta Pharmacol. Toxicol.* **46,** 270–277.

Ross S. B. and Renyi A. L. (1975a) Inhibition of the uptake of ^{3}H-dopamine and ^{14}C-5-hydroxytryptamine in mouse striatum slices. *Acta Pharmacol. Toxicol.* **36,** 56–66.

Ross S. B. and Renyi A. L. (1975b) Tricyclic antidepressant agents. I. Comparison of the inhibition of the uptake of ^{3}H-noradrenaline and ^{14}C-5-hydroxytryptamine in slices and crude synaptosome preparations of the midbrain-hypothalamus region of the rat brain. *Acta Pharmacol. Toxicol.* **36**, 382–394.

Ross S. B., Renyi A. L., and Ogren S.-O. (1972) Inhibition of the uptake of noradrenaline and 5-hydroxytryptamine by chlorphentermine and chloroimipramine. *Eur. J. Pharmacol.* **17**, 197–112.

Rudorfer M. V. and Potter W. Z. (1985) Metabolism of Drugs Used in Affective Disorders, in *Pharmacotherapy of Affective Disorders: Theory and Practice* (Dewhurst W. G. and Baker G. B., eds.) Croom Helm, London.

Saano V. (1982) Tofixopam selectively increases the affinity of benzodiazepine binding sites for [^{3}H]flunitrazepam but not for beta[^{3}H]carboline-3-carboxylic acid ethyl esther. *Pharmacol. Res. Commun.* **14**, 971–981.

Sachs C. and Jonsson G. (1972) Degeneration of central and peripheral noradrenaline neurons produced by 6-hydroxy-Dopa. *J. Neurochem.* **19**, 1561–1575.

Schoemaker H. and Nickolson V. J. (1982) Interaction between dopamine influx and efflux in rat striatal synaptosomes. *Life Sci.* **31**, 2455–2461.

Scott J. A. and Crews F. T. (1986) Down-regulation of serotonin$_2$, but not of beta-adrenergic receptors during chronic treatment with amitriptyline is independent of stimulation of serotonin$_2$ and beta-adrenergic receptors. *Neuropharmacology* **25**, 1301–1306.

Seeman P. (1980) Brain dopamine receptors. *Pharmacol. Rev.* **32**, 229–313.

Seeman P. (1982) Nomenclature of central and peripheral dopaminergic sites and receptors. *Biochem. Pharmacol.* **31**, 2563–2568.

Seeman P., Lee T., Chau-Wong M. and Wong K. (1976) Antipsychotic drug doses and neuroleptic/dopamine receptors. *Nature* **261**, 717–719.

Sethy V. H., Carlsson R. W., and Harris D. W. (1983) Effect of chronic antidepressant treatment on α_2-adrenoreceptors. *Drug Dev. Res.* **3**, 287–297.

Shaskan E. G. and Snyder S. H. (1970) Kinetics of serotonin accumulation into slices from rat brain. *J. Pharmacol. Exp. Ther.* **175**, 404–418.

Singer T. P., von Korff R. W., and Murphy D. L., eds. (1979) *Monoamine Oxidase: Structure, Function and Altered Functions* Academic, New York.

Skolnick P., Lock K. L., Paugh B., Marangos P., Windsor R., and Paul S. (1980) Pharmacologic and behavioral effects of EMD 28422: A novel purine which enhances [^{3}H]diazepam binding to brain benzodiazepine receptors. *Pharmacol. Biochem. Behav.* **12**, 685–689.

Snyder S. H. and Hendley E. D. (1968) A simple and sensitive fluorescence assay for monoamine oxidase and diamine oxidase. *J. Pharmacol. Exp. Ther.* **163**, 386–392.

Snyder S. H. and Peroutka S. J. (1984) Antidepressants and Neurotransmitter Receptors, in *Neurobiology of Mood Disorders* (Post R. M. and Ballenger J. C., eds.) Williams & Wilkins, Baltimore, Maryland.

Snyder S. H. and Yamamura H. T. (1977) Antidepressants and the muscarinic acetylcholine receptor. *Arch. Gen. Psychiatry* **34**, 236–239.

Snyder S. H., Kuhar M. J., Green A. I., Coyle J. T., and Shaskan E. G. (1970) Uptake and subcellular localization of neurotransmitters in the brain. *Int. Rev. Neurobiol.* **13**, 127–158.

Snyder S. H., U'Pritchard D. C., and Greenberg D. A. (1978) Neurotransmitter Receptor Binding in Brain, in *Psychopharmacology: A Generation of Progress* (Lipton M. A., DiMascio A., and Killam K. F., eds.) Raven, New York.

Snyder S. H., Banerjee S. P., Yamamura H. I., and Greenberg D. (1974) Drugs, neurotransmitters and schizophrenia. *Science* **184**, 1243–1253.

Southgate J. and Collins G. G. S. (1969) The estimation of monoamine oxidase using ^{14}C-labelled substrates. *Biochem. Pharmacol.* **17**, 109–119.

Sparatore A., Marchi M., Maura G., Paudice P., and Raiteri M. (1982) Effects of some rigid analogues of imipramine and amitriptyline on the uptake of noradrenaline, serotonin and choline in rat brain synaptosomes. *Pharmacol. Res. Commun.* **14**, 257–265.

Squires R. F. and Braestrup C. (1977) Benzodiazepine receptors in rat brain. *Nature* **266**, 732–734.

Stahl S. M. and Meltzer H. Y. (1978) A kinetic and pharmacologic analysis of 5-hydroxytryptamine transport by human platelets and platelet storage granules: Comparison with central serotonergic neurons. *J. Pharmacol. Exp. Ther.* **205**, 118–132.

Stoof J. C. and Kebabian J. W. (1984) Minireview: Two dopamine receptors: Biochemistry, physiology and pharmacology. *Life Sci.* **35**, 2281–2296.

Sugrue M. F. (1983) Do antidepressants possess a common mechanism of action? *Biochem. Pharmacol.* **32**, 1811–1817.

Sulakhe P. V., Gupta R. C., and Jagadeesh G. (1986) Brain Adenylate Cyclase, in *Neuromethods* vol. 5 *Neurotransmitter Enzymes* (Boulton A. A., Baker G. B., and Yu P. H., eds.) Humana, Clifton, New Jersey.

Sulser F. (1982) Regulation and adaptation of central norepinephrine receptor systems: Modification by antidepressant treatments. *Psychiatr. J. Univ. Ottawa* **7**, 196–203.

Sulser F. (1986) Update on neuroreceptor mechanisms and their implication for the pharmacotherapy of affective disorders. *J. Clin. Psychiatry* **47**, 13–18.

Tabor C. W., Tabor H., and Rosenthal S. M. (1954) Purification of amine oxidase from beef plasma. *J. Biol. Chem.* **208**, 645–661.

Takahashi H. and Takahara S. (1968) A sensitive fluorometric assay for monoamine oxidase based on the formation of 4,6-quinolinediol from 5-hydroxykynurenamine. *J. Biochem.* (Tokyo) **64,** 7–11.

Tenne M., Finberg J. P. M., Youdim M. B. H., and Utilzur S. (1985) A new rapid and sensitive bioluminescene assay for monoamine oxidase activity. *J. Neurochem.* **44,** 1378–1384.

Thomas P. C. and Jones R. B. (1977) The effects of clomipramine and desmethylclomipramine on the *in vitro* uptake of radiolabeled 5-HT and noradrenaline into rat brain cortical slices. *J. Pharm. Pharmacol.* **29,** 562–563.

Thompson J. M., Whitaker J. R., and Joseph J. A. (1981) [^{3}H]Dopamine accumulation and release from striatal slices in young, mature and senescent rats. *Brain Res.* **224,** 436–440.

Tipton K. F. (1969) A sensitive fluorometric assay for monoamine oxidase. *Anal. Biochem.* **28,** 318–325.

Tipton K. F. and Dawson A. P. (1968) The distribution of monoamine oxidase and a α-glycerophosphate dehydrogenase in pig brain. *Biochem. J.* **108,** 95–99.

Tipton K. F. and Youdim M. B. H. (1976) Assay of monoamine oxidase. *Ciba Foundation Symposium* (New Ser.) **39,** 393–403.

Tipton K. F., Dostert P., and Strolin-Benedetti M., eds. (1984) *Monoamine Oxidase and Disease: Prospects for Therapy with Reversible Inhibitors* Academic, London.

Tipton K. F., Fowler C. J., McCropdden J. M., and Strolin-Benedetti M. (1983) The enzyme-activated irreversible inhibition of type B MAO by 3-(4-[(3-chlorophenyl)methoxy]phenyl)-5-[(methylamino)methyl]-2-oxazolidinone methane sulphonate (compound MD780236) and the enzyme-catalysed oxidation of this compound as competing reactions. *Biochem. J.* **209,** 235–242.

Tran V. T., Chang R. S. L., and Snyder S. H. (1978) Histamine H$_1$ receptors identified in mammalian brain membranes with ^{3}H-mepyramine. *Proc. Natl. Acad. Sci. USA* **75,** 6290–6294.

Tuomisto J. and Tuomisto L. (1979) The accumulation of dopamine into striatal synaptosomes, and its inhibition: Comparison of crude, partially purified and purified synaptosomal preparations. *Med. Biol.* **57,** 238–245.

Tuomisto J. and Walaszek E. J. (1974) Inhibition of histamine and 5-hydroxytryptamine uptake by trans- and cis-2-phenylcyclopropylamine in rabbit blood platelets. *Med. Biol.* **52,** 255–259.

Turkish S., Yu P. H., and Greenshaw A. J. (1987) Monoamine oxidase inhibition: A comparison of *in vivo* and *ex vivo* measures of reversible effects *J. Neural. Transm.* (in press).

Udenfriend S. (1962) *Fluorescence Assay in Biology and Medicine* Academic, New York.

Udenfriend S., Weissbach H., and Clark C. T. (1955) The estimation of 5-hydroxytryptamine (serotonin) in biological tissue. *J. Biol. Chem.* **215**, 337–344.

Uji A., Kawai S., and Nagatsu T. (1980) Gas chromatographic determination of monoamine oxidase activity using mixed substrates. *J. Chromatogr.* **221**, 155–160.

U'Prichard D. C., Greenberg D. A., Sheehan P. P., and Snyder S. H. (1978) Tricyclic antidepressants: Therapeutic properties and affinity for alpha-noradrenergic receptor binding sites in the brain. *Science* **199**, 197–198.

Urwyler S. and von Wartburg J. P. (1980) Uptake and metabolism of catecholamines in rat brain synaptosomes: Studies on the contribution of monoamine oxidase. *Biochem. Pharmacol.* **30**, 2777–2785.

Vetulani J. and Sulser F. (1975) Action of various antidepressant treatments reduces reactivity of noradrenergic cyclic AMP generating systems in limbic forebrain. *Nature* **257**, 495–496.

Waldmeier P. C. (1982) Effects of antidepressant drugs on dopamine uptake and metabolism. *J. Pharm. Pharmacol.* **34**, 391–394.

Waldmeier P. C., Baumann P. A., and Maitre L. (1979) CGP 6085 A, a new, specific, inhibitor of serotonin uptake: Neurochemical characterization and comparison with other serotonin uptake blockers. *J. Pharmacol. Exp. Ther.* **211**, 42–49.

Waldmeier P. C., Felner A., and Lauber J. (1984) Reversible Monoamine Oxidase Inhibitors: Relation Between Effects on Enzymatic Activity Measured *Ex Vivo* and on Amine Activity, in *Monoamine Oxidase and Disease: Prospects for Therapy with Reversible Inhibitors* (Tipton K. F., Dostert P., and Strolin-Benedetti M., eds.) Academic, London.

Weissbach H., Smith T. E., Daly J. W., Witkop B., and Udenfriend S. (1960) A rapid spectrophotometric assay of monoamine oxidase based on the disappearance of kyneuramine. *J. Biol. Chem.* **235**, 1160–1163.

West D. P. and Fillenz M. (1980) Storage and release of noradrenaline in hypothalamic synaptosomes. *J. Neurochem.* **35**, 1323–1328.

Wheeler D. D. (1978) Some problems inherent in transport studies in synaptosomes. *J. Neurochem.* **30**, 109–120.

White T. D. (1976) Evidence that the rapid binding of newly accumulated noradrenaline within synaptosomes invokes synaptic vesicles. *Brain Res.* **108**, 87–96.

White T. D. and Paton D. M. (1972) Effects of external Na^+ and K^+ on the initial rates of noradrenaline uptake by synaptosomes prepared from rat brain. *Biochim. Biophys. Acta* **266**, 116–127.

White K. and Simpson G. (1981) Combined MAOI-tricyclic antidepressant treatment: A reevaluation. *J. Clin. Psychopharmacol.* **5**, 264–282.

Whittaker V. P. (1984) The Synaptosome, in *Handbook of Neurochemistry* vol. 7 (2nd Edn.) (Lajtha A., ed.) Plenum, New York.

Williams M. (1983) Anxioselective anxioytics. *J. Med. Chem.* **26,** 619–628.

Williams M. (1984) Molecular aspects of the action of benzodiazepine and non-benzodiazepine anxiolytics. *Prog. Neuropsychopharmacol. Biol. Psychiatry* **8,** 209–247.

Williams L. T. and Lefkowitz R. J. (1978) *Receptor Binding Studies in Adrenergic Pharmacology* Raven, New York.

Williams M. and U'Prichard D. C. (1984) Drug discovery at the molecular level: A decade of radioligand binding in retrospect. *Ann. Rep. Med. Chem.* **19,** 283–292.

Williams M. and Wood P. L. (1986) Receptor Binding in Drug Discovery and Development, in *Neuromethods* vol. 4 *Receptor Binding* (Boulton A. A., Baker G. B., and Hrdina P. D., eds.) Humana, Clifton, New Jersey.

Wolfe B. B., Harden T. K., Sporn J. R., and Molinoff P. B. (1978) Presynaptic modulation of beta-adrenergic receptors in rat cerebral cortex after treatment with antidepressants. *J. Pharmacol. Exp. Ther.* **207,** 446–457.

Wood K. and Wyllie M. G. (1983) Critical assessment of noradrenaline uptake in synaptosomal preparations. *Naunyn Schmiedebergs Arch. Pharmacol.* **322,** 129–135.

Wood P. L., Loo P., Braunwalder A., Yokoyama N., and Cheney D. L. (1984) *In vitro* characterization of benzodiazepine receptor agonists, antagonists, inverse agonists and agonist/antagonists. *J. Pharmacol. Exp. Ther.* **231,** 572–576.

Wood P. L., McPherson S. E., and Braunwalder A. (1984b) In vivo flunitrazepam binding: Actions of buspirone and clozapine. *Soc. Neurosci. Abstr.* **7,** 117–12.

Wood K., Whiting K., and Coppen A. (1986) Lymphocyte beta-adrenergic receptor density of patients with recurrent affective illness. *J. Affective Disord.* **10,** 3–8.

Wu P. H. and Dyck L. E. (1975) Microassay for the estimation of monoamine oxidase activity. *Anal. Biochem.* **72,** 637–642.

Wurtman R. J. and Axelrod J. (1963) A sensitive and specific assay for the estimation of monoamine oxidase. *Biochem. Pharmacol.* **12,** 1439–1441.

Yamada H., Suzuki H., and Ogura Y. (1972) Amine Oxidases from Microorganism: Stoichiometry of the Reaction Catalyzed by Amines Oxidase of *Aspergillus niger,* in *Advances in Biochemistry and Psychopharmacology* vol. 5 (Costa E. and Greengard P., eds.) Raven, New York.

Yamamura H. I. and Enna S. J. (1981) *Neurotransmitter Receptors* part 2 *Biogenic Amines* Chapman and Hall, London.

Yamamura H. I., Enna S. J., and Kuhar M. J. (eds.) (1978) *Neurotransmitter Receptor Binding* Raven, New York.

Yang H.-Y. T. and Neff N. H. (1974) The monoamine oxidases of brain: Selective inhibition with drugs and the consequences for the metabolism of the biogenic amines. *J. Pharmacol. Exp. Ther.* **189,** 733–740.

Yokoyama N., Ritter B., and Neubert A. (1982) 2-Arylpyrazolo-[4,3c]quinolin-3-ones: Novel agonist, partial agonist and antagonist of benzodiazepines. *J. Med. Chem.* **25,** 337–341.

Youdim M. B. H. and Finberg J. P. M. (1985) Monoamine oxidase inhibitor antidepressants. *Psychopharmacology* (Amsterdam) **2,** 35–70.

Youdim M. B. H. and Paykel E. S., eds. (1981) *Monoamine Oxidase Inhibitors: The State of the Art* Wiley, Chichester.

Yu P. H. (1986) Monoamine Oxidase, in *Neuromethods* vol. 5 *Neurotransmitter Enzymes* (Boulton A. A., Baker G. B., and Yu P. H., eds.) Humana, Clifton, New Jersey.

Yu J. H. and Smith C. B. (1977) Effect of cocaine and desmethylimipramine on the uptake, retention and metabolism of ^{3}H-5-hydroxytryptamine in rat brain slices. *Pharmacology* **15,** 242–253.

Yu P. H., Bailey B. A., and Durden D. A. (1986) High performance liquid chromatography of aldehydes and acids in monoamine oxidase catalyzed reactions. *Anal. Biochem.* **152,** 160–166.

Zeller V., Ramachander G., and Zeller E. A. (1965) Amine oxidases. XXI. A rapid method for the determination of the activity of monoamine oxidase and monoamine oxidase inhibitors. *J. Med. Chem.* **8,** 440–450.

Animal Models for Assessing Anxiolytic, Neuroleptic, and Antidepressant Drug Action

Andrew J. Greenshaw, Tuong Van Nguyen,
and David J. Sanger

1. Criteria for Animal Models

The preclinical behavioral analysis of effects of psychiatric drugs may be seen from two viewpoints. On the one hand it is necessary to identify drugs that may be useful for treating psychiatric disorders. With this approach the basic question is: "Does a drug fit into the particular class of drugs that is useful in treating a specific condition?" It should be noted that in this context the only necessary feature of a screening procedure is the ability to identify a clinically effective compound. In this sense there does not have to be any similarity between the behavior measured in the laboratory animal and the psychiatric disorder in question. Alternatively there is the attempt to understand mechanisms of action of these psychotherapeutic drugs. In this context it is necessary to attempt the measurement of those features of drug action that are critical for the alleviation of psychiatric symptoms. The first goal, i.e., identifying drugs that may be clinically useful, and the second goal, i.e., analyzing the mechanisms of action of psychotherapeutic drugs with animal behavior, are in fact intertwined. This chapter is written from a pragmatic viewpoint utilizing currently acceptable ideas to illustrate major approaches to the problem of using or developing an animal model. The complexities of defining an adequate animal model have been clearly discussed by Carlton (1978, 1983); a useful discussion of animal models in relation to depression has also been recently provided by Willner (1984).

The validity of animal models in relation to psychiatric disorders has traditionally been assessed using a set of criteria proposed by McKinney and Bunney (1969). According to this set of criteria, the *animal model* should correspond to the psychiatric disorder in terms of *biochemical profile, symptoms, treatment*, and, more optimistically, *etiology*. It seems clear that with the present state of understanding these criteria are rather ambitious. In fact, with regard to the initially proposed goal of an animal model as described above, it is apparent that these criteria suggested by McKinney and Bunney (essentially criteria related to what is called *"face validity,"* i.e., the obvious similarities between the model and the disorder) are not necessary conditions for a useful animal model. A clear example of this is found with the yohimbine potentiation test as an animal model of antidepressant drug action. In this case the observation that antidepressants may potentiate the effects of yohimbine (either in terms of a measure of lethality as a function of dose in mice or in terms of the effects of this compound on blood pressure in dogs) has led to the use of the yohimbine potentiation test as a screening test for antidepressant drugs; indeed, although this test has no face validity, it does have a fair degree of predictive validity, i.e., it may be used to identify likely antidepressant compounds. Thus in this case, the animal "model," i.e., the measure of an effect of antidepressant drugs, is useful, although it does not provide any real model of a psychiatric disorder. Carlton has pointed out that in relation to "screening", face validity may have no relevance as a test criterion. In relation to the use of animal models to elucidate mechanisms of action of psychotherapeutic drugs, there is clearly a need for models that parallel psychiatric disorders as closely as possible.

This issue of "what constitutes an acceptable or desirable animal model of psychotherapeutic drug action?" is a contentious one. In the present chapter, the term animal model is used in the broadest possible sense in relation to analyzing drug action. It must be stressed, however, that both "screening" and assessing modes of action of drugs are seen as equally important goals. The following section considers some fundamental features of animal models.

1.1. Predictive Validity

Predictive validity is the most important criterion for establishing an animal model of drug action. Predictive validity is an index of the accuracy of an animal model in relation to the identification of a particular class of drug. There are two main features of predictive validity. The first is obvious: if the model clearly determines whether or not a compound has the ability to alleviate symptoms of the disorder in humans, it has high predictive validity. In relation to this it is necessary to determine to what extent the model will identify what may be called false positives and false negatives. *False positives* are compounds that *do not have the ability to alleviate the clinical symptoms,* but *do come out as effective compounds in the model. False negatives* are compounds that *do have the ability to allevi-ate the symptoms of the clinical disorder,* but are *not* identified by the model. A further factor that determines predictive validity in an animal model is the degree to which the potency of a compound in the animal test correlates with clinical potency. In other words, a comparison of the clinical potency of a wide range of compounds against their potency in the animal test should yield a highly significant correlation if the test has high predictive validity (*see* Table 1).

1.2. Face Validity

Although it has been stated that face validity is not a necessary condition for the definition or the development of an animal model of drug action, it remains an important consideration. An example of this is seen in relation to antidepressant drugs; typically clinical responsiveness to antidepressant drug action is delayed relative to the onset of drug administration (Lapierre, 1985). Thus the degree to which the onset of drug action in the animal model correlates with the onset of antidepressant drug action in the clinical setting is clearly of great interest. As suggested earlier, this becomes more important when one is considering the analysis of the mechanisms of action of psychotherapeutic drugs and relatively less important when one is concerned simply with identifying compounds that may be clinically effective.

1.3. Construct Validity

Construct validity corresponds to an assessment of the degree to which the behavior in the model and the behavior in (i.e., symptomology of) the disorder are similar or homologous. Essentially what this means is, does the animal model actually entail equivalent features or symptoms of the disorder to those seen in humans? This is a rather difficult form of validity to assess objectively; it is clearly dependent on our understanding of the disorder both in empirical and theoretical terms. Most researchers only consider predictive validity and face validity in assessing animal models of psychiatric drug action; indeed, this is the approach that will be adopted in the present review.

1.4. Profiles of Drug Action

In considering the analysis of psychotherapeutic effects of drugs it is important to realize that the complexity of drug action in the clinic is mirrored in the preclinical laboratory setting. It would

Table 1
Potency Relation for Anxiolytic Drugs in the Geller-Seifter Model[a]

Drug	Clinical dose[b], mg	MED[c], mg/kg
Diazepam	20 (1)	0.62 (1)
Chlordiazepoxide	40 (2)	2.20 (3)
Oxazepam	49 (3)	1.25 (2)
Phenobarbital	115 (4)	4.50 (4)
Amobarbital	175 (5)	5.00 (5)
Tybamate	1212 (6)	62.50 (6)
Meprobamate	1410 (7)	80.00 (7)

[a]*See* section 2.5.2. (from Carlton, 1983, with permission).
[b]Average daily oral doses, from Cook and Davidson (1973), except for tybamate, from Shader (1975).
[c]Data from Cook and Davidson, 1973.

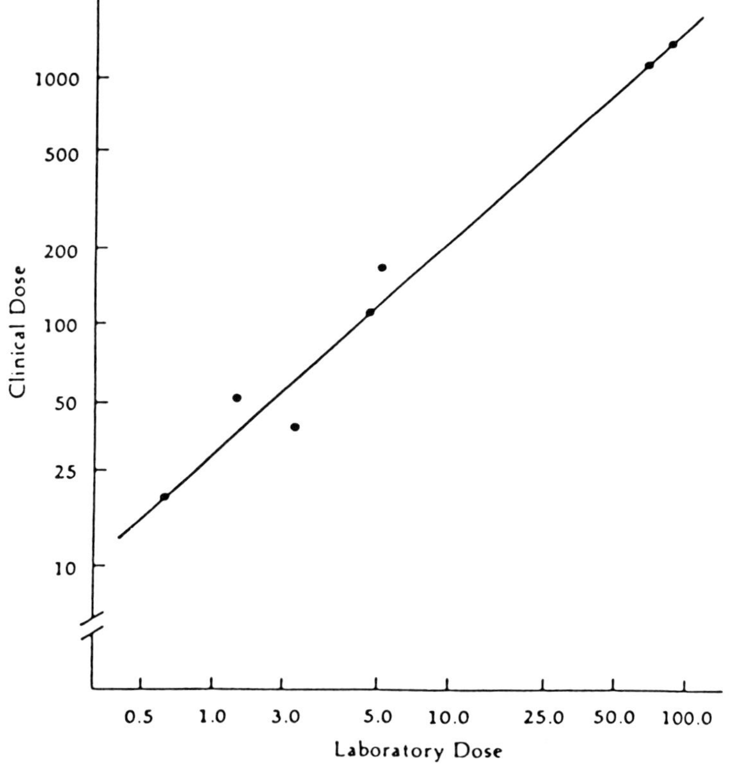

be convenient if one assay represented a perfectly accurate measure of psychotherapeutic potential, but obviously this is far from reality. The analysis of behavior is a complex affair, and whether we are considering drug action in the clinic or in the animal laboratory, the interplay of drugs and behavior presents us with a very complex interface. The aim of drug analysis in terms of both of the goals outlined earlier (*see* section 1) is for complete and parsimonious prediction, ultimately in the clinic.

This chapter describes a range of basic procedures that has contributed considerably to the development and study of anxiolytic, neuroleptic (antipsychotic), and antidepressant drugs. As Porsolt (1985) has stated, many behavioral tests have developed serendipitously—an indication of the interplay between fortuitous clinical advances and the development of screening tests. The current perspective is one from which general profiles of drug classes are visible, but atypical compounds are not uncommon. Indeed, such drugs often afford possibilities to achieve greater understanding of the neural basis of mental disorders and to develop new families of drugs. In subsequent sections of this chapter, the anxiolytics, neuroleptics, and antidepressants are considered in a general way. For a detailed analysis the reader is encouraged to explore specific articles describing the assays.

1.5. A Note on Drug Discrimination Learning

One area of research that is extremely important in this context is that of drug discrimination by infrahuman species. This approach assesses animals' subjective responses to drugs by measuring trained responses that animals have learned are rewarded (or not punished) dependent on the presence or absence of a drug state(s). This field has developed at a rapid rate over recent years and represents a powerful form of preclinical analysis; particularly for anxiolytic drugs. Because it is such a specialized and detailed area, the reader is referred to other extensive sources for further details (*see* Colpaert and Slangen, 1982; Jarbe, 1987).

2. Anxiolytic Drugs

2.1. General Appraisal

Anxiolytic drugs are in lay terms compounds that are effective in the reduction of feelings of anxiety. For the present purpose this is taken to mean drugs that are judged to be clinically effective in reducing formally described symptoms of anxiety. These issues of clinical definition will not be dealt with in the present chapter; the reader is referred to the *Diagnostic and Statistical Manual of Mental Disorders* (DSMIIIR) (1987) for a comprehensive treatment. Animal tests that are useful for identifying these compounds generally concern increases in behavior that is inhibited by novelty or by punishment (i.e., delivery of a noxious stimulus that is effective in reducing behavioral output). Some tests are relatively naturalistic (e.g., the social interaction test); others are more obviously contrived (e.g., conflict procedures). The value of these tests is not quantifiable in intuitive terms (i.e. face validity), but must be judged in terms of predictive validity—a fact that will emerge strongly through this chapter.

Nonspecific effects of drugs relative to the focus of their analysis (i.e., anxiolytic action in this case) are important in that (1) they may be undesirable or (2) they may confound the results of the test analysis. The latter point is borne out well by a consideration of factors such as sedation, under conditions in which decreases in responding are measured and increased feeding, under conditions whereby food deprivation is used to motivate animals to respond in the test situtation. The following section provides a brief description of some interesting "nonanxiolytic" effects or "side effects" of anxiolytic drugs.

2.2. Assessment of Side Effects of Anxiolytics

Anxiolytic drugs may possess considerable potency as sedatives and as anticonvulsants. In relation to anticonvulsant action, measures of protection against seizures induced either by electrical (trans-pinnal) shock or by injections of pentylenetetrazol are used as standard assays. Seizure induction in this context has also been achieved by repeated electrical stimulation of subcortical structures, a process known as kindling. This latter, chronic preparation has been used in attempts to develop models of seizure disorders in humans. Nevertheless, as pointed out by Hamor and Martin (1983), there are considerable difficulties in predicting possibly clinically useful anticonvulsants from animal tests in view of the multiplicity and complexity of convulsive disorders in humans (Millichap, 1969).

The relative sedative or muscle-relaxant properties of these compounds are obviously of considerable interest in the clinical context. Muscle relaxant effects are defined in terms of loss of muscle tone or motor coordination, whereas sedative effects are sometimes more specifically attributed to loss of vigilance. Such effects are generally assessed with catalepsy tests (*see* section 3) and with the rotarod test in which the animal's ability to maintain balance on a slowly rotating rod is assessed. Possible analgesic effects of these compounds are assessed using a variety of tests for nociceptive sensitivity (*see* Franklin and Abbott, 1987, for a review).

2.3. Aggressive Behavior

Anxiolytic drugs may selectively reduce aggressive behavior under some conditions and, under certain conditions, may increase aggressive responses (*see* Dantzer, 1987; Miczek and Winslow, 1987, for recent reviews). In view of the critical dependence of such variable effects on species, experimental protocol, and duration of drug treatment, this does not represent a particularly useful aspect of behavior for assessing anxiolytic drug action.

2.4. Exploratory Behavior

2.4.1. Staircase Test

The staircase test is a simple and rapid test that was proposed in 1973 as a screening procedure for anxiolytic drugs. Thiebot et al. (1973) have used this test to study the effects of drugs on two components of exploratory behavior in the rat. Conceptually this test is based on the assumption that anxiety is in some way equivalent to a fear of novelty.

For the test, a naive rat is placed in front of a five-step "staircase." The number of rearings and the number of steps climbed by the rat are counted separately. The number of steps climbed is an index of exploratory behavior, whereas the number of rearings reflects the level of "anxiety" of the animal. Thiebot et al. suggested that anxiolytic drugs could be distinguished from other classes of drugs based on their effect on the number of steps climbed and on rearing. Thus benzodiazepines, meprobamate, amobarbitone, and ethanol all decreased rearing at doses that left the number of steps climbed unchanged. At high doses a parallel decrease of both measures was seen. In contrast, amphetamine at low doses increased the rearing score only. These authors have suggested that the apparent "anxiolytic" effects were only partly attributable to muscle relaxant action of the drugs. It is interesting to note that the increase in rearing in response to amphetamine has been proposed as an anxiogenic action. An important recent appraisal has raised questions concerning the specificity of this test. Pollard and Howard (1986) have demonstrated a failure for the purported anxiogenics FG 7142 and pentylenetetrazol to produce significant effects in this test. More importantly, however, the anxiolytic alprazolam and the novel anxiolytic buspirone were identified as false negatives with this test, whereas morphine was a clear false positive.

2.4.2. The Plus-Maze

A further test of anxiolytic drug action involving exploration has been proposed using the plus-maze, so named because of its shape (Pellow et al., 1985). This device has two open arms and two closed arms. The effects of drugs are assessed on spontaneous activity of rodents in this apparatus. In a recent series of experiments, Lister (1987) has assessed the sensitivity to various drugs of behavior in the plus-maze (with mice), using a factor analytic approach. Anxiolytic drugs such as chlordiazepoxide and sodium pentabarbital are reported to selectively increase time spent in the open arms of the maze. Purported anxiogenics such as picrotoxin, caffeine, and FG7142 decrease this measure, whereas amphetamine and imipramine do not induce a selective change (Lister, 1987). Although Lister's work has confirmed the prior analysis of Pellow et al. (1985) using rats, an important feature of this test is that prior experience of the maze may attenuate the measured "anxiolytic" effect (Lister, 1987). The advantage of this test in terms of procedural ease may be offset by the possible limitations for repeated testing. Habituation is, of course, an important consideration with many tests of spontaneous behavior.

2.4.3. Social Interaction Test

The social interaction test has been proposed by File and Hyde (1978) as a useful method for studying anxiolytic drug action. As with many other tests of anxiolytic drug action, this procedure is based on the idea that anxiety will be reflected in terms of the inhibition of behavioral responses. Thus animals that have been socially isolated show a decrease in social interaction when exposed to a novel partner, relative to the level of interaction exhibited by unisolated controls.

In this test male rats are housed individually for a week with free access to food and water. They are then assigned to one of four test conditions. Half the rats are assigned to the familiar test condition and the other half are assigned to the unfamiliar test condition (familiarity being operationally defined in terms of habituation). For each of these sets of rats, half are assigned to low or high levels of illumination, respectively. In the familiar condition, rats are habituated to a test condition by being placed individually in the test box for 10 min on two consecutive days before the social interaction test. These familiarization sessions take place under the appropriate light level (*see* File and Hyde, 1978, for further details). Each rat is then tested for social interaction (i.e., certain behaviors are observed and rated) with an unknown, i.e., novel, test partner of the same weight. Pairs of rats are tested in the box for a 10-min screening period during which their behavior is videotaped for subsequent rating. The time rats spend in active contact is scored by two independent observers. Scored behaviors include sniffing, nipping, grooming, following, mounting, kicking, boxing, and crawling under or over the partner (*see* File and Hyde, 1978, for details). Maximum active interaction is observed when rats are tested under the low light in familiar conditions. If the light is increased or if the box is unfamiliar, active social interaction decreases.

Acute treatment with chlordiazepoxide reduces social interaction in this test; with chronic drug treatment (5 mg/kg/d for 5 d), however, sedative effects diminish and anxiolytic effects remain. Anxiolytic effects are reflected by an increase in social interaction. Various controls have shown that this may be a useful test for anxiolytic drug screening. For example, exploratory behavior decreases in the same manner as social interaction with changing test conditions in this procedure. Furthermore, as social interaction decreases, defecation and freezing increase. Thus in relation to previous literature describing responses of animals to novelty, the social interaction parameters change in a somewhat equivalent way. Control studies by File and Hyde (1978) indicate that a decrease in social interaction may not be simply attributed to olfactory changes in the partner. Nevertheless, it is possible that olfactory cues do play some role in this test (*vide infra*). As chronic treatment with an anxiolytic drug exhibits tolerance to sedative effects, but not to the anxiolytic effects, it is possible that this test may be useful for assessing the relative sedative/anxiolytic effects of drugs.

A variation of this test is the "colony intruder" paradigm in which an animal introduced into a colony is normally the object of intense physical attack from resident males. Chronic treatment of the intruder with chlordiazepoxide before introduction of the animal to the colony results in diminished aggression from the resident. This is despite the fact that the intruders submit less and initiate more interactions with residents (File, 1982). As Dantzer (1987) has pointed out, however, these effects may be caused by a decrease in submissive behavior on the part of the intruder, but may also be indirect; in other words caused by a change in the stimulus properties of the intruder. In a recent study, Dixon (1983) observed that diazepam resulted in increased aggression toward a treated male placed into a cage of a male resident mouse. This effect was apparently the result of olfactory cues, since applying urine from drugged animals on the fur of nondrugged animals resulted in an enhanced attack by the resident. In relation to these kinds of data, Dantzer (1987) has pointed out that further work is needed to determine at what level anxiolytic drugs act in these kinds of tests.

Nevertheless, if the social interaction test has high predictive validity, these considerations are only important when we are considering mechanisms of the anxiolytic action of these compounds.

2.5. Punished Behavior

2.5.1. Experimentally Induced Conflict

By far the most commonly used assays for anxiolytic drug action assess the ability of these compounds to increase responding that is suppressed by the delivery of noxious stimuli (e.g., electric shocks) that are contingent on responding (i.e., punishment). Clearly these procedures entail considerable face validity in that they involve experimentally induced conflict between a positively motivated response and concurrent (e.g., Geller-Seifter) or associated (e.g., conditioned suppression) aversive consequences. Nevertheless, caution must be excercised in this view since all anxiolytics are not identified by all conflict procedures. For example, buspirone is a novel atypical anxiolytic that is reported to be clinically effective (Rickels et al., 1982). This compound is sometimes detected by the Vogel conflict procedure, but generally not by the Geller-Seifter procedure *(vide infra)* with rats, although anticonflict effects have been observed with pigeons (Witkin and Barrett, 1986). The reasons for this are unclear and certainly not predictable on intuitive grounds (Weissman et al., 1984; Witkin and Barrett, 1986). The following section describes several of these procedures.

2.5.2. Geller-Seifter Conflict Test and Related Procedures

Geller and Seifter (1960) developed what is one of the most widely used screening tests for anxiolytic drug action. Their procedure is based on the observation that animals working on an intermittent schedule of reinforcement (i.e., a schedule whereby operant behavior such as lever pressing in rats results in a reward or reinforcer such as food by programmed intermittent delivery; *see* Sanger, 1987) will exhibit response suppression in periods of the schedule under which response contingent shock is presented. Thus animals working on a stable operant baseline will exhibit periods of sustained responding followed by periods of response suppression that are associated with times at which response contingent shock is programmed. The original procedure used liquid food reward. Currently most tests use food pellets as a reinforcer, although this procedure is also effective when electrical brain stimulation is substituted for food as a reward or reinforcer (Gomita and Ueki, 1981). In the original work, Geller and Seifter trained animals on a multiple schedule of variable interval 2 min (VI 2 min) and continuous reinforcement [i.e., every 5 min the VI schedule would change to a continuous reinforcement (CRF) schedule for a period of 3 min, the CRF schedule being signaled by a distinct auditory stimulus]. When these contingencies were well established, conflict was induced. This was done by punishing every lever response made in the presence of the tone with a shock delivered through the grid floor. Thus in the conflict periods rats were simultaneously rewarded with food and punished with shock for every lever response. After extended training this procedure induced a baseline of sustained responding interspersed with effective response suppression during periods of tone presentation that signalled response-contingent shock. The basic procedure is illustrated by the cumulative records displayed in Fig. 1. In the original report, meprobamate was effective in releasing punished behavior [i.e., the animals responded more in the CRF (tone and shock) period]. Qualitatively similar effects were observed with pentobarbital and phenobarbital. In contrast, promazine decreased the number of shocks the animals accepted in the conflict situation; trifluoperazine exhibited similar effects to those of promazine. *d*-Amphetamine under these conditions also reduced the tendency to accept shock in conflict situations. The relative effects of the anxiolytic chlordiazepoxide and of

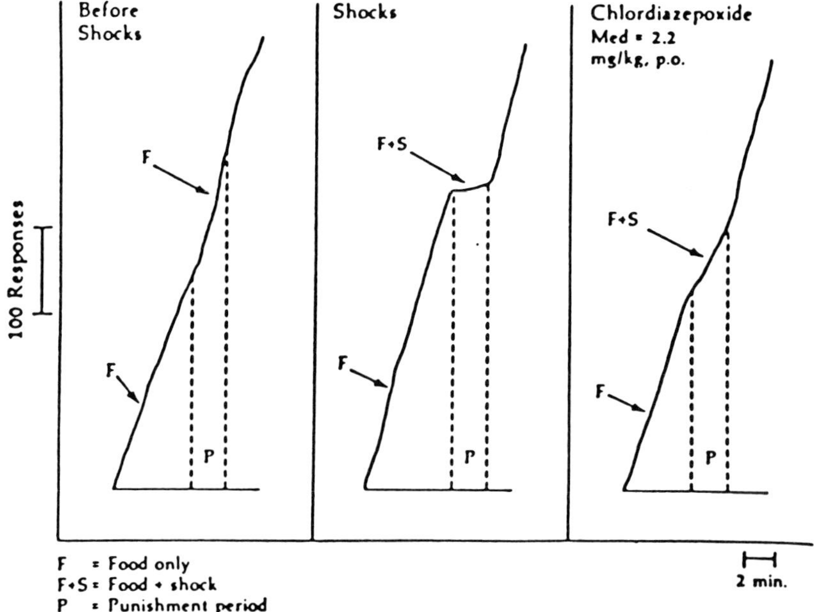

Fig. 1. The Geller-Seifter conflict "model" cumulative records of responding maintained by a multiple schedule involving punishment (from Carlton, 1983, reproduced with permission from author and publisher).

the neuroleptic chlorpromazine in this test are illustrated in Figs. 2 and 3, respectively.

In summary, with this procedure conflict behavior is induced by simultaneously rewarding with food and punishing with shock every lever response made in the presence of a tone. Anxiolytic drugs such as meprobamate, phenobarbital, and pentobarbital increase the number of shocks received by the animal in the conflict period of the schedule. In contrast, promazine, trifluoperazine, and amphetamine decrease the number of shocks. Geller and Seifter suggested that this technique might permit a separation of anxiolytic from nonanxiolytic drug effects. It is interesting to note that Geller and Seifter at this stage observed that compounds that were effective in releasing punished behavior were classified as depressant drugs that had anticonvulsant properties. It is currently evident that a wide range of clinically effective anxiolytics including benzodiazepines are able to restore or increase responding that is suppressed by punishment in this procedure. The validity of the

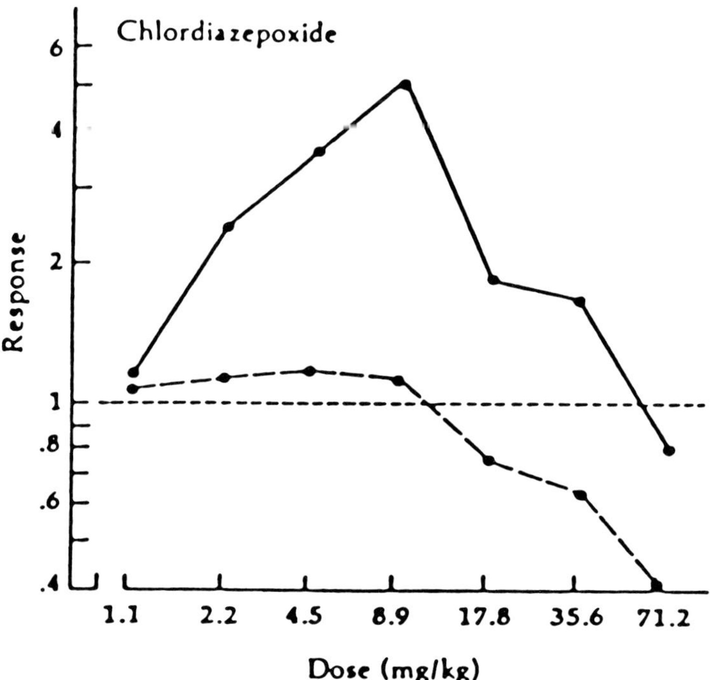

Fig. 2. Dose–response function describing the effects of chlordiazepoxide on punishment (solid line) and unpunished (dashed line) responding (from Cook and Davidson, 1973, with permission of authors and publisher).

Geller-Seifter Test for predicting anxiolytic efficacy has been emphasized by Cook and Davidson (1973) and subsequent authors. It is currently widely used as a screen for anxiolytic drug action. Furthermore, this test appears to have a considerable degree of face validity.

Nevertheless, this procedure is not without its difficulties, as Howard and Pollard (1977) have pointed out. There seem to be at least three problems:

1. Within and among animals there is a high degree of inconsistency of drug effects.
2. Baseline response rates seem to vary considerably particularly in the component in which responding is punished.
3. This procedure requires extensive training of test subjects.

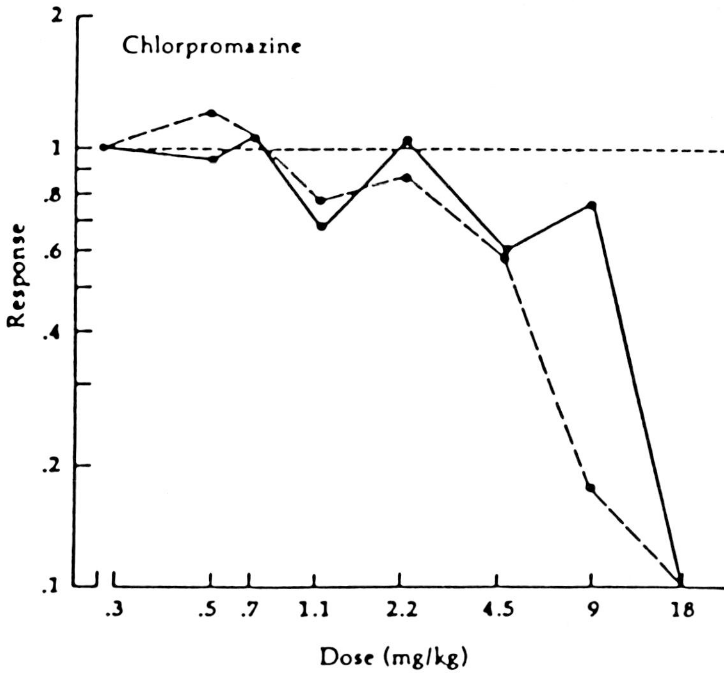

Fig. 3. Dose–response function describing the effect of chlorpromazine on punished (solid line) and unpunished (dashed line) responding
(from Cook and Davidson, 1973, with permission of authors and publisher).

Nevertheless, in view of its high predictive validity, this test
seems to be one of choice for measuring anxiolytic drug action.
Nonspecific effects such as effects of anxiolytic drugs on food
intake have been excluded as a critical feature of drug action in this
situation (Cook and Davidson, 1973). In these studies it has been
clearly demonstrated that increases in food-related motivation
(achieved with increasing food deprivation) do not mimic the
anticonflict effects of benzodiazepines. The marked increases in
food intake induced by these compounds (Cooper and Turkish,
1987) are therefore apparently not the basis for increases in food-
reinforced responding under the conflict test conditions. Furthermore it is evident that, in contrast to certain other classes of drugs,
the effects of benzodiazepines do not appear to be rate-dependent
(Sanger and Blackman, 1981).

2.5.3. Vogel Conflict

The suppression of water licking by electric shock has been used extensively to study the anxiolytic action of various compounds. Vogel et al. (1971) initially described a simple and rapid method for suppressing licking that is sensitive to the effect of a wide range of anxiolytic drugs. With this method naive rats are randomly allocated to different groups and water-deprived for 48 h prior to testing, food being available at all times in the home cages. Typically, animals are tested in a clear plexiglass box with a black plexiglass compartment attached to one wall. An opening leads from the large box to the small compartment. The entire apparatus has a stainless steel grid floor. A water bottle with a metal drinking spout is fitted to the outside of the small compartment so that the tube extends through a hole into the box. A lickometer circuit (*see* Bures et al., 1983) serves to measure licking, and shock is administered to the subject through the drinking tube and/or grid floor. After injection with a test compound, the subject is placed in the apparatus to find the drinking tube. If the subject completes 20 licks, shock is then administered for a period of up to 2 s. The subject can shorten the shock duration by withdrawing from the tube. A 3-min timer automatically starts at the termination of the first shock. During the 3-min period shocks are delivered following each twentieth lick and the number of shocks delivered during the 3-min session is recorded for each subject. With this procedure shock is very effective in inhibiting licking in animals that are not treated with drugs. In the original study benzodiazepines, pentobarbital, and probamate were very effective in releasing punished behavior (i.e., animals tested with these compounds received significantly more shocks than animals in the no-drug condition). There are certain advantages of this test in that a naive subject is tested in one session, and the test requires little time to complete and no training. Nevertheless, disadvantages are the lack of reference to baseline performance for each subject and a tendency for this test to detect false positives (Cook and Sepinwall, 1975a). The simplicity and low variability of data obtained with this procedure make it a reliable screening test. A number of false positives make it less attractive as a singular test of anxiolytic drug action, however.

Petersen and Lassen (1981) have described antipunishment effects of drugs in trained drug-experienced rats using weekly Vogel conflict testing. In this study the authors report that weekly exposure of water-deprived rats to this water-lick conflict test rapidly reduces undrugged responding to a very low and stable baseline. With this procedure rats were trained for up to 35 weekly test sessions under 48 h. of water deprivation. Suppression of licking was attenuated by clorazepam, diazepam, phenobarbital, and meprobamate. The anticonvulsant drug sodium valproate was also active. Nevertheless, serotonin antagonists, methysergide, cyproheptadine, cinanserin, and the serotonin depletor *p*-chlorophenylalanine (PCPA), were all inactive, as were other distinct psychotropic drugs, including propanolol, clonidine, THIP, theophylline, chlorpromazine, paroxetine, and ethanol. It is interesting to note that the antiserotonin compounds in this Petersen and Lassen study were active in the original Vogel conflict and also in the Geller conflict test (*see* section 2.3.1; Sepinwall and Cook, 1978), but, since they are not known to have clinical anxiolytic properties, should presumably be considered as false positives. Thus this modified Vogel procedure may have some advantages in terms of specificity of drug response, i.e., less false positives.

2.6. Conditioned Suppression

Conditioned suppression refers to the phenomenon whereby behavior is suppressed by a stimulus that has previously been associated with a noxious event. This is equivalent to pavlovian fear conditioning. In relation to procedures such as these it is important to note that conditioning may occur on the baseline or off the baseline. In other words, if an animal is working in the presence of a stimulus and a noxious event is then introduced (as in training for the Geller-Seifter conflict), this would be on-baseline conditioning. If, however, a stimulus is paired with a noxious event in a situation that is different from the test situation, this conditioning would be referred to as off-baseline. Gray (1977), in reviewing anxiolytic effects, has suggested that conditioned suppression literature shows that anxiolytic agents release conditioned suppression when pavlovian fear conditioning is on the baseline, but not when conditioning is off the baseline. There are various interpretations of these findings in terms of proposals ranging from adventitious punishment (Kelleher and Morse, 1968) to the ability of animals to discriminate between contingent and noncontingent events (Herrnstein, 1966). These issues are beyond the scope of the present treatment. Nevertheless it is important to note that *on-baseline* conditioning procedures such as the Geller-Seifter test are more sensitive to anxiolytics and are therefore procedures of choice in this context.

Interestingly, a number of experiments has shown that anxiolytics can interfere with learning. Several studies have analyzed the effects of these compounds with passive-avoidance or conditioned suppression procedures (Berger and Stein, 1968; Sanger and Joly, 1985; Broekkamp et al., 1984; Jensen et al., 1979; Komiskey et al., 1981; Oishi et al., 1972; Scobie and Garske, 1970). These experiments indicate that benzodiazepines when administered before fear conditioning trials induce poor performance during a subsequent retention test. It seems clear that anxiolytic drugs may disrupt the acquisition of avoidance responses at doses that neither produce gross motor impairment nor alter sensitivity to electric shocks (Patel et al., 1979). In a typical experiment to assess the effects of such drugs on learning, mice receive electric shocks in one side of a two-compartment box. Retention of conditioned fear is evaluated during the second trial, 24 h after the first, when the animals are returned to the box. The direct behavioral effects of these compounds are assessed by measuring motor activity during the first trial before presentation of the shock. In this test one compartment has transparent walls (light walls); the other has dark walls. The floor of both compartments is a metal grid and the animal can pass from one compartment to the other through a small hole in the dividing wall. The animals are shocked in the dark compartment, and on the retest both latency to enter and time spent in the dark compartment are measured. Using this procedure, Sanger and Joly (1985) have reported that diazepam and zopiclone induce dose-related decreases in motor activity; these same doses disrupt fear conditioning by decreasing latency to enter and increasing time spent in the dark (shock compartment).

3. Neuroleptic Drugs

3.1. Basic Screening Tests

In general, compounds that are seen to induce the following effects on animal behavior may be considered neuroleptics:

1. induction of catalepsy
2. induction of ptosis

3. inhibition of amphetamine- and apomorphine-induced behavior, i.e., stereotyped responses and locomotor activation
4. inhibition of apomorphine-induced emesis
5. inhibition of adrenaline or noradrenaline toxicity
6. inhibition of exploratory locomotor activity
7. inhibition of positively reinforced operant behavior
8. selective inhibition of active avoidance

One of the most widely used tests that is apparently selective for detecting neuroleptic activity is that of selective inhibition of active avoidance (Cook and Weidley, 1957). The inhibition of positively reinforced operant behavior by neuroleptic compounds is currently the subject of much research; its validity as a specific screening test for neuroleptics is unclear, however. Inhibition of brain self-stimulation fits into the same general category as inhibition of positively reinforced operant behavior, and although this test provides a sensitive assay for neuroleptic action, compounds other than neuroleptics may also inhibit self-stimulation.

Janssen et al. (1965) have pointed out that effects in each of the above tests may relate to particular properties of neuroleptic compounds in terms of critical features of action. Thus, antagonism of locomotor responses induced by apomorphine or amphetamine may predict antiagitation and antimanic effects of these compounds. Antagonism of stereotyped behavior induced by apomorphine and amphetamine may predict antipsychotic, antihallucinatory and antimanic activity. The inhibition of epinephrine (adrenaline) or norepinephrine (noradrenaline) toxicity is used as a predictor of autonomic side effects such as hypotension and tachycardia. The induction of catalepsy is used as an index of extrapyramidal effects of neuroleptics and may provide useful information concerning nonspecific motor dysfunction induced by these compounds in patients. Antagonism of emesis (i.e., vomiting) induced by apomorphine in dogs is predictive of similar effects in humans. Janssen has pointed out that this test may be useful to predict the onset and duration of critical antipsychotic effects. In this case it is interesting to note that many of the tests, including emesis, have, of course, no face validity in terms of antipsychotic action.

A description of the most common screening tests has been provided by Niemegeers and Janssen (1974). A brief description of these tests follows.

3.2. Catalepsy and Palpebral Ptosis

Rats are treated at 1-min intervals and placed in small individual observation cages. In the same order each rat is briefly observed for catalepsy and palpebral ptosis after *hourly* intervals of 1–10 h by two "blind" observers. Several doses of each compound and five to ten animals per dose are used. Catalepsy is assessed by placing each animal in an abnormal position. One observer attempts to hang the rat's forepaws on the outer edge of the wall of the observation cage. Both observers then assess whether or not they observe catalepsy according a simple rating scale.

After having handled the rats (subsequent to catalepsy testing), the two observers independently assess the degree of palpebral ptosis, again according to a simple rating scale.

3.3. Amphetamine Antagonism

Wistar rats are injected with *d,l*-amphetamine intraveneously at a dose of 10 mg/kg (free base). A "blind" observer assesses the intensity of chewing behavior and locomotor activity according to standard rating scales.

In a typical experiment, each dose of each compound is studied in separate rats, and control rats are, of course, included. The animals are pretreated with various subcutaneous doses of neuroleptic drugs and challenged with amphetamine after 1, 3, 7, 15, and 31 h. The inhibition of amphetamine-induced chewing and locomotor activity is measured at 1, 2, 4, 8, 16, and 32 h. after administration of the neuroleptic.

For further details of control data, the reader is referred to Niemegeers and Janssen (1974).

3.4. Norepinephrine Antagonism

Wistar rats are injected with norepinephrine intravenously with a dose of 1.25 mg/kg (0.2 mL/100 g). This high dose, three times the intravenous LD_{50}, kills all control rats within 15 min. Rats are pretreated with various subcutaneous doses of neuroleptic drugs 1, 2, 4, 8, 16, and 32 h before the challenging dose of norepinephrine.

Twenty-four hour survival after norepinephrine injection is adopted as the criterion for the neuroleptic drug effect.

3.5. Interpretation of the Results of Catalepsy and Palpebral Ptosis Tests and Antagonism of Amphetamine and Norepinephrine Tests

In relation to predicting sedative properties of neuroleptic drugs, Niemegeers has pointed out that the ratio of palpebral ptosis to catalepsy (i.e., in terms of relative potency) is a useful index of the relative sedative versus neurologic effects, compounds with a low palpebral ptosis to catalepsy ratio being the most sedative.

As stated earlier, the blockade of amphetamine and norepinephrine effects may relate to the relative antipsychotic versus autonomic effects of neuroleptic compounds. Niemegeers has again pointed out that the relative norepinephrine antagonism to amphetamine antagonism potency is a useful index of the autonomic versus antipsychotic activity of neuroleptics, compounds with a low norepinephrine antagonism to amphetamine antagonism ratio having the highest potential for autonomic effects.

Ljungberg and Ungerstedt (1985) have recently proposed what is essentially an extension of the original amphetamine antagonism test as a screening procedure for the simultaneous assessment of relative limbic and striatal actions of neuroleptics. In this test, animals are tested in a modified open-field apparatus after injection of 2 mg/kg of d-amphetamine; in a 10-min test, the relative effects of the neuroleptic drug on measures of d-amphetamine-induced locomotor activity and weak stereotypy are assessed. Locomotor activity changes are proposed to reflect potential antipsychotic action, whereas inhibition of the weak stereotypy is interpreted in terms of potential extrapyramidal effects of the neuroleptic being tested. These authors have validated their procedure with haloperidol, chlorpromazine, and thioridazine as compounds with significant extrapyramidal side effects, in contrast to clozapine and sulpiride, which are considered to give rise to fewer extrapyramidal side effects. This is contentious, since thioridazine has been characterized as inducing little or no extrapyrimidal symptomatology in the clinic (Sovner and DiMascio, 1978). The reader is referred to Ljungberg and Ungerstedt (1985) for the specific protocol. At this stage their test appears to possibly represent a convenient and rapid refinement for screening, although the utility of this procedure must be established in future experiments.

3.6. Apomorphine-Induced Emesis in Dogs

Adult beagles weighing 15–20 kg are given apomorphine subcutaneously at a dose of 0.31 mg/kg. This high standard dose (four times the ED_{95}) induces emesis (vomiting) in all control animals (*see* Niemegeers and Janssen, 1974). As with the other tests, the animals are pretreated with various subcutaneous doses of neuroleptics at 0.25, 0.50, 1, 2, 4, 8, 16, 32, and 64 h. before being challenged with apomorphine. Complete protection against vomiting for 1 h. after apomorphine is adopted as the criterion for drug effect.

3.7. Jumping Box Test

The jumping box for dogs is a large cage divided into two compartments by a hurdle. The floor of each compartment is a grid that can be electrically charged. Beagles (body weight as above) are fully trained to avoid shocks from the grid by jumping into the alternate compartment at the sound of a buzzer (i.e., a conditioned avoidance test). These dogs are treated with various subcutaneous doses of neuroleptics; 1, 5, and 25 h. after administration of the drug, beagles undergo a 10 min/10 trial session in a jumping box. Loss of avoidance (i.e., failure to avoid electric shock within 15 s. of the buzzer) more than once in any of the three post-drug sessions is adopted as a criterion for the drug effect.

3.8. Prediction of Neurological Side Effects

The ratio of the lowest ED_{50} [i.e., from each time interval] in the jumping box test and the lowest ED_{50} in the apomorphine antagonism test is proposed as a useful index of relative neurologic versus antipsychotic activity, compounds with a low ratio having highest neurologic potential (the ratio being 2.2/0.7, 0.66/0.024 and 0.091/0.012 for chlorpromazine, haloperidol and pimozide respectively). Neurological side effects may be predicted with chlorpromazine at 3.14 times the lowest effective dose; with haloperidol at 3.30 times the lowest effective dose; and with pimozide at 7.58 times the lowest effective dose. Neurological side effects, as Niemegeers has pointed out, are difficult to avoid with neuroleptics that have no difference between the lowest antipsychotic dose and the dose which induces neurological reactions. In this regard, with clinical use of neuroleptics, side effects may be a problem with compounds that have rapid onset and short duration of action. In this case a clinically *useful* dose will far exceed the lowest effective dose in the view of the need for maintenance of drug levels in the patient.

3.9. Overview of These Tests in Relation to the Prediction of Neuroleptic Specificity

Niemeegers and Janssen (1974) have pointed out that compounds with a low palpebral ptosis to catalepsy ratio and/or norepinephrine to amphetamine antagonism ratio will induce profound sedation and/or autonomic effects at doses that are at or below the threshold for antipsychotic action. Compounds with high values of the above ratios may be considered specific neuroleptic compounds that will be quite inactive as sedative agents that induce autonomic dysfunction.

3.10. Active Avoidance Tests

As stated earlier, a widely used and apparently sensitive test of neuroleptic drug action is that of active avoidance responding. This procedure is based on the use of an operant conditioning schedule originally described by Sidman (1957) referred to as active or Sidman avoidance. Under the test conditions, animals are trained to perform a response within a certain time to avoid delivery of a noxious electric shock. Failure to respond results in the repeated delivery of shocks at predetermined intervals, the shock-shock interval. If the animal responds, a time-out will be initiated that will delay the delivery of the next shock for a certain fixed period of time, the response-shock interval. Thus, animals may respond at relatively constant rates and successfully avoid all shocks. This nondiscriminated avoidance schedule is quite sensitive to the behavioral effects of neuroleptics. Indeed, it has been suggested that, at least under certain conditions, Sidman avoidance is a more sensitive procedure than conditioned (discriminated) avoidance schedules. Rats are commonly used in these experiments; primates are sometimes employed, however. The particular details of test procedures in terms of the intensity of shock and the shock-shock and response-shock intervals vary from laboratory to laboratory and are, therefore, not described here. The reader is, however, referred to Liebman et al. (1981) and Niemegeers et al. (1969) as examples of such protocols.

The use of more than one species may be valuable in assessing neuroleptic drug action. This is particularly important in determining potency, since the potency of one compound may vary considerably among tests with rats and tests with certain primate species (Setler et al., 1976; Liebman et al., 1981).

To improve avoidance testing as a screen for neuroleptic drugs, antagonists have been used to assess the specificity of the drug response. Benztropine is an antagonist that has been used in this context. Benztropine is a compound that may reverse the effect of certain neuroleptics on avoidance schedules. It does not, however, block the depressant effects of other compounds that may disrupt avoidance behavior. For example, the depressant effects of benzodiazepines or barbiturate drugs may be facilitated by benztropine. Indeed, antagonism of neuroleptic effects by benztropine in animals clearly relates to similar effects that have been reported for human psychotics. If benztropine is used for extrapyramidal dysfunction, typically this treatment is accompanied by loss of the therapeutic action of certain antipsychotic neuroleptic drugs. Benztropine antagonism of neuroleptic effects is apparently limited to shock avoidance tests, since it does not attenuate the decrease in brain self-stimulation responding induced by these compounds, although it is effective as an inhibitor of dyskinesias induced by neuroleptics in primates (Neale et al., 1984). In this respect benztropine antagonism may be useful for detecting possible extrapyramidal side effects of neuroleptics in primate avoidance tests (Liebman et al., 1981). It is interesting to note that analgesic narcotics may mimic neuroleptics in certain tests of avoidance responding, although it may be relatively easy to distinguish between such narcotic responses and neuroleptic responses by using an antagonist such as naloxone. Thus, naloxone reverses the decrease in avoidance responding induced by morphine, but not that induced by neuroleptic drugs.

Discriminated or conditioned avoidance responding is also a useful test for neuroleptic drug action (Denenberg et al., 1959; Bignami, 1978; Houser, 1978; Kuribara and Tadokoro, 1981; Niemeegers et al., 1969). One particularly important observation is that antipsychotic or neuroleptic drugs disrupt the ability of animals to avoid shocks at doses that do not prevent escape from shocks. A variety of other centrally acting drugs, including hypnotics, sedatives, and anxiolytics, however, may disrupt avoidance and escape responding at similar doses. The selective effects of neuroleptics on avoidance responding are illustrated in Fig. 4.

The attenuation of avoidance (with antagonism by benztropine) provides good evidence for neuroleptic drug action. In particu-

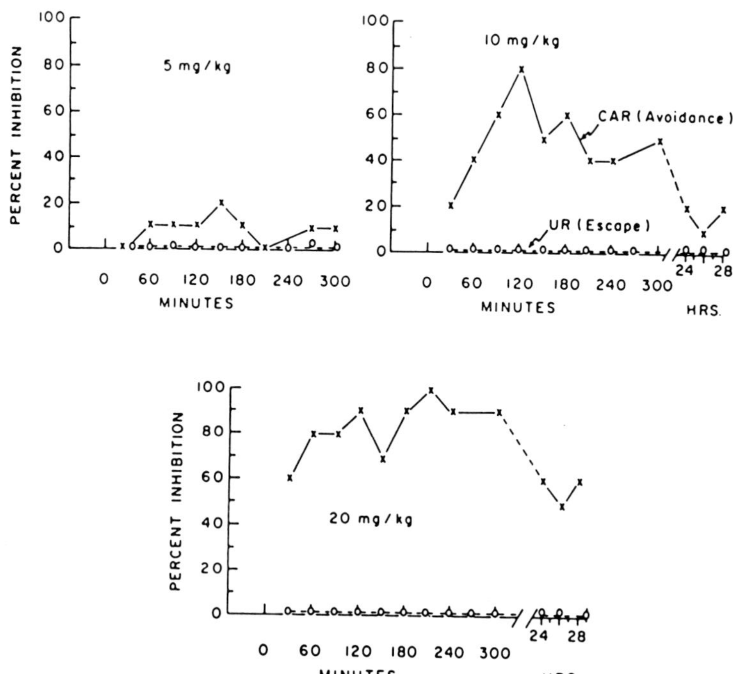

Fig. 4. Effect of chlorpromazine on discrete conditioned pole-climb avoidance response in rats. Ten Charles River (CD albino) rats were assigned to each treatment group and trained to consistently avoid or escape footshock (0.8 mA, 350 V.A.C.) before treatments. Solutions of chlorpromazine hydrochloride were administered orally at 0 time, on the abscissa. During testing, trials were spaced 30 min apart. Each point represents the percentage of animals that failed to make either the conditioned avoidance response (CAR — solid line) or the unconditioned escape response (UR — — dashed line) on a given trial. Chlorpromazine was effective in selectively blocking the CAR at doses which did not alter the UR (from Cook and Sepinwall, 1975b, with permission of authors and publisher).

lar, a selective deficit in avoidance, but not escape responding, also provides a useful index of neuroleptic activity; nevertheless some antipsychotic drugs do not clearly exhibit this profile of action (Sanger, 1985, 1986).

3.11. Tests with Brain Self-Stimulation

Since the mid-1950s, it has been established that animals will perform responses that result in electrical stimulation of certain brain areas. It has been proposed that depression of self-stimulation behavior is a powerful indicator for neuroleptic activity. This is currently a subject of intense controversy and debate. Although it is apparent that neuroleptic drugs are certainly very effective in attenuating self-stimulation, it is clear that other compounds may have this effect. Thus attenuation of self-stimulation is certainly not a specific feature of neuroleptic drug action.

It is interesting to note, however, that certain researchers have suggested that the attenuation of self-stimulation by neuroleptic drugs is functionally equivalent to behavior seen in extinction, i.e., when the reward or reinforcer is no longer available to the animal. This, of course, has important implications in relation to current theories of psychosis and mechanisms of antipsychotic drug action (*see* Wise, 1982). Nevertheless, although it is clear that under certain conditions neuroleptics may have extinction-like effects in this test, an extinction-like response is not always seen, as illustrated in Fig. 5 (*see* Sanger, 1986). The reader is referred to recent reviews by Wise (1978, 1982) and Liebman (1988) for further details.

3.12. Self-Administration of Cocaine

It is well established that many neuroleptic drugs, at low doses, increase the rate of self-administration of psychomotor stimulant drugs (*see* Wise, 1978). This technique is technically rather demanding and, although it has been successfully employed in many laboratories, requires more initial specialization of skills than most other approaches. The reader is referred to Roberts and Goeders (1988) for a detailed appraisal of techniques in this area. Nevertheless, recent experiments indicate that the self-administration procedure may be particularly useful for detecting atypical compounds that are effective antipsychotics, but are false negatives in many tests. For example, some of these atypical compounds induce only mild catalepsy and weakly antagonize amphetamine-induced behavioral activation (Stille et al., 1971; Jenner et al., 1978).

Roberts and Vickers (1984) have recently demonstrated that a number of atypical neuroleptics will increase cocaine self-administration and report a high correlation ($r = 0.94$) between potency (ED_{125}) in this test and recommended daily clinical dose. This phenomenon is illustrated by the data of Fig. 6. Nevertheless, a notable false negative in this test is clozapine, an effective but atypical antipsychotic agent. In terms of general screening for

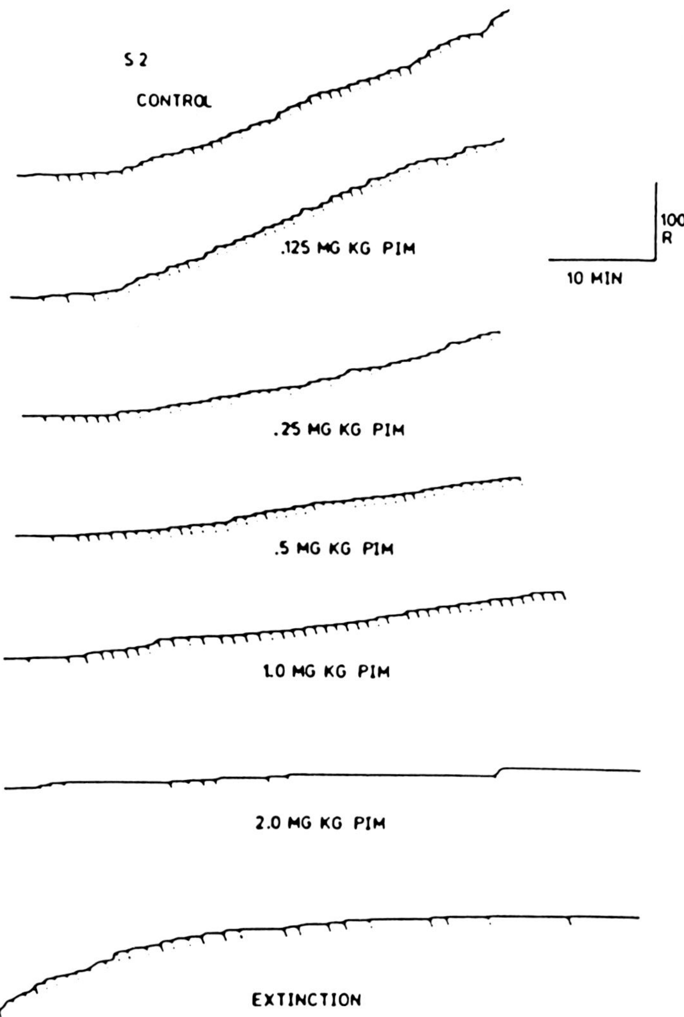

Fig. 5. Cumulative records illustrating the performance of one rat on the FI 60-s schedule under control conditions, during a reward omission session (extinction) and after administration of each dose of pimozide. Extinction produced an increasing decline in responding over time, whereas pimozide reduced responding evenly over the session up to the 2 mg/kg dose, which markedly disrupted responding in this animal (from Greenshaw et al., 1981, with permission of authors and publisher).

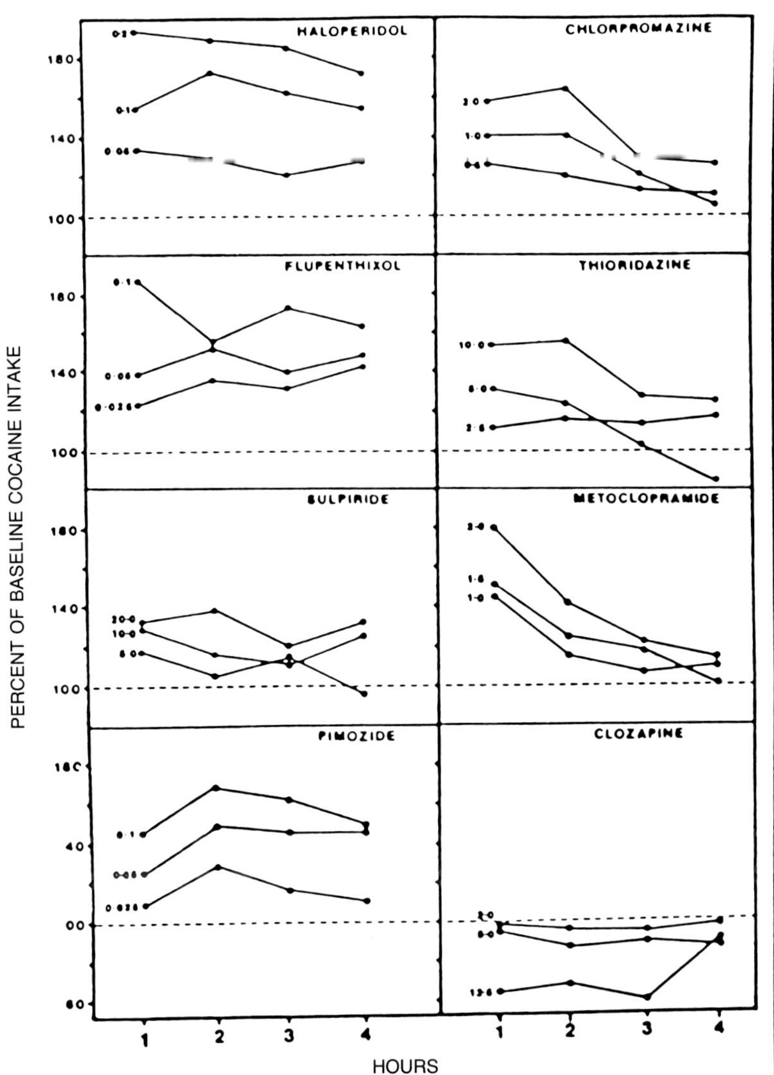

Fig. 6. The effect of various neuroleptics on the hourly intake of self-administered cocaine in rats. Each point represents the average hourly intake of cocaine expressed as a percentage of baseline intake over the 4-h test session. The range of baseline intake varied less than 10%. Doses are in mg/kg ip (from Roberts and Vickers, 1984, with permission of authors and publisher).

neuroleptics, this seems to be a promising approach. Roberts and Vickers have suggested that it is advantageous in terms of sensitivity to low doses of neuroleptics; in terms of the fact that it permits a clear analysis of the time course of neuroleptic effects and that the effects (i.e., increases in rate) are clearly distinguishable from motor impairment.

4. Antidepressant Drugs

4.1. Present Status

Tests of antidepressant drug action have recently been extensively discussed by Willner (1984), with particular attention being paid to each test in terms of validity. In the present chapter only tests with significant validity (i.e., at least significant predictive validity as outlined earlier) will be discussed. The models or tests identified and assessed by Willner (1984) are illustrated in Table 2. The reader is encouraged to consult the above review, since it represents a clear and extensive analysis of this topic.

As may be seen from Table 2, the following tests showed high validity: behavioral despair, chronic stress, separation, and self-stimulation. The following tests showed reasonable validity: yohimbine potentiation in dogs, chronic isolation, exhaustion stress, circadian rhythms. Other tests showing some degree of predictive validity were muricide, olfactory bulbectomy, and learned helplessness. The remainder of the tests or models have either very poor or questionable validity and are, therefore, not discussed in any detail in the present chapter. For references to these tests, the reader is referred to Willner's excellent review (Willner, 1984) as a source of literature. The authors realize that by deleting some of these tests, classical models of antidepressant drug action such as reserpine reversal will not be dealt with; on balance, however, the validity of some of these tests is questionable enough to justify their omission in the present view.

4.2. Muricide

An early observation in the study of antidepressant drug action was that imipramine (a tricyclic antidepressant) impaired mouse-killing behavior (muricide) in rats (Horovitz, 1967). This effect of imipramine is also exhibited by a wide range of antidepressants. It is important to note that electroconvulsive shock, which is also an effective antidepressant treatment in the clinic, is effective in inhibiting muricide (Ueki, 1982). Nevertheless in some cases antidepressants block muricide only at doses that cause significant motor impairment (Sofia, 1969). Furthermore, many false positives such as psychomotor stimulants, antihistamines, and certain anticholinergics also inhibit muricide. This high incidence of false positives makes muricide a relatively unattractive test for antidepressant drug action.

4.3. Yohimbine Potentiation

Antidepressant drugs are also found to potentiate the effects of systemically administered yohimbine. There are two versions of this test; in one version the increased potency of yohimbine in toxicity testing with mice (i.e., decreased LD_{50}) is used (Quinton, 1963). In another version, yohimbine-induced increases in blood pressure and behavioral changes (body tremors and increased motor activity) in dogs are assessed (Lang and Gershon, 1962). In the mouse version of this test, electroconvulsive shock is ineffective, and, as in the case of muricide, false positive responses are given by psychomotor stimulants, anticholinergics, and antihistamines.

In the dog test, various tricylcic compounds and nialamide (a monoamine oxidase inhibitor) are effective. Nevertheless atypical antidepressants and electroconvulsive shock have apparently not been tested. The false positives from the previous tests, i.e., psychomotor stimulants, anticholinergics, and antihistamines, are ineffective in this context. Thus, the yohimbine potentiation test in dogs seems to have a reasonable amount of predictive validity, although, as Willner (1984) has pointed out, it needs to be assessed with a wider range of drugs before its usefulness can be fully determined.

4.4. Olfactory Bulbectomy

Olfactory bulbectomy, i.e., bilateral ablation of the brain olfactory bulbs, in rats induces a variety of behavioral effects. Olfactory bulbectomized rats are hyperreactive, and a prominent feature of these animals is that they show a deficit in passive avoidance learning. In the passive avoidance test, animals are typically placed on a raised platform surrounded by an electrified grid. If they step down on the electrified grid they receive an electric shock. Control

Table 2
Assessment of Models of Antidepressant Drug Action Against Various Validating Criteria[a]

Models	Responsive to antidepressants	Wide range tested	No false positives	No false negatives	Correlation of potencies	Time course	Similarity of symptoms	Coherence of symptoms	No dissimilarities	Specificity to depression	Clear interpretation of model	Clear interpretation of features	Homology	Empirical relationship to depression	Theoretical relationship to depression	Summary P	F	C	Total
Muricide	+	+ +	−	−?	−?											+			1
Yohimbine (mice)	+	+ +	=	−	−														0

Model	1	2	3	4	5	6	7	8	9	10	11	12	13	14	15	P	F	C	Total
Yohimbine (dogs)	+	+	+?	+												++			2
Dopa	+	+	−	=															0
Kindling	+		−	=															0
Reserpine	+	++	−	=	+?	−	+		−										0
Amphetamine	+	+	−	=		−	−												0
5-HTP	+	+	−	−		−	+												0
Bulbectomy	+	++	−	−		?	++	?		?						+		+	2
Isolation	+	++	+?	+	+?	−	+			−						+++			3
Exhaustion	+					+	+									+	+		2
Rhythms	+	+	+	+?		+	+									++	+		3
Helplessness	+	++	+	+		+	++	−?	−	−	+?	+?	+?	?	?	++	+	+	4
"Despair"	+	++	?	−?	+	+?										++	+		3
Chronic stress	+	++		+	−	+	++	+		−	+		?			++	++	+	5
Separation	+					+	++	+	−		+	+	+	+?	+	+	+	++	4
Disengagement											+	+	+	+?	+		+	+	2
ICSS	+	+	+?	+		+				+	+	+?	+?	+?	+	++	+	++	5

[a]The main body of the table estimates the extent to which a model meets each criterion (+ or ++, model does well; − or =, model does badly). The summary columns on the right estimate predictive (P), face (F), and construct (C) validity on a four-point scale (blank, +, ++, +++): these scores are summed to give a grand total at the far right. (from Willner, 1984, with permission).

animals on retesting exhibit an increased latency to step down; typically olfactory bulbectomized rats show a deficit in this kind of avoidance learning. This syndrome induced by olfactory bulbectomy is apparently reversed by chronic antidepressant drug treatments (Cairncross et al., 1978). These animals also show characteristically high levels of plasma corticosteroids. Antidepressant drugs are not the only compounds that reverse this response to these lesions. Neuroleptic and anxiolytic drugs may decrease plasma corticosteroid levels and also reverse hyperreactivity. As Willner (1984) has pointed out, it is not clear whether the whole syndrome induced by olfactory bulbectomy or only the passive avoidance deficit is sensitive to antidepressant drugs (Joly and Sanger, 1986). Reversal of passive/avoidance deficit in olfactory bulbectomized animals is a response that has been demonstrated for a variety of antidepressant drugs. Nevertheless it is apparent that very few nonantidepressant drugs have been tested in this context, although the psychomotor stimulant amphetamine and the anticholinergic atropine are ineffective in reversing passive avoidance deficit. Interestingly quipazine, a 5-HT agonist, was effective (Lloyd et al., 1982), whereas tranylcypromine, a monoamine oxidase inhibitor with known antidepressant action, was ineffective in reversing this passive avoidance deficit (Cairncross et al., 1978). With the exception of this peculiar failure to identify a monoamine oxidase-inhibiting antidepressant (but *see* Jesberger and Richardson, 1986), this model appears to be sensitive to various antidepressant drugs.

4.5. Diurnal Rhythms

Rats typically show a diurnal activity pattern that is high at night and low during the day, being nocturnal animals. After reversal of the light/dark cycle, readjustment to a normal cycle of locomotor activity is facilitated by antidepressant treatment administered prior to and after the shift in light/dark cycle. This test may be insensitive to anxiolytics, neuroleptics, psychomotor stimulants, and reserpine (Baltzer and Weiskrantz, 1973). Goodwin et al. (1982) have subsequently reported adjustment of circadian rhythms with hamsters shifted from a normal light/dark cycle to constant darkness. Willner has pointed out that since disturbance of circadian rhythms appears to be a characteristic feature of depression, studies such as this may provide an interesting model with which to screen or investigate the actions of antidepressant drugs.

In relation to circadian rhythms and the observation that a variety of sleep deprivation procedures may be effective as antidepressant treatments (Gillin, 1983), it is interesting to note that antidepressants may suppress REM sleep in cats and rats (Scherschlicht et al., 1982). In terms of this measure, amphetamine and morphine suppressed both REM and non-REM sleep, although its specificity is reduced by the fact that phenobarbital had an effect similar to that of antidepressants.

4.6. Chronic Isolation

Rats that are exposed to social isolation from the age of about 2–3 wk exhibit a marked increase in spontaneous motor activity relative to group-reared control animals (Einon et al., 1975; Sakakian et al., 1975, 1977). Isolation of this kind is not accompanied by any sign of aggression toward other organisms, and it seems that the time of initial isolation may be critical, since isolation of adult animals results in aggression, but not hyperactivity. The isolation-induced hyperactivity described above is apparently abolished by acute treatment with various antidepressant drugs (*see* Willner, 1984). Neuroleptics and anxiolytics apparently do not abolish this hyperactivity response. The effects of anticholinergic drugs on this test are unknown. Norepinephrine receptor antagonists may be effective blockers of hyperactivity. Also, chronic amphetamine appears to abolish hyperactivity, although it should be noted that the isolated animals are hypersensitive to acute administration of amphetamine.

4.7. Learned Helplessness

Learned helplessness is a term coined by Seligman (1975) to describe the phenomenon whereby exposure to uncontrollable stress produces severe deficits in learning tasks. An important point is that these deficits are only seen when stress is uncontrollable; subjects exposed to identical treatments, but with the provision that they may exert control over the stressful stimulus, do not display learned helplessness. Learned helplessness deficits are apparently reversed by antidepressant treatment. A wide variety of antidepressants has been tested in this context (*see* Willner, 1984), Acute antidepressant treatment is apparently ineffective and, importantly, chronic treatment with neuroleptics, psychomotor stimulants, sedatives, and anxiolytics is ineffective in reversing learned helplessness. There are some reports that learned helplessness may be reversed by stimulation of catecholamine receptors and by the anticholinergic drug scopolamine. This learned helplessness phenomenon seems to be a sensitive model for testing antidepressant drug action. There are apparently no false negatives with this test (i.e., no known antidepressant drug has failed to reverse learned helplessness).

4.8. Behavioral "Despair"

A rather convenient and simple test of antidepressant drug action was developed by Porsolt (Porsolt et al., 1977, 1978, 1979). In this test, if mice or rats are placed in a confined space and are forced to swim, they initially show high levels of swimming activity and then, after a certain latency, maintain a relatively passive or immobile posture. Repeated exposure of the animal to this situation results in a decrease in the latency to assume immobility. Behavioral immobility after forced swimming has been described as behavioral "despair." Although this description is rather anthropromorphic in assuming that the animals have given up hope or despaired of escaping, this is by no means a necessary assumption for interpretation of the data. Latency to immobility is delayed significantly by pretreatment with a wide range of antidepressant drugs, including tricyclics, monoamine oxidase inhibitors, and atypical compounds. Electroconvulsive shock and REM sleep deprivation are also effective in delaying the onset of behavioral "despair" (Porsolt, 1985). Indeed it is interesting and important to note that there is a significant correlation between the clinical potency of antidepressant drugs and the potency of these compounds in the behavioral "despair" test. This relationship has to date not been reported for any other test of antidepressant drug action (Willner, 1984).

This test suffers, nevertheless, from its lack of specificity, since several false positives are evident. Furthermore, some antidepressants are not effective, and false positive responses have been reported for psychomotor stimulants, convulsant drugs, anticholinergics, antihistamines, barbiturates, opiates, and a variety of other compounds, although it is notable that neuroleptic drugs and anxiolytic drugs are apparently not effective in this test. In relation to the false positive effect of psychomotor stimulants and anticholinergics, it has been suggested that prolonging the immobility test may allow the dissociation of false positives from antidepressants because the stimulants and anticholinergics may act through an indiscriminate stimulation of motor activity (Kitada et al., 1981). A recent refinement of this test has been reported (Cools, 1980; Thornton et al., 1985). In the latter study a modified version of the test, which included an escape component, was shown to be sensitive to nomifensine. Interestingly, although the response to antidepressants in this test is potentiated by chronic treatment, tolerance develops to the antihistamine effect (Kitada et al., 1981). Notable false negatives in this test are some atypical neuroleptics that have significant antidepressant action (Gorka and Janus, 1985).

4.9. Tail Suspension Test: A Novel Method for Screening Antidepressant Drug Action

As a refinement or an extension of the original test described by Porsolt, a new procedure has been proposed for testing antidepressant drugs. This test has been called the tail suspension test (Steru et al., 1985). This method was based on the observation that a mouse suspended by the tail exhibits alternate periods of activity and immobility.

In this study imipramine, desipramine, and amitriptyline each induced a significant dose-related reduction of immobility. A peripherally active analog of imipramine had no effect on this measure. Mianserin and viloxazine decreased the duration of immobility, as did nomifensine. High doses of the anticholinergic drug atropine reduced immobility, as did amphetamine at high doses. Nevertheless amphetamine increased immobility at low doses, as did the anxiolytic drug diazepam.

These authors claim that this may be a specific test of antidepressant drug action and does not merely measure locomotor stimulation. There are two arguments; first, sedative antidepressants decrease immobility (i.e., increase the activity of the animal) at doses that induce sedation. Second, antidepressants with stimulant activity decrease the duration of immobility at doses that are lower than those reported to induce locomotor stimulant responses. One limitation of this model seems to be that acute administration of drugs is particularly effective, and there is no evidence that the antidepressant response emerges as a function of chronic exposure to drug treatment. The authors suggest that the tail suspension test may have distinct advantages over the behavioral "despair" test on two counts. The immersion test induces hypothermia, and this is avoided in the tail suspension test; this is possibly a confounding factor in the antidepressant drug responses in the behavioral "despair" test. Also, recording of behavior in the tail suspension test may be more precise than human observation, which is the basis for the immobility measure in the behavioral "despair" test. The second point seems rather trivial since immobility is relatively easy to identify in rodents in the swimming test. In any event, it is not difficult to conceive of ways in which to automate measures of activity in the swimming test. A more important point may be that the tail suspension test is apparently sensitive to lower doses of antidepressant drugs and, on the basis of preliminary results, may provide clearer dose–response analyses. Furthermore, one group of antidepressant drugs that are false negatives in the behavioral despair test, inhibitors of 5-HT uptake such as zimelidine, are active in this test (Steru et al., 1985). Nevertheless this test is very new and it will be necessary to subject it to fairly rigorous analysis in a number of laboratories.

4.10. Chronic Stress

In relation to models that involve the exposure of laboratory animals to stressful events, such as the behavioral "despair" model discussed above, chronic unpredictable stress has been used to assess effects of antidepressant drugs. Of course, the definition of stress remains problematic [*see* Anisman and Zacharko (1982) for a recent discussion of this topic]. For the present purposes, stress is referred to in operational terms, i.e., stress refers to the exposure of animals to defined stimuli. In a recent series of experiments, Katz and colleagues (*see* Katz, 1982) have assessed the effects of chronic exposure of rats to a variety of "stressful" stimuli. This class of stimuli includes electric shocks, cold water immersion, and reversal of the light/dark cycle. At the end of exposure to this operationally defined stress over a period of about 3 wk, the animals are tested in the open field for activity responses to loud noises and bright lights. Basically, Katz and colleagues observed that in unstressed animals the open-field test exposure to noise and light resulted in an increase in locomotor activity; this effect was not seen in animals exposed to chronic stress. Daily administration of antidepressants during the chronic stress period, however, resulted in a restoration of the activating effect. This restoration was observed with pargyline, tricyclic antidepressants, atypical antidepressants, and electroconvulsive shock. Neuroleptics, anxiolytics, antihistamines, and anticholinergics were ineffective, as were psychomotor stimulants such as amphetamine. Nevertheless it is notable that the antidepressant monoamine oxidase inhibitor tranylcypromine was also ineffective in restoring this activation. As Willner (1984) has pointed out, this chronic stress treatment is reported to increase plasma corticosteroid levels, an effect that shows the same pharmacological sensitivity (i.e., it is reversed by chronic treatment with antidepressant drugs). It is notable in this case that the anticholinergic drug scopolamine was also effective in reversing this increase in plasma corticosteroids. These effects have been reported for both rats and mice. A further effect of chronic stress observed by Katz (1982) is a failure for animals to show a taste preference (i.e., increased consumption for fluid when saccharin is added to drinking water). This effect was partially attenuated by imipramine. Speculation may arise as to the validity of this model, since investigations of possible relationships between stress and depression have not yielded a clear result. Nevertheless as pointed out earlier, if such a model has high predictive validity for detecting antidepressant drugs, the issue of to what degree stress precipitates depression remains important only in relation to attempting to elucidate the mechanisms of action of antidepressant drugs. It is interesting to note that in relation to this chronic stress model, recent studies with intracranial self-stimulation (see later section) have yielded similar behavioral data. In Katz's model, antidepressant action is seen to be chronic, as in the case with human clinical studies, whereas in some self-stimulation studies to be discussed later, the deficits are reversed by acute treatment with antidepressants. Again, although the difference between acute and chronic responsiveness is not a critical feature in terms of predictability, it is an important consideration in attempting to analyze the mechanisms of antidepressant drug action.

4.11. Separation Models

The observation that separation of organisms from their conspecifics may lead to a profound behavioral syndrome of withdrawal and aberrant social behavior ("despair") is significant in relation to analyzing possible animal models of depression. As Willner (1984) has stated, "the evolutionary proximity of primate species seems to afford intuitive insights into their behavior which are lacking in less closely related animals." Thus it is not surprising that a number of authors propose that the only true "animal model of depression" is concerned with separation phenomena in infrahuman primates. Although primates are not the only species to exhibit this response to separation (Katz, 1981; McKinney and Bunney, 1969), it is easy to understand why most researchers have focused on primates in this context. Infant monkeys respond to maternal separation by an initial "protest" stage followed by despair after 24 or 48 h. The protest stage is characterized by increased activity, disruption of sleep patterns, and vocalization. The despair phase is characterized by decrease in overall activity, food intake, and social interaction and play, eventually resulting in the assumption of a withdrawn posture and "sad" facial expression (Suomi, 1976). Similar behavioral responses are observed when group-reared animals are subsequently isolated from their conspecifics (Suomi, 1976). It is somewhat surprising that few published studies have examined the efficacy of antidepressant drug treatment in relation to "despair" induced by separation in primates. Hrdina et al. (1979) have assessed the effects of chronic desmethylimipramine. This compound is reported to decrease stress vocalization and to increase social contact in infant macaques separated from their mothers. Similar effects have been reported in response to chronic administration of imipramine with respect to self-clasping in peer-separated Rhesus monkeys. Nevertheless, other behavioral changes induced by separation were apparently unaffected (Suomi et al., 1978). Electroconvulsive shock has also been reported to have therapeutic effects in isolated Rhesus monkeys (Lewis and McKinney, 1966). Some compounds that are not antidepressant have been found to be ineffective in changing the social unresponsiveness of isolated chimpanzees (Menzel et al., 1963; Turner et al., 1969), although chlorpromazine is reported to have had some therapeutic effects in isolated Rhesus monkeys (McKinney et al., 1973). It is apparent that this model may provide a very interesting and possibly, in some respects, equivalent set of behaviors for the analysis of antidepressant drug action. Nevertheless insufficient studies of different drugs are available, and much further work needs to be done in order to characterize or attempt to characterize a clear antidepressant profile for this separation syndrome. The possible homology between primate separation response and some forms of human depression, particularly anaclitic (i.e., related to maternal loss) depression, is intuitively appealing, but remains contentious (*see* Willner, 1984, for a discussion).

4.12. Intracranial Self-Stimulation

An assessment of the subjective responses to electrical stimulation of local brain areas in human subjects during neurosurgery (Heath, 1983; Heath et al., 1968; Valenstein, 1973) has indicated clearly that electrical stimulation of the brain may evoke pleasurable sensations in humans. One hypothesis concerning depression is that depression is a consequence of a reduction of activity in neural systems of the brain involved in the perception of the hedonic value of environmental stimuli. In relation to this, it is particularly interesting that many species of laboratory animals will perform operant responses for stimulation of local brain areas. In these experiments animals are implanted chronically with electrodes located in a chosen brain area and are trained to lever-press in response to electrical stimulation of the brain. A number of researchers have studied the effects of antidepressant drugs administered to animals self-stimulating on an operant schedule, i.e., while the animals were lever-pressing or nose-poking on a schedule whereby these responses resulted in the delivery of electrical brain stimulation to a specified brain area. In one such series of experiments, it has been demonstrated that following chronic amphetamine treatment, on withdrawal of amphetamine, response rates are markedly reduced and the threshold current for brain stimulation reinforcement is elevated for a period of weeks (Barrett and White, 1980; Kokkinidis and Zacharko, 1980; Leith and Barrett, 1976, 1980; Simpson and Annau, 1977). A period of 2-d treatment with imipramine or amitriptyline attenuated this response deficit. With continued chronic treatment with these drugs, the deficit was reversed (Kokkinidis et al., 1980).

In other studies a reduction of self-stimulation has been reported following lesions of the internal capsule. Such response deficits are alleviated by tricyclics and some atypical antidepressant drugs. Other nonantidepressant drugs such as diazepam, yohimbine, and propranolol are apparently ineffective in reversing this deficit (Cornfeldt et al., 1982; Szewczaic et al., 1982).

Other studies have involved antidepressant drug treatment in the absence of some imposed or induced deficit in self-stimulation behavior. In one such study, Wauquier (1980) reported that antidepressant drug treatment increased the "breaking point" on progressive ratio schedules of brain stimulation reinforcement. This is a particularly interesting finding; it has not subsequently been substantiated, however. Chronic administration of desmethyl-imipramine (Fibiger and Phillips, 1981) or clorgyline (Aulakh et al., 1983) are also reported to increase rates for self-stimulation progressively. In the Fibiger and Phillips study, threshold currents for brain stimulation reinforcement were reduced following chronic desmethylimipramine treatment. In the clorgyline study by Aulakh et al. (1981), no threshold measures were reported. In a subsequent study (O'Regan et al., 1987), however, these facilitatory effects of chronic clorgyline treatment were accompanied by a decrease in the threshold for brain stimulation reinforcement. Nevertheless, further studies are necessary to validate this method in view of the fact that only desmethylimipramine and one monamine oxidase inhibitor have yielded such threshold effects to date.

In other studies, Zacharko and colleagues (Zacharko et al., 1984; Kokkinidis et al., 1980) have reported that, following repeated exposure to uncontrollable shocks or chronic ampethamine treatment, animals exhibit deficits in self-stimulation responding. These effects are reported to be attenuated by acute treatment with antidepressant drugs such as desmethylimipramine. These latter studies clearly involve features of both the self-stimulation model and the chronic stress model.

5. Conclusion

This chapter has described basic approaches to the preclinical behavioral analysis of psychotherapeutic drug action. From the nature of the described procedures, it is evident that the main emphasis of this type of analysis is predictive. Predictive validity is undoubtedly the most important feature of these tests. There is a wide divergence of approaches to this area, exemplified by ongoing attempts to discover more accurate means of identifying antidepressant drugs and to determine their critical actions on brain and behavior. With anxiolytics and neuroleptics, the situation is somewhat clearer. The discrepancies between tests with respect to identifying psychotherapeutic drugs in these classes do, however, serve as a prominent reminder of the limitations of available "behavioral technology" in this field. At the present time it is apparent that no single preclinical test will yield a realistic estimate of clinical efficacy. To generate a degree of confidence in preclinical analysis, it is necessary to use a number of tests and to look for concordant results. This necessity may, for some readers at least, present this field as something of an art. Nevertheless, in all fields of science, whether at an early or advanced stage, careful sampling with different measuring devices is necessary to avoid false conclusions (either false positives or false negatives). In the complex field of assessing psychotherapeutic drug action, behavioral analysis is necessary. The use of infrahuman species is intrinsically fraught with pitfalls. Continuing careful analysis comparing behavioral and biochemical data from these studies with limited data available from clinical studies will, however, eventually lead to an advanced and clear understanding of the biological bases and treatment of psychiatric disorders.

References

Anisman H. A. and Zacharko R. M. (1982) Depression: The predisposing influence of stress. *Behav. Brain Sci.* **5,** 89–137.

Aulakh C. S., Cohen R. M., Pradhan S. N., and Murphy D. L. (1983) Self-stimulation responses are altered following long-term but not short-term treatment with clorgyline. *Brain Res.* **270,** 383–386.

Baltzer V. and Weiskrantz L. (1973) Antidepressant agents and reversal of diurnal activity cycles in the rat. *Biol. Psychiatry* **10,** 199–209.

Barrett R. J. and White D. I. C. (1980) Reward system depression following chronic amphetamine antagonism by haloperidol. *Pharmacol. Biochem. Behav.* **13,** 555–559.

Berger B. D. and Stein L. D. (1968) Assymetrical dissociation of learning between scopolamine and Wy 4036, a new benzodiazepine tranquilizer. *Psychopharmacologia* **14,** 351–358.

Bignami G. (1978) Effects of Neuroleptics, Ethanol, Hypnotic-Sedatives, Tranquilizers, Narcotics and Minor Stimulants in Aversive Paradigms, in *Psychopharmacology of Aversively Motivated Behaviour* (Anisman H. and Bignami G., eds.) Plenum, New York.

Broekkamp C. L., LePichan M., and Lloyd K. G. (1984) The comparative effects of benzodiazepines, progabide and PK 9084 on acquisition of passive avoidance in mice. *Psychopharmacology* **82,** 122–125.

Bureš J., Burešová O., and Huston J. P. (1983) *Basic Techniques and Experiments in Brain Behaviour Research* Elsevier, Amsterdam.

Cairncross K. D., Cox B., Forster C., and Wren A. F. (1978) A new model for the detection of antidepressant drugs: Olfactory bulbectomy in the rat compared with existing models. *J. Pharmacol. Meth.* **1,** 131–143.

Carlton P. L. (1978) Theories and Models in Psychopharmacology, in *Psychopharmacology: A Generation of Progress* (Lipton M. A., DiMascio A., and Killam K. F., eds.) Raven, New York.

Carlton P. L. (1983) *Behavioural Pharmacology* McGraw Hill, New York.

Colpaert F. C. and Slangen J. L. (1982) *Drug Discrimination: Applications in CNS Pharmacology* Elsevier, Amsterdam.

Cook L. and Davidson A. B. (1973) Effects of Behaviourally Active Drugs in a Conflict-Punishment Procedure in Rats, in *The Benzodiazepines* (Garattini S., Mussini E., and Randall O., eds.) Raven, New York.

Cook L. and Sepinwall J. (1975a) Behavioural Analysis of Effects and Mechanisms of Action of Benzodiazepines, in *Mechanism of Action of Benzodiazepines* (Costa E. and Greengard P., eds.) Raven, New York.

Cook L. and Sepinwall J. (1975b) Psychopharmacological Parameters of Emotion, in *Emotions—Their Parameters and Measurement* (Levi L., ed.) Raven, New York.

Cook L. and Weidley E. (1957) Behavioural effects of some psychopharmacological agents. *Ann. NY Acad. Sci.* **66,** 740–752.

Cools A. R. (1980) Role of the neostriatal dopaminergic activity in sequencing and selecting behavioural strategies: Facilitation of processes involved in selecting the best strategy in a stressful situation. *Behav. Brain Res.* **1,** 361–378.

Cooper S. J. and Turkish S. (1987) Psychopharmacology of Food and Water Intake, in *Experimental Psychopharmacology: Concepts and Methods* (Greenshaw A. J. and Dourish C. T., eds.) Humana, Clifton, New Jersey.

Cornfeldt M., Fisher B., and Fielding S. (1982) Rat internal capsule lesion: A new test for detecting antidepressants. *Fed. Proc.* **41,** 1006.

Dantzer R. (1987) Behavioral Analysis of Anxiolytic Drug Action, in *Experimental Psychopharmacology: Concepts and Methods* (Greenshaw A. J. and Dourish C. T., eds.) Humana, Clifton, New Jersey.

Denenberg V. H., Ross S., and Ellsworth J. (1959) Effects of chlorpromazine on acquisition and extinction of a conditioned response in mice. *Psychopharmacologia* **1,** 59–64.

Diagnostic and Statistical Manual of Mental Disorders (1987) 3rd Edn. Revised, American Psychiatric Association, Washington, DC.

Dixon A. K. (1983) Indirect drug effects on behavior. Second Symposium of the European Society for Research on Aggression, October, Zeist.

Einon D. F., Morgan M. J., and Sahakian B. J. (1975) The development of intersession habituation and emergence in socially reared and isolated rats. *Dev. Psychobiol.* **8,** 553–559.

Fibiger H. C. and Phillips A. G. (1981) Increased intracranial self-stimulation in rats after long-term administration of desipramine. *Science* **214,** 683–685.

File S. E. (1982) Colony aggression: Effects of benzodiazepines on intruder behavior. *Physiol, Psychol.* **10,** 413–416.

File S. E. and Hyde J. R. G. (1978) Can social interaction be used to measure anxiety? *Br. J. Pharmacol.* **62,** 19–74

Franklin K. B. J. and Abbott F. V. (1988) Techniques for Assessing the Effects of Drugs on Nociceptive Responses, in *Neuromethods: Psychopharmacology* (Boulton A. A., Baker G. B., and Greenshaw A. J., eds.) Humana, Clifton, New Jersey.

Geller I. and Seifter J. (1960) The effects of meprobamate, barbiturates, *d*-amphetamine and promazine on experimentally-induced conflict in the rat. *Psychopharmacologia* **1,** 482–492.

Gillin J. C. (1983) The sleep therapies of depression. *Prog. Neuropsychopharmacol.* **7,** 351–364.

Gomita Y. and Ueki S. (1981) "Conflict" situation based on intracranial self-stimulation behavior and the effect of benzodiazepines. *Pharmacol. Biochem. Behav.* **14,** 219–222.

Goodwin F. K., Wirz-Justice A., and Wehr T. A. (1982) Evidence that Pathophysiology of Depression and the Mechanism of Action of Antidepressant Drugs both Involve Alteration in Circadian Rhythms, in *Typical and Atypical Antidepressants: Clinical Practice* (Costa E. and Racagni G., eds.) Raven, New York.

Gorka Z. and Janus K. (1985) Effects of neuroleptics displaying antidepressant activity on the behavior of rats in the forced swimming test. *Pharmacol. Biochem. Behav.* **23,** 203–206.

Gray J. A. (1977) *The Neuropsychology of Anxiety* Oxford University Press, Oxford.

Greenshaw A. J., Sanger D. J., and Blackman D. E. (1981) The effects of pimozide and of reward omission on fixed-interval behavior of rats maintained by food and electrical brain stimulation. *Pharmacol. Biochem. Behav.* **15,** 227–233.

Hamor T. A. and Martin I. L. (1983) The benzodiazepines. *Prog. Med. Chem.* **20,** 158–223.

Heath R. G. (1983) Electrical self-stimulation of the brain in man. *Am. J. Psychiat.* **120,** 571–577.

Heath R. G., John S. B., and Fontana C. J. (1968) The Pleasure Response: Studies by Stereotaxic Techniques in Patients, in *Computer and Electrical Devices in Psychiatry* (Kline N. S. and Laska E., eds.) Grune & Stratton, New York.

Herrnstein R. J. (1966) Superstition, in *Operant Behavior: Areas of Research and Application* (Honig W. K., ed.) Appleton-Century-Crofts, New York.

Horovitz Z. P. (1967) Selective block of mouse-killing by antidepressants. *Science* **135,** 375–377.

Houser V. P. (1978) The Effects of Drugs on Behavior Controlled by Aversive Stimuli, in *Contemporary Research in Behavioral Pharmacology* (Sanger D. J. and Blackman D. E. eds.) Plenum, New York.

Howard J. L. and Pollard G. T. (1977) The Geller-conflict Test: A Model of Anxiety and a Screening Procedure for Anxiolytics, in *Animal Models in Psychiatry, and Neurology* (Hanin I. and Usdin E., eds.) Oxford, Pergamon.

Hrdina P. D., von Kulmiz P., and Stretch R. (1979) Pharmacological modification of experimental depression in infant macaques. *Psychopharmacology* **64,** 89–93.

Janssen P. A. J., Niemegeers C. J. E., and Schellekens K. H. L. (1965) Is it possible to predict the clincal effects of neuroleptic drugs (major tranquilizers) from animal data? 1. Neuroleptic activity spectra for rats. *Arzneimittelforsch.* **15,** 104–117.

Jarbe T. U. C. (1987) Drug Discrimination Learning, in *Experimental Psychopharmacology: Concepts and Methods* (Greenshaw A. J. and Dourish C. T., eds.) Humana, Clifton, New Jersey.

Jenner P., Chow A., Reavill C., Theoporou A., and Marsden C. D. (1978) A behavioural and biochemical comparison of dopamine receptor blockade produced by haloperidol with that produced by substituted benzamide drugs. *Life Sci.* **23,** 545–550.

Jensen R. A., Martinez J. L., Vasquez B. L., and McGaugh J. L. (1979) Benzodiazepines alter acquisition and retention of an inhibitory avoidance response in mice. *Psychopharmacology* **64,** 125–125.

Jesberger J. A. and Richardson J. S. (1986) Effects of antidepressant drugs on the behavior of olfactory bulbectomized and sham-operated rats. *Behav. Neurosci.* **100** 256–274.

Joly D. and Sanger D. J. (1986) The effects of fluoxetine and zimelidine on the behavior of olfactory bulbectomized rats. *Pharmacol. Biochem. Behav.* **24,** 199–204.

Katz R. J. (1981) Animal models and human depresive disorders. *Neurosci. Biobehav. Rev.* **5,** 231–246.

Katz R. J. (1982) Animal model of depression: Pharmacological sensitivity of a hedonic deficit. *Pharmacol. Biochem. Behav.* **16,** 965–968.

Kelleher R. J. and Morse W. H. (1968) Determinants of the behavioral specificity of drugs. *Ergeb. Physiol.* **60,** 1–56.

Kitada Y., Miyauchi T., Satoh A., and Satoh S. (1981) Effects of antidepressants in the rat forced swimming test. *Eur. J. Pharmacol.* **72,** 145–152.

Kokkinidis L. and Zacharko R. M. (1980) Response sensitization and depression following chronic amphetamine treatment in a self-stimulation paradigm. *Psychopharmacology* **68,** 73–76.

Kokkinidis L., Zacharko R. M., and Predy, P. A. (1980) Post-amphetamine depression of self-stimulation responding from the substantia nigra: Reversed by tricyclic antidepressants. *Pharmacol. Biochem. Behav.* **13**, 379–383

Komiskey H. L., Cook T. M., Lin C.-F., and Hayton W. I. (1981) Impairment of learning or memory in the mature and old rat by diazepam. *Psychopharmacology* **73**, 304–305.

Kuribara H. and Tadokoro S. (1981) Correlation between antiavoidance activities of anti-psychotic drugs in rats and daily clinical doses. *Pharmacol. Biochem. Behav.* **14**, 181–192.

Lang W. and Gershon S. (1962) Effects of psychoactive drugs on yohimbine-induced responses in conscious dogs. A study of anti-depressant drugs. *Med. Exp.* (Basel) **7**, 125–134.

Lapierre Y. D. (1985) Course of clinical response to antidepressants. *Prog. Neuropsychopharmacol. Biol. Psychiatry* **9**, 503–507.

Leith N. J. and Barrett R. J. (1976) Amphetamine and the reward system: Evidence for tolerance and post-drug depression. *Psychopharmacologia* **46**, 19–25.

Leith N. J. and Barrett R. J. (1980) Effects of chronic amphetamine on self-stimulation responding: Animal model of depression? *Psychopharmacology* **72**, 9–15.

Lewis J. K. and McKinney W. T. (1966) Effects of electroconvulsive shock on the behaviour of normal and abnormal Rhesus monkeys. *Behav. Psychiat.* **37**, 687–693.

Liebman J. M. (1988) Drug Effects on Behaviors Maintained by Electrical Brain Stimulation, in *Neuromethods: Psychopharmacology* (Boulton A. A., Baker G. B., and Greenshaw A. J., eds.) Humana, Clifton, New Jersey.

Liebman J. M., Neale R., Noreika L., and Braunwalder A. (1981) Differential reversal of various dopamine antagonists by anticholinergics in Sidman avoidance: Possible relationship to adrenergic blockade. *Psychopharmacology* **72**, 248–253.

Lister R. G. (1987) The use of a plus-maze to measure anxiety in the mouse. *Psychopharmacology* **92**, 180–185.

Ljungberg T. and Ungerstedt U. (1985) A rapid and simple behavioural screening method for simultaneous assessment of limbic and striatal blocking effects of neuroleptic drugs. *Pharmacol. Biochem. Behav.* **23**, 479–485.

Lloyd K. G., Garrigan D., and Broekkamp C. L. E. (1982) The Action of Monoaminergic Cholinergic and Gabaergic Compounds in the Olfactory Bulbectomized Rat Model of Depression, in *New Vistas in Depression* (Langer S. Z., Takahashi R., Segawa T., and Briley M., eds.) Pergamon, New York.

McKinney W. T. and Bunney W. E. (1969) Animal model of depression: Review of evidence and implications for research. *Arch. Gen. Psychiat.* **29,** 490–494.

McKinney W. T., Young L. D., and Suomi S. J. (1973) Chlorpromazine treatment of disturbed monkeys. *Arch. Gen. Psychiatry* **29,** 490–494.

Menzel E. W., Davenport R. K., and Rogers C. M. (1963) Effects of environmental restriction upon the chimpanzee's responsiveness to objects. *J. Comp. Physiol. Psychol.* **56,** 78–85.

Miczek K. A. and Winslow J. T. (1987) Psychopharmacological Research on Aggressive Behavior, in *Experimental Psychopharmacology: Methods and Concepts* (Greenshaw A. J. and Dourish C. T., eds.) Humana, Clifton, New Jersey.

Millichap J. C. (1969) Relation of laboratory evaluation to clinical effectiveness of antiepileptic drugs. *Epilepsia* **10,** 315–328.

Neale R., Gerhardt S., and Liebman J. M. (1984) Effects of dopamine agonists, catecholamine depletors and cholinergic and GABAergic drugs on acute dyskinesias in squirrel monkeys. *Psychopharmacology* **82,** 20–26.

Niemegeers C. J. E. and Janssen P. A. J. (1974) A systematic study of the pharmacological properties of dopamine antagonists. *Life Sci.* **24,** 2201–2216.

Niemegeers C. J. E., Verbruggen F. J., and Janssen P. A. J. (1969) The influence of various neuroleptic drugs on shock avoidance responding in rats: Nondiscriminanted Sidman avoidance procedure. *Psychopharmacologia* **16,** 161–174.

Oishi H., Iwahara S., Yang K.-M., and Yogi A. (1972) Effects of chlordiazepoxide on passive avoidance responses in rats. *Psychopharmacologia* **23,** 373–385.

O'Regan D., Kwok R., Bailey B. A., Yu P. H., Greenshaw A. J., and Boulton A. A. (1987) A behavioural and neurochemical analysis of chronic and selective monoamine oxidase inhibition. *Psychopharmacology* **92,** 42–47.

Patel J. B., Ciofalo V. B., and Ioro L. C. (1979) Benzodiazepine blockade of passive avoidance task in mice: A state-dependent phenomenon. *Psychopharmacology* **61,** 25–28.

Petersen E. N. and Lassen J. B. (1981) A water-lick conflict paradigm using drug experienced rats. *Psychopharmacologia* **75,** 236–239.

Pollard G. T. and Howard J. L. (1986) The staircase test: Some evidence of non-specificity for anxiolytics. *Psychopharmacology* **89,** 14–19.

Porsolt R. D. (1985) Animal Models of Affective Disorders, in *Pharmacotherapy of Affective Disorders* (Dewhurst W. G. and Baker G. B., eds.) New York University Press, New York.

Porsolt R. D., Bertin A., and Jalfre M. (1977) Behavioural despair in mice: A primary screening test for antidepressants. *Arch. Int. Pharmacodyn. Ther.* **229,** 327–336.

Porsolt R. D., Arton G., Biaret N., Daniel M., and Jalfre M. (1978) Behavioural despair in rats, a new model sensitive to antidepressant treatments. *Eur. J. Pharmacol.* **47,** 379–381.

Porsolt R. D., Bertin A., Blavet N., Daniel M., and Jalfre M. (1979) Immobility induced by forced swimming in rats: Effects of agents which modify central catecholamine and serotonin activity. *Eur. J. Pharmacol.* **57,** 210.

Pellow S., Chopin P., File S. E., and Briley R. M. (1985) Validation of open-closed arm entries in an elevated plus-maze as a measure of anxiety in the rat. *J. Neurosci. Meth.* **14,** 149–167.

Quinton R. M. (1963) The increase in the toxicity of yohimbine induced by imipramine and other drugs in mice. *Br. J. Pharmacol.* **21,** 51–66.

Rickels K., Weisman K., Norstad N., Singer M., Stoltz D., Brown A., and Dunton J. (1982) Buspirone and diazepam in anxiety: A controlled study. *J. Clin. Psychiatry* **43,** 81–82.

Roberts D. C. S. and Goeders N. (1988) Self Administration Methodology, in *Neuromethods: Psychopharmacology* (Boulton A. A., Baker G. B., and Greenshaw A. J., eds.) Humana, Clifton, New Jersey.

Roberts D. C. S. and Vickers G. (1984) Atypical neuroleptics increase self-administration of cocaine: An evaluation of a behavioural screen for antipsychotic activity. *Psychopharmacology* **82,** 135–139.

Sahakian B. J., Robbins T. W., and Iversen S. D. (1977) The effects of isolation rearing on exploration in the rat. *Anim. Learning Behav.* **5,** 193–198.

Sahakian B. J., Robbins T. W., Morgan M. J., and Iversen S. D. (1975) The effects of psychomotor stimulants on stereotypy and locomotor activity in socially deprived and control rats. *Brain Res.* **84,** 195–205.

Sanger D. J. (1987) Effects of Drugs on Schedule Controlled Behavior, in *Experimental Psychopharmacology: Methods and Concepts.* Humana, Clifton, New Jersey.

Sanger D. J. (1986) Response decrement patterns after neuroleptics and non-neuroleptic drugs. *Psychopharmacology* **89,** 98–104.

Sanger D. J. (1985) The effects of clozapine on shuttle box avoidance responding in rats: Comparisons with haloperidol and chlordiazepoxide. *Pharmacol. Biochem. Behav.* **23,** 231–236.

Sanger D. J. and Blackman D. E. (1981) Rate-Dependence and The Effects of Benzodiazepines, in *Advances in Behavioral Pharmacology* (Thompson T. and Dews P., eds.) Academic, New York.

Sanger D. J. and Joly D. (1985) Anxiolytic drugs and the acquisition of conditioned fear in mice. *Psychopharmacology* **85,** 284–288.

Schershlicht R., Palc P., Schneeberger J., Steiner M., and Haefely W. (1982) Selective Suppression of Rapid Eye Movement Sleep (REMS) in Cats by Typical and Atypical Antidepressants, in *Typical and Atypical Antidepressants: Molecular Mechanisms* (Costa E. and Racagni G., eds.) Raven, New York.

Scobie S. and Garske G. (1970) Chlordiazepoxide and conditioned suppression. *Psychopharmacologia* **16**, 272–280.

Seligman M. E. P. (1975) *Helplessness: On Depression, Development and Death* Freeman, San Francisco.

Sepinwall J. and Cook L. (1978) Behavioral Pharmacology of Antianxiety Drugs, in *Handbook of Psychopharmacology* vol. 13 (Iversen, L. L., Iversen, S. D., and Snyder, F. H.) *Biology of Mood and Antianxiety Drugs* Plenum, New York.

Setler P., Sarau H., and Mckenzie G. (1976) Differential attenuation of some effects of haloperidol in rats given scopolamine. *Eur. J. Pharmacol.* **39**, 117–126.

Shader R. I. (1975) *Manual of Pediatric Therapeutics* Little, Brown, Boston.

Sidman M. (1957) Avoidance conditioning with brief shock and no exteroceptive warning signal. *Science* **118**, 157–158.

Simpson D. M. and Annau Z. (1977) Behavioural withdrawal following several psychoactive drugs. *Pharmacol. Biochem. Behav.* **7**, 59–64.

Sofia R. D. (1969) Structural relationship and potency of agents which selectively block mouse killing (muricide) behaviour in rats. *Life Sci.* **8**, 1201–1210.

Sovner R. and DiMascio A. (1978) Extrapyramidal Syndromes and Other Neurological Effects of Psychotropic Drugs, in *Psychopharmacology: A Generation of Progress* (Lipton M. A., DiMascio A., and Killam K. F., eds.) Raven, New York.

Steru L., Chermat R., Thierry B., and Simon P. (1985) The tail suspension test: A new method for screening antidepressants in mice. *Psychopharmacology* **85**, 367–370.

Stille G., Lauener H., and Eichenberger E. (1971) The pharmacology of 8-chloro-11(4-methyl-2-piperazinyl)-5H-dibenzo[b,e]-diazepine (clozapine). *Il Farmaco* **26**, 603–625.

Suomi S. J. (1976) Factors Affecting Responses to Social Separation in Rhesus Monkeys, in *Animal Models in Human Psychopharmacology* (Serban, G. and Kling, A., eds.) Plenum, New York.

Suomi S. J., Seaman S. F., Lewis J. K., Delizio R. D., and McKinney W. T. (1978) Effects of imipramine treatment on separation-induced social disorders in Rhesus monkeys. *Arch. Gen. Psychiat.* **35**, 321–325.

Szewczaic M. R., Fielding S., and Cornfeldt M. (1982) Rat internal capsule lesion: Further characterization of antidepressant screening potential. *Soc. Neurosci. Abstr.* **8**, 465.

Thiebot M. H., Soubrie P., Simon P., and Boissier J. R. (1973) Dissociation de deux composantes du comportement chez le rat sous l'effect de psychotropes. *Psychopharmacol.* **31**, 77–90.

Thornton E. W., Evans J. A. C., and Harris C. (1985) Attenuated response to nomifensine in rats during a swim test following lesion of the habenula complex. *Psychopharmacology* **87**, 81–85.

Turner C., Davenport R., and Rogers C. (1969) The effect of early deprivation of the social behaviour of adolescent chimpanzees. *Am. J. Psychiatry* **125**, 1531–1536.

Ueki S. (1982) Mouse Killing Behaviour (Muricide) in the Rat and the Effect of Antidepressants, in *New Vistas in Depression* (Langer S. Z., Takahashi R., Syama T., and Briley M., eds.) Pergamon, New York.

Valenstein E. S. (1973) *Brain Control: A Critical Examination of Brain Stimulation and Psychosurgery* Wiley, New York.

Vogel J. R., Beer B., and Clody D. E. (1971) A simple and reliable conflict procedure for testing anti-anxiety agents. *Psychopharmacologia* **21**, 1–7.

Wauquier A. (1980) The Influence of Psychoactive Drugs on Brain Self-Stimulation in Rats: A Review, in *Brain Stimulation Reward* (Wauquier A. and Rolls E. T., eds.) Elsevier/North Holland, Amsterdam.

Weissman B. A., Barrett J. E., Brady L. S., Witkin J. M., Mendelson W. B., Paul S. M., and Skolnick P. (1984) Behavioural and neurochemical studies on the anticonflict actions of buspirone. *Drug. Dev. Res.* **4**, 83–93.

Willner P. (1984) The validity of animal models of depression. *Psychopharmacology* **83**, 1–16.

Wise R. A. (1978) Minireview: Neuroleptic attenuation of intracranial self-stimulation: Reward or performance deficits? *Life Sci.* **22**, 535–542.

Wise R. A. (1982) Neuroleptics and operant behaviour: The anhedonia hypothesis. *Behav. Brain Sci.* **5**, 39–87.

Witkin J. M. and Barrett J. E. (1986) Interaction of buspirone and dopaminergic agents on punished behavior of pigeon. *Pharmacol. Biochem. Behav.* **24**, 751–756.

Zacharko R. M., Bowers W. J., and Anisman H. (1984) Responding for brain stimulation: Stress and desmethylimipramine. *Prog. Neuropsychopharmacol. Biol. Psychiatry* **8**, 601–606.

High-Affinity Binding of Antidepressants to Platelets and Brain Tissue

Hans Schoemaker and Salomon Z. Langer

1. Introduction

Although antidepressant drug treatment is a well-established therapeutic approach in manic-depressive disorders, the mechanism by which such drugs ameliorate the depressive syndromes remains an area of active research. The mechanism of action of antidepressants most likely has two components. The clinical manifestation of their therapeutic effects generally is seen after a 1- to 2-wk treatment, thus implying that a degree of neuronal adaptation under the effect of antidepressants is required for clinical efficacy. The development of neuronal sub- or supersensitivity following subchronic antidepressant treatment, however, is thought to be secondary to the direct interaction of these drugs with specific receptors and/or enzymes in the brain. To characterize this interaction of antidepressants with such receptors and/or enzymes, radioligand binding studies using mostly ^{3}H-labeled antidepressants have been used extensively. Foremost among these is [^{3}H]-imipramine, a tricyclic antidepressant that inhibits neuronal serotonin (5-hydroxytryptamine, 5-HT) and norepinephrine (noradrenaline, NA) transport. Other radiolabeled antidepressants whose high-affinity binding sites have been studied in the brain and peripheral tissues include [^{3}H]-desipramine, [^{3}H]-nomifensine, [^{3}H]-indalpine, [^{3}H]-paroxetine, [^{3}H]-cyanoimipramine, [^{3}H]-amitriptyline, [^{3}H]-doxepin, [^{3}H]-mianserin, and [^{3}H]-rolipram (Table 1). A discussion of high-affinity antidepressant binding sites in the brain and peripheral tissues and their associated methodological aspects, however, is facilitated by discussing the molecular structures labeled by these drugs whenever possible. Indeed, the majority of radiolabeled antidepressants have been shown to label the presynaptic serotonin, NA, dopamine (DA), or epinephrine (adrenaline, A) transporter in the brain and peripheral tissues.

Table 1
Antidepressants as High-Affinity Radioligands

[³H]-Antidepressant	Putative identity of its high-affinity recognition site
[³H]-Imipramine [³H]-Cyanoimipramine [³H]-Indalpine [³H]-Paroxetine [³H]-Norzimelidine	Serotonergic transporter
[³H]-Desipramine [³H]-Nomifensine	Noradrenergic transporter
[³H]-Nomifensine	Dopaminergic transporter
[³H]-Desipramine	Adrenergic transporter
[³H]-Amitriptyline	Muscarinic and histamine (H_1) receptor
[³H]-Doxepin	Histamine (H_1) receptor
[³H]-Mianserin	Histamine (H_1) and serotonin ($5\text{-}HT_2$) receptors
[³H]-Rolipram	cAMP phosphodiesterase

2. High-Affinity Antidepressant Binding Sites and the Serotonergic Transporter

Among the first antidepressants for which a high-affinity binding site could be demonstrated was [³H]-imipramine (Langer et al., 1980a,b, 1981a,b; Raisman et al. 1979). Over the years, a substantial body of evidence has accumulated to support the hypothesis that the high-affinity [³H]-imipramine binding site identified in the brain and platelets of several species is located within the plasmolemmal serotonin transporter complex and possesses most of the properties of a pharmacological receptor. Current evidence suggests that the [³H]-imipramine binding site may not be identical with the substrate recognition site of the serotonin transporter, but may represent a novel presynaptic receptor that functions to modulate serotonin transport and can therefore control the synaptic availability of this indolamine neurotransmitter (for reviews, *see* Davis, 1984a; Langer, 1984; Langer and Raisman, 1983; Langer and Briley, 1981; Langer and Schoemaker, 1987; Langer et al., 1986a).

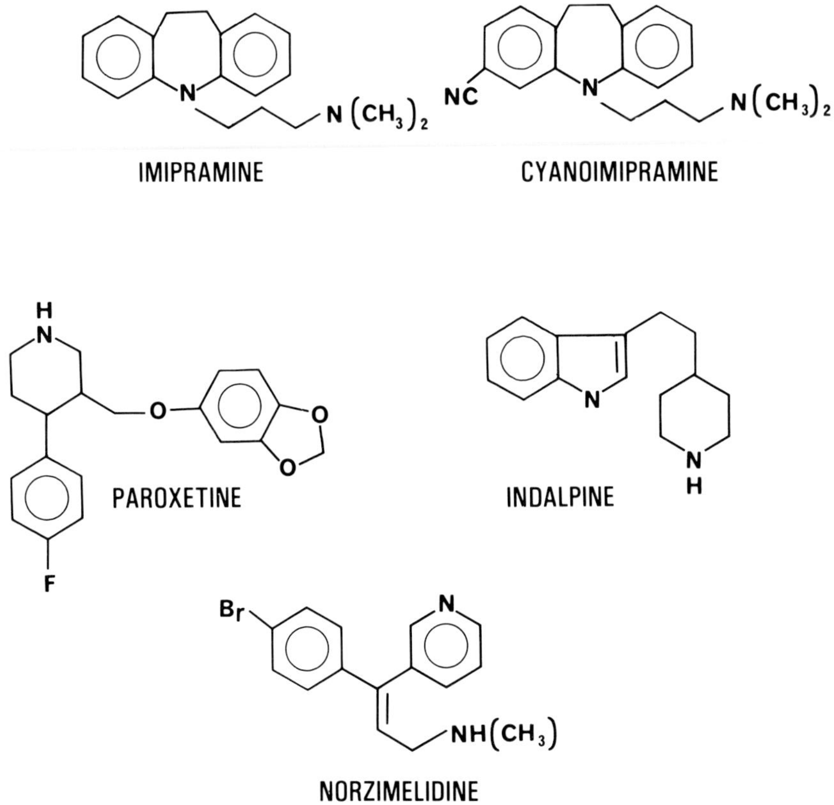

Fig. 1. Structural formula of some antidepressants used as radioligands to label the serotonergic transporter.

Furthermore, high-affinity binding sites for other tricyclic ([³H]-cyanoimipramine) and nontricyclic ([³H]-paroxetine, [³H]-indalpine, [³H]-norzimelidine) antidepressants (Fig. 1) are thought to be associated with the neuronal serotonin transporter.

2.1. [³H]-Imipramine Binding in Brain and Platelets

Most studies on [³H]-imipramine binding to membrane fractions from the brain have been performed using methods identical or closely related to the original protocol described by Raisman et al. (1980). As with many binding studies, the tissue is dissected and homogenized with the use of a polytron homogenizer in 50 vol of ice-cold 50 mM Tris-HCl buffer containing 100 mM NaCl and 5 mM KCl (pH 7.4). The advantage of this homogenization procedure is a minimalization of the formation of vesicular structures (e.g., synaptosomes following a "Teflon-in-glass" homogenization in $0.32M$ sucrose) that could accumulate the radioligand under study. Following homogenization in Tris-HCl buffer (100 mM NaCl, 5 mM KCl), the homogenate is centrifuged at 30000g for 10 min (4°C), and the resulting pellet is resuspended in buffer as above, and recentrifuged. After repeating this procedure a third time, the final pellet is resuspended in 33.3 vol of ice-cold buffer. This membrane homogenate can be used freshly or stored at −80°C for later use. Aliquots of this suspension (final tissue concentration, 30 mg original wet weight/mL) are then incubated with [^{3}H]-imipramine in a final vol of 250 μL buffer for 60 min at 0°C. Following incubation, a 100-μL aliquot is diluted into 5 mL ice-cold buffer and rapidly filtered over Whatman GF/F glass fiber filters. Filters are then washed with three 5-mL vol of buffer and counted for radioactivity. Nonspecific binding is defined as residual binding observed in the presence of 100 μM desipramine.

[^{3}H]-Imipramine binding in the brain has also been studied recently using autoradiographic methodologies (*see* section 2.2.1). Apart from the preparation of the tissue under study, the experimental conditions employed are closely related to those described above for the membrane preparation.

In studying the binding of [^{3}H]-imipramine to platelet membranes, most authors employ a protocol closely related to that of Langer et al. (1980b). In this case, venous blood is withdrawn into plastic tubes containing Na-citrate (0.38% final concentration). Platelet-rich plasma (PRP) is prepared by centrifugation at room temperature (250g for 10 min), and platelets isolated by centrifugation at 16000g for 10 min (4°C). Platelets are then washed twice with buffer (5 mM Tris-HCl, pH 7.5, 20 mM EDTA, 120 mM NaCl) and submitted to hypotonic lysis in 5 mM Tris-HCl containing 5 mM EDTA (pH 7.5). Membranes are collected by centrifugation at 39000g for 10 min, washed with 70 mM Tris-HCl buffer (pH 7.5), and finally resuspended in 50 mM Tris-HCl buffer (pH 7.4) containing 120 mM NaCl and 5 mM KCl. [^{3}H]-Imipramine binding is then determined as above.

Since many studies on platelet [^{3}H]-imipramine binding have been done using outdated human PRP, a readily available "source" of [^{3}H]-imipramine binding sites, it is important to note that the characteristics of the [^{3}H]-imipramine binding site appear not to be affected by the storage of PRP at low temperatures, even for prolonged periods.

Although this protocol may be expected to be devoid of intact platelets, Friedl et al. (1983) reported that, at least in their hands, samples prepared according to Langer et al. (1980b) contained varying proportions of intact platelets that would not have released intracellular protein stores that do not contribute any [³H]-imipramine binding sites. Consequently, the density of [³H]-imipramine binding sites, often expressed on the basis of "membraneous" protein content, may vary to some extent depending on the reproducibility and completeness of platelet lysis.

Using the methodology described above, high-affinity [³H]-imipramine binding sites have been demonstrated in brain and platelets. The dissociation constant (K_d) of [³H]-imipramine for its binding site in the brain or platelets of different species is usually given as 0.6–5 nM, depending on the experimental conditions employed. A contributing factor to this relatively large variation of the affinity of [³H]-imipramine may be a heterogeneity of [³H]-imipramine binding sites reported by several authors (*see* section 2.2.4.). In addition, the K_d of [³H]-imipramine, at least in human platelet membranes, depends to a certain extent on the membraneous protein concentrations (Arora et al., 1985; Barkai et al., 1985), as has previously been shown for, e.g., [³H]-quinuclidinyl benzylate (QNB) binding to the muscarinic cholinergic receptor (Fields et al. 1978). Such artifactual phenomena can most likely be attributed in part to a depletion of the "free" [³H]-imipramine concentration caused by "nonspecific" protein binding.

The maximal binding density (B_{max}) of [³H]-imipramine in human platelet membranes is generally estimated to be approximately 1000 fmol/mg protein, as determined by the "rapid-filtration" technique (Langer et al., 1980b). However, an equilibrium dialysis method to determine [³H]-imipramine binding to human platelet membranes indicated a significantly higher maximal binding density (Lui et al., 1984).

Platelet [³H]-imipramine B_{max} values also appear to be species-dependent. Thus, as compared to human platelets, higher B_{max} values are observed in rabbit (A. Segonzac, H. Schoemaker, and S. Z. Langer, unpublished observations) and cat platelets (Briley et al., 1982).

Using the methodologies as outlined above, studies can be conducted to determine whether the high-affinity [³H]-imipramine binding site has the characteristics of a pharmacological receptor. Several criteria must and can be met to reach this conclusion. The high-affinity [^{3}H]-imipramine binding site must, among other things, show a discrete regional and cellular distribution, and its pharmacology must correspond to that of a functional process.

2.2. Characterization of the Imipramine Recognition Site

2.2.1. Regional and Cellular Localization of the Imipramine Recognition Site

[^{3}H]-Imipramine binding sites are distributed heterogenously in the brain, being more dense in the hypothalamus and amygdala than in the cerebellum. Both in the rat and human brain, this regional localization of [^{3}H]-imipramine binding sites corresponds closely to the regional density of serotonergic innervation, but not to that of noradrenergic innervation (Palkovits et al., 1981; Langer et al., 1981a). Higher-definition autoradiographic studies show a selective localization to the cingulate cortex, anterior thalamic nuclei, CA$_3$ of the hippocampus, lateral and basal nuclei of the amygdala, thalamic nuclei, and substantia nigra, areas with a high density of serotonergic nerve terminals (Dawson and Wamsley, 1983; Fuxe et al., 1983; Grabowsky et al., 1983; Rainbow et al., 1982; Biegon, 1986). [^{3}H]-Imipramine binding sites were also observed in the dorsal raphe (Dawson and Wamsley, 1983; Fuxe et al., 1983; Biegon, 1986), where they are synthetized in the cell body of serotonergic neurons and subsequently transported down the axon to the nerve terminals (Dawson et al., 1985). Denervation of serotonergic neurons using the neurotoxin 5,7-dihydroxytryptamine or by electrolytic lesions of the dorsal raphe nucleus results in a decrease in the density of [^{3}H]-imipramine binding, indicating that the [^{3}H]-imipramine binding site is localized to serotonergic neurons (Paul et al., 1981b; Sette et al., 1981; Gross et al., 1981; Brunello et al., 1982; Luine et al., 1983; Dawson et al., 1985; Hrdina et al., 1985b).

A majority of [^{3}H]-imipramine binding sites is localized to serotonergic nerve terminals as indicated by subcellular fractionation studies. Thus, [^{3}H]-imipramine binding sites were recovered predominantly from the synaptosomal fraction (Rehavi et al., 1983; Sette et al., 1983b; Stockert et al., 1984), although some [^{3}H]-imipramine binding sites in the rat cerebral cortex may also be localized postsynaptically (Stockert et al., 1984).

In addition to the brain, [^{3}H]-imipramine binding sites have been identified in several peripheral tissues, such as the lung (Raisman and Langer, 1983), retina (Borbe and Zube, 1983), and, most importantly, platelets (Briley et al., 1979; Langer et al., 1980b; Paul et al., 1980; Wennogle et al., 1981).

2.2.2. Pharmacological Profile of the Imipramine Recognition Site

[^{3}H]-Imipramine binding in the brain is stereospecific (Langer et al., 1980c) and shows a distinct pharmacological profile that closely correlates with that of inhibition of serotonin transport into the presynaptic serotonergic terminal (Langer et al., 1980a). Similarly, [^{3}H]-imipramine binding to membranes from blood platelets, often taken as a model for serotonergic neurons in the brain, follows a pharmacological profile identical to that established for the serotonin transporter complex in the brain or for the inhibition of [^{3}H]-serotonin uptake into platelets (Paul et al., 1980; Langer et al., 1980b; Schoemaker et al., 1986; Segonzac et al., 1987). Table 2 shows the conservation of the pharmacological profile of the [^{3}H]-imipramine recognition site in platelets between different species (rabbit and humans) and between platelets and the (rat) cortex.

As is already well established for [^{3}H]-serotonin transport in the brain or platelets, [^{3}H]-imipramine binding to the serotonin transporter also critically depends on the presence of Na^+ and Cl^- ions (Briley and Langer, 1981). The monovalent cations K^+ and Li^+ do not substitute for Na^+, whereas the divalent cations Mg^{2+}, Ca^{2+}, and Zn^{2+} are either without effect or inhibit [^{3}H]-imipramine binding (Severson et al., 1986). Other halide ions, however, substitute for Cl^- to a certain extent, with a efficacy rank order of $Br^- > F^- = I^-$ (Severson et al., 1986).

Although the pharmacological profile of [^{3}H]-imipramine binding in brain and platelets shows a statistically significant correlation with the drug inhibitory profile of the serotonin transporter, a significant discrepancy exists between the ability of individual drugs to affect [^{3}H]-imipramine binding and [^{3}H]-serotonin uptake (Langer et al., 1980a; Laduron, 1984a,b; Segonzac et al., 1987). Most relevant, imipramine itself shows an affinity (K_d) of binding to human platelet membranes of 0.79 nM, but only inhibits [^{3}H]-serotonin uptake into human platelets with an IC_{50} of 30 nM (Fig. 2; Segonzac et al., 1987). Differences between drug inhibitory potency for [^{3}H]-imipramine binding, studied at 0°C, and inhibitory affinity for [^{3}H]-serotonin transport at physiological temperatures of 37°C have been used as an argument against the hypothesis that [^{3}H]-imipramine labels the serotonin transporter complex (Laduron, 1984a,b). However, it seems likely that thermodynamic considerations explain to a large degree these

Table 2

Comparison of the Pharmacological Profile of [3H]-Imipramine Binding to the Rat Cerebral Cortex and to Human and Rabbit Platelets[a]

| | [3H]-Imipramine binding | | |
| | K_i, nM | | IC_{50}, nM |
Drug	Human[b] platelets	Rabbit[c] platelets	Rat[d] cortex
Imipramine	3.0	1.6	11.5
Chlorimipramine	3.2	5.0	23.0
Indalpine	1.1	0.8	1.15
Fluoxetine	4.5	1.2	25.0
Citalopram	6.7	2.8	34.0
Femoxetine	26.3	5.3	66.0
Fluvoxamine	13.2	4.1	47.5
Pyrilamine	125.0	62.8	380.0
Chlorpromazine	48.8	30.5	267.0
5-Methoxytryptoline	24.7	11.3	150.0
5-Hydroxytryptoline	275.0	114.0	580.0
Tryptoline	939.0	470.0	1800.0
Serotonin	1999.0	561.0	3183.0
Tryptamine	2002.0	5372.0	4800.0

[a]Drug K_i values were determined on [3H]-imipramine binding at 0°C to human or rabbit platelet membranes. The correlation coefficient of K_i values in these two membranes sources is 0.97 ($p < 0.001$). For drug inhibition of [3H]-imipramine binding to the rat cerebral cortex, only IC_{50} values are shown, as the Hill coefficients for inhibition were less than unity for some of these drugs (cf., Sette et al., 1983a).

[b]Data from Segonzac et al., 1987.

[c]Data from Segonzac et al., unpublished observations.

[d]Data from Langer et al., unpublished observations.

differences in drug affinity between incubation temperatures of 0 and 37°C (Schoemaker et al., 1986; Segonzac et al., 1987). Thus, [3H]-imipramine binding to brain (Plenge and Mellerup, 1984; Reith et al., 1984) and platelet membranes (Davis, 1984b; Davis et al., 1983, 1985) is temperature-dependent. [3H]-Imipramine binding is of high affinity at 0°C, and of progressively lower affinity at incubation temperatures of 20 and 37°C (Davis et al., 1983; Schoemaker et al., 1986; Segonzac et al., 1987). This decrease in the

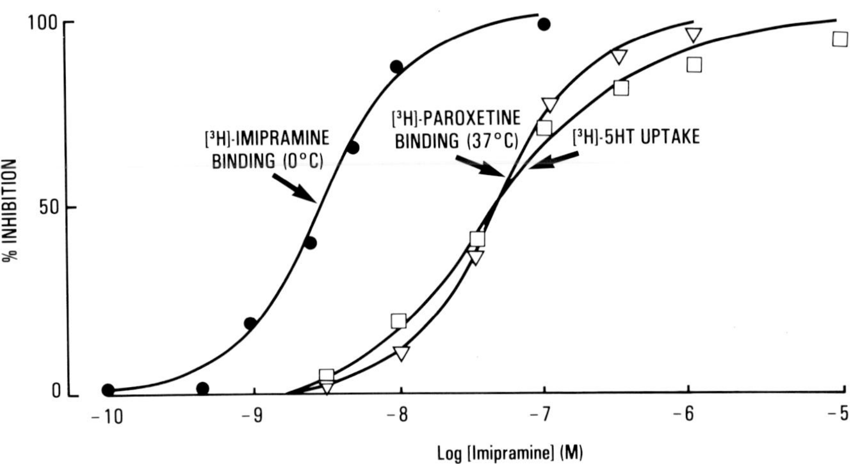

Fig. 2. Temperature-dependence of the interaction of imipramine with the serotonin transporter. The interaction of imipramine with the serotonergic transporter in human platelets was studied as its inhibition of [³H]-imipramine binding at 0°C (0.6 nM, 90 min) (●), [³H]-paroxetine binding at 37°C (0.1 nM, 180 min) (▽), and [³H]-serotonin uptake at 37°C (0.6 μM, 30 s) (□). Shown is the percentage inhibition evoked by different imipramine concentrations of specific radioligand binding or radiosubstrate uptake (A. Segonzac, H. Schoemaker, and S. Z. Langer, unpublished observations).

[³H]-imipramine binding affinity is caused, at least in part, by a significant increase in the rate of ligand–receptor dissociation with increasing temperature (Segonzac et al., 1987). Consequently, [³H]-imipramine binding cannot be determined reliably at the temperature at which serotonin uptake occurs (i.e., 37°C). Radioligand binding studies to the serotonergic transporter at 37°C have now become possible, using a new, selective, and nontricyclic serotonin uptake inhibitor, [³H]-paroxetine (Habert et al., 1985; Langer et al., 1984c; Mellerup et al., 1983, 1985; Plenge and Mellerup, 1984; Segonzac et al., 1987). As shown in Fig. 3, [³H]-paroxetine binds pseudoirreversibly to the serotonin transporter complex at 0°C (Plenge and Mellerup, 1984; Segonzac et al., 1987), but at higher incubation temperatures, [³H]-paroxetine binding is readily reversible and can be used in equilibrium binding analyses (Mellerup et al., 1983; Segonzac et al., 1987). [³H]-Paroxetine binding is of high affinity at 20°C (K_d = 0.073 nM). Moreover, there exists a significant correlation (r = 0.97; p < 0.01) among the K_i values of 14 tricyclic and nontricyclic drugs to inhibit [³H]-paroxetine binding

(0°C)

PAROXETINE (100 μM)
SPECIFIC [³H]-PAROXETINE BINDING (%)
100
50
0
250
500
1 200
TIME (min)

(37°C)

PAROXETINE (100μM)
SPECIFIC [³H]-PAROXETINE BINDING (%)
100
50
0
100
200
300
400
TIME (min)

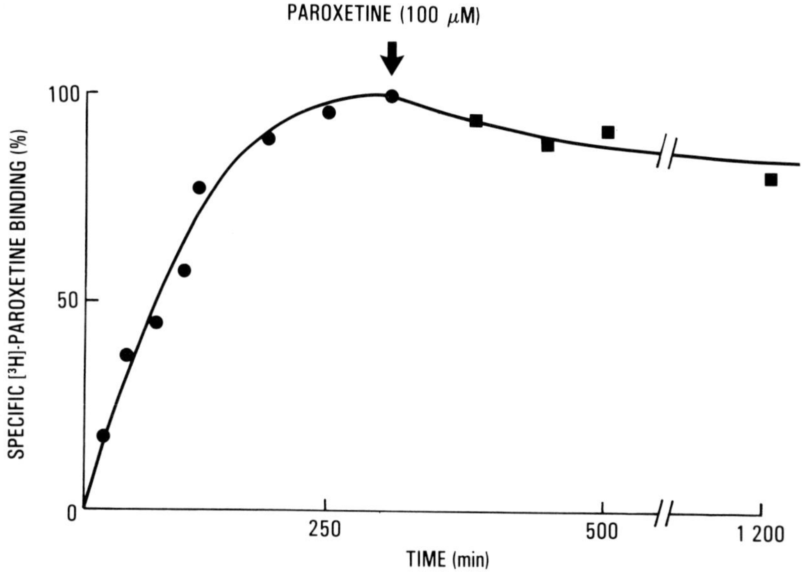

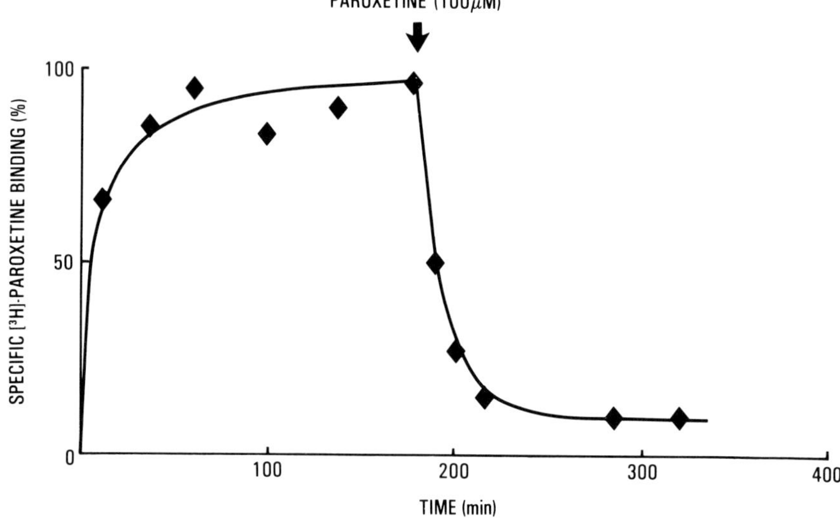

to human platelet membranes at 20°C and [^{3}H]-serotonin uptake into human platelets at 37°C (Segonzac et al., 1987). Similarly, there exists a highly significant correlation between drug inhibition of [^{3}H]-paroxetine binding to the rat cortex and inhibition of synaptosomal serotonin uptake (Habert et al., 1985). The close agreement between K_i values observed for these drugs on [^{3}H]-paroxetine binding and [^{3}H]-imipramine binding in human platelets, both performed at 20°C (Fig. 4), indicates that the [^{3}H]-paroxetine and [^{3}H]-imipramine binding sites within the human platelet serotonin transporter complex are at least closely related, if not identical (Segonzac et al., 1987).

[^{3}H]-Paroxetine binding to human platelet membranes remains of high affinity (K_d = 0.20 nM), even at 37°C (Fig. 5), and is ideally suited for studying the binding characteristics of serotonin-uptake inhibitors at temperatures at which their functional effects on serotonin transport are usually evaluated (Mellerup et al., 1983; Schoemaker et al., 1986; Segonzac et al., 1987). The pharmacological profile of [^{3}H]-paroxetine binding to human platelet membranes at 37°C is significantly correlated (r = 0.97; p < 0.001) with that of [^{3}H]-serotonin-uptake inhibition. The correlation of K_i values for drug inhibition of [^{3}H]-serotonin uptake with K_i values obtained on [^{3}H]-paroxetine binding at 37°C is significantly better (p < 0.05) than that with drug K_i values measured at 0°C using [^{3}H]-imipramine as the radioligand (Segonzac et al., 1987). The affinity of nontricyclics such as paroxetine, citalopram, and indalpine for the serotonin transporter complex is largely temperature-independent. Conversely, but with the exception of chlorimipramine, the affinity of tricyclic serotonin-uptake inhibitors such as imipramine and 5-methoxytryptoline is strongly influenced by the incubation temperature, the K_i values being lower

---

Fig. 3. Dissociation kinetics of [^{3}H]-paroxetine binding to human platelet membranes at 0 and 37°C. [^{3}H]-Paroxetine binding (0.1 nM) to human platelet membranes was studied at 0 (A) and 37°C (B) as a function of the incubation time (cf., Segonzac et al., 1987). Maximal binding (i.e., apparent equilibrium binding) was reached after 300 min at 0°C and 180 min at 37°C. Dissociation kinetics were then studied after the addition of 100 µM unlabeled paroxetine as indicated. At 0°C (A) dissociation proceeds with a half-life greater than 300 min, indicating a pseudoirreversible binding. At 37°C (B), [^{3}H]-paroxetine binding is rapidly reversible with a half-life of 10 min.

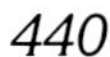

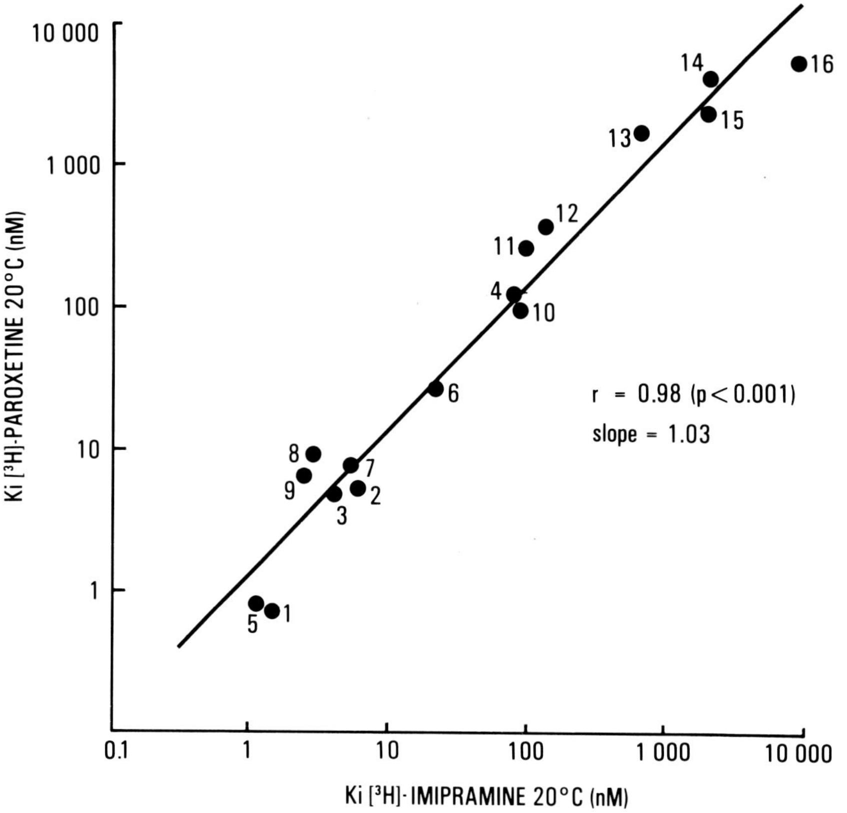

Fig. 4. Comparison of the pharmacological profile of [³H]-imipramine and [³H]-paroxetine binding to human platelet membranes at 20°C. The inhibition of specific [³H]-imipramine (0.6 nM, 180 min) and [³H]-paroxetine binding (0.1 nM, 180 min) to human platelet membranes by different drugs was studied at 20°C, as described by Segonzac et al. (1987). Shown is a correlation analysis of drug K_i values obtained under these conditions (data from Segonzac et al., 1987). Drugs are identified as (1) chlorimipramine, (2) desmethylchlorimipramine, (3) imipramine, (4) desipramine, (5) paroxetine, (6) femoxetine, (7) fluoxetine, (8) citalopram, (9) indalpine, (10) chlorpromazine, (11) pyrilamine, (12) 5-methoxytryptoline, (13) 5-hydroxytryptoline, (14) tryptoline, (15) serotonin, and (16) tryptamine.

at 0°C than at 37°C (Segonzac et al., 1987). Thus, [³H]-imipramine derives its high affinity for the serotonin transporter in human platelets from the temperature-dependence of its interaction with the transporter complex. Whereas in human platelets, imipramine inhibits both [³H]-serotonin uptake and [³H]-paroxetine binding at 37°C with a K_i of 30 nM (Fig. 2), its affinity for the transporter

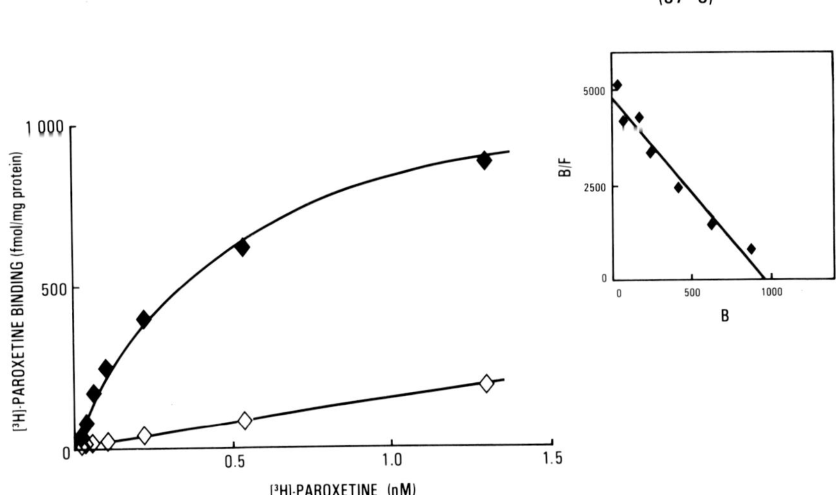

Fig. 5. Saturation analysis of [³H]-paroxetine binding to human platelet membranes at 37°C. [³H]-Paroxetine binding to human platelet membranes was studied at 37°C and an incubation time of 180 min as described by Segonzac et al. (1987). Shown is a saturation analysis of specific (◆) and nonspecific (◇) binding at different radioligand concentrations. The inset represents a Scatchard analysis of specific [³H]-paroxetine binding (B: fmol/mg protein) versus the ratio B/F, in which F is the free radioligand concentration (from A. Segonzac, H. Schoemaker, and S. Z. Langer, unpublished observations).

complex increases with a decrease in incubation temperature, and at 0°C [³H]-imipramine labels the serotonin transporter with a K_d of 0.79 nM (Segonzac et al., 1987). In contrast, the interaction of imipramine with the noradrenergic transporter appears to be temperature-independent (Javitch et al., 1984). Imipramine inhibits [³H]-norepinephrine uptake into rat cortical synaptosomes at 37°C with a K_i of 58 nM, while inhibiting [³H]-mazindol binding (*see* section 3.2) to the noradrenergic transporter in rat cortical membranes at 0°C with a K_i of 95 nM. Thus, a low incubation temperature favors the selectivity of [³H]-imipramine as a ligand for the serotonin transporter complex, and explains why imipramine, at 37°C, inhibits the uptake of serotonin and NA with similar potency, but at 0°C labels only the serotonin transporter complex. Similarly, the affinity of imipramine for the serotonin transporter complex at 0°C is much higher than its affinity for the noradrenergic transporter, and for histamine H_1, α_1-adrenergic, and S_2 serotonin receptors (Laduron, 1984b).

The imipramine receptor density in human platelet membranes, measured using either [³H]-imipramine or [³H]-paroxetine, however, is independent of the incubation temperature. Platelet [³H]-imipramine B_{max} values at 0°C, as determined, e.g., in clinical studies, therefore accurately reflect the density of serotonin-uptake modulatory sites at a physiological temperature (Segonzac et al., 1987).

2.2.3. The Imipramine Binding Site: A Novel Modulatory Receptor?

Several arguments are in favor of the hypothesis that, in platelets, the [³H]-imipramine and substrate recognition sites of the transporter are not identical. Serotonin allosterically affects [³H]-imipramine binding, and, in particular, produces a pronounced decrease in the rate of dissociation of [³H]-imipramine from its receptor (Fig. 6), as compared to the rate of [³H]-imipramine dissociation in the presence of fluoxetine (*see also* Wennogle and Meyerson, 1983; Segonzac et al., 1985a; Plenge and Mellerup, 1985). This property is shared by tryptamine, a substrate for the serotonin transporter in rabbit and human platelets (Segonzac et al., 1984, 1985a). Thus, the imipramine recognition site might be a novel type of receptor that modulates the serotonin transporter in platelets. A similar function of the imipramine receptor in relation to serotonin transport was postulated in the brain (Langer and Raisman, 1983; Sette et al., 1983a,b; Plenge and Mellerup, 1985; Kim and Reith, 1986). Recently, Plenge and Mellerup (1985) observed a similar interaction of serotonin with the [³H]-paroxetine binding site in platelet and neuronal membranes. It follows from this concept that an endogenous ligand for the [³H]-imipramine-labeled modulatory site may exist, and that it is different from serotonin. A factor extracted from the rat brain that inhibits both [³H]-imipramine binding and serotonin uptake was recently reported by Barbaccia et al. (1983a, 1986) and Rehavi et al. (1985). Similarly, a fraction from human plasma inhibits serotonin uptake into rat brain synaptosomes (Angel and Paul, 1984). Because 5-methoxytryptoline inhibits both serotonin uptake (IC_{50} = 860 nM) and [³H]-imipramine binding in the brain (IC_{50} = 179 nM), Langer and coworkers (1984a,b, 1985) suggested that this tricyclic indoleamine derivative or a closely related analog may be an endogenous ligand for the imipramine receptor. 5-Methoxytryptoline is more active as an inhibitor of [³H]-imipramine binding to human platelet membranes at 0°C (K_i = 25 nM) than as an inhibitor of [³H]-serotonin uptake at 37°C into human platelets (K_i = 335 nM;

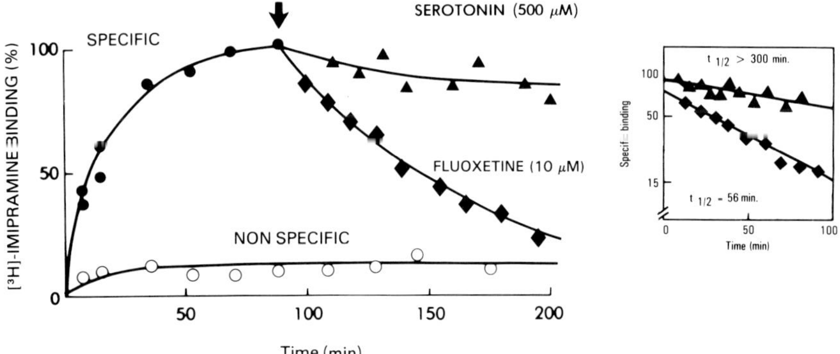

Fig. 6. Dissociation kinetics of [³H]-imipramine binding to human platelet membranes in the presence of fluoxetine or serotonin. Human platelet membranes were incubated with [³H]-imipramine (0.6 nM), and specific (●) and nonspecific (○) binding was measured. At equilibrium (90 min), dissociation kinetics of specific [³H]-imipramine binding were studied after the addition (↓) of fluoxetine (10 μM, ◆) or serotonin (500 μM, ▲). The inset shows a first-order rate plot of the [³H]-imipramine dissociation kinetics in the presence of fluoxetine (10 μM, ◆) or serotonin (500 μM, ▲). The half-life ($t_{1/2}$) of [³H]-imipramine dissociation in the presence of serotonin and fluoxetine was >300 and 56 min, respectively (cf., Segonzac et al., 1985a).

Segonzac et al., 1985b, 1987). Again, this apparent discrepancy is caused by the thermodynamics of the interaction of 5-methoxytryptoline with the serotonin transporter complex; at 37°C, its affinity for inhibition of [³H]-paroxetine binding is the same as that for inhibition of [³H]-serotonin uptake (Segonzac et al., 1987). In spite of its structural similarity to serotonin, 5-methoxytryptoline does not act through the serotonin recognition site of the transporter, but behaves as a competitive inhibitor of [³H]-imipramine binding (Segonzac et al., 1985b), as shown by dissociation experiments.

We recently confirmed that, although [³H]-5-methoxytryptoline can be accumulated in several tissues, it is not actively transported through the rabbit platelet membrane (A. Segonzac, H. Schoemaker, and S. Z. Langer, unpublished observations) nor into rat hypothalamic slices (A. M. Galzin and S. Z. Langer, unpublished observations). The temperature-dependent accumulation of [³H]-5-methoxytryptoline into rabbit platelets is not Na^+-dependent nor inhibited by ouabain, but is most likely related to differences in the rate of diffusion of [3H]-5-methoxytryptoline at 0 and 37°C.

2.2.4. Heterogeneity of [3H]-Imipramine Binding

Most of the studies discussed above, which indicate that the [3H]-imipramine binding site is of high affinity and satisfies most criteria of a pharmacological receptor associated with the serotonin transporter, have assumed or demonstrated that [3H]-imipramine labels a single class of binding sites (i.e., homogeneity of [3H]-imipramine binding). However, depending on the experimental conditions, this may not always be the case.

Saturation analysis of [3H]-imipramine binding to human platelet membranes over a very wide concentration range (0.1–250 nM; Ieni et al., 1985) allows for the identification of a second high-capacity, low-affinity (K_d = 293 nM) binding site for [3H]-imipramine of unknown pharmacological relevance. The observation that most drugs show sigmoidal inhibition curves on [3H]-imipramine binding to human platelet membranes with Hill coefficients close to unity (Segonzac et al., 1987) indicates that the presence of a low-affinity [3H]-imipramine binding site does not interfere with the study of its high-affinity recognition site when measured at low nanomolar radioligand concentrations.

The presence of saturable [3H]-imipramine binding sites on glass fiber filters used in binding assays poses a more significant problem. Thus, it has been noted by several authors that, under some experimental conditions, saturable but low-affinity [3H]-imipramine binding to glass fiber filter may be demonstrated (Phillips et al., 1984; Reith, 1985, 1986). Since this phenomenon appears to depend on the filtration equipment used, and the batch and type of glass fiber filter (Reith, 1986), individual assay conditions need to be evaluated carefully. Presoaking of the filter material in 0.1% polyethyleneimine may also be helpful to decrease filter binding (Reith, 1986).

[3H]-Imipramine binding heterogeneity appears more prominent in the brain. A number of serotonin-uptake inhibitors displace [3H]-imipramine binding to cerebral cortical membranes in a complex manner with Hill slopes less than unity (Sette et al., 1983a; Marcusson et al., 1985). Similarly, [3H]-imipramine saturation analyses in this tissue are indicative of the presence of both low- and high-nanomolar affinity sites (Conway and Brunswick, 1985; Hrdina, 1982, 1984; Reith, 1986). A further indication of [3H]-imipramine binding heterogeneity is the observation that binding that remains after degeneration of the presynaptic serotonergic nerve terminals in the rat cerebral cortex is of lower affinity than that seen under control conditions (Sette et al., 1983b). On the other hand, this phenomenon is not observed when [^{3}H]-paroxetine is used to label the serotonin transporter complex (Habert et al., 1985), indicating that [^{3}H]-paroxetine may be a more specific radioligand than [^{3}H]-imipramine in the brain. [^{3}H]-Cyanoimipramine, which labels the same number of high-affinity binding sites in human platelets as [^{3}H]-imipramine, but only half of them in the brain (Dumbrille-Ross and Tang, 1983), may represent a similar advantage; raphe lesions reduced [^{3}H]-cyanoimipramine binding and endogenous serotonin levels in the rat cortex to a similar extent (Dumbrille-Ross and Tang, 1983). However, the pseudoirreversible binding noted with [^{3}H]-cyanoimipramine at 0°C limits its use in equilibrium binding studies (Dumbrille-Ross and Tang, 1983).

An additional degree of [^{3}H]-imipramine binding heterogeneity in the rat cerebral cortex was revealed by Marcusson et al. (1985). Thus, these workers demonstrated that, in this tissue, [^{3}H]-imipramine binds to both protease- and norzimelidine-sensitive sites, as well as protease- and norzimelidine-insensitive sites. [^{3}H]-Imipramine binding to both sites was displaceable by unlabeled desipramine. Only [^{3}H]-imipramine binding to its protease-sensitive recognition site appears related to the neuronal serotonin transporter (Marcusson et al., 1986).

It is unclear whether the [^{3}H]-imipramine binding heterogeneity, as discussed above, corresponds to that reported for [^{3}H]-2-nitroimipramine binding to human platelet membranes by Wennogle et al. (1985), but not by Rehavi et al. (1983).

Finally, the question of a possible heterogeneity among [^{3}H]-antidepressant binding sites within the serotonin transporter complex has been addressed by several authors. Thus, whereas the pharmacological profile of [^{3}H]-paroxetine and [^{3}H]-imipramine binding to human platelet membranes is significantly correlated (*see* section 2.2.2), a difference in the molecular size of their respective binding polymers was reported by Mellerup et al. (1985). However, it was not established whether this apparent difference in molecular size of the [^{3}H]-paroxetine and [^{3}H]-imipramine binding sites could have been caused by a difference in the incubation temperature used to study the binding of these radioligands (cf., Segonzac et al., 1987). Similarly, [^{3}H]-indalpine binding to the serotonin transporter in the rat brain (Benavides et al., 1985; Savaki et al., 1985) may not be fully equivalent to [^{3}H]-imipramine binding. Thus, as opposed to [^{3}H]-imipramine binding (Sette et al., 1983a), [^{3}H]-indalpine binding is competitively inhibited by serotonin (Benavides et al., 1985). Furthermore, [^{3}H]-indalpine binding appears more sensitive to the presence of sodium ions than that of [^{3}H]-imipramine (Benavides et al., 1985). Finally, whereas serotonin is only a weak inhibitor of [^{3}H]-imipramine binding in the rat cortex (Sette et al., 1983a), serotonin inhibits [^{3}H]-paroxetine (Habert et al., 1985) and [^{3}H]-indalpine binding (Benavides et al., 1985) in this tissue with a K_i value (698 and 280 nM, respectively) similar to that of its affinity for the neuronal transporter. The binding of [^{3}H]-norzimelidine to the serotonin transporter (Hall et al., 1982) appears even more sensitive to serotonin (IC_{50} = 43 nM). It remains to be examined if the pharmacological characteristics of the binding of these nontricyclic [^{3}H]-antidepressants correspond more closely to those of the protease-sensitive [^{3}H]-imipramine binding site in the rat cerebral cortex (Marcusson et al., 1985, 1986).

2.2.5. Clinical Studies on the Imipramine Recognition Site

Although it was demonstrated that several ^{3}H-labeled antidepressants label the serotonin transporter complex, clinical research has centered almost exclusively on [^{3}H]-imipramine binding to platelets as a biological marker in depression (Langer and Briley, 1981; Langer and Raisman, 1983; Langer et al., 1981b, 1982; Briley et al., 1980; Raisman et al., 1981, 1987).

2.2.5.1. DRUG STUDIES IN HEALTHY VOLUNTEERS   In an attempt to evaluate the effect of subchronic antidepressant treatment on platelet [^{3}H]-imipramine binding, we recently treated healthy volunteers for 1 wk with a low dose of chlorimipramine (50 mg/day), a tricyclic antidepressant that is more potent at inhibiting serotonin than NA uptake (Poirier et al., 1984, 1987). [^{3}H]-Imipramine B_{max} values in platelets were decreased by 63% by the end of the 1 wk chlorimipramine treatment and throughout the first week of washout (Poirier et al., 1984). These results clearly show that platelet [^{3}H]-imipramine binding can be down-regulated in humans. The B_{max} values of platelet [^{3}H]-imipramine binding remained decreased compared to pretreatment values for up to 3 wk after the end of treatment with this tricyclic antidepressant. In contrast, serotonin uptake into platelets was completely inhibited during this treatment, but was fully recovered after 1 wk of washout. These results again support the view that the [^{3}H]-imipramine recognition site is associated with, but not identical to, the substrate recognition site of the serotonin transporter. However, in healthy volunteers, the changes in [³H]-imipramine binding induced by treatment with chronic antidepressants that affect serotonin uptake may differ depending on the antidepressants used. Amitriptyline increases the B_{max} and K_d values of platelet [³H]-imipramine binding (Braddock et al., 1984), whereas imipramine causes no change at all (Suranyi-Cadotte et al., 1985c). There is at present no explanation for such differences, although it is of interest to note that chlorimipramine, but not imipramine, also "downregulates" [³H]-imipramine binding to human platelet membranes in vitro (Mellerup and Plenge, 1986). Similar 1-wk treatments of healthy volunteers with maprotiline, an antidepressant inhibiting NA uptake, and amineptine, which acts on dopaminergic mechanisms, did not significantly affect the parameters of [³H]-serotonin uptake or [³H]-imipramine binding in platelets (Poirier et al., 1987). These results support our previous suggestion (Poirier et al., 1984) that a washout period of at least 4 wk may be required for studies on [³H]-imipramine binding in platelets from depressed patients, if they were previously medicated with chlorimipramine or other drugs that inhibit serotonin uptake.

2.2.5.2. STUDIES IN DEPRESSIVE AND NONDEPRESSIVE DISORDERS   Sixteen out of 26 studies published to date on platelet [³H]-imipramine binding in depression reported a significant decrease in the B_{max} of [³H]-imipramine binding in the platelets of depressed, untreated patients, without concomitant changes in the K_d values (Table 3). In contrast, one study (Mellerup et al., 1982) reported a small but statistically significant increase in the B_{max} of [³H]-imipramine binding in the platelets of depressed manic melancholic patients, whereas seven studies failed to detect any differences in B_{max} of platelet [³H]-imipramive binding between depressed untreated patients and control volunteers (Table 3). Discrepancies among these studies may be caused by differences in patient population or diagnostic criteria, the length of the washout drug-free period, or methodological differences in the determination of platelet [³H]-imipramine binding (*see* section 2.1).

Changes in the binding of [³H]-imipramine in human platelets may be a reflection of those occurring in the brain (Briley et al., 1982; Langer et al., 1982). In cats treated chronically with imipramine, the B_{max} of [³H]-imipramine binding was decreased to a similar extent in the hypothalamus and platelets (Briley et al., 1982). Although the validity of this correlation between platelets and brain remains to be established in humans, it was, however,

Table 3

[^{3}H]-Imipramine Binding Density (B_{max}) in Platelets from Untreated
Depressed Patients in Comparison to Age and
Sex-Matched Volunteers: Summary of Clinical Observations

Study	Platelet [^{3}H]-imipramine binding, % change from control[a]
Decreased [^{3}H]-imipramine binding	
Briley et al. (1980)	−54
Asarch et al. (1981)	−22
Paul et al. (1981a)	−29
Raisman et al. (1981)	−48
Suranyi-Cadotte et al. (1982)	−54
Raisman et al. (1982b)	−32
Langer and Raisman (1983)	−43
Suranyi-Cadotte et al. (1985a)	−34
Suranyi-Cadotte et al. (1985b)	−26
Langer et al. (1986b)	−54
Lewis and McChesney (1985)	−40
Schneider et al. (1985)	−20
Tanimoto et al. (1985)[b]	−21
Wagner et al. (1985)	−10
Poirier et al. (1986)	−47
Baron et al. (1986)[c]	−20
Increased [^{3}H]-imipramine binding	
Mellerup et al. (1982)	+9
No significant differences	
Baron et al. (1983)	NS
Egrise et al. (1983)	NS
Whitaker et al. (1984)	NS
Gentsch et al. (1985)	NS
Tang and Morris (1985)	NS
Hrdina et al. (1985a)	NS
Baron et al. (1986)[c]	NS
Carstens et al. (1986)	NS
Muscettola et al. (1986)	NS

[a]Shown is the percent change in the B_{max} of [^{3}H]-imipramine binding in platelets from untreated depressed patients when compared with control volunteers matched for age and sex for those studies in which significant differences were reported. Those studies that failed to demonstrate a significant difference between untreated depressed patients and their appropriate controls are shown as NS (nonsignificant). Note that 16 out of 26 studies reported a significant decrease in B_{max} of [^{3}H]-imipramine binding in platelets from depressed untreated patients.

[b]Tanimoto et al. (1985) measured [^{3}H]-imipramine binding at a single concentration of 1 nM.

[c]Baron et al. (1986) reported a significant decrease in platelet [^{3}H]-imipramine B_{max} values in euthymic bipolar, but not in unipolar, depression.

demonstrated that the B_{max} of [³H]-imipramine binding in the postmortem brain of suicides and depressed patients was significantly lower than that of appropriate controls in two studies (Perry et al., 1983; Stanley et al., 1982). Different results were, however, reported in a recent study of Owen et al. (1986).

Some of the results suggest that the decreased density of [³H]-imipramine binding sites in platelets may be a "state-independent," rather than a "state-dependent" biological marker in depression and may reflect a vulnerability factor or a genetic susceptibility to depression. Thus, short-term longitudinal studies indicate that the density of [³H]-imipramine binding in platelets from depressed patients does not change during treatment with tricyclic antidepressant drugs, in spite of a major improvement in the psychiatric rating (Raisman et al., 1981). In addition, severely depressed patients treated with the selective NA-uptake inhibitor maprotiline did not show any change in the low B_{max} value for platelet [³H]-imipramine binding, in spite of a successful clinical recovery (Raisman et al., 1981). Moreover, we recently reported that after at least six sessions of electroconvulsive therapy and without other antidepressant medication, the B_{max} value for [³H]-imipramine binding in platelets from severely endogenously depressed patients did not differ significantly from the corresponding values obtained before treatment, in spite of the clinical improvement (Langer et al., 1986b).

This observation may be of considerable interest, since it is generally agreed that a genetic predisposition exists in a substantial proportion of patients with manic-depressive illness. Indeed, a study by Paul et al. (1981a) reported that the B_{max} and K_d values of [³H]-imipramine binding in platelets from pairs of monozygotic twins were practically identical, suggesting that [³H]-imipramine binding in platelets may be a genetically determined parameter.

In one of our most recent studies, however, depressed patients treated by electroconvulsive shock therapy were followed longitudinally for up to 18 mo. It was of interest to note that over this 18-mo period, the platelet imipramine receptor density increased toward the range of normal values (Langer et al., 1986b). A delayed return to control [³H]-imipramine binding values in platelets was also reported for depressed patients under treatment with imipramine (Suranyi-Cadotte et al., 1985b). These data would support the view that [³H]-imipramine binding in platelets may be a state-dependent biological marker in depression, but that it recovers to normal B_{max} values some time after the clinical improvement, probably at a time when the clinical remission is well consolidated. Clearly, additional studies in patients that were formerly depressed, but are currently euthymic and unmedicated, are required to clarify whether [^{3}H]-imipramine binding is a state-dependent or a state-independent biological marker in depression.

A much more limited number of studies has examined whether changes in platelet [^{3}H]-imipramine binding parameters are specific to depression. Thus, in platelets from cirrhotic or hypertensive patients, a significant decrease in the V_{max} of [^{3}H]-serotonin uptake was reported, whereas there was no modification in platelet [^{3}H]-imipramine binding parameters (Ahtee et al., 1981; Kamal et al., 1984).

Platelet [^{3}H]-imipramine binding parameters were reportedly unaffected in schizophrenia (Wood et al., 1983), Alzheimer's disease (Suranyi-Cadotte et al., 1985a), adolescent affective disorders (Rehavi et al., 1984), or adult females immediately postpartum, despite an increased rating of depressive mood (Katona et al., 1985).

Whereas platelet [^{3}H]-imipramine binding was not affected in Parkinson's disease (Suranyi-Cadotte et al., 1985a), the density of [^{3}H]-imipramine binding sites was reduced in postmortem samples of the Parkinsonian putamen and prefrontal cortex, as indicated in a preliminary study by Cash et al. (1985). It remains to be established whether these findings are related to a primary pathophysiological lesion in central serotonergic neurons and/or associated with the occurrence of depression often encountered in Parkinson's disease.

Although these studies generally show so far that a decrease in [^{3}H]-imipramine binding density may be specific for depression, several recent reports indicate that changes in platelet [^{3}H]-imipramine binding may also be found in other diseases not primarily associated with affective disorders. Thus, high-affinity [^{3}H]-imipramine binding was decreased in platelets from enuretic children and adolescents compared to healthy age-matched controls (Weizman et al., 1985). Subjects with familial or nonfamilial enuresis, who often benefit from imipramine treatment, could not be discriminated on the basis of the B_{max} values of platelet [^{3}H]-imipramine binding.

A further preliminary study by Bloomfield et al. (1985) shows that platelet [^{3}H]-imipramine binding is also significantly reduced in Cushing's disease. The authors speculated that the reduced platelet [^{3}H]-imipramine binding density in Cushing's disease may be related to altered cortisol levels and suggest that this adrenocorticosteroid may also contribute to the changes reported in the B_{max} of platelet [^{3}H]-imipramine binding in depression.

A recent interesting observation was made by Weizman et al. (1986) in adolescent females with anorexia nervosa. When compared to normal controls, a highly significant 29% decrease in platelet [^{3}H]-imipramine binding density was observed in anorexia nervosa. Moreover, anorexia nervosa appears to represent the counterpart of diseases like essential hypertension or alcoholic cirrhosis in that, in contrast to the latter two conditions, [^{3}H]-imipramine recognition sites were decreased in density, whereas platelet [^{3}H]-serotonin uptake was normal (Weizman et al., 1986).

Taken together, these results suggest that the site of [^{3}H]-imipramine binding and the site of [^{3}H]-serotonin uptake, although associated, are not identical, and can be regulated independently. Moreover, although the uptake of [^{3}H]-serotonin into platelets is modified in both psychiatric and nonpsychiatric disorders, changes in density of platelet [^{3}H]-imipramine binding appear to be predominantly associated with depressive states.

2.3. Conclusions

Although the high-affinity binding sites observed for several antidepressants like [^{3}H]-cyanoimipramine, [^{3}H]-nitroimipramine, [^{3}H]-paroxetine, [^{3}H]-indalpine, and [^{3}H]-norzimelidine appear related to the serotonin transporter, most evidence thus far in favor of such an association has been presented for [^{3}H]-imipramine. In the case of [^{3}H]-imipramine, certain data indicate that it binds, not to the substrate recognition site of the transporter, but possibly to a novel presynaptic receptor that serves to modulate serotonin transport. The density of the imipramine receptor in human platelets is decreased in untreated depressed patients, as indicated by the majority of studies on this subject, although it remains to be established whether this decrease is a state- or trait-dependent marker in depression. The hypothesis that the imipramine receptor is a novel modulatory site for the serotonin transporter suggests that an endogenous ligand for the imipramine receptor may exist. The isolation and identification of such an endocoid, and information concerning its regulation and involvement in serotonergic neurotransmission, is likely to significantly advance our understanding of depression.

3. High-Affinity Antidepressant Binding Sites and the Noradrenergic Transporter

Similar to the case of [³H]-imipramine, the binding of [³H]-desipramine, a tricyclic antidepressant that preferentially inhibits neuronal NA uptake, has been studied in the brain and peripheral tissues. Whereas high-affinity [³H]-imipramine receptors are associated with the serotonergic transporter, high-affinity binding sites for [³H]-desipramine and [³H]-nomifensine (Fig. 7) appear linked to the neuronal noradrenergic transporter.

3.1. Regional and Cellular Distribution of the [³H]-Desipramine Recognition Sites

The presence of [³H]-desipramine binding sites in the central and peripheral nervous system is closely associated with the density of noradrenergic innervation (Langer et al., 1981c; Raisman et al., 1982a). [³H]-Desipramine binding is of high affinity (K_d = 1.1–1.4 nM; Raisman et al., 1982a), and is localized to noradrenergic neurons. Thus, [³H]-desipramine binding was decreased after surgical denervation (Langer and Raisman, 1983) or after 6-hydroxydopamine (Hrdina et al., 1981) or DSP4 (Lee et al., 1982) lesioning of noradrenergic neurons in the rat brain. Similarly, sympathetic denervation of the rat heart by stellate ganglionectomy or of the submaxillary gland by superior cervical ganglionectomy, resulted in a parallel decrease in endogenous NA levels and

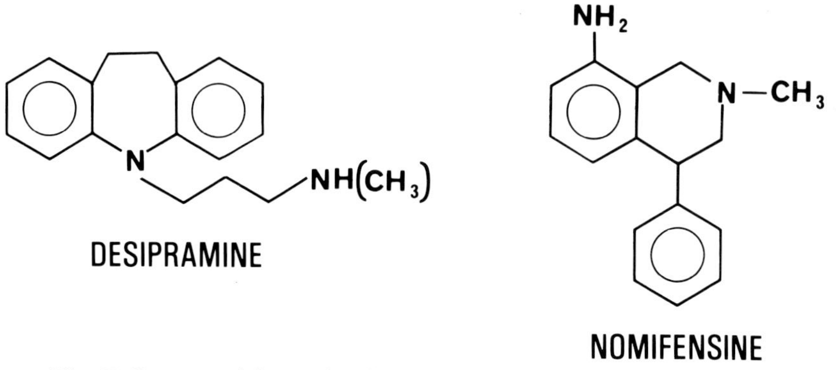

DESIPRAMINE

NOMIFENSINE

Fig. 7. Structural formula of antidepressants used as radioligands to label the noradrenergic transporter. Note that [³H]-desipramine can also be used to label the adrenergic transporter (*see* section 5), whereas [³H]-nomifensine additionally labels the dopaminergic transporter (*see* section 4).

of [³H]-desipramine binding (Raisman et al., 1982a). In the rat, the highest [³H]-desipramine binding density is found in the vas deferens (B_{max} = 1015 fmol/mg protein; Raisman et al., 1982a; Langer et al., 1984d).

[³H]-Desipramine binding sites associated with the noradrenergic transporter have also been demonstrated in rat pheochromocytoma cells (PC 12 cells) by Bonish and Harder (1986). These authors have also solubilized the [³H]-desipramine binding site in PC 12 cells (Schomig and Bonish, 1986).

3.2. Pharmacological Profile of the [³H]-Desipramine Recognition Site

The pharmacological profile of [³H]-desipramine binding in various tissues, including PC 12 cells, significantly correlates with that of inhibition of [³H]-NA uptake (Langer et al., 1984d; Raisman et al., 1982a; Hrdina, 1982; Bonish and Harder, 1986). The noradrenergic uptake inhibitor oxaprotiline stereospecifically affects [³H]-desipramine binding, (+)-oxaprotiline being 275 times more potent than (–)-oxaprotiline (Raisman et al., 1982a). Significantly, NA is virtually inactive as an inhibitor of [³H]-desipramine binding (Lee et al., 1982; Raisman et al., 1982a), although a significant qualitative correlation exists between the inhibitory effects of transporter substrates for inhibition of [³H]-desipramine binding and [³H]-NA transport in PC 12 cells (Bonish and Harder, 1986). It has been suggested (Raisman et al., 1982a) that, in analogy with [³H]-imipramine binding to a modulatory site of the serotonin transporter (*see* section 2.2.3), [³H]-desipramine may label a site different from the NA recognition site of the transporter, but allosterically coupled to it (Langer et al., 1984d). Although this proposal remains an open question, Lee and coworkers (1982) noted that NA does not affect the rate of dissociation of [³H]-desipramine, as was observed for the effect of serotonin on the dissociation of [³H]-imipramine (*see* section 2.2.3).

We recently reexamined the question of the substrates of NA uptake with respect to [³H]-desipramine and sodium-dependent [³H]-mazindol binding to the rat vas deferens and salivary gland (Langer et al., 1986c; De Oliveira et al., in preparation). In these tissues, sodium-dependent [³H]-mazindol binding is also associated with the NA transporter (Javitch et al., 1984, 1985). In this study, it was observed that the affinity of NA and other substrates such as A, DA, and metaraminol for [³H]-desipramine or [³H]-

Table 4

Effect of the Incubation Temperature on the Pharmacological
Profile of [^{3}H]-Mazindol Binding to the Noradrenergic
Transporter in the Rat Salivary Gland

Drug	[^{3}H]-Mazindol binding[a]	
	IC$_{50}$, μM 0°C	IC$_{50}$, μM 25°C
Uptake inhibitors		
Nisoxetine	0.0117	0.0148
Maprotiline	0.100	0.100
Uptake substrates		
(–)-NA	1.95	100
(–)-A	25.1	110
Metaraminol	0.603	50.1

[a][^{3}H]-Mazindol binding was studied at a ligand concentration of 2 nM and
incubation temperatures of 0 or 25°C (data from Langer et al., 1986c).

mazindol binding varies as a function of the incubation temperature. When binding assays are carried out at 0°C as opposed to the
usual incubation temperature of 20°C (cf., Raisman et al., 1982a),
these substrates of the noradrenergic transporter have significantly higher affinity (in some cases 100-fold) for inhibition of radioligand binding (Table 4; Langer et al., 1986c).

3.3. Conclusions

Most evidence available would suggest that [^{3}H]-desipramine
binding sites are associated with the noradrenergic transporter in
the brain and peripheral tissues. Many of the intricate details of the
[^{3}H]-desipramine binding site, such as its molecular relationship to
the substrate recognition site and its involvement in the etiology
and mechanism of action of antidepressant drugs, remain to be
elucidated.

4. High-Affinity Antidepressant Binding Sites and the Dopaminergic Transporter

High-affinity, sodium-dependent binding sites (K_d = 80 nM)
for the nontricyclic antidepressant [^{3}H]-nomifensine have been

identified in membranes from the rat and rabbit striatum (Dubocovich and Zahniser, 1985).

4.1. Cellular Localization of [³H]-Nomifensine Binding

[³H]-Nomifensine binding is localized to dopaminergic terminals in the rat striatum, as shown by a decrease in their density, but not their affinity following 6-hydroxydopamine lesions (Dubocovich and Zahniser, 1985). Quantitative autoradiography of [³H]-nomifensine binding sites in the rat and human brain (Scatton et al., 1985, 1986) confirms their localization to dopaminergic nerve terminals, although the NA transporter also appeared to be labeled.

4.2. Pharmacology of Nomifensine Binding

The pharmacological profile of [³H]-nomifensine binding was highly significantly correlated with that previously established for inhibition of [³H]-DA uptake into rat striatal synaptosomes, indicating that striatal [³H]-nomifensine binding sites are associated with the dopaminergic transporter. The enantiomers of nomifensine stereoselectively inhibit [³H]-nomifensine binding, in agreement with their effects on [³H]-DA uptake (Schacht and Leven, 1984). As in the case of [³H]-antidepressant binding to the serotonergic and noradrenergic transporters (*see* section 3.2), the substrate of the transporter, DA, inhibited [³H]-nomifensine binding with lower affinity than its K_m for transport into the nerve terminal (Dubocovich and Zahniser, 1985).

It should be noted that the DA transporter can also be labeled using other radioligands such as [³H]-cocaine (Kennedy and Hanbauer, 1983; Schoemaker et al., 1985; Pimoule et al., 1983), [³H]-mazindol (Javitch et al., 1984, 1985), [³H]-GBR-12935 (Berger et al., 1985), [³H]-GBR-12783 (Bonnet et al., 1986), and [³H]-threo-(±)-methylphenidate (Janowsky et al., 1985). Among the different radioligands used to label the dopaminergic transporter, [³H]-GBR-12783 and [³H]-GBR-12935 appear to be of highest affinity (K_d = 1–2 nM).

4.3. Conclusions

Antidepressant binding sites associated with the dopaminergic transporter were only lately identified. The characterization of [³H]-antidepressant binding sites associated with the dopaminergic transporter provides a means of studying this carrier at the molecular level, and under conditions in which the functional process of transport can no longer be evaluated, such as in the postmortem human brain. Thus, binding of [^{3}H]-cocaine as a radioligand to label the dopaminergic transporter has been shown to be decreased in the human putamen from Parkinson's disease patients (Pimoule et al., 1983; Schoemaker et al., 1985), but was unaffected in schizophrenia (Pimoule et al., 1985). The role of the dopaminergic transporter in depressive states remains an open question.

5. High-Affinity Binding and the Adrenergic Transporter

Whereas high affinity binding sites for [^{3}H]-antidepressants associated with the neuronal transporter for serotonin, NA, and DA have been well identified and characterized, the identification of [^{3}H]-antidepressant binding associated with the neuronal transporter for A has only been reported recently (Langer et al., 1987).

5.1. [^{3}H]-Desipramine Binding in the Frog Heart

In the identification of [^{3}H]-desipramine binding to the neuronal transporter for A, use was made of the fact that the sympathetic transmitter in the frog heart is A (Azuma et al., 1965), and consequently, the frog heart may be assumed to contain exclusively the neuronal transporter for A. [^{3}H]-Desipramine binding in the frog heart is of high affinity (K_d = 1.94 nM), and shows a B_{max} of 816 fmol/mg protein. As shown for inhibition of [^{3}H]-desipramine binding to the noradrenergic transporter in the rat heart, [^{3}H]-desipramine binding to the adrenergic transporter in the frog heart is potently inhibited by unlabeled desipramine, nisoxetine, and (+)-oxaprotiline (Table 5). Interestingly, however, imipramine and the atypical antidepressants mianserin and iprindole are 20–50 times more potent as inhibitors of [^{3}H]-desipramine binding to the adrenergic than noradrenergic transporter (Table 5). It would thus appear that, although species differences cannot be excluded, the pharmacology of the neuronal transporter for A may be different from that for NA (Langer et al., 1987). Significantly, iprindole was reported to antagonize the 6-hydroxydopa-induced fall in endogenous A, but not NA levels in the mouse brain in vivo (VonVoigtlander and Losey, 1978). The implications of these

Table 5
Comparison of the Pharmacology of [³H]-Desipramine
Binding to the Neuronal Epinephrine Transporter in the Frog Heart
and the Norepinephrine Transporter in the Rat Heart

| | [³H]-Desipramine binding, IC$_{50}$, μM | |
Drug	Frog heart[a]	Rat heart[b]
Desipramine	0.0025	0.006
Nisoxetine	0.013	0.007
(+)-Oxaprotiline	0.040	0.020
Imipramine	0.004	0.220
Mianserin	0.050	0.950
Iprindole	0.250	10.0

[a]Data from Langer et al. (1987)
[b]Data from Lee et al. (1982) and Raisman et al. (1982a)

observations for the mechanism of antidepressant effects of these compounds remain attractive possibilities that deserve additional work.

6. High-Affinity Binding of Other Radiolabeled Antidepressants

6.1. [³H]-Amitriptyline

[³H]-Amitriptyline was the first antidepressant used in binding studies in an attempt to define the site of action of tricyclic antidepressants. However, [³H]-amitriptyline binding was predominantly to muscarinic cholinergic and histamine H$_1$ receptors (Rehavi and Sokolovsky, 1978).

6.2. [³H]-Doxepin

High-affinity [³H]-doxepin binding has been demonstrated by Tran et al. (1981). These binding sites were found to represent the histamine H$_1$ receptor. In addition, these authors reported the presence of a low-affinity site (K_d = 25.8 nM) of unknown pharmacology, for which several antidepressants had moderate (IC$_{50}$ values from 30 to 600 nM) affinity.

6.3. [³H]-Mianserin

High-affinity binding sites in the brain for the [³H]-labeled antidepressant mianserin were first described by Dumbrille-Ross et al. (1980) and Whitaker and Cross (1980). The dissociation constant (K_d) of [³H]-mianserin binding to the rat cerebral cortex was given by Dumbrille-Ross et al. (1980) as 1.4 nM. Although several tricyclic antidepressants were moderately active as inhibitors of the [³H]-mianserin binding site, these authors concluded that the [³H]-mianserin binding site was not identical to the [³H]-imipramine receptor, but possibly related to a serotonin receptor. Whitaker and Cross (1980) reached a similar conclusion, although noting that tricyclic, tetracyclic, and other antidepressants potently affected [³H]-mianserin binding to the calf caudate. Peroutka and Snyder (1981) reported that [³H]-mianserin effectively labels the serotonin 5-HT$_2$ receptor, but also the histamine H$_1$ receptor. However, as opposed to radioligand binding to the 5-HT$_2$ receptor, [³H]-mianserin binding in the rat hippocampus is up-regulated following destruction of its serotonergic innervation, whereas it is not affected following chronic imipramine treatment (Barbaccia et al., 1983b), indicating that [³H]-mianserin may not exclusively label the 5-HT$_2$ receptor, even in the presence of a histamine H$_1$ blocker. In addition, in contrast to [³H]-imipramine binding, [³H]-mianserin binding is up-regulated following chronic *p*-chlorophenylalanine treatment (Brunello et al., 1982), showing that [³H]-mianserin recognition sites are under the regulation of endogenous serotonin in vivo. More recent evidence indicates that mianserin also has high affinity for a novel serotonin receptor, the 5-HT$_{1C}$ receptor (Pazos et al., 1984). It remains to be established whether [³H]-mianserin also labeled the 5-HT$_{1C}$ receptor in the above studies.

6.4. [³H]-Rolipram

Preliminary data indicate that [³H]-rolipram binds with high affinity ($K_d = 1.3$ nM) to a soluble fraction of the pig frontal cortex ($B_{max} = 400$ fmol/mg protein) or pons ($B_{max} = 25$ fmol/mg protein) (Schneider, 1982; Schneider et al., 1986). The binding of this selective cAMP phosphodiesterase inhibitor was stereospecific and sensitive to inhibition by other phosphodiesterase inhibitors, such as ICI 63197 and Ro 20-1724. High-affinity [³H]-rolipram binding, however, remains to be characterized in more detail.

7. Conclusions

Radioligand binding studies have identified high-affinity binding sites for [³H]-antidepressants that are thought to contribute to their therapeutic efficacy in depressive disorders. Among these, [³H]-imipramine has been studied most extensively. Its binding site would appear to be a novel pharmacological receptor associated with the serotonergic transporter in brain and platelets, but different from the substrate recognition site of the carrier. The identification and characterization of a putative endocoid acting through the imipramine receptor may significantly advance our knowledge of serotonergic homeostasis and provide new insights into the etiology of depressive disorders, for which the [³H]-imipramine receptor is thought to be a biological marker.

In analogy with the association of the [³H]-imipramine receptor and the serotonergic transporter, binding sites of [³H]-desipramine and [³H]-nomifensine are associated with the noradrenergic and dopaminergic transporter systems under the proper experimental conditions and allow for their study at the molecular level. However, our understanding of these receptors and their relation to the catecholamine recognition site of the transporter is much less advanced than that of the [³H]-imipramine receptor associated with the serotonergic transporter.

It may be concluded that the study of high-affinity binding sites for classical as well as some of the newer antidepressants has proven fruitful and holds promise for the future in the identification and characterization of the mechanism of drug antidepressant action.

Acknowledgment

The authors gratefully acknowledge Colette Villebeuf for the preparation of this manuscript.

References

Ahtee L., Briley M. S., Raisman R., Lebrec D., and Langer S. Z. (1981) Reduced uptake of serotonin but unchanged ³H-imipramine binding in the platelets from cirrhotic patients. *Life Sci.* **29,** 2323–2329.

Angel I. and Paul S. M. (1984) Inhibition of synaptosomal 5-[^{3}H]-hydroxytryptamine uptake by endogenous factor(s) in human blood. *FEBS Lett.* **171**, 280–284.

Arora R. C., Wunnicke V., and Meltzer H. (1985) Effect of protein concentration on kinetic constants (K_d and B_{max}) of ^{3}H-imipramine binding in blood platelets. *Biol. Psychiatry* **20**, 116–119.

Asarch K. B., Shih J. C., and Kubesar A. (1981) Decreased ^{3}H-imipramine binding in depressed males and females. *Commun. Psychopharmacol.* **4**, 425–432.

Azuma T., Binia A., and Visscher M. B. (1965) Adrenergic mechanisms in the bullfrog and turtle. *Am. J. Physiol.* **209**, 1287–1294.

Barbaccia M. L., Gandolfi O., Chuang D. M., and Costa E. (1983a) Modulation of neuronal serotonin uptake by a putative endogenous ligand of imipramine recognition sites. *Proc. Natl. Acad. Sci. USA* **80**, 5134–5138.

Barbaccia M. L., Gandolfi O., Chuang D. M., and Costa E. (1983b) Differences in the regulatory adaptation of the 5HT$_2$ recognition sites labeled with ^{3}H-mianserin. *Neuropharmacology* **22**, 123–126.

Barbaccia M. L., Melloni P., Pozzi O., and Costa E. (1986) [^{3}H]Imipramine displacement and 5HT uptake inhibition by tryptoline derivatives: In rat brain 5-methoxytryptoline is not the autacoid for [^{3}H]imipramine recognition sites. *Eur. J. Pharmacol.* **123**, 45–52.

Barkai A. I., Kowalik S., and Baron M. (1985) Effect of membrane protein concentration on binding of ^{3}H-imipramine in human platelets. *Biol. Psychiatry* **20**, 215–219.

Baron M., Barkai A., Gruen R., Kowalik S., and Quitkin F. (1983) ^{3}H-Imipramine platelet binding sites in unipolar depression. *Biol. Psychiatry* **18**, 1403–1409.

Baron M., Barkai A., Gruen R., Peselow M. D., Fieve R. R., and Quitkin F. (1986) Platelet ^{3}H-imipramine binding in affective disorders: Trait versus state characteristics. *Am. J. Psychiat.* **143**, 711–717.

Benavides J., Savaki H. E., Malgouris C., Laplace C., Margelidon C., Daniel M., Courteix J., Uzan A., Gueremy C., and Le Fur G. (1985) Quantitative autoradiography of [^{3}H]-indalpine binding sites in the rat brain: I. Pharmacological characterization. *J. Neurochem.* **45**, 514–520.

Berger P., Janowsky A., Vocci F., Skolnick P., Schweri M. M., and Paul S. M. (1985) [^{3}H]-GBR-12935: A specific high affinity ligand for labelling the dopamine transport complex. *Eur. J. Pharmacol.* **107**, 289–290.

Biegon A. (1986) Effect of chronic desipramine treatment on dihydroalprenolol, imipramine and desipramine binding sites: A quantitative autoradiographic study in the rat brain. *J. Neurochem.* **47**, 77–80.

Bloomfield J. G., Elliott J. M., and Evans D. J. (1985) Platelet [³H]-imipramine and [³H]-yohimbine binding are reduced in Cushing's disease. *Br. J. Pharmacol.* **86,** 402P.

Bonish H. and Harder R. (1986) Binding of ³H-desipramine to the neuronal noradrenaline carrier of rat phaeochromocytoma cells (PC-12 cells). *Naunyn Schmiedebergs Arch. Pharmacol.* **334,** 403–411.

Bonnet J. J., Protais P., Chagraoui A., and Costentin J. (1986) High-affinity [³H]GBR 12783 binding to a specific site associated with the neuronal dopamine uptake complex in the central nervous system. *Eur. J. Pharmacol.* **126,** 211–222.

Borbe H. O. and Zube I. (1983) The detection of specific [³H]imipramine binding in bovine retina. *Brain Res.* **264,** 178–180.

Braddock L. E., Cowen P. J., Eliott J. M., Fraser S., and Stump K. (1984) Changes in the binding to platelets of [³H]imipramine and [³H]yohimbine in normal subjects taking amitriptyline. *Neuropharmacology* **23,** 285–286.

Briley M. S. and Langer S. Z. (1981) Sodium-dependency of ³H-imipramine binding in rat cerebral cortex. *Eur. J. Pharmacol.* **72,** 377–380.

Briley M. S., Raisman R., and Langer S. Z. (1979) Human platelets possess high-affinity binding sites for ³H-imipramine. *Eur. J. Pharmacol.* **58,** 347–348.

Briley M. S., Langer S. Z., Raisman R., Sechter D., and Zarifian E. (1980) Tritiated imipramine binding sites are decreased in platelets of untreated depressed patients. *Science* **209,** 303–305.

Briley M. S., Raisman R., Arbilla S., Casadamont M., and Langer S. Z. (1982) Concomitant decrease in (³H)-imipramine binding in cat brain and platelets after chronic treatment with imipramine. *Eur. J. Pharmacol.* **81,** 305–314.

Brunello M., Chuang D., and Costa E. (1982) Different synaptic localizations of mianserin and imipramine binding sites. *Science* **215,** 1112–1115.

Carstens M. E., Engelbrecht A. H., Russel V. A., Aalbers C., Gagiano C. A., Chalton D. O., and Taljaard J. J. F. (1986) Imipramine binding sites on platelets of patients with major depressive disorder. *Psychiatry Res.* **18,** 333–342.

Cash R., Raisman R., Ploska A., and Agid Y. (1985) High and low affinity [³H]imipramine binding sites in control and parkinsonian brains. *Eur. J. Pharmacol.* **117,** 71–80.

Conway P. G. and Brunswick D. J. (1985) High- and low-affinity binding components for [³H]-imipramine in rat cerebral cortex. *J. Neurochem.* **45,** 206–209.

Davis A. (1984a) Molecular aspects of the imipramine "receptor". *Experientia* **40,** 782–794.

Davis A. (1984b) Temperature-sensitive conformational changes in [³H]imipramine binding sites and the involvement of sulphur-containing bonds. *Eur. J. Pharmacol.* **102**, 341–347.

Davis A., Morris J. M., and Tang S. W. (1983) Temperature-sensitive conformational changes in membrane-bound and solubilized [³H]imipramine binding sites. *Eur. J. Pharmacol.* **88**, 407–410.

Davis A., Morris J. M., and Tang S. W. (1985) Partial characterization of solubilized platelet imipramine binding sites using a new probe, [³H]3-cyanoimipramine ([³H]Ro 11-2465). *Eur. J. Pharmacol.* **109**, 97–104.

Dawson T. M. and Wamsley J. K. (1983) Autoradiographic localization of [³H]-imipramine binding sites: Association with serotonergic neurons. *Brain Res. Bull.* **11**, 325–334.

Dawson T. M., Gehlert D. R., Snowhill E. W., and Wamsley J. K. (1985) Quantitative autoradiographic evidence for axonal transport of imipramine receptors in the central nervous system of the rat. *Neurosci. Lett.* **55**, 261–266.

Dubocovich M. L. and Zahniser N. (1985) Binding characteristics of the dopamine uptake inhibitor [³H]-nomifensine to striatal membranes. *Biochem. Pharmacol.* **32**, 1137–1144.

Dumbrille-Ross A. and Tang S. W. (1983) Binding to a subpopulation of ³H-imipramine binding sites by ³H-Ro 11-2465, a possible irreversible ligand. *Mol. Pharmacol.* **23**, 607–613.

Dumbrille-Ross A., Tang S. W., and Seeman P. (1980) High-affinity binding of [³H]-mianserin to rat cerebral cortex. *Eur. J. Pharmacol.* **68**, 395–396.

Egrise D., Desmedt D., Schoutens A., and Mendlewicz J. (1983) Circannual variations in the density of tritiated imipramine binding sites on blood platelets in man. *Neuropsychobiology* **10**, 101–102.

Fields J. Z., Roeske W. R., Morkin E., and Yamamura H. I. (1978) Cardiac muscarinic cholinergic receptors. *J. Biol. Chem.* **9**, 3251–3258.

Friedl W., Propping P., and Weck B. (1983) ³H-Imipramine binding in platelets: Influence of varying proportions of intact platelets in membrane preparations of binding. *Psychopharmacology,* **80**, 96–99.

Fuxe K., Calza L., Benfenati F., Zini I., and Agnati L. F. (1983) Quantitative autoradiographic localization of [³H]-imipramine binding sites in the brain of the rat: Relationship to ascending 5-hydroxytryptamine neuron systems. *Proc. Natl. Acad. Sci. USA* **80**, 3836–3840.

Gentsch C., Lichtsteiner M., Gastpar M., Gastpar G., and Feer H. (1985) ³H-Imipramine binding sites in platelets of hospitalized psychiatric patients. *Psychiatry Res.* **14**, 177–187.

Grabowsky K. L., McCabe R. T., and Wamsley J. K. (1983) Localization of [³H]-imipramine binding sites in rat brain by light microscopic autoradiography. *Life Sci.* **32**, 2355–2361.

Gross G., Gothert M., Ender H. C., and Schumann H. J. (1981) ^{3}H-imipramine binding sites in the rat brain: Selective localization on serotoninergic neurons. *Naunyn Schmiedebergs Arch. Pharmacol.* **317,** 310–314.

Habert E., Graham D., Tahraoui L., Claustre Y., and Langer S. Z. (1985) Characterization of ^{3}H]paroxetine binding to rat cortical membranes. *Eur. J. Pharmacol.* **118,** 107–114.

Hall H., Ross R., Ogren S. O., and Gawell L. (1982) Binding of a specific 5HT uptake inhibitor, ^{3}H-norzimelidine, to rat brain homogenates. *Eur. J. Pharmacol.* **80,** 281–282.

Hrdina P. D. (1982) Tricyclic Antidepressants: Differences in High Affinity Binding of ^{3}H]-Desipramine and ^{3}H]-Imipramine in the Brain, in *Advances in the Biosciences* (Langer S. Z., Takahashi R., and Briley M., eds.) Pergamon, Oxford.

Hrdina P. D. (1984) Differentiation of two components of specific ^{3}H]-imipramine binding in rat brain. *Eur. J. Pharmacol.* **102,** 481–488.

Hrdina P. D., Elson-Hartman K., Roberts D., and Pappas B. (1981) High affinity ^{3}H]-desipramine binding in rat cerebral cortex decreases after selective lesion of noradrenergic neurons with 6-hydroxydopamine. *Eur. J. Pharmacol.* **73,** 375–376.

Hrdina P. D., Lapierre Y. D., Horn E. R., and Bakish D. (1985a) Platelet ^{3}H-imipramine binding: A possible predictor of response to antidepressant treatment. *Prog. Neuropsychopharmacol. Biol. Psychiat.* **9,** 619–623.

Hrdina P. D., Pappas B. A., Roberts D. C. S., Bialik R. J., and Ryan C. L. (1985b) Relationship between levels and uptake of serotonin and high affinity ^{3}H]-imipramine recognition sites in the rat brain. *Can. J. Physiol. Pharmacol.* **63,** 1239–1244.

Ieni J. R., Zukin S. R., and Van Praag H. M. (1985) Human platelets possess multiple ^{3}H]imipramine binding sites. *Eur. J. Pharmacol.* **106,** 669–672.

Janowsky A., Schweri M. M., Berger P., Long R., Skolnick P., and Paul S. M. (1985) The effects of surgical and chemical lesions on striatal ^{3}H]-threo-(±)-methylphenidate binding: Correlation with ^{3}H]-dopamine uptake. *Eur. J. Pharmacol.* **108,** 187–191.

Javitch J. A., Blaustein R. O., and Snyder S. H. (1984) ^{3}H]Mazindol binding associated with neuronal dopamine and norepinephrine uptake sites. *Mol. Pharmacol.* **26,** 35–44.

Javitch J. A., Strittmatter S. M., and Snyder S. H. (1985) Differential visualization of dopamine and norepinephrine uptake sites in rat brain using ^{3}H]-mazindol autoradiography. *J. Neurosci.* **5,** 1513–1521.

Kamal L. A., Raisman R., Meyer P., and Langer S. Z. (1984) Reduced V_{max} of ^{3}H-serotonin uptake but unchanged ^{3}H-imipramine binding in the platelets of untreated hypertensive subjects. *Life Sci.* **34,** 2083–2088.

Katona C. L. E., Theodorou A. E., Missouris C. G., Bourke M. P., Horton R. W., Moncrieff D., Paykel E. S., and Kelly J. S. (1985) Platelet ^{3}H-imipramine binding in pregnancy and the puerperium. *Psychiatry Res.* **14**, 33–37.

Kennedy L. T. and Hanbauer I. (1983) Sodium-sensitive cocaine binding to rat striatal membrane: Possible relationship to dopamine uptake sites. *J. Neurochem.* **41**, 172–178.

Kim S. S. and Reith M. E. A. (1986) Effect of serotonin on the dissociation of high-affinity binding of [3H]-imipramine in mouse cerebral cortex. *Neurosci. Lett.* **67**, 123–128.

Laduron P. M. (1984a) In reply to S. Z. Langer. *Trends Pharmacol. Sci.* **5**, 52–53.

Laduron P. M. (1984b) Criteria for receptor sites in binding studies. *Biochem. Pharmacol.* **33**, 833–839.

Langer S. Z. (1984) [3H]Imipramine and [3H]desipramine binding: Nonspecific displaceable sites or physiologically relevant sites associated with the uptake of serotonin and noradrenaline? *Trends Pharmacol. Sci.* **5**, 51–52.

Langer S. Z. and Briley M. S. (1981) High affinity ^{3}H-imipramine binding: A new biological tool for studies in depression. *Trends Neurosci.* **4**, 28–31.

Langer S. Z. and Raisman R. (1983) Binding of [3H]-imipramine and [3H]-desipramine as biochemical tools for studies in depression. *Neuropharmacology* **22**, 407–413.

Langer S. Z. and Schoemaker H. (1987) Platelet Imipramine Binding in Depression, in *Receptors and Ligands in Psychiatry and Neurology*, (Sen A. K. and T. Lee, eds.) Cambridge University Press, Cambridge (in press).

Langer S. Z., Moret C., Raisman R., Dubocovich M. L., and Briley M. S. (1980a) High-affinity ^{3}H-imipramine binding in rat hypothalamus is associated with the uptake of serotonin but not norepinephrine. *Science* **210**, 1133–1135.

Langer S. Z., Briley M. S., Raisman R., Henry J. F., and Morselli P. L. (1980b) Specific ^{3}H-imipramine binding in human platelets: Influence of age and sex. *Naunyn Schmiedebergs Arch. Pharmacol.* **313**, 189–194.

Langer S. Z., Raisman R., and Briley M. S. (1980c) Stereoselective inhibition of ^{3}H-imipramine binding by antidepressant drugs and their derivatives. *Eur. J. Pharmacol.* **64**, 89–90.

Langer S. Z., Javoy-Agid F., Raisman R., Briley M. S., and Agid Y. (1981a) Distribution of specific high-affinity binding sites for imipramine in human brain. *J. Neurochem.* **37**, 267–271.

Langer S. Z., Zarifian E., Briley M. S., Raisman R., and Sechter D. (1981b) High-affinity binding of ^{3}H-imipramine in brain and platelets and its relevance to the biochemistry of affective disorders. *Life Sci.* **29,** 211–218.

Langer S. Z., Raisman R., and Briley M. S. (1981c) High-affinity [³H]-DMI binding is associated with neuronal noradrenaline uptake in the periphery and the central nervous system. *Eur. J. Pharmacol.* **72,** 423–424.

Langer S. Z., Zarifian E., Briley M. S., Raisman R., and Sechter D. (1982) High-affinity ³H-imipramine binding: A new biological marker in depression. *Pharmacopsychiatry* **15,** 3–10.

Langer S. Z., Lee C. R., Segonzac A., Tateishi T., Esnaud H., Schoemaker H., and Winblad B. (1984a) Possible endocrine role of the pineal gland for 6-methoxytetrahydro-β-carboline, a putative endogenous neuromodulator of the [³H]imipramine recognition site. *Eur. J. Pharmacol.* **102,** 379–380.

Langer S. Z., Raisman R., Tahraoui L., Scatton B., Niddam R., Lee C. R., and Claustre Y. (1984b) Substituted tetrahydro-β-carbolines are possible candidates as endogenous ligand of the [³H]imipramine recognition site. *Eur. J. Pharmacol.* **98,** 153–154.

Langer S. Z., Schoemaker H., and Segonzac A. (1984c) Comparative studies on [³H]-paroxetine and [³H]-imipramine binding to human platelets. *Br. J. Pharmacol.* **84,** 59P.

Langer S. Z., Tahraoui L., Raisman R., Arbilla S., Najar M., and Dedek J. (1984d) ³H-Desipramine Labels a Site Associated with the Neuronal Uptake of Noradrenaline in the Peripheral and Central Nervous System, in *Neuronal and Extraneuronal Events in Autonomic Pharmacology* (Fleming W. W., Langer S. Z., Graefe K. H., and Weiner N., eds.) Raven, New York.

Langer S. Z., Lee C. R., Schoemaker H., Segonzac A., and Esnaud H. (1985) 5-Methoxytryptoline and Close Analogs as Candidates for the Endogenous Ligand of the ³H-Imipramine Recognition Site, in *Endocoids* (Lal H., Labella F., and Lane J., eds.) Alan R, Liss, New York.

Langer S. Z., Galzin A. M., Lee C. R., and Schoemaker H. (1986a) Antidepressant Binding Sites in Brain and Platelets, in *Antidepressants and Receptor Function,* Ciba Foundation Symposia 123, John Wiley, New York.

Langer S. Z., Sechter D., Loo H., Raisman R., and Zarifian E. (1986b) Electroconvulsive shock therapy and maximum binding of platelet ³H-imipramine binding in depression. *Arch. Gen. Psychiatry* **43,** 949–952.

Langer S. Z., De Oliveira A. M., and Schoemaker H. (1986c) Temperature-dependence of substrate interaction with ³H-mazindol binding to the noradrenaline transporter. *Br. J. Pharmacol.* **89,** 630 P.

Langer S. Z., Pimoule C., and Schoemaker H. (1987) In the frog heart, [³H]-desipramine labels the neuronal transporter for adrenaline. *Br. J. Pharmacol.* **90,** 285P.

Lee C. M., Javitch J. A., and Snyder S. H. (1982) Characterization of [^{3}H]-desipramine binding associated with neuronal norepinephrine uptake sites in rat brain membranes. *J. Neurochem.* **2**, 1515–1525.

Lewis D. A. and McChesney C. (1985) Tritiated imipramine binding distinguishes among subtypes of depression. *Arch. Gen. Psychiat.* **42**, 485–488.

Lui N., Almon R. R., and Jusko W. J. (1984) Comparison of filtration and equilibrium dialysis methods for [^{3}H]-imipramine binding to human platelets. *Anal. Biochem.* **139**, 42–57.

Luine V. N., Frankfurt M., Rainbow T. C., Biegon A., and Azmitia E. (1983) Intrahypothalamic 5,7-dihydroxytryptamine facilitates feminine sexual behavior and decreases [^{3}H]-imipramine binding and 5-HT uptake. *Brain Res.* **264**, 344–348.

Marcusson J., Fowler C. J., Hall H., Ross S. B., and Winblad B. (1985) "Specific" binding of [^{3}H]-imipramine to protease-sensitive and protease-resistant sites. *J. Neurochem.* **44**, 705–711.

Marcusson J. O., Backstrom I. T., and Ross S. B. (1986) Single-site model of the neuronal 5-hydroxytryptamine uptake and imipramine binding site. *Mol. Pharmacol.* **30**, 121–128.

Mellerup E. T., Plenge P., and Rosenberg R. (1982) ^{3}H-Imipramine binding sites in platelets from psychiatric patients. *Psychiatry Res.* **7**, 221–227.

Mellerup E. T., Plenge P., and Engelstoft M. (1983) High affinity binding of [^{3}H]paroxetine and [^{3}H]imipramine to human platelet membranes. *Eur. J. Pharmacol.* **96**, 303–309.

Mellerup E. T., Plenge P., and Nielsen M. (1985) Size determination of binding polymers for [^{3}H]imipramine and [^{3}H]paroxetine in human membranes. *Eur. J. Pharmacol.* **106**, 411–413.

Mellerup E. T. and Plenge P. (1986) Chlorimipramine, but not imipramine, rapidly reduces [^{3}H]-imipramine binding in human platelet membranes. *Eur. J. Pharmacol.* **126**, 155–158.

Muscettola G., Di Lauro A., and Gianini C. P. (1986) Platelet ^{3}H-imipramine binding in bipolar patients. *Psychiatry Res.* **18**, 343–353.

Owen F., Chambers D. R., Cooper S. J., Crow T. J., Johnson J. A., Lofthouse R., and Poulter M. (1986) Serotonergic mechanisms in brains of suicide victims. *Brain Res.* **362**, 185–188.

Palkovits M., Raisman R., Briley M. S., and Langer S. Z. (1981) Regional distribution of [^{3}H]-imipramine binding in rat brain. *Brain Res.* **210**, 493–498.

Paul S. M., Rehavi M., Skolnick P., and Goodwin F. K. (1980) Demonstration of specific "high affinity" binding sites for [^{3}H]imipramine on human platelets. *Life Sci.* **26**, 953–959.

Paul S. M., Rehavi M., Skolnick P., Ballenger J. C., and Goodwin F. K. (1981a) Depressed patients have decreased binding of ^{3}H-imipramine to the platelets serotonin "transporter". *Arch. Gen. Psychiatry* **38**, 1315–1317.

Paul S. M., Rehavi M., Rice K. C., Ittah Y., and Skolnick P. (1981b) Does high-affinity [^{3}H]-imipramine binding label serotonin reuptake sites in brain and platelet? *Life Sci.* **28**, 2753–2760.

Pazos A., Hoyer D., and Palacios J. (1984) The binding of serotonergic ligands to the porcine choroid plexus: Characterization of a new type of serotonin recognition site. *Eur. J. Pharmacol.* **106**, 539–546.

Peroutka S. J. and Snyder S. H. (1981) [^{3}H]-Mianserin: Differential labelling of serotonin$_2$ and histamine$_1$ receptors in rat brain. *J. Pharmacol. Exp. Ther.* **216**, 142–148.

Perry E. K., Marshall E. F., Blessed G., Tomlinson B. E., and Perry R. H. (1983) Decreased imipramine binding in the brain of patients with depressive illness. *Br. J. Psychiat.* **142**, 188–192.

Phillips O. M., Wood K. M., and Williams D. C. (1984) Binding of [^{3}H]-imipramine to human platelet membranes with compensation for saturable binding to filters and its implication for binding studies with brain membranes. *J. Neurochem.* **43**, 479–486.

Pimoule C., Schoemaker H., Javoy-Agid F., Scatton B., Agid Y., and Langer S. Z. (1983) Decrease in [^{3}H]-cocaine binding to the dopamine transporter in Parkinson's disease. *Eur. J. Pharmacol.* **95**, 145–146.

Pimoule C., Schoemaker H., Reynolds G. P., and Langer S. Z. (1985) ^{3}H-SCH 23390 labeled D$_1$ dopamine receptors are unchanged in schizophrenia and Parkinson's disease. *Eur. J. Pharmacol.* **114**, 235–237.

Plenge P. and Mellerup E. T. (1984) Imipramine binding site. Temperature dependence of the binding of [^{3}H]-labeled imipramine and [^{3}H]-labeled paroxetine to human platelet membrane. *Biochem. Biophys. Acta* **770**, 22–28.

Plenge P. and Mellerup E. T. (1985) Antidepressive drugs can change the affinity of [^{3}H]-imipramine and [^{3}H]-paroxetine binding to platelet and neuronal membranes. *Eur. J. Pharmacol.* **119**, 1–8.

Poirier M. F., Loo H., Benkelfat C., Sechter D., Zarifian E., Galzin A. M., Schoemaker H., Segonzac A., and Langer S. Z. (1984) [^{3}H]Imipramine binding and [^{3}H]5HT uptake in human blood platelets: Changes after one week chlorimipramine treatment. *Eur. J. Pharmacol.* **106**, 629–633.

Poirier M. F., Benkelfat C., Loo H., Sechter D., Zarifian E., Galzin A. M., and Langer S. Z. (1986) Reduced B$_{max}$ of [^{3}H]-imipramine binding to platelets of depressed patients free of previous medication with 5HT uptake inhibitors. *Psychopharmacology* **89**, 456–461.

Poirier M. F., Galzin A. M., Loo H., Pimoule C., Segonzac A., Benkelfat C., Sechter D., Zarifian E., Schoemaker H., and Langer S. Z. (1987) Dissociation between changes in [^{3}H]-5-HT uptake and [^{3}H]-

imipramine binding in blood platelets following one week administration of chlorimipramine to healthy volunteers: Comparison with one week administration of maprotiline and amineptine. *Biol. Psychiatry* **22** 287–302.

Rainbow T. C., Biegon A., and McEwen B. S. (1982) Autoradiographic localization of imipramine binding in rat brain. *Eur. J. Pharmacol.* **77,** 363–364.

Raisman R. and Langer S. Z. (1983) Specific high-affinity ³H-imipramine binding sites in rat lung are related with a non-neuronal uptake site for serotonin. *Eur. J. Pharmacol.* **94,** 345–348.

Raisman R., Briley M. S., and Langer S. Z. (1979) Specific tricyclic antidepressant binding sites in rat brain. *Nature* **281,** 148–150.

Raisman R., Briley M. S., and Langer S. Z. (1980) Specific tricyclic antidepressant binding sites in rat brain characterised by high-affinity [³H]-imipramine binding. *Eur. J. Pharmacol.* **61,** 373–380.

Raisman R., Sechter D., Briley M. S., Zarifian E., and Langer S. Z. (1981) High affinity ³H-imipramine binding in platelets from untreated and treated depressed patients compared to healthy volunteers. *Psychopharmacology* **75,** 368–371.

Raisman R., Sette M., Pimoule C., Briley M., and Langer S. Z. (1982a) High-affinity [³H]-desipramine binding in the peripheral and central nervous system: A specific site associated with the neuronal uptake of noradrenaline. *Eur. J. Pharmacol.* **78,** 345–351.

Raisman R., Briley M. S., Bouchami F., Sechter D., Zarifian E., and Langer S. Z. (1982b) ³H-Imipramine binding and serotonin uptake in platelets from untreated depressed patients and control volunteers. *Psychopharmacology* **77,** 332–335.

Rehavi M. and Sokolovsky M. (1978) Multiple binding sites of tricyclic antidepressant drugs to mammalian brain receptors. *Brain Res.* **149,** 525–529.

Rehavi M., Tracer H., Rice K., Skolnick P., and Paul S. M. (1983) [³H]2-nitroimipramine: A selective "slowly-dissociating" probe of the imipramine binding site ("serotonin transporter") in platelets and brain. *Life Sci.* **32,** 645–653.

Rehavi M., Weizman R., Carel C., Apter A., and Tyano S. (1984) High-affinity ³H-imipramine binding in platelets of children and adolescents with major affective disorders. *Psychiatry Res.* **13,** 31–39.

Rehavi M., Ventura I., and Sarne Y. (1985) Demonstration of endogenous "imipramine like" material in rat brain. *Life Sci.* **36,** 687–693.

Reith M. E. A. (1985) Possible artefacts in imipramine binding assays. *J. Neurochem.* **45,** 661.

Reith M. E. A. (1986) Binding of [³H]-imipramine to mouse cerebrocortical membranes and to glass fiber filters. *J. Neurochem.* **46,** 760–766.

Reith M. E. A., Sershen H., and Lajtha A. (1984) Thermodynamics of the interactions of tricyclic drugs with binding sites for [^{3}H]imipramine in mouse cerebral cortex. *Biochem. Pharmacol.* **33**, 4101–4104.

Savaki H., Malgouris C., Benavides J., Laplace C., Uzan A., Gueremy C., and Le Fur G. (1985) Quantitative autoradiography of [^{3}H]-indalpine binding sites in the rat brain. II. Regional distribution. *J. Neurochem.* **45**, 521–526.

Scatton B., Dubois A., Dubocovich M. L., Zahniser N. R., and Fage D. (1985) Quantitative autoradiography of ^{3}H-nomifensine binding sites in rat brain. *Life Sci.* **36**, 815–822.

Scatton B., Dubois A., Camus A., Zahniser N. R., Dubocovich M. L., Agid Y., and Cudennec A. (1986) Autoradiographic Visualization and Quantification of Dopamine Receptors and Uptake Sites in the Rat and Human C.N.S., in *Dopaminergic Systems and Their Regulation* (Woodruff G. N., Poat J. A., and Roberto P. G., eds.) Macmillan, London.

Schacht U. and Leven M. (1984) Stereoselective inhibition of synaptosomal catecholamine uptake by nomifensine. *Eur. J. Pharmacol.* **98**, 275–277.

Schneider H. H. (1982) High affinity binding of the selective cyclic adenosin monophosphate (cAMP) phosphodiesterase (PDE) inhibitor rolipram to an isolated soluble PDE from rat brain. Abstract to the 1st Symposium on Cyclic Nucleotide Phosphodiesterases, Houston, Texas.

Schneider H. H., Schmiechen R., Brezinski M., and Siedler J. (1986) Stereospecific binding of the antidepressant rolipram to brain protein structures. *Eur. J. Pharmacol.* **127**, 105–115.

Schneider L. S., Severson J. A., and Sloane R. B. (1985) Platelet ^{3}H-imipramine binding in depressed elderly patients. *Biol. Psychiatry* **20**, 1234–1237.

Schoemaker H., Pimoule C., Arbilla S., Scatton B., Javoy-Agid F., and Langer S. Z. (1985) Sodium-dependent [^{3}H]-cocaine binding associated with dopamine uptake sites in the rat striatum and human putamen: Decrease after dopaminergic denervation and in Parkinson's disease. *Naunyn Schmiedebergs Arch. Pharmacol.* **329**, 227–235.

Schoemaker H., Segonzac A., and Langer S. Z. (1986) Characterization of ^{3}H-paroxetine binding to the 5HT transportrer in human platelets. *Symposia in Neuroscience; Modulation of Central and Peripheral Transmitter Function* (Biggio G., Spano P. F., Toffano G., and Gessa G. L., eds.) Liviana Press, Padova.

Schomig E. and Bonisch H. (1986) Solubilization and characterization of the [^{3}H]-desipramine binding site of rat phaeochromocytoma cells (PC-12 cells). *Naunyn Schmiedebergs Arch. Pharmacol.* **334**, 412–417.

Segonzac A., Tateishi T., and Langer S. Z. (1984) Saturable uptake of [^{3}H]-tryptamine in rabbit platelets is inhibited by 5-hydroxytryptamine uptake blockers. *Naunyn Schmiedebergs Arch. Pharmacol.* **328,** 33–37.

Segonzac A., Raisman R., Tateishi T., Schoemaker H., Hicks P. E., and Langer S. Z. (1985a) Tryptamine, a substrate for the serotonin transporter in human platelets, modifies the dissociation kinetics of [^{3}H]imipramine binding: Possible allosteric interaction. *J. Neurochem.* **44,** 349–356.

Segonzac A., Schoemaker H., Tateishi T., and Langer S. Z. (1985b) 5-Methoxytryptoline, a competitive endocoid acting at [^{3}H]imipramine recognition sites in human platelets. *J. Neurochem.* **45,** 249–256.

Segonzac A., Schoemaker H., and Langer S. Z. (1987) Temperature-dependence of drug interaction with the platelet 5HT transporter: A clue to the imipramine selectivity paradox. *J. Neurochem.* **48,** 331–339.

Sette M., Raisman R., Briley M. S., and Langer S. Z. (1981) Localization of tricyclic antidepressant binding sites on serotonin nerve terminals in rat hypothalamus. *J. Neurochem.* **37,** 40–42.

Sette M., Briley M. S., and Langer S. Z. (1983a) Complex inhibition of ^{3}H-imipramine binding by serotonin and non-tricyclic serotonin uptake blockers. *J. Neurochem.* **40,** 622–628.

Sette M., Ruberg M., Raisman R., Scatton B., Zivkovic B., Agid Y., and Langer S. Z. (1983b) [^{3}H]-Imipramine binding in subcellular fractions of the rat cerebral cortex after chemical lesion of serotoninergic neurons. *Eur. J. Pharmacol.* **95,** 41–51.

Severson J. A., Woodward J. J., and Wilcox R. E. (1986) Subdivision of mouse brain [^{3}H]-imipramine binding based on ion dependence and serotonin sensitivity. *J. Neurochem.* **46,** 1743–1754.

Stanley M., Virgilio S., and Gershon S. (1982) Tritiated imipramine binding sites are decreased in the frontal cortex of suicides. *Science* **216,** 1337–1339.

Stockert M., Veiga A., Butman S., and De Robertis E. (1984) Pre- and post-synaptic localization of ^{3}H-imipramine binding sites in rat cerebral cortex. *J. Recep. Res.* **4,** 755–771.

Suranyi-Cadotte B., Wood P. L., Nair N. P. V., and Schwartz G. (1982) Normalization of platelet (^{3}H)imipramine binding in depressed patients during remission. *Eur. J. Pharmacol.* **85,** 357–358.

Suranyi-Cadotte B. E., Gauthier S., Lafaille F., Deflores S., Dam T. V., Nair N. P. V., and Quirion R. (1985a) Platelet ^{3}H-imipramine binding distinguishes depression from Alzheimer dementia. *Life Sci.* **37,** 2305–2311.

Suranyi-Cadotte B. E., Quirion R., Nair N. P. V., Lafaille F., and Schwartz G. (1985b) Imipramine treatment differentially affects platelets ^{3}H-imipramine binding and serotonin uptake in depressed patients. *Life Sci.* **36,** 795–799.

Suranyi-Cadotte B., Lafaille F., Schwartz G., Nair N. P. V., and Quirion R. (1985c) Unchanged platelet ^{3}H-imipramine binding in normal subjects after imipramine administration. *Biol. Psychiatry* **20,** 1237–1240.

Tang S. W. and Morris J. M. (1985) Variation in human platelet ^{3}H-imipramine binding. *Psychiatry Res.* **16,** 141–146.

Tanimoto K., Maeda K., and Terada T. (1985) Alteration of platelet [^{3}H]imipramine binding in mildly depressed patients correlates with disease severity. *Biol. Psychiatry* **20,** 340–343.

Tran V. T., Lebovitz R., Toll L., and Snyder S. H. (1981) [^{3}H]-Doxepin interactions with histamine H_1-receptors and other sites in guinea pig and rat brain homogenates. *Eur. J. Pharmacol.* **70,** 501–509.

VonVoigtlander P. F. and Losey E. G. (1978) 6-Hydroxydopa depletes both brain epinephrine and norepinephrine: Interactions with antidepressants. *Life Sci.* **23,** 147–150.

Wagner A., Aberg-Wistedt A., Asberg M., Ekqvist B., Martensson B., and Montero D. (1985) Lower ^{3}H-imipramine binding in platelets from untreated depressed patients compared to healthy controls. *Psychiatry Res.* **16,** 131–139.

Weizman R., Carel C., Tyano S., and Rehavi M. (1985) Decreased high affinity ^{3}H-imipramine binding in platelets of enuretic children and adolescents. *Psychiatry Res.* **14,** 39–46.

Weizman R., Carmi M., Tyano S., Apter A., and Rehavi M. (1986) High affinity [^{3}H]imipramine binding and serotonin uptake to platelets of adolescent females suffering from anorexia nervosa. *Life Sci.* **38,** 1235–1242.

Wennogle L. P. and Meyerson L. R. (1983) Scrotonin modulates the dissociation of [^{3}H]-imipramine from human platelets recognition sites. *Eur. J. Pharmacol.* **86,** 303–307.

Wennogle L. P., Beer B., and Meyerson L. R. (1981) Human platelet imipramine recognition sites: Biochemical and pharmacological characterization. *Pharmacol. Biochem. Behav.* **15,** 975–982.

Wennogle L. P., Ahston R. A., Schuster D. I., Murphy R. B., and Meyerson L. R. (1985) 2-Nitroimipramine: A photoaffinity probe for the serotonin uptake/tricyclic binding site complex. *EMBO J.* **4,** 971–977.

Whitaker P. M. and Cross A. J. (1980) ^{3}H-Mianserin binding in calf caudate: Possible involvement of serotonin receptors in antidepressant drug action. *Biochem. Pharmacol.* **29,** 2709–2712.

Whitaker P. M., Warsh J. J., Stancer H. C., Persad E., and Vint C. K. (1984) Seasonal variation in platelet ^{3}H-imipramine binding: Comparable values in control and depressed populations. *Psychiatry Res.* **11,** 127–131.

Wood P. L., Suranyi-Cadotte B. E., Nair N. P. V., Lafaille F., and Schwartz G. (1983) Lack of association between [^{3}H]imipramine binding sites and uptake of serotonin in control, depressed and schizophrenic patients. *Neuropharmacology* **22,** 1211–1214.

Structure–Activity Relationships at the Benzodiazepine Receptor

Pier A. Borea, Thomas A. Hamor, and Ian L. Martin

1. Introduction

There can be little doubt that the benzodiazepines represent one of the major successes of the pharmaceutical industry. Since their introduction into clinical practice in 1960, they have become the most frequently prescribed of all psychotropic drugs; at present there are over 35 such compounds available in various countries throughout the world.

Their discovery owes much to serendipity. Sternbach (1982) recounts that in the mid-1950s considerable emphasis was being placed on the putative clinical value of tranquilizers. The group at Roche engaged themselves in a search for compounds with such properties. Largely because of the interesting chemical possibilities, work was initiated on a series of benzheptoxdiazines, compounds with which Sternbach had some experience as a result of his involvement with dyestuff chemistry some 20 years previously. Modifications were made to the structure of these compounds such that basic side chains could be introduced into the heterocyclic ring by their reaction with secondary amines. In the course of such studies it was discovered, however, that the compounds were not heptoxdiazines at all, but quinazoline-3-oxides, a completely new class of compound. Unfortunately, the compounds possessed no interesting biological profile. Some time later it was found that reaction of the quinazoline-*N*-oxides with primary rather than secondary amines led to a heterocyclic ring enlargement, and, instead of the expected product of such a reaction being a six-membered pyrimidine ring, a seven-membered diazepine ring was formed. Subsequent pharmacological studies showed that this compound was indeed of considerable pharmacological interest. Further chemical studies with this skeleton allowed patent application to be filed in 1958, and this was subsequently granted one year later. The first of the benzodiazepines, chlordiazepoxide, was introduced into the market place in 1960, just three and a half years after the initial pharmacological studies had been carried out (Randall, 1982).

Since that time an enormous amount of effort has been devoted to the study of the chemistry of these compounds. Over 2000 analogs have been synthesized, and their pharmacology investigated, by Hofmann-La Roche alone, whereas compounds with heterocyclic rings fused to the "a" and "d" faces of the basic skeleton have been investigated by Upjohn (Hester et al., 1971). The compounds possess an interesting and unique pharmacology. They are anticonvulsant, anxiolytic, sedative/hypnotic, and muscle relaxant; they exhibit a high therapeutic index and have proved to be extremely safe in overdose. The compounds are essentially devoid of peripheral side effects, and it is these attributes that have made the compounds such a clinical and commercial success.

It is clear from investigations that have been carried out to date that these compounds produce their effects through interaction with specific receptors in the mammalian central nervous system. It is the structure–activity relationships at this site to which this review is devoted. The pharmacology of the benzodiazepine receptor has been addressed in this series previously (Martin, 1986), and the following section is included to contain only those essential details to allow the reader to appreciate the main subject matter of this chapter. Following this, detailed consideration is given to the structural characteristics of ligands for the benzodiazepine receptor before the relationships between structure and biological activity are addressed.

2. Mechanisms of Action at the Benzodiazepine Receptor

The first indication concerning the possible mechanism of action of the benzodiazepines was obtained from the electrophysiological studies of Schmidt et al. (1967). They found that diazepam was able to increase primary afferent depolarization (PAD) in the cat spinal cord. It is now known that presynaptic inhibition underlies PAD, and this in turn is dependent on the major inhibitory amino acid transmitter gamma-aminobutyric acid (GABA). Since that time it has become clear that the benzodiazepines increase GABA-mediated transmission in many areas of the mammalian CNS (*see* Haefely, 1984). By the mid-1970s, there was an accumulating body of electrophysiological, biochemical, and behavioral evidence to suggest that the whole gamut of the pharmacological actions of the benzodiazepines was mediated through facilitation of the actions of GABA acting at postsynaptic receptors (Costa et al., 1975; Haefely et al., 1975).

GABA produces its effects by interaction with two different types of receptor (Bowery et al., 1984). The effects of the benzodiazepines are restricted to those effects of GABA that are mediated by the so called GABA-A receptor subtype. The direct consequence of activation of this receptor on the cell membrane is to increase the probability of anion channels opening in that membrane. As a result chloride flows down its electrochemical gradient and, in general, causes membrane hyperpolarization. The effect of the benzodiazepines appears to be to increase the frequency of such channel opening in response to a given GABA stimulus (Study and Barker, 1981). The benzodiazepines themselves are not able to increase the probability of this channel opening in the absence of a GABA stimulus; they do not interact directly with the recognition site for the amino acid, but with a different site, on the same protein complex, that is allosterically linked to it.

In 1977 specific high-affinity binding sites for the benzodiazepines were discovered in the mammalian CNS (Squires and Braestrup, 1977; Mohler and Okada, 1977). The distribution and ligand recognition properties of these sites indicated that they were indeed part of the pharmacological receptor through which these compounds produced their effects (Braestrup and Squires, 1978; Mohler and Okada, 1978). Binding sites for certain benzodiazepines were also initially found in peripheral tissues, but the recognition properties of these sites were quite distinct from those identified in the CNS. These so-called "peripheral type" binding sites have since been found to occur in the CNS, though their function, in both locations, remains unclear.

The initial studies with the benzodiazepine receptor indicated that it recognized only those compounds of the benzodiazepine class. Further, the interaction of these ligands with this receptor appeared to obey simple mass action kinetics, as expected from a homogenous receptor population. These initial conclusions turned out to be falacious. The first nonbenzodiazepine compound found to displace the benzodiazepines from their binding sites on CNS membranes was the triazolopyridazine Cl 218872. Although this compound was able to completely displace either diazepam or flunitrazepam from rat CNS membranes, the Hill coefficient indicated a complexity in the interaction (Squires et al., 1979). Subsequently another chemical class, the esters of β-carboline-3-carboxylic acid, have also been shown to exhibit similar displacement characteristics in certain regions of the mammalian CNS (Nielsen and Braestrup, 1980). The hypothesis put forward to explain this data suggested that there were two populations of benzodiazepine binding sites, both of which were able to bind the classical benzodiazepines with the same affinity. Both the triazolopyridazines and the β-carboline-3-carboxylic acid esters, however, could differentiate between these two populations of sites, since they exhibited different affinities for each. The argument was then completed by assuming that the different subtypes of receptor, Bz1 and Bz2, were differentially located in the mammalian CNS; the cerebellum consisted almost entirely of the Bz1 subtype, whereas the hippocampus comprised roughly equal proportions of both types of site. It is important to realize, however, that alternative explanations of this biochemical data are equally plausible, and additional evidence for the multiple receptor hypothesis must be sought (Martin et al., 1983; Chiu and Rosenberg, 1983).

Despite the arguments that have continued over the intervening years, there are few that would now dispute that the receptor population in the mammalian CNS is heterogeneous, in the biochemical sense at least. Ligand binding characteristics already mentioned, photoaffinity labeling experiments (Sieghart and Karobath, 1980), and the dissociation kinetics of Ro 15-1788 (Brown and Martin, 1984b), all suggest that the receptor population can be distinguished in certain areas of the CNS. There remains, however, no evidence that the functional characteristics of these putative receptor subtypes can be distinguished.

The nonbenzodiazepine ligands produced yet more surprises, which are unrelated to the problem of receptor heterogeneity. The classical benzodiazepines have a specific pharmacodynamic profile; they all exhibit anticonvulsant, anxiolytic, and sedative/hypnotic activity. Ethyl β-carboline-3-carboxylate, however, was shown to be proconvulsant and anxiogenic (Tenen and Hirsch, 1980; Cowan et al., 1981; Dorow et al., 1983). It is now clear that within the β-carboline series compounds exist that span the whole range of activities from the full agonist ZK 93423, with a spectrum of activity similar to that of the classical benzodiazepines, through to the potent convulsant DMCM (see later section for structures). In the middle of the range lies a compound with essentially no intrinsic activity—ZK 93426 (Braestrup et al., 1983; 1984).

The data have been rationalized in terms of the three-state model that was put forward independently by several groups (Polc et al., 1982; Nutt et al., 1982; Jensen et al., 1983; Prado de Carvalho et al., 1983). It assumes that the benzodiazepine receptor exists in three energy states. The first, when occupied, results in no effect being observed on GABA activation of chloride channel opening: the antagonist state. The second, occupied by the so-called agonists, results in an increased coupling between GABA and its effector mechanism. The third state is preferred when the receptor is occupied with inverse agonists; it results in a decreased coupling between the amino acid and the ion channel, which produces its effects on neuronal excitability. The concept can be extended to include the possibility of partial agonists and partial inverse agonists.

Of particular concern to us here, however, are the recognition properties of the binding site(s) for these ligands, and one series of observations has proved of considerable value in this regard. Incubation of flunitrazepam with CNS membrane preparations, followed by irradiation of the material with UV light, results in the irreversible attachment of a fragment of flunitrazepam to the protein matrix in the vicinity of the recognition site (Battersby et al., 1979; Mohler et al., 1980). This procedure of "photoaffinity labeling" of the benzodiazepine receptor has proved extremely useful to probe further the mechanisms by which the different ligand classes interact with the receptor complex.

This photoaffinity labeling of the benzodiazepine receptor causes a drastic modification of the recognition properties of the binding site for the classical agonist benzodiazepines; up to 90% of the high-affinity sites are lost. Saturation analysis with [^{3}H]-ethyl β-carboline-3-carboxylate, the propyl ester, or [^{3}H]-Ro 15-1788, subsequent to photoaffinity labeling, suggest that the recognition site for these ligands is not so compromised. If the modified receptor population is labeled with one of these radioligands, it is clear that the classical benzodiazepine can still cause displacement, but with a much reduced affinity compared to that found in membranes not so modified (Brown and Martin, 1982; Gee and Yamamura, 1982; Hirsch, 1982; Karobath and Supavilai, 1982; Mohler, 1982).

This observation has led to some interesting speculation about the recognition site characteristics of the different ligand classes. The benzodiazepines can be classed as agonists at their receptor; they cause a clear functional response. The interaction of an agonist with its receptor can be envisaged as an initial recognition process of the ligand, and then subsequently, or in some concerted fashion, a conformational change occurs in the protein as the first stage in the initiation of the effector sequence of events that leads to an overt response (Franklin, 1980). A marked loss in the affinity of the agonist benzodiazepines, as a result of photoaffinity modification of the receptor population, could be the result of changes in the recognition site process or the effector sequence. It has been proposed that the increase in the affinity of compounds for the benzodiazepine receptor, seen in vitro on the addition of GABA to the incubation medium, is indicative of an agonist profile (Braestrup and Nielsen, 1981; Mohler and Richards, 1981; Doble et al., 1982; Fujimoto et al., 1982; Skolnick et al., 1982). It has now been shown that the GABA facilitation of benzodiazepine binding remains unimpaired after photoaffinity labeling, suggesting that the modification that occurs is the result of the modification of the recognition site process and not its effector sequence (Brown and Martin, 1984a).

The recognition site for the benzodiazepines is present on the GABA-A receptor-ionophore complex, and it is now clear that this complex exhibits several additional accessory binding sites that are able to modulate its activity (Olsen, 1981). Biochemical interactions between these sites can be demonstrated in vitro, and presumably reflect the complicated allosteric manner in which they control chloride channel gating in vivo. The sites with which these diverse compounds interact are distinct and separate from those that recognize GABA or the benzodiazepines.

In parallel with the pharmacological characterization of the receptor complex, attempts were being made by the biochemists to purify the receptor protein involved. The evidence from the neurochemical studies in which the plethora of interactions could be demonstrated indicated that the binding sites for many of these compounds were present within a single protein complex (Sigel and Barnard, 1984). Radiation inactivation studies in CNS membranes indicated that the molecular weight of the total complex was about 230,000 daltons (Chang and Barnard, 1982).

The receptor complex has now been solubilized from the CNS of a number of mammalian species and subsequently purified using affinity column chromatographic procedures (for review, *see* Stephenson and Barnard, 1986). The purified material appears to be a multisubunit complex consisting of at least two subunits, alpha and beta, with molecular weights of 53 and 55 kdalton, respectively (Sigel et al., 1983; Olsen et al., 1984; Barnard et al., 1984; Schoch et al., 1984). The binding site for the benzodiazepines appears to be on the alpha subunit. There have been no reports of a successful functional reconstitution of the purified receptor protein into artificial lipid bilayers, though the receptor protein has been successfully cloned and the amino acid sequence of the sub-units determined (Barnard, personal communication).

It must be clear from the brief summary given above that the benzodiazepines produce their effects by interaction with specific binding sites that are located on a complicated membrane receptor protein. The way in which ligands acting at the benzodiazepine binding site modulate the activity of the total complex is still poorly understood. Here we restrict ourselves to a discussion of the relationship between the structural characteristics of these ligands and their biological activity. To do this we have limited our discussion largely to those benzodiazepines that are marketed in the UK and a small number of other compounds that are particularly important in the clarification of these structure–activity relationships.

3. Structural Characteristics

3.1. Introduction

Currently 20 benzodiazepines [*see* formulae (1), (2.1)–(2.11), (3.1), (3.2), and (4)–(9) in Table 1] are marketed in the UK for clinical use (MIMS, 1987). Considering these as representative of the pharmacologically most useful compounds, it is instructive to consider commonalities in their chemical and molecular structures.

All except bromazepam (6) have a phenyl substituent at the 5-position; all except clobazam (8) are 1,4-benzodiazepines, and all except chlordiazepoxide (1), alprazolam (3.1), triazolam (3.2), loprazolam (4), midazolam (5), and medazepam (7) have a C2 ketonic oxygen atom. Apart from these eight and ketazolam (9), the remaining eleven compounds [(2.1)–(2.11)] are all 5-phenyl-1,4-benzodiazepin-2-ones with different substituents at the 1-, 3-, and 7-positions and at the *ortho* position of the 5-phenyl ring.

The nitrogen atom at the 1-position generally carries either hydrogen or a methyl group, but flurazepam (2.5) and prazepam (2.10) have larger groups at the 1-position. C3 is generally un-substituted, but clorazepate potassium (2.3) has a carboxylate group, and lorazepam (2.6), lormetazepam (2.7), oxazepam (2.9), and temazepam (2.11) have a hydroxy substituent at this site. The 7-substituent is Cl in 15 cases, NO_2 in 4, and Br in 1. The *ortho* position of the 5-phenyl ring is unsubstituted in 11 of the 20

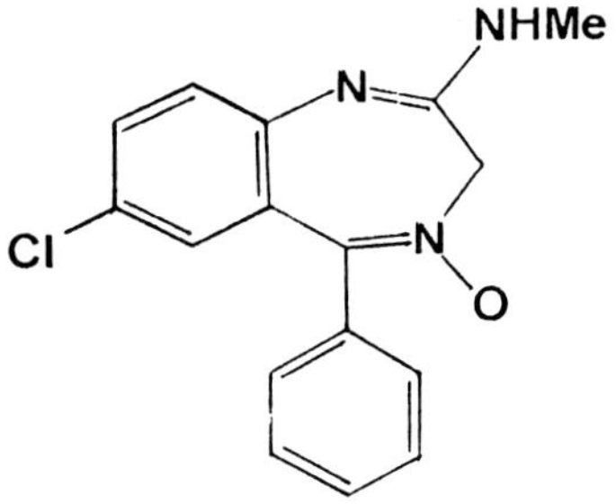

(1) Chlordiazepoxide (figure 11).

Table 1 *(continued)*

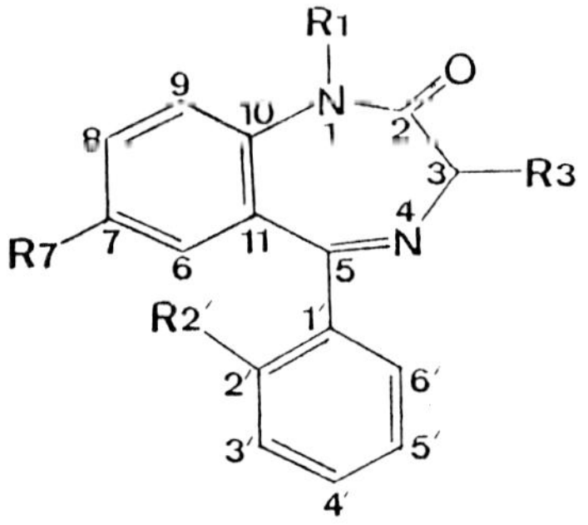

(2)

ompound ɔ.	Fig. no. in text	Drug name	R_1	R_3	R_7	R_{21}
1)	Fig. 1	Diazepam	Me	H	Cl	H
2)	Fig. 2	Clonazepam	H	H	NO_2	Cl
3)		Clorazepate	H	COOK	Cl	H
4)	Fig. 3	Flunitrazepam	Me	H	NO_2	F
5)		Flurazepam	$(CH_2)_2NEt_2$	H	Cl	F
6)	Fig. 4	Lorazepam	H	OH	Cl	Cl
7)		Lormetazepam	Me	OH	Cl	Cl
8)	Fig. 5	Nitrazepam	H	H	NO_2	H
9)	Fig. 6	Oxazepam	H	OH	Cl	H
10)	Fig. 7	Prazepam	$CH_2.(c\text{-}C_3H_5)$	H	Cl	H
11)	Fig. 8	Temazepam	Me	OH	Cl	H
12)			Me	Me	NO_2	F
13)			H	Me	Cl	H
14)			CH(Ph)Me	Me	Cl	H
15)			H	$CHMe_2$	Cl	H
16)			$(CH_2)_2Cl$	OEt	Cl	H
17)			$CH_2CONHMe$	H	Cl	H
18)			H	H	Cl	H
19)			$CHMe_2$	H	Cl	H
20)			CH_2Ph	H	Cl	H
21)			Et	H	Cl	H
22)			Me	H	NO_2	H
23)			CH_2OMe	H	NO_2	H

Table 1 *(continued)*

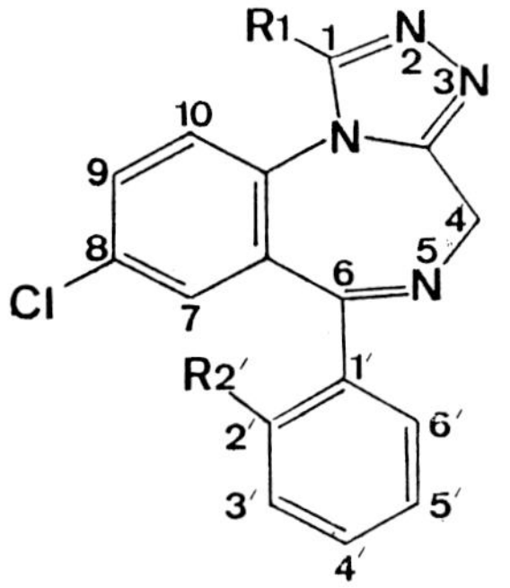

(3.1) Alprazolam R1=Me, R2'=H

(3.2) Triazolam R1=Me, R2'=Cl

(3.3) (Fig.10) Estazolam R1=H, R2'=H

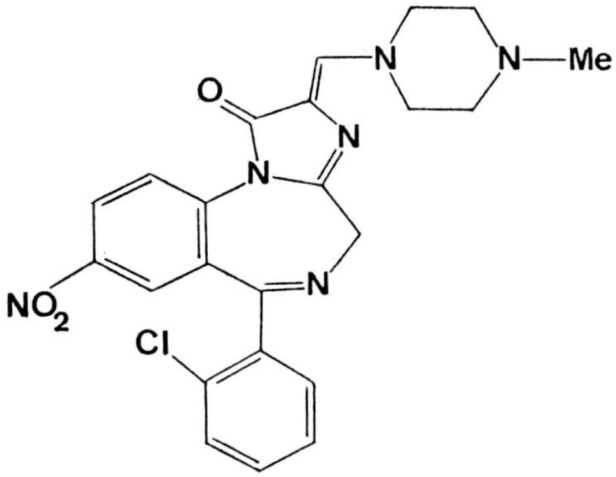

(4) Loprazolam

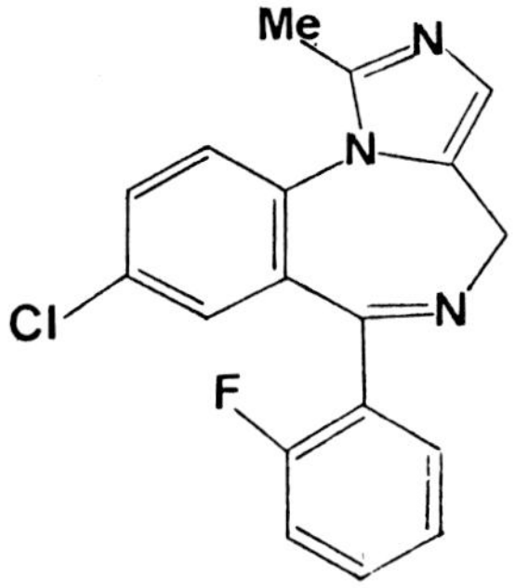

(5) Midazolam

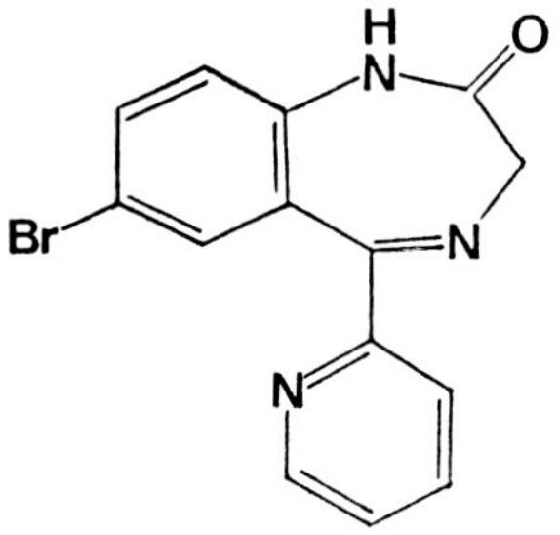

(6) Bromazepam (Fig. 9)

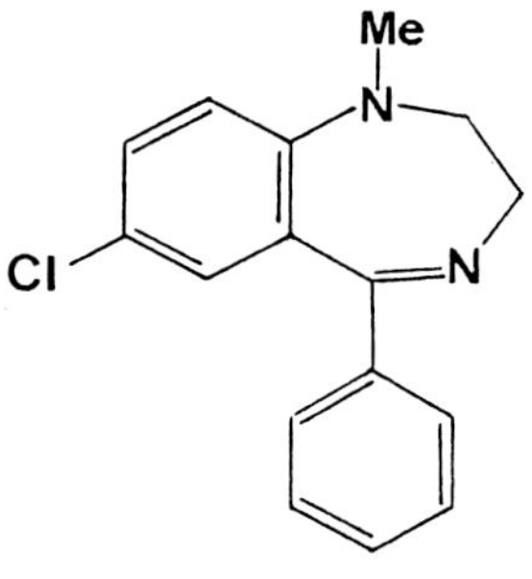

(7) Medazepam (Fig. 12)

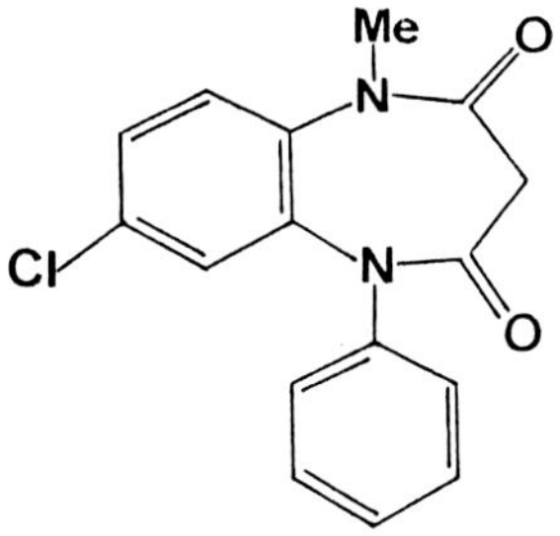

(8) Clobazam (Fig. 13)

Table 1 *(continued)*

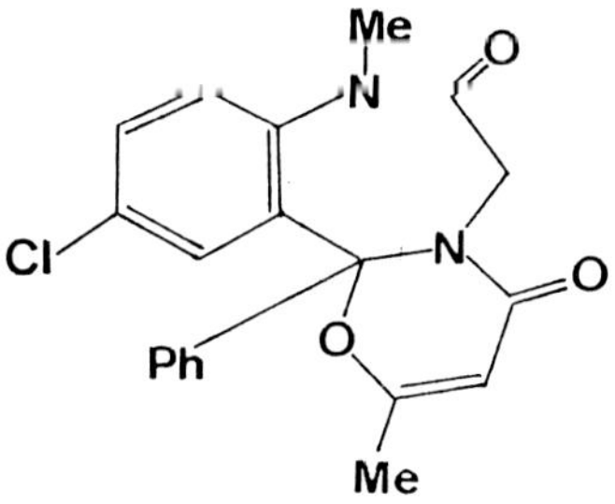

(9) Ketazolam

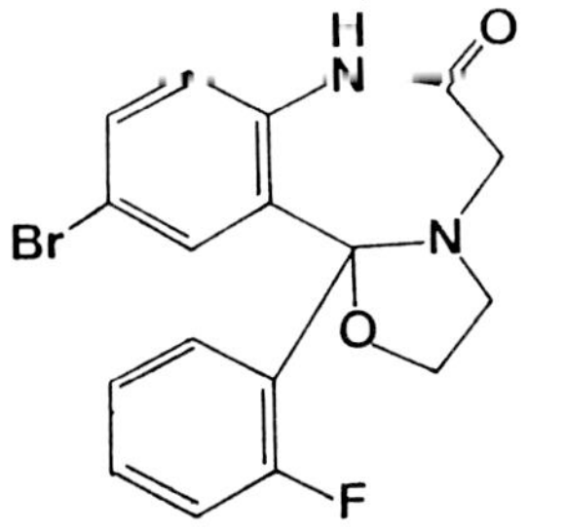

(10) (Fig. 14)

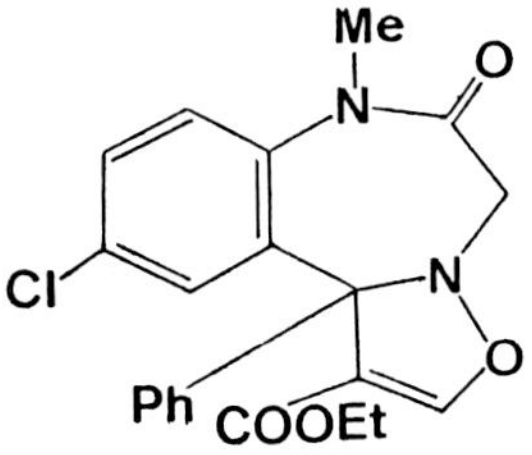

(11) (Fig. 15)

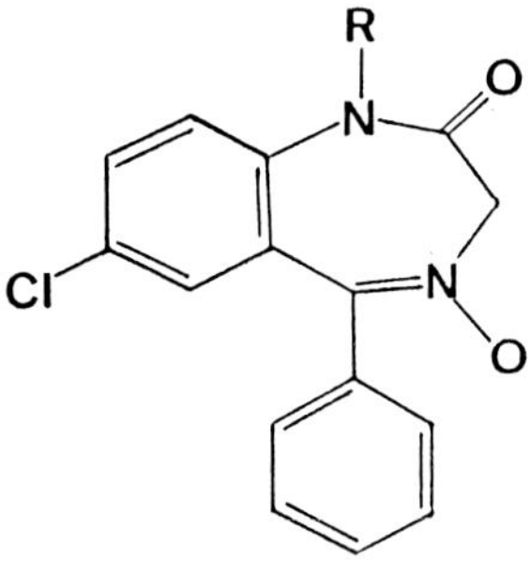

(12) $R = (Ch_2)_2\text{-}O\text{-}CO\text{-}(CH_2)_2\text{-}COOH$

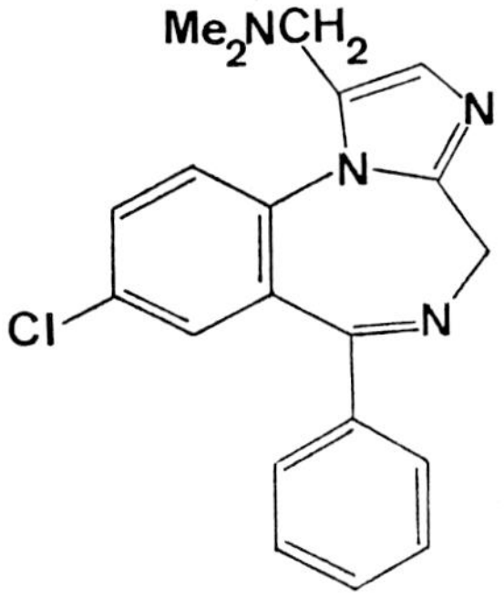

(13)

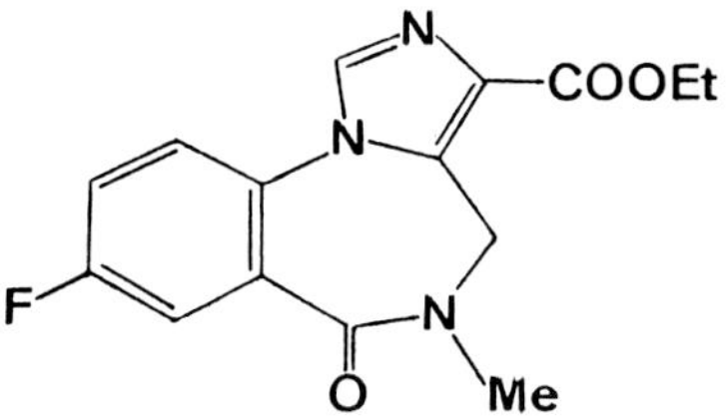

(14) Ro15-1788

Table 1 *(continued)*

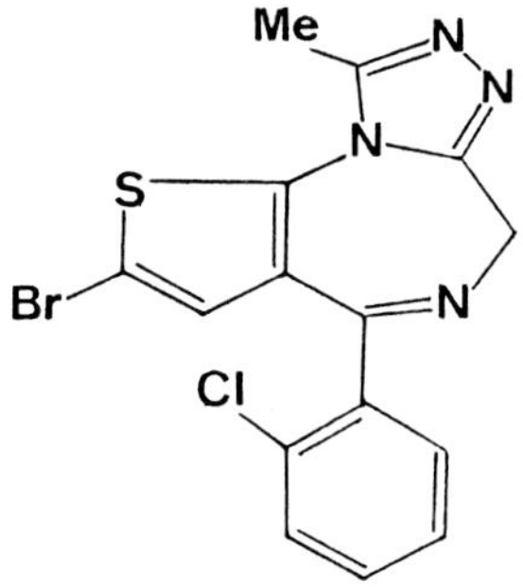

(15) Brotizolam (Fig. 16)

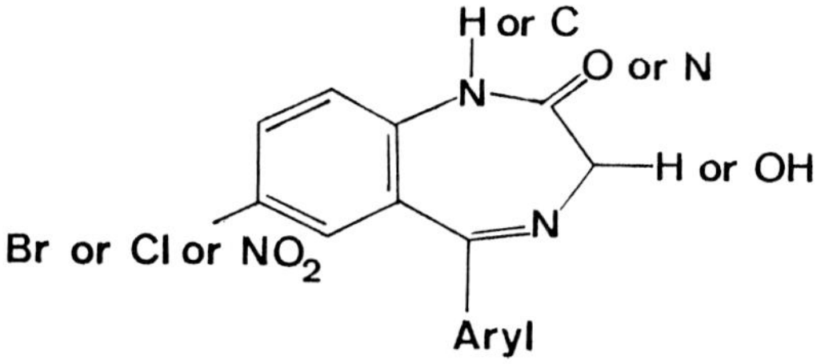

(X)

compounds, contains Cl in 5 compounds, and F in 3. Bromazepam (6) has a 5-pyridyl ring such that the heterocyclic nitrogen atom is in the *ortho* position.

Alprazolam (3.1), triazolam (3.2), loprazolam (4), and midazolam (5) contain an additional ring fused to the N1-C2 bond, the a-face of the benzodiazepine system. Ketazolam (9) also contains an additional ring, with fusion, however, being to the d-face.

The basic moiety (X) is thus common to all the clinically used (MIMS, 1987) benzodiazepines, excepting only chlordiazepoxide (1), midazolam (5), medazepam (7), clobazam (8), and ketazolam (9).

Benzodiazepines containing cyclic structures on the b- or c-faces have markedly different pharmacological properties and are not considered in this review.

By the end of 1985 the solid-state structures of 14 of the listed benzodiazepines determined by X-ray crystallography had been reported in the literature (Allen et al., 1979), as had those of a considerable number of experimental compounds covering a wide range of pharmacological activity. Many of these compounds have also been studied in solution by NMR spectroscopy, an environment that is closer to the in vivo conditions under which a drug interacts with its receptor. In the solid state, crystal packing forces may affect the geometry of flexible parts of the molecule in a different way from that because of solvent effects in solution. Confidence in the validity of the solid-state structural results is, however, gained from the fact that in all cases in which both X-ray solid-state data and NMR solution data are available for the same compound, the results agree reasonably well. Further, in a number of cases, the crystal structure contains two, three, or even four independent molecules. Chemically identical molecules are therefore exposed to different packing forces, so that the effect of these can be assessed. In general only the orientation of the aromatic ring at C5 is affected to an appreciable extent (*see* Table 2).

3.2. Solid-State Structure

X-ray structures are available for 9 of the 15 type (X) benzodiazepines listed by MIMS (1987). Computer drawings (Motherwell and Clegg, 1978) of these molecules based on the X-ray results are shown in Figs. 1–9 for diazepam (2.1) (Camerman and Camerman, 1972), clonazepam (2.2) (Chananont et al., 1979), flunitrazepam (2.4) (Butcher et al., 1983), lorazepam (2.6) (Bandoli and Clemente,

Table 2
Structural Parameters from X-Ray Crystallographic Analysis of Benzodiazepines

	$\theta 1^a$	$\theta 2^b$	$\theta 3^b$	$^c\Delta$	$^d l(N1\text{-}C2)$	$^d l(C5\text{=}C1')$	$^d t(C11\text{-}C5\text{-}C1'\text{-}C2')$	$^d t(N4\text{-}C5\text{-}C1'\text{-}C6')$
(1)	68	34	58	3.7	1.30	1.48	45	48
	70	37	57	5.8	1.30	1.47	40	40
	61	33	56	3.8	1.30	1.48	40	43
	63	33	55	7.7	1.30	1.48	45	45
(2.1)	55	38	58	9.3	1.37	1.49	25	23
(2.2)	84	34	58	4.4	1.36	1.50	66	63
	78	33	59	2.4	1.36	1.51	61	57
(2.4)	75	37	61	3.9	1.37	1.48	48	43
(2.6)	73	35	62	3.1	1.35	1.48	52	52
	81	33	58	2.9	1.31	1.49	58	58
(2.8)	62	32	60	1.3	1.36	1.49	35	34
(2.9)	54	33	63	2.3	1.36	1.49	30	29
	66	33	63	1.6	1.36	1.49	42	40
(2.10)	71	37	62	4.6	1.37	1.49	44	43
	67	40	60	5.3	1.37	1.49	29	25
(2.11)	66	34	63	2.8	1.38	1.49	40	39
	67	36	64	2.1	1.37	1.50	37	35
	59	40	63	3.5	1.37	1.49	24	21
(2.13)	53	37	59	4.6	1.33	1.50	19	15
	58	34	62	5.1	1.37	1.46	22	24
	64	34	62	4.3	1.34	1.49	30	28

(2.14)	63	39	59	8.7	1.38	1.50	33	31
(2.15)	68	31	62	3.0	1.36	1.50	47	46
(2.16)	68	38	62	2.0	1.37	1.49	32	29
(2.17)	62	36	61	4.8	1.37	1.50	32	30
(3.3)	75	31	54	3.2	1.34	1.53	48	52
(6)	60	31	60	1.9	1.35	1.50	40	40
(7)	63	44	57	15.1	1.46	1.49	26	25
	55	46	56	17.1	1.46	1.49	14	16
(8)[f]	85	41	59	5.7	1.36	1.45	63	60
(10)	85	24	65	22.4	1.37	1.52	69	119
(11)	85	51	51	18.3	1.36	1.53	−81	−143
(12)	64	42	60	3.5	1.37	1.48	30	33
(13)	66	35	56	2.2	1.36	1.49	33	30
(14)	—	35	56	8.8	1.39	—	—	—
(15)	72	29	57	2.9	1.38	1.50	−88	−86
(1)H$^+$	67	42	58	8.1	1.32	1.47	27	30
(7)H$^+$	67	46	51	23.1	1.46	1.48	35	34

[a]$\theta 1$ is the angle between the planes of the 5-phenyl ring and the fused benzo moiety.

[b]$\theta 2$ and $\theta 3$ are the stern and bow angles of the boat-shaped seven-membered ring (see text).

[c]Δ is the "deviation parameter" (see text).

[d]"l" and "t" refer to lengths and torsion angles, respectively. Estimated standard deviations are generally 0.005–0.01 Å for lengths and 0.3–1.0° for angles.

[e]The numbering of the atoms is specified in formula (2) (for certain compounds containing additional fused rings, this does not correspond to standard chemical numbering.)

[f]In clobazam (8), atom 4 is carbon and atom 5 is nitrogen.

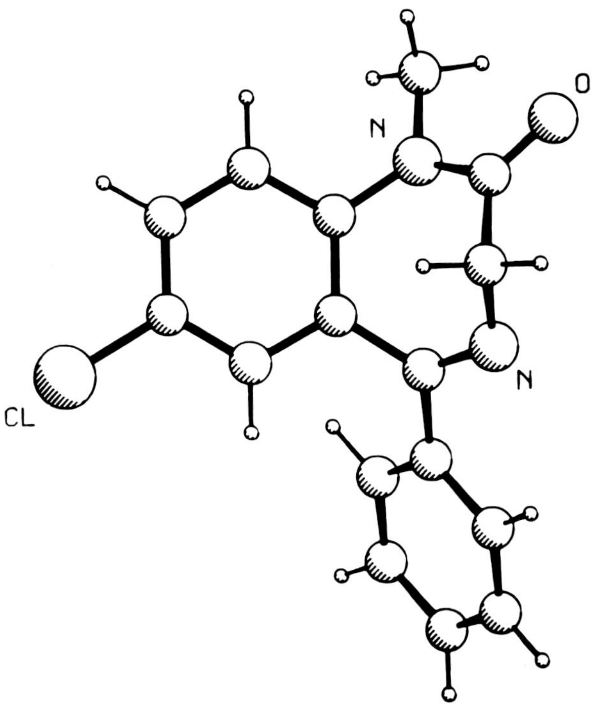

Fig. 1. Diazepam [(2.1) in Table 1)] (Camerman and Camerman, 1972).

1976), nitrazepam (2.8) (Gilli et al., 1977a), oxazepam (2.9) (Gilli et al., 1978a), prazepam (2.10) (Brachtel and Jansen, 1981), temazepam (2.11) (Galdecki and Glowka, 1980), and bromazepam (6) (Butcher et al., 1983). In each case the molecule is viewed in a direction perpendicular to the plane of the fused benzo ring.

It is noteworthy that in lorazepam (2.6), oxazepam (2.9), and temazepam (2.11), which have C3 substituted unsymmetrically, the larger hydroxy group is oriented quasi-equatorially in the solid state, as illustrated in Figs. 4, 6, and 8. (It may be noted that "flipping," or inversion of the seven-membered ring interchanges equatorial and axial substituents at C3.) A literature search reveals that in benzodiazepines of this type a C3 substituent is invariably oriented equatorially in the solid state, e.g., (2.12) (Blount et al., 1983), (2.13) (Sikirica and Vickovic, 1982), (2.14) (Ruzic-Toros et al., 1982), (2.15) (Dvorkin et al., 1982), and (2.16) (Bertolasi et al., 1982a). These results are in agreement with solution studies.

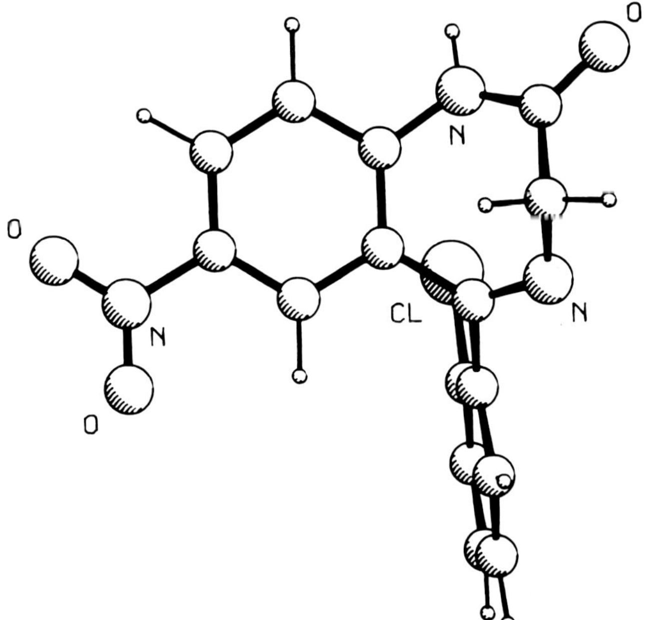

Fig. 2. Clonazepam (2.2) (Chananont et al., 1979).

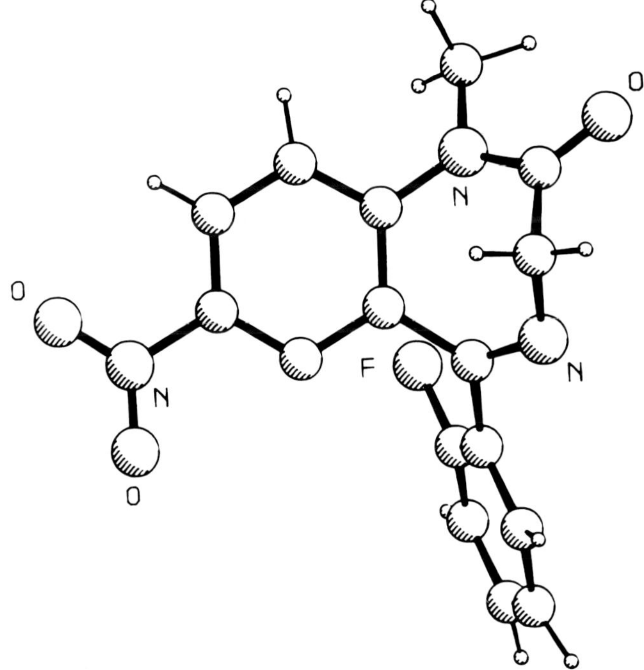

Fig. 3. Flunitrazepam (2.4) (Butcher et al., 1983).

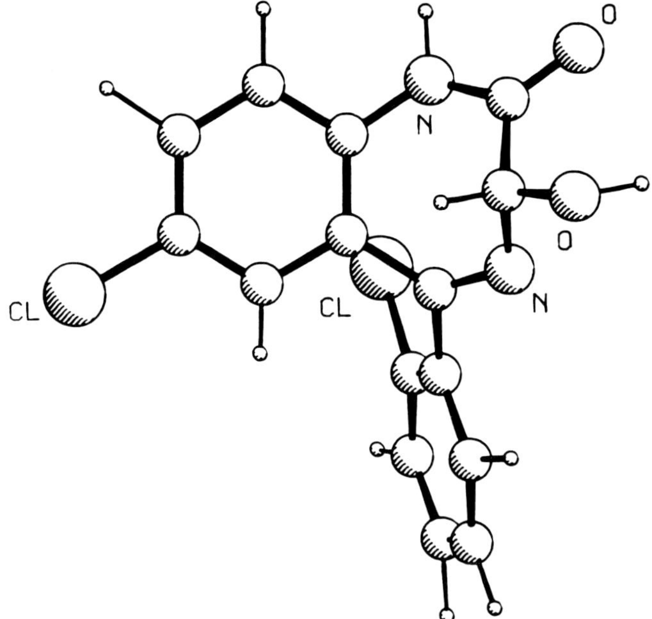

Fig. 4. Lorazepam (2.6) (Bandoli and Clemente, 1976).

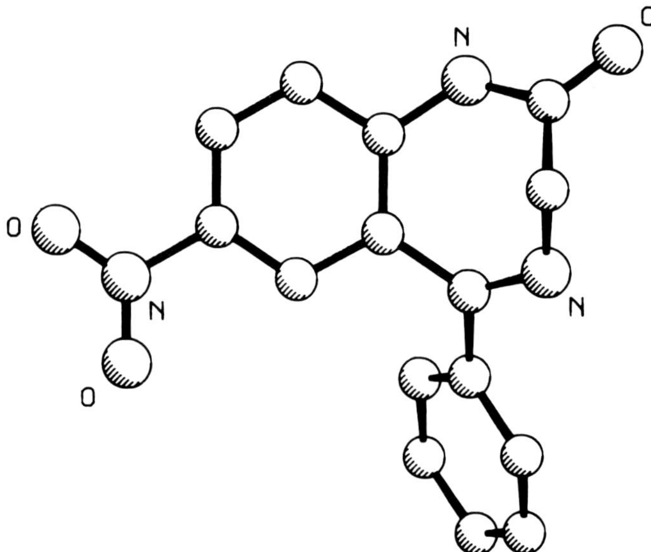

Fig. 5. Nitrazepam (2.8) (Gilli et al., 1977a) (H-atom positions not published).

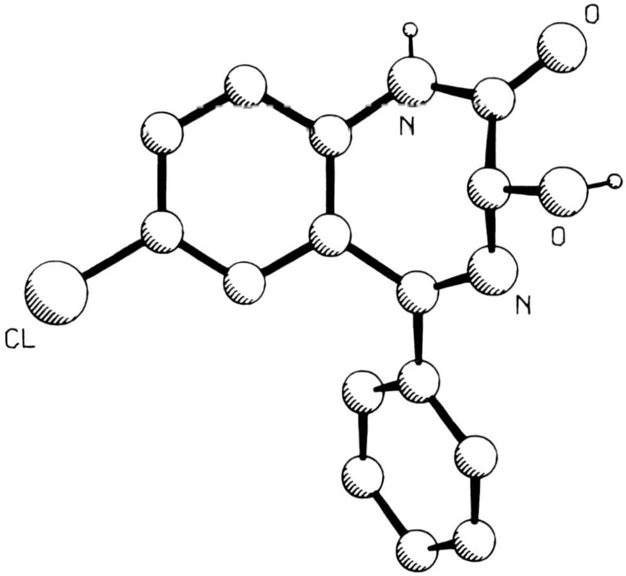

Fig. 6. Oxazepam (2.9) (Gilli et al., 1978a) (H-atom positions not published apart from those of N1—H and 3-hydroxy groups).

C3-substituted compounds are resolvable, and in a study of 3-methyl R,S pairs, in every case the S-enantiomer was found to be pharmacologically more active than the corresponding R-enantiomer (Blount et al., 1983). Our figures, therefore, depict the molecules in the configuration corresponding to the pharmacologically active 3S-enantiomer of the methyl derivatives.

At first sight the C,N framework (X) is essentially identical for all the depicted molecules, differing only in the orientation of the

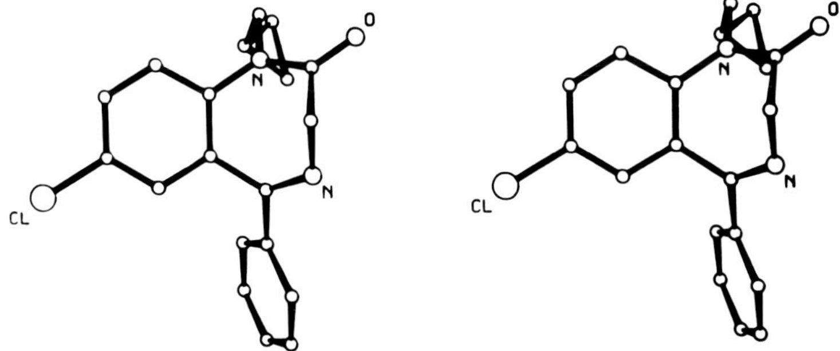

Fig. 7. Stereoscopic view of prazepam (2.10) (Brachtel and Jansen, 1981) (H-atom positions not available).

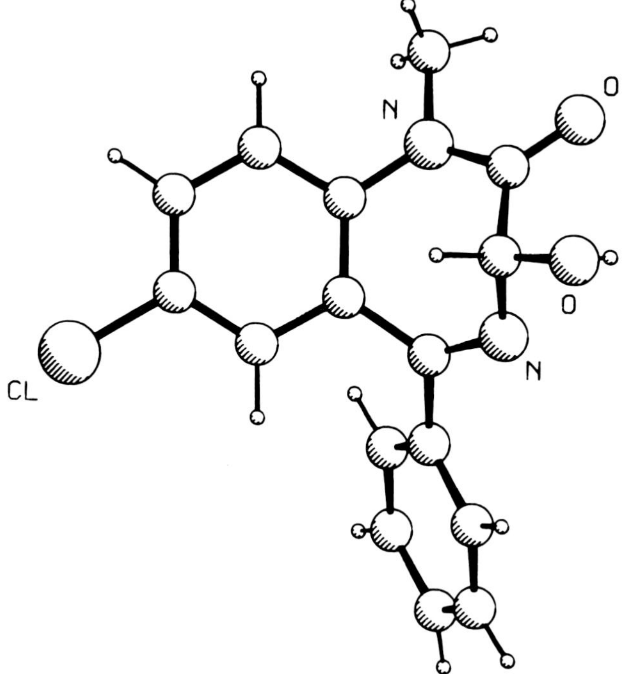

Fig. 8. Temazepam (2.11) (Galdecki and Glowka, 1980).

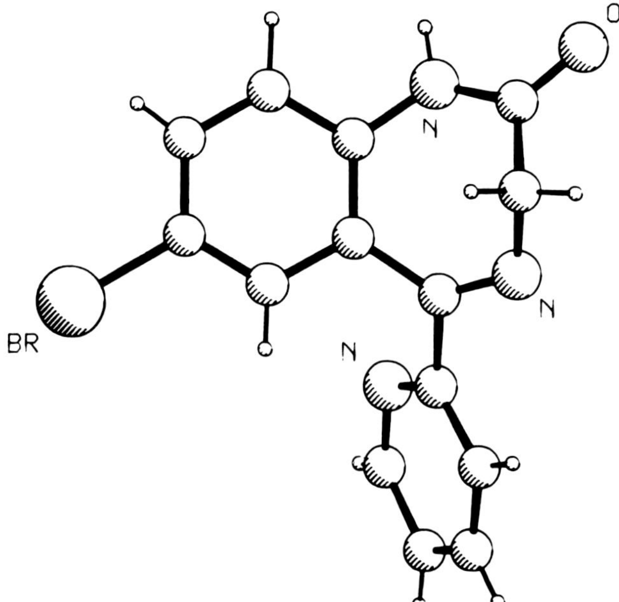

Fig. 9. Bromazepam (6) (Butcher et al., 1983).

5-aryl ring. The seven-membered heterocyclic ring adopts a cyclo-heptatriene-like boat conformation. The third double bond (in addition to the formal N4=C5 double bond and the shared aromatic bond, C10—C11) is the amide bond N1—C2. This bond partici-pates in electron delocalization between the nitrogen lone pair and the carbonyl oxygen atom and is shortened to approximately 1.36Å, about half-way between the C—N pure single and double bond lengths. The disposition of bonds at N1 is near planar, and the overall geometry of the N1—C2 bond approximates that of a normal double bond.

Molecules of this type (X) can, therefore, be described in terms of four planes and three interplanar angles. The planes are the "stern" of the boat (atoms N1, Cl0, C11, and C5), which is es-sentially coplanar with the fused benzo moiety (i.e., atoms C6, C7, C8, and C9 also lie in this plane, or very close to this plane, as does a C7-substituent), the central plane of the boat (atoms N1, C2, N4, and C5), the "bow" of the boat (atoms C2, C3, and N4), and the 5-phenyl ring [or pyridyl ring in bromazepam (6)].

Considering first the 5-aryl ring, this is oriented at angles within the range 54–84° to that of the benzo (or boat stern) plane. If the 5-phenyl ring carries an *ortho*-substituent (chloro or fluoro), in every case the substituent "points" toward the "benzo end" of the molecule, in such a way that it lies on the opposite side of this plane from C3 (*see* Figs. 2–4). Unsubstituted rings are oriented similarly, but less steeply inclined to the benzo plane (interplanar angles 54–71° for unsubstituted rings and 73–84° for substituted rings). The orientation of the 5-phenyl ring can be defined more precisely in terms of the torsion angles C11—C5—C1'—C2' and N4—C5—C1'—C6', where C2' is the atom that carries an *ortho*-substituent (if present), and C6' is the other *ortho*-carbon atom of the 5-phenyl ring. These and other selected geometric parameters are listed in Table 2. Multiple entries for a compound occur when the crystal contains more than one independent molecule.

The angles between the central plane of the boat and the stern and bow planes, respectively, fall within narrower ranges: 31–40° for the stern angles and 58–64° for the bow angles (*see* Table 2).

The three interplanar angles ($\theta1$, $\theta2$, $\theta3$, in Table 1) and the two torsion angles discussed above define the geometry of the 5-aryl-1,4-benzodiazepine system (X) to a good approximation. In all cases, however, the seven-membered ring deviates to a greater or lesser extent from the "ideal" cycloheptatriene boat conformation with mirror (C_s) symmetry. In the ideal conformation the ring torsion angles about the N1—C2, N4—C5, and C10—C11 "double" bonds are zero, and the other four torsion angles related in pairs by the mirror plane (i.e., those about C2—C3 and C3—C4, and those about C5—C11 and C10—N1) are equal in magnitude, but of opposite sign. A quantitative measure of the deviation from ideal mirror symmetry is given by the "deviation parameter," Δ, based on the asymmetry parameter of Duax et al. (1976):

$$\Delta = [1/5(T1^2 + T4^2 + T6^2 + (T2 + T3)^2 + (T5 + T7)^2)]^{1/2}$$

where T1–T7 are the ring torsion angles about the bonds N1—C2, C2—C3, C3—N4, N4—C5, C5—C11, C11—C10, and C10—N1, respectively.

The calculated values of Δ are listed in Table 2. In general, Δ is small and approaches the ideal value of zero. Compounds with a methyl group at N1 tend to show a slightly larger value (mean for five molecules 4.3°) than the N1-H compounds (mean for eight molecules 2.5°). Prazepam (2.10) (*see* Fig. 7), with a cyclopropyl-methyl group at N1, has Δ = 4.6 and 5.3° for the two independent molecules in the crystal.

Of the listed type (X) compounds containing an additional fused ring, only the crystal structure of alprazolam (3.1) has been determined (Hester et al., 1971), but structural parameters are not available. Structural parameters have, however, been reported (Kamiya et al., 1973) for the closely analogous estazolam (3.3) (*see* Fig. 10). The seven-membered ring is essentially a cycloheptatriene boat with stern and bow angles of 31 and 54°, and deviation parameter Δ = 3.2°. The bow angle is thus relatively small (see Table 2). The angle between the 5-phenyl ring and the fused benzo moiety is 75°, a slightly larger angle than is found in type (2) compounds with unsubstituted 5-phenyl rings (54–71°).

Considering the listed benzodiazepines that are not of type (X), the seven-membered ring of chlordiazepoxide (1) contains three double bonds, but moderately large deviations from mirror symmetry occur, within the range 3.7–7.7° for the four independent molecules in the crystal (Bertolasi et al., 1982b) (*see* Table 2 and Fig. 11). Medazepam (7) does not have an unsaturated function at C2, and its seven-membered ring adopts a cyclohepta-diene-like pseudo-boat conformation (Gilli et al., 1978b). The cyclo-heptatriene "deviation parameter" is, as might be expected, large (15.1 and 17.1° for two independent molecules), and the length of the N1—C2 bond is that of a single bond (*see* Table 2 and Fig. 12).

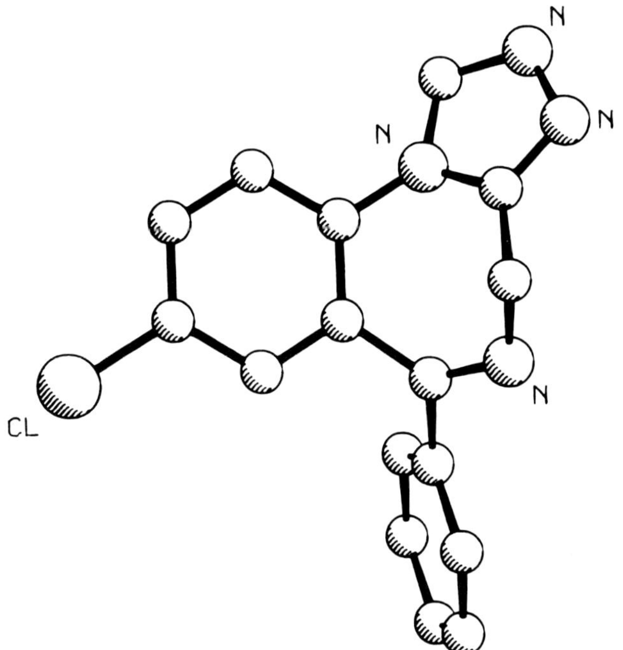

Fig. 10. Estazolam (3.3) (Kamiya et al., 1973) (H-atom positions not published).

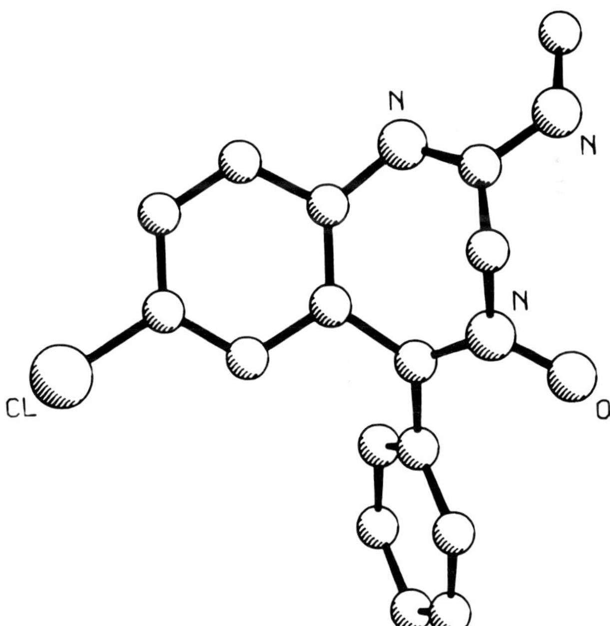

Fig. 11. Chlordiazepoxide (1) (Bertolasi et al., 1982b) (H-atom positions not published).

495

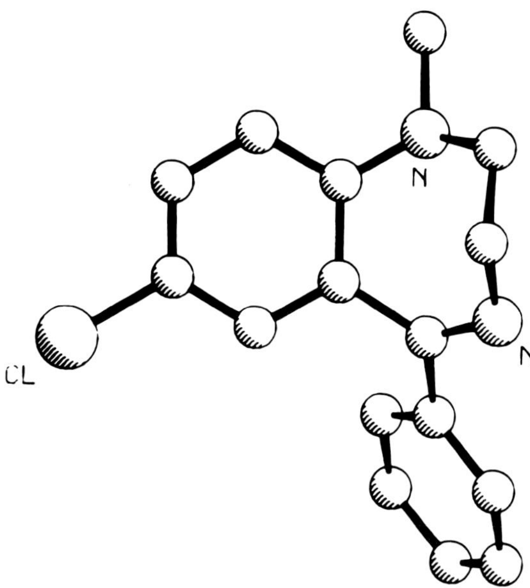

Fig. 12. Medazepam (7) (Gilli et al., 1978b) (H-atom positions not published).

The seven-membered ring of the 1,5-benzodiazepine clobazam (8) (Fig. 13) contains two amide functions, and these together
with a shared aromatic ring bond again give rise to a cycloheptatriene-like boat conformation with a marginally large stern angle
and typical bow angle and deviation parameter as shown in Table 2
(Butcher and Hamor, 1985a). The angle $\theta 1$ between the 5-phenyl
ring and the benzo moiety is 85°, which is larger than is generally
found when the 5-phenyl ring is unsubstituted. Angles of this
magnitude are, however, characteristic of compounds in which the
ring carries a chlorine substituent in the *ortho*-position.

The crystal structure of the d-face-fused ketazolam (9) has also
been determined (Szmuszkovicz et al., 1971), but structural
parameters are not available. In the analogous compounds (10) and
(11), the seven-membered ring contains only two "double" bonds,
and the deviation parameter Δ, as defined above, is large (Okada et
al., 1983; Freeman et al., 1982). Details are given in Table 2. Views
of the molecules are shown in the stereo-diagrams in Figs. 14 and
15. In both compounds the phenyl ring is steeply inclined to the
benzo plane (85°), but the geometry of substitution differs. The
disposition of bonds at C5 is essentially tetrahedral; in compound

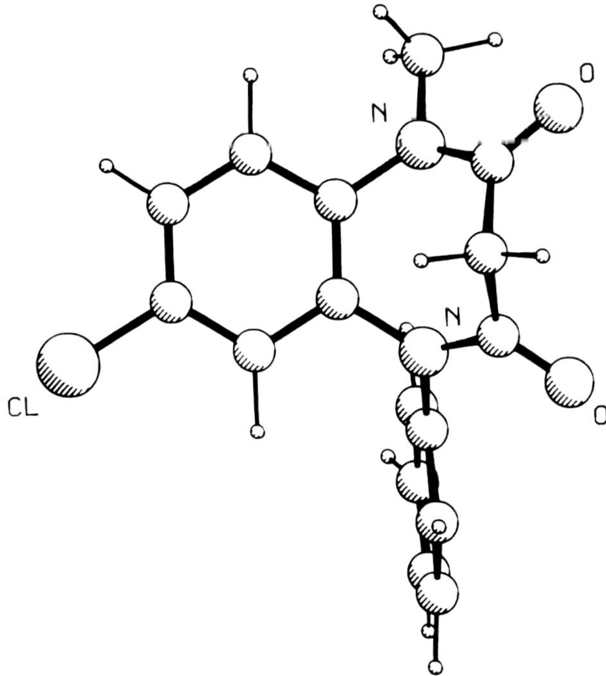

Fig. 13. Clobazam (8) (Butcher and Hamor, 1985a).

(10) the fluorophenyl ring is quasi-axial, whereas in (11) the phenyl ring is quasi-equatorial.

In addition to the structural studies on the clinically approved benzodiazepines described above, a considerable number of experimental compounds, covering a large range of pharmacological activity, have been examined by spectroscopic and x-ray crystallographic methods. The literature up to the end of 1981 has been reviewed previously (Hamor and Martin, 1983), and some of the more recent results have also, in passing, been cited above.

Further examples of recent X-ray structures include that of compound (12) (Ferretti et al., 1984), which carries a large N1 substituent. The geometry of the molecular framework does not seem to be affected to an appreciable extent, although the stern angle is relatively large (*see* Table 2). In general, large substituents at N1 do not affect the basic geometry apart from a slight increase in Δ [*see* entries for (2.10), (2.14), (2.16), and (2.17) (Chananont et al., 1980a) in Table 2]. C3 substitution also does not affect the basic framework geometry to an appreciable extent [*see* entries for (2.6), (2.9), (2.11), and (2.13)–(2.16) in Table 2].

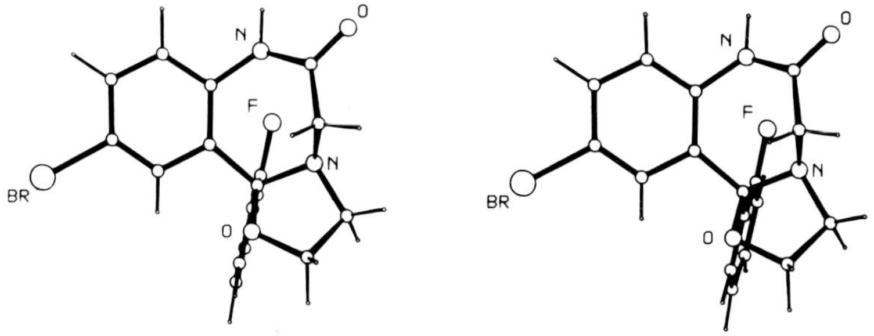

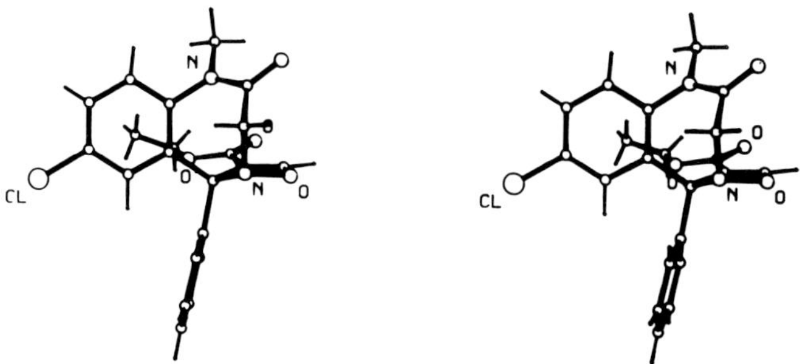

Fig. 14. Stereoscopic view of compound (10) (Okada et al., 1983).

Structural parameters have been reported for the two a-face-fused benzodiazepines (13) (Butcher and Hamor, 1984) and (14) (Codding and Muir, 1985). In (13), the geometry of the basic framework (X) is normal, apart from a marginally small (56°) bow angle, similar to that in the analogous estazolam (3.3) (*see* Table 2). Compound (14) is noteworthy in that it does not contain the usual 5-aryl ring. The N1—C2 and N4—C5 formal single bonds are both involved in N→O electron delocalization, and have lengths of 1.39 and 1.36 Å, respectively, with near planar disposition of valencies. The seven-membered ring is essentially a cycloheptatriene-like boat, but with a relatively large deviation parameter (*see* Table 2). Compound (14), although binding to the benzodiazepine receptor, has no biological effect, and is classed as a benzodiazepine antagonist.

Also containing an additional ring fused to the a-face is brotizolam (15). This is a "heterodiazepine," containing a thieno ring instead of the usual fused benzene ring, and has found use as a hypnotic (Weber et al., 1978). The thieno ring is planar, and the seven-membered ring is in the usual cycloheptatriene-like boat

Fig. 15. Stereoscopic view of compound (11) (Freeman et al., 1982).

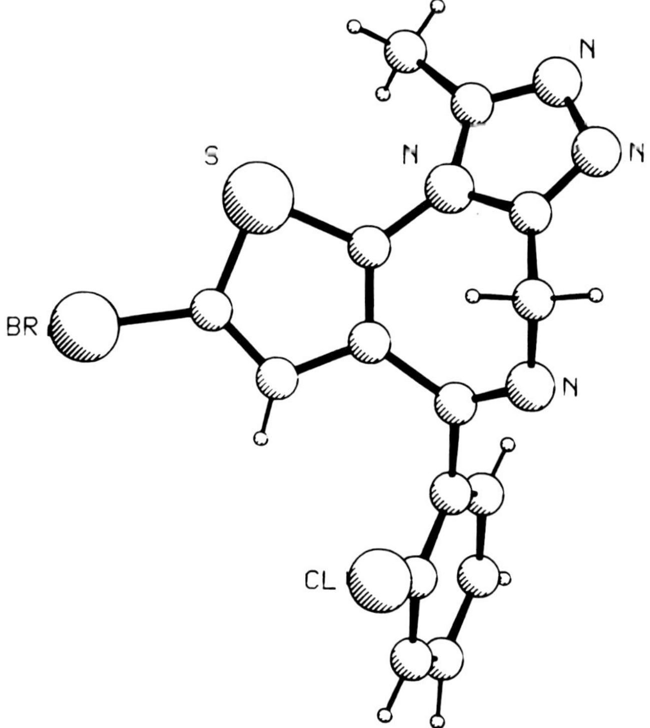

Fig. 16. Brotizolam (15) (Butcher and Hamor, 1985b).

conformation (Butcher and Hamor, 1985b), which is, however, slightly flatter than commonly found in benzodiazepines (*see* Table 2 and Fig. 16). The angle between the thieno and chlorophenyl rings is 72°, slightly smaller than found in *ortho*-chlorophenyl-1,4-benzodiazepines. It may be noted that the chloro substituent lies on the *same* side of the thieno plane as C3, whereas normally in type (X) 1,4-benzodiazepines, the 5-phenyl ring is oriented such that an *ortho*-substituent lies on the side of the benzo plane remote from C3. This difference is highlighted by the *negative* sign associated with the torsion angles C11—C5—C1'—C2' and N4—C5—C1'—C6' compared with positive values in all other cases except in the d-face fused compound (11). The torsion angles about C5—C1' for compound (10) are not directly comparable because of the different geometry of substitution of the phenyl ring. A somewhat similar situation to that in brotizolam occurs in the crystal structure of the hydrochloride salt of the 5-(2-fluorophenyl)-1,4-benzodiazepine opioid, tifluadom (Petcher et al., 1985), in which

the fluorophenyl ring is rotated by 180° from the normal orientation, with torsion angles about C5—C1', as defined above, both
–131°. Here the ring is, however, affected by rotational disorder. In
contrast, the *p*-toluenesulphonate salt of tifluadom has the normal
conformation about C5—C1', with corresponding torsion angles
60 and 56°, respectively (Petcher et al., 1985).

For two of the listed benzodiazepines, chlordiazepoxide (1)
and medazepam (7), the crystal structures of the hydrochloride
salts have also been reported (Herrnstadt et al., 1979; Chananont et
al., 1980b) [*see* Table 2 under (1)H$^+$ and (7)H$^+$, and Figs. 17 and 18].
Protonation occurs at N1 in chlordiazepoxide, but at N4 in
medazepam [NMR measurements on solutions by Kuwayama and
Yashiro (1985) indicate that protonation at N4 occurs also in diazepam (2.1)]. Moderately large changes occur in the geometry of
the seven-membered ring on protonation. Thus in chlordiazepoxide the endocyclic angle at N1 increases by approximately 2°, and
those at C2, C5, and C10 decrease by 5, 3, and 3.5°, respectively,
compared with the mean values for the four independent molecules in the crystal of the free base. Protonation also causes the

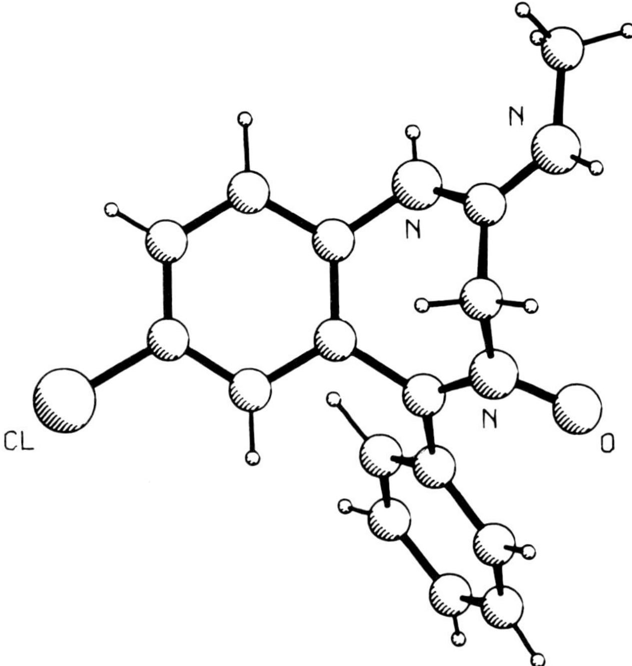

Fig. 17. Chlordiazepoxide cation (1)H$^+$ (Herrnstadt et al., 1979).

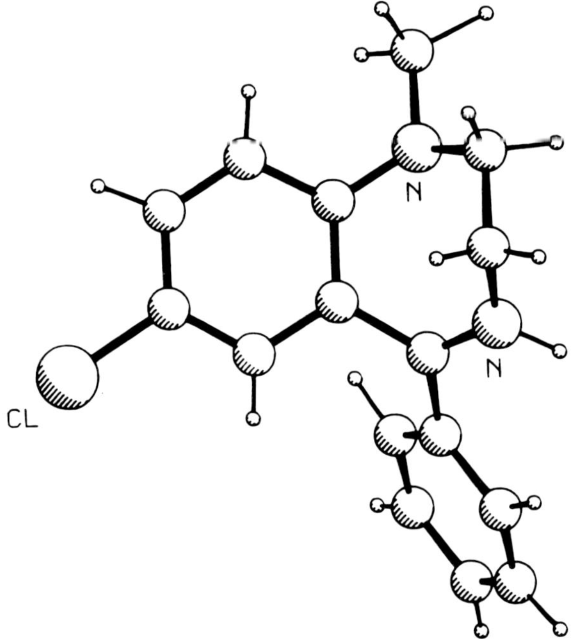

Fig. 18. Medazepam cation (7)H$^+$ (Chananont et al., 1980b).

C10—N1 and N1—C2 bonds to increase in length by 0.04 and 0.025 Å, respectively, and the exocyclic C2—N bond to decrease by 0.035 Å. Other bond lengths and angles are not affected significantly. The stern angle of the "boat" at 42° is larger by 5° than that in any of the four independent molecules of the base (*see* Table 2). In medazepam, the bond angle at N4 increases by 6.7° and that at C5 decreases by 4° on protonation, compared with the mean values for the two independent molecules of the free base. Other bond angles and the bond lengths are not affected significantly. Thus in both compounds the valence angle at the site of protonation increases by a small amount. The bow angle of the pseudoboat-shaped cycloheptadiene-like ring of medazepam decreases by 5–6° on protonation (*see* Table 2). Overall, however, the free base and cation look quite similar (compare Figs. 11 and 17 for chlodiazepoxide and Figs. 12 and 18 for medazepam).

The crystal structure analysis of alprazolam (3.1) (Hester et al., 1971) was carried out on the hemihydrobromide salt, and the crystal contains both protonated and unprotonated species; protonation occurs at N2 of the triazolo ring. In agreement with this is a ^{15}N NMR study of alprazolam (Scahill and Smith, 1983a) that indicates that N2 is the most basic nitrogen in the molecule. On the other hand a recent proton NMR analysis of estazolam (3.3) in 0.5*M* DCl solution indicates "protonation" of the imino nitrogen atom of the seven-membered heterocyclic ring (Kuwayama et al., 1986).

3.3. Solution Structure

In the first detailed structural investigation to be reported, Linscheid and Lehn (1967) studied a number of type (2) compounds [diazepam (2.1), its N1-H analog (2.18), and the N1-substituted derivatives (2.19)–(2.21)] by proton NMR spectroscopy over a range of temperatures both in deuteropyridine and in hexachlorobutadiene solution. At ambient temperatures, the seven-membered heterocyclic ring was found to undergo inversion between two boat-shaped enantiomeric, energetically equivalent, forms[†], which, on the NMR time-scale, is rapid for (2.18) and slow for (2.1), (2.19)–(2.21). The free energy of activation (ΔG^*) for the inversion was calculated to be 51.9 kJ/mol for (2.18) and in the range 72.4–79.6 kJ/mol for the other compounds, which have N1 substituted by methyl or larger groups. These values are not greatly affected by the nature of the solvent. Unless stated otherwise, values of ΔG^* refer to a temperature of 310K; when other temperatures are quoted they are normally the C3 methylene group coalescence temperatures.

These results were confirmed by further NMR studies on diazepam and nitrazepam (2.8) (Bley et al., 1968) and on nitrazepam, clonazepam (2.2), flunitrazepam (2.4), and compounds (2.22) and (2.23) (Sarrazin et al., 1975). The energy barrier to inversion for nitrazepam is given as 50.2 kJ by Bley et al. (1968) or 51.1 kJ by Sarrazin et al. (1975), similar to that found for (2.18) (51.9 kJ/mol) by Linscheid and Lehn (1967), so that the nature of the C7 substituent appears to have little effect on the inversion barrier. A recent ^{13}C NMR study has, however, indicated a through-space interaction between the seven-substituent and the C2 carbonyl function, which increases with increasing electronegativity of the substituent (Paul et al., 1983). Also, Unterhalt (1981) and Scahill and Smith (1983b) find from ^{15}N NMR that replacing Cl by NO_2 at C7 produces a downfield shift of the N1 resonance.

[†]The inversion to the enantiomeric form of the *molecule* involves also a rotation of the 5-phenyl ring about the C5—C1' bond such that the torsion angles about C5—C1' change sign, but remain equal in magnitude. In solution, the energy barrier to this is small (see below).

A major change in the energy barrier is caused by replacing the hydrogen atom at N1 by a methyl group, the energy barriers for the N1-methyl analogs of (2.8) and (2.18), i.e., (2.22) and (2.1), being 72.4 kJ/mol (Sarrazin et al., 1975) and, depending on solvent, 72.4–73.7 kJ/mol (Linscheid and Lehn, 1967), respectively, an increase of some 22 kJ. Larger N1 substituents cause only a relatively small increase over the N1-Me value, maximum for the above compounds 79.6 kJ/mol in (2.20). The value for prazepam (2.10), with a cyclopropylmethyl substituent, is slightly higher, 82.8 kJ/mol at 405K (Cazaux et al., 1983) or 81.0kJ/mol at 398K (Kovar et al., 1983). This effect is considered to be probably caused by a *peri* interaction of the N1 substitutent with the C9 hydrogen atom in the planar transition state of the ring inversion. The energy barrier in chlordiazepoxide (1) is 63.2 kJ/mol (Bley et al., 1968), intermediate between the N1-H and N1-Me compounds of type (2).

The presence of an *ortho*-chloro or *ortho*-fluoro substituent in the 5-phenyl ring reduces the barrier to inversion by approximately 7 kJ/mol^{-1} (cf. (2.2) 43.5, (2.8) 50.2 kJ and (2.4) 65.7, (2.22) 72.4 kJ) (Bley et al., 1968; Sarrazin et al., 1975). A chloro substituent in the *para* position has no measurable effect (Linscheid and Lehn, 1967).

Fusion of a triazolo ring to the a-face of the 1,4-benzodiazepine nucleus increases the energy barrier significantly (Wade et al., 1979). Kovar et al., (1983) report 76.5/kJ/mol for triazolam (3.2) at 378K, and Scahill and Smith (1985) report approximately 88 kJ/mol for alprazolam (3.1) and estazolam (3.3) (temperature not given) from variable temperature measurements. The value for brotizolam (15) is 46.1 kJ/mol at 231K, the relatively low energy barrier being attributed to the reduced steric hindrance to inversion imposed by the thieno moiety compared with a benzo moiety (Kovar et al., 1983).

When one of the hydrogen atoms on C3 is substituted by a hydroxy or methyl group, the inversion is no longer between energetically equivalent enantiomers. The invertomers are diastereoisomeric and of unequal energy, one with the substituent quasi-equatorial and hydrogen quasi-axial and the other having the substituent quasi-axial (and hydrogen quasi-equatorial). The 3-hydroxy compounds oxazepam (2.9) and temazepam (2.11) and the 3-methyl derivative (2.13) are found (Sadee et al., 1973; Sunjic et al., 1979) to exist only in one conformation at 310K, with the larger group in the sterically more favorable quasi-equatorial position. This conformation is also found in the solid state (see above and, in particular, Figs. 4, 6, and 8).

Medazepam (7) and its N1—H analog both exhibit conformational equilibrium at room temperature, involving inversion between two pseudo-boat forms (Romeo et al., 1981) that would not be energetically equivalent (N1 substituent quasi-equatorial or quasi-axial). The energy barrier to inversion of pyramidal nitrogen is relatively low, however, so that the distinction between equatorial and axial substitution at N1 disappears and ring inversion is again between energetically-equivalent forms, as for the cycloheptatriene-like rings of the type (X) compounds. A later investigation leads to a reassignment of resonances and to a value of 55.5° for the N—C—C—N endocyclic torsion angle (Finner et al., 1985), in good agreement with the solid state values of 58° (mean over two independent molecules) in the free base (Gilli et al., 1978b) and 49° in the hydrochloride salt (Chananont et al., 1980b). The energy barrier to ring inversion is approximately 38 kJ/mol at 193K and in agreement with the solid-state structures (*see* Figs. 12 and 18), the N1—Me substituent is found to be predominantly quasi equatorial (Finner et al., 1985). The 3-methyl derivative of the N1-H analog of medazepam exists predominantly with the methyl substituent quasi equatorial (Decorte et al., 1983).

The 1,5-benzodiazepine clobazam (8) and a number of derivatives have been studied by ^{13}C NMR spectroscopy (Aversa et al., 1983) and by variable temperature ^{1}H NMR techniques (Chidichimo et al., 1984). The energy barrier to ring inversion for clobazam is 88 kJ/mol at 310K.

With regard to the orientation of the 5-phenyl ring, Haran and Tuchagues (1980), by analysis of the ^{13}C NMR spectra, postulate that at room temperature, if the ring is unsubstituted, there is rapid rotation of the ring about the C5—C1' bond. In flunitrazepam (2.4) there is, however, evidence of through-space H6 to F coupling (Haran and Tuchagues, 1980) and coupling between the fluorine atom and both C6 and C11, indicating a predominance of the conformer with the *ortho*-substituent atom "pointing" toward the benzo moiety (Finner et al., 1984a), as is generally found in the solid state (*see* Figs. 2–4). Nevertheless, it is considered that the rotational isomer with the fluorine atom pointing toward N4 (rotation of about 180° about C5—C1') also contributes to the equilibrium. In this context it is of interest that an *ortho*-chloro or -fluoro substituent has a deshielding effect on the ^{15}N resonance of N4, indicating a possible through-space interaction (Scahill and Smith, 1983b).

Such a substituent has also been found to reduce the energy barrier to inversion (Sarrazin et al., 1975) (see above). The possibility that this might be related to a reduction in electron delocalization across the C5—C1' bond brought about by the increased dihedral angle between the N4=C5 double bond and the C1'—C6' bond of the phenyl ring in the *ortho*-substituted compounds (*see* Table 2) has been suggested by Sarrazin et al. (1975). The X-ray crystallographic analyses show, however, that within the limits of experimental error, the length of the C5—C1' bond corresponds to that of a single bond, regardless of whether a halo substituent is present (*see* Table 2). Thus, considering the type (2) compounds, the mean length of the bond for the five molecules containing an *ortho*-substituted ring is 1.49 Å [compounds (2.2), (2.4), (2.6)], the same as for the 16 molecules containing an unsubstituted 5-phenyl ring. The X-ray results, therefore, indicate that there is no electron delocalization across the C5—C1' bond.

In bromazepam (6), the pyridine nitrogen atom is predominantly antiperiplanar to N4, "pointing" in the same direction as the fluoro-substituent in flunitrazepam (Finner et al., 1984b). This is the conformation found in the solid state (*see* Fig. 9). An earlier postulate of a C6—H . . . N(pyridyl) hydrogen bond (Sarrazin et al., 1980), for which there is no evidence in the solid state, may not be valid in solution either (Finner et al., 1984b).

A quantitative measure of the preferred orientation of the 5-phenyl ring has been obtained by Paul et al. (1982) by use of lanthanide-induced shifts to assign the ^{13}C resonances in the NMR spectra of diazepam (2.1), flurazepam (2.5), prazepam (2.10), and compound (2.18). There are considerable differences between the values obtained from the two shift reagents used [Yb(fod)$_3$ and Pr(fod)$_3$]. In the case of compounds (2.1) and (2.10), however, for which crystal structures have been determined (Camerman and Camerman, 1972; Brachtel and Jansen, 1981), the agreement between the solid-state and solution values is reasonably good. Expressed as the C11—C5—C1'—C2' torsion angle, for diazepam (2.1) the solution values are 18.1 and 35.5°, compared with 25.5° in the solid. For prazepam (2.10), the solution angles are 20.9 and 13.4°; in the crystal structure, which contains two independent molecules, the corresponding angle is 43.7° in one molecule and 28.8° in the other (*see* Table 2).

The conformation of the N1-cyclopropylmethyl substituent of prazepam has been investigated by Cazaux et al. (1983). Their results appear to agree with the crystal structure of Brachtel and Jansen (1981) (*see* Fig. 7), although both in the solid and in solution there is evidence of an equilibrium involving two or more conformations. Figure 7 shows the major conformer in the crystal structure.

4. Structure–Activity Studies

The original structure–activity studies with the classical 1,4-benzodiazepines were empirical and derived from the pioneering synthetic work of Sternbach (Sternbach et al., 1964; Sternbach, 1973; 1982), in concert with a series of standard pharmacological tests for sedative, muscle relaxant, taming, and anticonvulsant activity carried out by Randall (Randall et al., 1974). The results of the screening of a large number of analogs in the series were summarized as follows (*see* Formula 2):

Substitution in the fused benzene ring: Position 7 appeared to be the most important; here electron-withdrawing substituents (e.g., nitro-, chloro- or trifluoromethyl-) resulted in increases of activity amounting to several orders of magnitude. Alternative substitution at this position with electron-donating groups did not produce significant modification of the activity of the un-substituted parent. Substituents in other positions appeared unimportant.

Substitution in the 7-membered heterocyclic ring: Reduction of the N4-C5 double bond led to a marked reduction in activity, as did the removal of the carbonyl function at position 2. Methylation at position 1 increased activity, but larger groups led to a decrease. Fusion of additional heterocyclic rings across N1—C2 ("a" face) or N4—C5 ("d" face) also provided active compounds. Addition of a hydroxyl group at position 3 increased activity somewhat, though alternative substituents did not prove beneficial. The 5-phenyl ring was of paramount importance. It could be modified by ortho-chloro or ortho-fluoro substitution leading to increases of activity; additional substituents at other positions were forbidden, however.

The above studies, though derived from a large amount of data, were essentially qualitative in nature; compounds were classed as active or inactive. By making use of more detailed pharmacological measurements, attempts were later made to quantify the relationship between structure and activity. The first of these studies, with a relatively small number of benzodiaze-pines, concluded that although lipophilicity was not significantly correlated with anticonvulsant activity, the charge on N4 was of major importance (Sarrazin et al., 1976). In a much more detailed investigation, Blair and Webb (1977) carried out a study with 59 benzodiazepines in which they searched for structural rela-tionships with pharmacological potency of the compounds as anti-convulsants, sedatives, or muscle relaxants. Electronic parameters were calculated using CNDO/2 based on the crystal structure of diazepam, making the assumption that the molecular mod-ifications would not affect the skeleton structure. Though more recent data indicate that this is not strictly correct, the differences are relatively small. Significant correlations were found with the charge on the oxygen at C2 and the total molecular dipole moment; the introduction of the Hansch lipophilic constants for the sub-stituents did not improve the correlations. In another study using CNDO/2 molecular orbital calculations on a set of seven ben-zodiazepines, Gilli et al. (1977b) found significant correlations be-tween the antipentylenetetrazole activity of the compounds and the energy differences between their lowest unoccupied and high-est occupied molecular orbitals.

The Free-Wilson linear model has been applied to activity data obtained from antipentylenetetrazole, footshock, and inclined screen tests on 55 benzodiazepin-2-one derivatives variously sub-stituted in positions 7, 8 9, 1, 2', and 4' (Borea et al., 1979). Position 7 was found to be the most important in determining the biological activities in all three tests considered; activities were found to increase with the increasing π (Hansch hydrophobic constant) and σ (Hammett constant) values of the substituents. The low in-dividual group contributions for positions 1, 8, and 9 showed that substitution in these positions was of small practical interest. Posi-tion 2' was found to be the most important after position 7. The positive effect reached a maximum in the antipentylenetetrazole test, for which the observed order of values was $Cl > F > Br > NO_2 > CF_3 > H > OCH_3 > CH_3$. This order was rationalized by saying that biological activities increase with increasing electron-withdrawing properties of the substituents and a decrease in their steric hindrance. In the case of position 4', large negative values were obtained for all substituents tested regardless of their physi-cochemical characteristics.

Biagi et al. (1980), making use of the Hansch approach, studied the influence exerted by hydrophobic factors (measured by means of chromatographic R_m values) on the biological activity of 26 benzodiazepines as measured by exploratory behavior in rats. A significant correlation was found ($r = 0.878$), though it was found necessary to introduce an indicator variable to take account of halogen substitution at position 2'.

In another quantitative study using Hansch analysis, attention was focused on substitution at positions 7, 1, and 2' with the aid of 35 1,4-benzodiazepin-2-ones, the antipentylenetetrazole activity of which was known (Borea, 1981). The best regression equation found was:

$$\log(1/C) = -0.338 + 0.643 \; \Sigma \; \pi + 2.787 \; \sigma_7 + 2.962 \; F_{2'} +$$

$$0.638 \; E_{S2'} + 2.986 \; I_7$$

where $n = 35$; $r = 0.918$; $S = 0.421$; $F = 30.92$; $P < 0.005$. Here $\Sigma\pi$ is the hydrophobic substituent parameter and σ_7 is the Hammett constant that is a measure of the electronic effect of the substituent in position 7 on the reaction. $E_{S2'}$ and $F_{2'}$ indicate the Taft steric and field inductive constants, respectively, of the substituents in position 2'. This indicates that this particular measure of biological activity depends on the overall molecular lipophilicity and on the presence of strong electron-withdrawing substituents in positions 7 and 2'. In the latter position, moreover, the optimal activity is reached in substituents exhibiting small steric hindrance. The most striking feature of the equation is that it demonstrates, in an in vivo test, the strong dependence of the activities on σ_7 and $F_{2'}$, as shown by their large regression coefficients. A dimethyl amino substituent at position 7, however, produced derivatives that were more active than could be predicted by any equation in $\Sigma\pi$, σ_7, $F_{2'}$ and $E_{S2'}$. An indicator variable was used to include these compounds into the regression, but no explanation for the anomalous activity of the compounds could be given.

The above studies attempted to correlate various structural features of the benzodiazepines with their pharmacological potency in a number of in vivo tests. The interpretation of these correlations is fraught with danger. Before the compound in question arrives at its site of action, it must be absorbed and possibly suffer metabolic degradation or sequestration at various sites within the animal. Studies of this type are therefore of limited value, except in the delineation of the most global aspects of the structure–activity relationships. Subsequent to the discovery of the benzodiazepine binding site in the mammalian CNS, the affinity of individual compounds for this site was used as an alternative measure of its biological activity. Clearly, in this case the above problems of bioavailability, in the whole animal studies, are circumvented. These later investigations, using essentially the same approaches as those described previously, largely confirmed the earlier findings, excepting the cases in which the biological activity of the compound in vivo was caused by a metabolite of the parent (Borea and Bonora, 1983; Borea and Gilli, 1984; Borea, 1983; Hamor and Martin, 1984). The conclusions of these studies can be summarized as follows:

Affinities of 1-alkylated or dealkylated benzodiazepines are very similar [e.g., K_i diazepam = 8.9 nM, K_i demethyldiazepam = 8.9 nM (Squires and Braestrup, 1977)], showing that the main drug-receptor interaction is not mediated by the N1—H group.

N4 substituted benzodiazepines or benzodiazepines lacking the C2=O group show low or very low receptor binding affinities (e.g., K_i medazepam = 3850 nM, K_i chlordiazepoxide = 574 nM, suggesting that these two points can be considered the basic requirement for optimal benzodiazepine–receptor binding).

Previous structure–activity results are confirmed as regard the positive effects of overall lipophilicity (Borea and Bonora, 1983) of small electrophiles in position 2' and the strong negative effect of 4' substituents (Borea, 1983).

R and S stereoisomeric 3-substituted benzodiazepines display a dramatic difference in affinities (e.g., K_i Ro 11-6896,3S = 4.8 nM, K_i Ro 11-6893,3R = 1040 nM (Mohler and Okada, 1978; Blount et al., 1983).

It is of interest that one of the most strongly held tenets from the early structure–activity studies was that substituents in the 7 position had to be electron-withdrawing. Based on receptor binding affinity data, it is now known that compounds having electron-releasing groups in 7 (e.g., OCH_3 and OH) can display affinities in the nanomolar range (Haefely et al., 1985).

Based on these findings, Borea and Gilli (1984) postulated that the 1,4-benzodiazepines interact with their specific binding sites via two hydrogen bonds; the acceptor sites on the ligand being the 0 at the C2 position and the N4 of the seven-membered heterocycle. The 5-phenyl substituent would provide the third ligand point by interaction with a hydrophobic pocket on the binding protein. Similar importance was placed on electrostatic interactions within the ligand–receptor complex by Loew et al. (1984). By modeling the molecular electrostatic potential around a number of benzodiazepine analogs, they concluded that for high-affinity ligands, specific cationic interactions take place with the electron-withdrawing groups at C7, the 0 at C2 and N4.

The structure–activity studies discussed so far were restricted to the benzodiazepines, and a marked progression can be seen from the original relationships developed empirically, from the synthetic chemistry, to the delineation of structural parameters using molecular orbital calculations and the subsequent development of model receptor sites. In fact, at least in the case of the classical benzodiazepines, little new insight was provided despite the increasing sophistication of the techniques used to address the question of the importance of structural characteristics to biological activity.

By 1980, however, it was clear that the benzodiazepines could be displaced from their binding sites on CNS membranes by compounds with no obvious structural similarity to the benzodiazepines. This naturally led to attempts to search for structural congruence between these distinct chemical classes, on the assumption that all these compounds bound to the same site, though not necessarily with identical ligand–site interactions. Although it is quite possible that these additional ligands may bind to the same site, it is equally plausible to envisage completely separate sites; apparently competitive displacement of the benzodiazepines could occur through tightly coupled allosteric mechanisms; it is still not known which of these two alternatives is correct. In this respect it is salutary to note the proposal of Camerman and Camerman (1972), in which they suggested that the anticonvulsant behavior of several classes of drugs, including the benzodiazepines, barbiturates, hydantoins, and succinamides, were related to a common sterochemical pattern comprising two bulky hydrophobic groups and two electron-donating atoms in a fixed spatial arrangement. It is now clear that these agents do not produce their effects through interaction with a common recognition site, and it is difficult to envisage what meaning can be placed on such structural commonalities when the molecular mechanism of action of the compounds is different. This limitation must be kept in mind when attempts are made to study the relationship between structure and activity. In the case of the benzodiazepine receptor, we have a further complexity in that three classes of ligand are known: agonists, inverse agonists, and antagonists. In the final analysis the information obtained from structure–activity studies must not only incorporate data obtained from binding studies, which are likely to be dependent on the recognition properties of the site, but also the differentiation of the compounds in terms of their intrinsic activity.

The ligands that appear to interact directly with the benzodiazepine site fall mainly into the following chemical classes:

Cyclopyrolones: e.g., Zopiclone ([6-(5-chloro-2-pyridil)-6,7-dihydro-7-oxo-5,5 *H*-pyrrolo-[3,4-b] pyrazin-5-yl] 4-methyl-1-piperazine carboxylate) and Suriclone ([6-(7-chloro-1,8-napthyridin-2-yl)-7-oxo-2,3,6,7-tetrahydro-5 *H*-1,4 dithiino [2,3-c] pyrrol-5-yl]4-methyl-1-piperazine carboxylate).

Triazolopyridazines: e.g., Cl 218872 (3-methyl-6-[3-trifluoro-methylphenyl]-1,2-4-triazolo[4,3-b]pyridazine).

Phenylquinolines: e.g., PK 9084 (phenyl-2-([piperidinyl-3)-2-ethyl]-quinoline.

Pyrazoloquinolines: e.g., CGS 9896 (2-(4-chlorophenyl)-2,5-dihydpyrazolo[4,3-c] quinolin-3(3 *H*)-one).

Imidazobenzodiazepines: e.g., Ro 15-1788 (8-fluoro-3-carboethoxy-5 dihydro-5-methyl-6-oxo-4 *H* -imidazo[1,5-a]1, 4-benzodiazepine.

β-Carbolines: e.g., DMCM (3-carbomethoxy-4-ethyl-6,7-dimethoxy-β-carboline), β-CCM (3-carbomethoxy-β-carboline), β-CCE (3-carboethoxy-β-carboline) PrCC (3-carbo-*n*-propyloxy-β-carboline), ZK 93426 (3-carboethoxy-5-*i*-propyloxy-4-methyl-β-carboline), ZK 91296 (3-carboethoxy-5-benzyloxy-4-methoxymethyl-β-carboline, and ZK 93423 (3-carboexthoxy-6-benzyloxy-4-methoxymethyl-β-carboline.

Table 3 shows a synopsis of pharmacological and biochemical data for a selection of benzodiazepine receptor ligands. The chemical class of the ligands for the benzodiazepine recognition site, their potency in displacing the benzodiazepines from this site, and their efficacy classification are shown.

At the present time the beta-carboline class of compounds is the only one for which extensive structure-affinity studies have been reported (*see* Fig. 19).

The first structure–activity analysis on the beta-carbolines used the method of Free-Wilson and was performed by Schauzu and Mager (1983) with a series of compounds described by Cain and colleagues (1982). The small number of compounds used, however, did not allow definitive conclusions to be drawn. Loew (Loew et al., 1985), using methods of molecular electrostatic potential mapping, previously applied by the same group to the benzodiazepines, concluded that N2 and the substituent at C3 played an important part in the definition of the affinity of the ligands. These

Table 3
Synopsis of Pharmacological and Biochemical Data for a Selection of
Benzodiazepine Recognition Site Ligands

	Che-mical class[a]	IC_{50} nM^{b}	Pharmacological profile
Flunitrazepam	BDZ	5	Agonist
Diazepam	BDZ	16	Agonist
Oxazepam	BDZ	38	Agonist
Triazolam	tBDZ	2	Agonist
Zopiclone	CYP	36	Agonist
Suriclone	CYP	2	Agonist
Cl 218-872	TPA	140	Agonist
ZK 93423	BC	2	Agonist
Medazepam	BDZ	6800	Agonist
PK 9084	PHQ	459	Partial agonist
CGS 9896	PQ	0.6	Partial agonist
ZK 91296	BC	1.1	Partial agonist
Ro 15-1788	iBDZ	3.3	Antagonist
ZK 93426	BC	0.4	Antagonist
PrCC	BC	1.2	Antagonist
CGS 8216	PQ	0.3	Partial inverse agonist
βCCE	BC	7	Partial inverse agonist
βCCM	BC	8	Inverse agonist
DMCM	BC	11	Inverse agonist

[a]tBDZ, iBDZ, BDZ = Triazolo-, imidazo-, benzodiazepines; CYP = cyclopyr-rolones; TAP = triazolopyridizines; PHQ = phenylquinolines; PQ = pyrazolo-quinolines; BC = β-carbolines

[b]IC_{50} for inhibition of specific BDZ binding from rat synaptosomal membranes; taken from Squires and Braestrup (1977), Braestrup and Nielsen (1983), Petersen et al. (1984), and Jensen et al. (1984).

workers went on to develop hypotheses about the possible amino acid residues present in the receptor protein binding site that may be relevant to these interactions.

A sterochemical model rationalizing both the binding site affinities and the intrinsic activities of the compounds in this series has been developed by Borea and colleagues (Borea et al., 1985, 1987). In this work it was suggested that high-affinity binding was obtained only if there was an ester function substitution at the

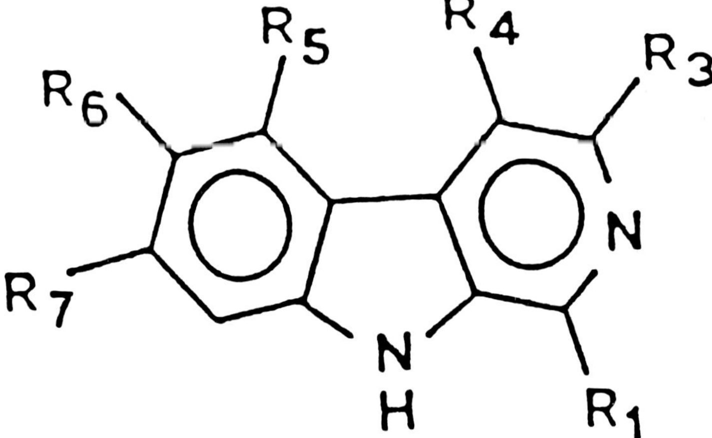

Fig. 19. Numbering scheme for β-carbolines

3-position and the three-ring system was totally aromatic (and therefore planar) with no substituent at position R1. This suggested that the recognition site may be a planar cleft with the ester oxygen atoms acting as hydrogen bond acceptors; the carboline N may also provide an additional hydrogen bonding acceptor site while the pyole N a donor site. A Free-Wilson analysis carried out on a series of 33 variously substituted β-carbolines agrees with these conclusions (Borea and Ferretti, 1986). Regarding the intrinsic activity of the compounds, extension of the alkyl group of the ester function at positions 3 would increase the agonist activity of the compound, as would the introduction of hydrophobic substituents at position 5. On the other hand, introduction of sterically bulky groups at positions 6 and 7 increases the inverse agonist activity of the ligands.

The first attempt to generate a model binding site to accommodate several of the chemical classes of ligand mentioned above was due to Crippen (1982). The study investigated 29 molecules with representatives from the benzodiazepines, the beta-carboline-3-carboxylates, and the cyclopyrrolones. Using the distance geometry approach, he was able to generate a set of 15 site points with five adjustable energy parameters that were able to predict the free energy of binding of 18 compounds from the total set. Crippen concluded that five atoms of each ligand could occupy corresponding points in the site constituting a possible pharmacophore.

Starting from the X-ray crystallographic data obtained for the active S-isomer of 3-methyl flunitrazepam (2.12), and a wealth of additional information concerning the structural requirements for activity in the benzodiazepine series, Fryer (1983) postulated a three-dimensional molecular model of ligands with four possible binding regions. In the case of the classical benzodiazepines, these would be the fused benzene ring, an electron-withdrawing substituent at position 7, and two systems associated with the carbonyl oxygen at C2 and the phenyl substituent at C5. By some modification to this scheme it was possible to accommodate several other ligands that were known to displace the benzodiazepines from their binding sites, namely zopiclone, Cl 218872, and the beta-carboline analogs. The final model suggested that there were five possible site points, three of which were required if a ligand was to display high-affinity binding. The relative orientation of the benzodiazepine agonists with Cl 218872 and zopiclone are essentially those shown in Figs. 20a, b, and c, respectively. The beta-carboline analogs were in a different orientation than that shown in Fig. 20e. Fryer suggested that the fused benzene rings of the carbolines and the classical benzodiazepines were essentially superimposed with the oxygen of the carboxy ethyl side chain of the former compound located in the vicinity of the carbonyl group at C2 of the benzodiazepines.

Codding and Muir (1985) determined the crystal structure of the benzodiazepine antagonist Ro 15-1788 and used this information to generate a model site that encompassed the agonist benzodiazepines, the beta-carboline-3-carboxylates, and CGS 8216. The latter two structures, together with Ro 15-1788, were found to have a similar arrangement of binding features, i.e., an aromatic ring, a carbonyl oxygen atom, and a hydrophobic side chain, whereas CGS 8216 and the beta-carboline analogs exhibited an additional hydrogen bonding donor not present in Ro 15-1788. The relative orientations of the beta-carboline-3-carboxylates, the CGS analog, and Ro 15-1788 are those shown in Figs. 20e, d, and f, respectively.

These latter two studies developed from the insight of experienced structural chemists and, as such, the matches are of a qualitative nature. The test of such hypotheses developed is by the synthesis of appropriate compounds. Fryer (1983) indicates that compounds were synthesized, on the basis of his model for ligand interactions, and that active compounds were obtained, though he also makes it clear that the model is insufficiently refined for accurate prediction at this stage.

The alternative to experience is to adopt a purely objective approach to search for commonalities between structurally differ-

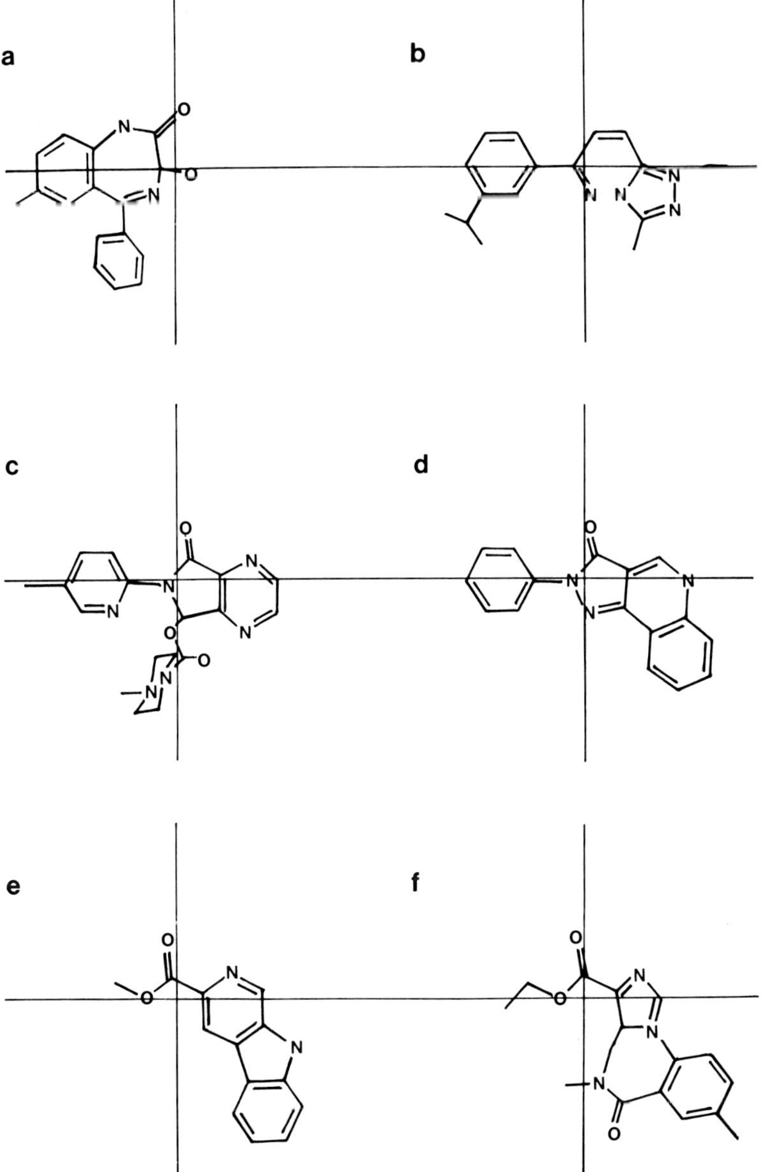

Fig. 20. Molecular structures of: (a) oxazepam (b) Cl 218872 (c) Zopiclone (d) CGS 9896 (e) methyl β-carboline-3-carboxylate, and (f) Ro 15-1788. The structure skeletons of the above compounds are shown in the superimposed orientations described in the text. The horizontal and vertical coordinate axes are the same throughout and are intended to allow the superimposition of all molecules. Hydrogen atoms have been omitted for clarity

ent molecules. Such a method is available using a combinatorial optimization procedure developed by Danziger and Dean (1985). This involves the allocation of certain properties to atoms or groups of atoms within the two molecules, which are then defined as the ligand points. A study has been carried out by Borea, Danziger, Dean, and Martin (results unpublished) with Ro 15-1788, the antagonist from the benzodiazepine series and ZK 93426, the antagonist from the beta-carboline series. The crystal structures for both these compounds were available from the literature (Codding and Muir, 1985; Bertolasi et al., 1984). This information allowed the putative ligand points to be defined in three-dimensional space, and these were then used in the efficient algorithm of Danziger and Dean (1985) to obtain the best match of all possible matches between the two molecules, as defined by the putative ligand points described above. Using this matched orientation, accessible surfaces and molecular electrostatic potentials were then superimposed on the molecules and compared directly using a procedure developed by Namasivayam and Dean (1986). Statistical analysis of this data suggested that, on the best matching face, the Spearman rank correlation coefficient for the surfaces was 0.957 and, for the electrostatic potentials, 0.847. The superimposition of these two molecules obtained using this approach is very similar to that suggested in the study of Codding and Muir mentioned above. No similar data are yet available to suggest how the agonist compounds from these two structurally dissimilar series may be incorporated into the scheme.

Most recently Borea et al. (1987) have attempted to use the crystallographic data from representative compounds of each chemical class that appear to interact with the benzodiazepine receptor, in concert with an accumulation of the pharmacological data on these compounds, to generate a general stereochemical model to rationalize both the binding characteristics and the pharmacological activities of these molecules. Each class of molecule is oriented relative to the other as shown in Fig. 20. Figure 21 is taken from the above reference and is designed to illustrate the essential features of this model, which can be summarized as follows:

A unique recognition site for all ligands is assumed. The overall envelope of all compounds studied is essentially planar, but with a definite bump (marked in Fig. 21 as out-of-plane displacement zones 00P1 and 00P2) in the middle. The thickness of the planar part is the usual value for aromatic rings (3.54 Å). The bump is asymmetrical, being more extended in the lower than in the upper part, where the maximum height is nearly 3.68 Å from the mean ligand plane at the extreme limit of the van der Waals radius of the hydrogen attached to the C3 atom of benzodiazepines. This point can be considered a zone of steric confinement of the ligand (marked S3 in the figure) able to discriminate R and S enantiomers of 3-substituted benzodiazepines. The lower part of the bump is much larger and is associated, typically, with the 5-phenyl group of benzodiazepines or the N-carboxypiperazine moiety in zopiclone. In its horizontal plane, the overall molecular envelope is an ovoid of about 12 × 18 Å. As Fig. 21 has been drawn on the contour of the largest ligands, the figure can be considered to map the internal borders of the overall binding site.

Specific binding determining drug–receptor interactions would be B1 and B2 for benzodiazepines and CGS compounds, B1 and B3 for carbolines, Ro 15-1788 and zopiclone, and only B2 for Cl 218872. The forces responsible for the binding would probably originate from hydrogen bonds accepted by the ligands. Drugs not capable of binding at B1 display IC_{50} values higher by one or two orders of magnitude. The different ligands bind to the same site, but in different positions, still having a superimposition zone responsible for their ability to displace one another. The different location in the site delineates the different kinds of intrinsic activity.

Attempts to correlate structural characteristics with biological activity are not only intellectually intriguing, but possibly of great commercial value. There can be few examples in which the conditions have been more favorable for the delineation of structure–activity relationships than in the benzodiazepine series. Structurally the compounds are conformationally very restricted and are ideal candidates for the use of X-ray crystallographic techniques; few would argue that packing forces in the crystal seriously affect the structure. There are literally thousands of chemical analogs whose gross pharmacology has been investigated; many hundreds of the compounds have been screened to determine their affinity for the benzodiazepine binding sites in the mammalian CNS, which is thought to be the pharmacological receptor through which these compounds have their effects. These measurements are capable of considerable accuracy. It is thus possible to determine both sides of the structure–activity equation with very reasonable precision. We have additional factors in the observation that compounds of quite distinct chemical structure appear to interact with this binding site; a fact that should allow a greater

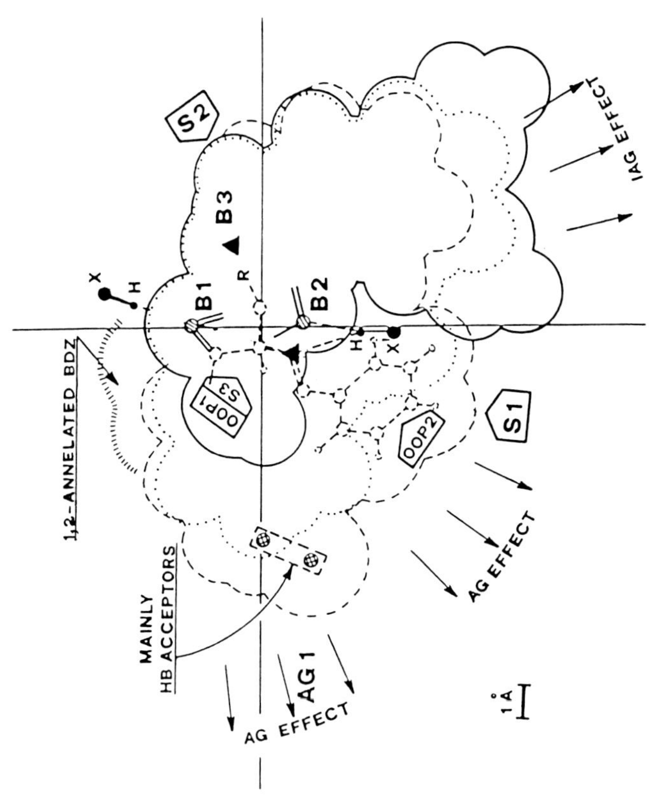

S2
B3
B1
B2
R
H
X
OOP1
S3
S1
OOP2
1,2-ANNELATED BDZ
MAINLY
HB ACCEPTORS
AG1
AG EFFECT
AG EFFECT
IAG EFFECT
1Å

Fig. 21. A sketch of the model proposed by Borea et al. (1987). The van der Waals envelope for DMCM is represented by a continuous line, that for a combination of oxazepam and the partial agonist CGS 9896, by the dashed line, and similarly for PrCC and Ro 15-1788 by a dotted line. The binding site is essentially planar except for the zones of out-of-plane displacement marked as OOP1 and OOP2. B1, B2, and B3 are the main binding points; S1 and S3 on one side and S2 on the other are regions of steric hindrance for the binding of benzodiazepines and β-carbolines, respectively; X-H and X'-H are hypothesized hydrogen bonding donor groups. The rectangular area indicated as "mainly HB acceptors" marks the position of strong electrophilic groups (C1, F, NO_2, . . .) in the agonist molecules. The arrows show that the bound molecule causes agonist or inverse agonist effect by pushing the receptor macromolecule in the directions marked as "AG EFFECT" or "IAG EFFECT." Perfect balancing of these opposite forces produces antagonist behavior.

definition of the recognition properties of the site. There are inevitably disadvantages. These revolve around the fact that the benzodiazepine site appears to modulate the GABA-A receptor. It is a type of fine-tuning mechanism for the actions of GABA at this particular receptor subtype. Because of this, ligands can be classified as agonists, inverse agonists, and antagonists. It is essential that if we are to make sense of the structural requirements for activity at this important recognition site, we must proceed with caution. Initially we must compare antagonists with antagonists across the different classes that we surmise interact at the same site. Subsequently we may superimpose the agonists and inverse agonists. We have a wealth of data in this arena, not only to understand the detailed molecular information that defines intrinsic activity, but also, with the recently elucidated amino acid sequence of the GABA-benzodiazepine receptor protein (Schofield et al., 1987), to probe the molecular events that underlie the actions of these clinically important agents.

Acknowledgments

Extensive use has been made of the Cambridge Crystallographic Database (Allen et al., 1979). The authors would like to thank Mrs. G. Carter for her considerable efforts in typing the manuscript.

References

Allen F. II., Bellard S., Brice M. D., Cartwright B. A., Doubleday A., Higgs A. Hummelink T., Hummelink-Peters B. G., Kennard O., Motherwell W. D. S., Rodgers J. R., and Watson D. G. (1979) The Cambridge Crystallographic Data Centre: computer-based search, retrieval, analysis and display of information. *Acta Crystallogr.* **B35,** 2331–2339.

Aversa M. C., Ferlazzo A., and Giannetto P. (1983) A carbon-13 NMR assignment study of a series of 2,3,4,5-tetrahydro-1-methyl-1H-1,5-benzodiazepin-2,4-diones. *J. Heterocycl. Chem.* **20,** 1641–1644.

Bandoli G. and Clemente D. A. (1976) Crystal, molecular and electronic structure of an antianxiety agent: 7-Chloro-5-(2-chlorophenyl)-1,3-dihydro-3-hydroxy-1,4-benzodiazepin-2-one. *J. Chem. Soc. Perkin Trans.* **2,** 413–418.

Barnard E. A., Stephenson F. A., Sigel E., Mamalaki C., Bilbe G., Constanti A., Smart T. E., and Brown D. A. (1984) Structure and Properties of the Brain GABA/Benzodiazepine Receptor Complex, in *Neurotransmitter Receptors: Mechanism of Action and Regulation* (Kito S., Segawa T., Kuriyama K., Yamamura H. I., and Olsen R. W., eds.) Plenum, New York.

Battersby M. K., Richards J. G., and Mohler H. (1979) Benzodiazepine receptor: Photoaffinity labelling and localisation. *Eur. J. Pharmacol.* **57**, 277–278.

Bertolasi V., Ferretti V., Gilli G., and Borea P. A. (1982a) 7-Chloro-1,3-dihydro-1-chloroethyl-3-ethoxy-5-phenyl-2H-1,4-benzodiazepin-2-one. *Cryst. Struct. Commun.* **11**, 1481–1486.

Bertolasi V., Sacerdoti M., Gilli G., and Borea P. A. (1982b) Structure of 7-chloro-2-methylamino-5-phenyl-3H-1,4-benzodiazepine-4-oxide (chlordiazetoxide) *Acta Crystallogr.* **B38**, 1768–1772.

Bertolasi V., Ferretti V., Gilli G., and Borea P. A. (1984) Structure of methyl beta-carboline-3-carboxylate (beta-CCM). $C_{13}H_{10}N_2O_2$. *Acta Crystallogr.* **C40**, 1981–1983.

Biagi G. L., Barbaro A. M., Guerra M. C., Babbini M., Gaiardi M., Bartoletti M., and Borea P. A. (1980) Rm values and structure–activity relationships of benzodiazepines. *J. Med. Chem.* **23**, 193–201

Blair T. and Webb G. A. (1977) Electronic factors in the structure–activity relationship of some 1,4-benzodiazepin-2-ones. *J. Med. Chem.* **20**, 1206–1210.

Bley W., Nuhn P., and Benndorf G. (1968) NMR spectroscopic study of drugs. II. Conformation and ring inversion of diazepam, nitrazepam and chlordiazepoxide. *Arch. Pharm.* (Weinheim) **301**, 444–451.

Blount J. F., Fryer R. I., Gilman N. W., and Todaro L. J. (1983) Quinazolines and 1,4-benzodiazepines. 92. Conformational recognition of the receptor by 1,4-benzodiazepines. *Mol. Pharmacol.* **24**, 425–428.

Borea P. A. (1981) Structure activity relationships in 1,4-benzodiazepines. *Boll. Soc. It. Bol. Sper.* **57**, 627–632

Borea P. A. (1983) De novo analysis of receptor binding affinity data of benzodiazepines. *Arzneimittelforsch.* **33**, 1086–1088.

Borea P. A. and Bonora A. (1983) Brain receptor binding and lipophilic character of benzodiazepines. *Biochem. Pharmacol.* **32**, 603–607.

Borea P. A., and Ferretti V. (1986) De novo analysis of receptor binding affinity data of beta-carbolines. *Biochem. Pharmacol.* **35**, 2836–2839.

Borea P. A. and Gilli G (1984) The nature of 1,4-benzodiazepine-receptor interactions. *Arzneimittelforsch.* **34**, 649–652.

Borea P. A., Ferretti V., and Gilli G. (1985) A stereochemical model accounting for receptor binding affinity and intrinsic activity of beta-carbolines. *Br. J. Pharmacol.* **86**, 666P.

Borea P. A., Bertolasi V., Ferretti V., and Gilli G (1987) Stereochemical features controlling binding and intrinsic activity properties of benzodiazepine-receptor ligands. *Mol. Pharmacol.*, **31**, 334–344.

Borea P. A., Gilli G., and Bertolasi V. (1979) Application of the Free-Wilson model to the analysis of three different pharmacological activity tests in benzodiazepines. *Farmaco. Ed. Sci.* **34**, 1073–1082.

Bowery N. G., Hill D. R., Hudson A. L., Price G. W., Turnbull M. J., and Wilkin G. P. (1984) Heterogeneity of Mammalin GABA Receptors, in *Actions and Interactions of GABA and Benzodiazepines* (Bowery N. G., ed.) Plenum, New York.

Brachtel G. and Jansen M. (1981) Prazepam. *Cryst. Struct. Commun.* **10**, 669–672.

Braestrup C. and Nielsen M. (1981) GABA reduces the binding of ^{3}H-methyl beta-carboline-3-carboxylate to brain benzodiazepine receptors. *Nature* (Lond.) **294**, 472–474.

Braestrup C. and Nielsen M. (1983) Benzodiazepine Receptors, in *Handbook of Psychopharmacology* vol. 17 (Iversen L. L., Iversen S. D., and Snyder S. H., eds.) Plenum, New York.

Braestrup C. and Squires R. F. (1978) Brain specific benzodiazepine receptors. *Br. J. Psychiat.* **133**, 249–260.

Braestrup C., Nielsen M., Honore T., Jensen L. H., and Petersen E. M. (1983) Benzodiazepine receptor ligands with positive and negative efficacy. *Neuropharmacology* **22**, 1451–1457.

Braestrup C., Honore T., Nielsen M., Petersen E. N. and Jensen L. H. (1984) Ligands for benzodiazepine receptors with positive and negative efficacy. *Biochem. Pharmacol.* **33**, 859–862.

Brown C. L. and Martin I. L. (1982) Photoaffinity labelling of the benzodiazepine receptor does not occlude the beta-CCE binding site. *Br. J. Pharmacol.* **77**, 312P.

Brown C. L. and Martin I. L. (1984a) Photoaffinity labelling of the benzodiazepine receptor compromises the recognition site but not its effector mechanism. *J. Neurochem.* **43**, 272–273.

Brown C. L. and Martin I. L. (1984b) Kinetics of Ro 15-1788 binding to membrane bound rat brain benzodiazepine receptors. *J. Neurochem.* **42**, 918–923.

Butcher H. J. and Hamor T. A. (1984) Structure of 8-chloro-1-((dimethylamino)methyl)-6-phenyl-4H-imidazo(1,2-a)(1,4)-benzodiazepine. *Acta Crystallogr.* **C40**, 848–850.

Butcher H. J. and Hamor T. A. (1985a) Structure of 7-chloro-1-methyl-5-phenyl-1H-1,5-benzodiazepine-2,4(3H,5H)-dione (Clobazam). *Acta Crystallogr.* **C41**, 1081–1083.

Butcher H. J. and Hamor T. A. (1985b) Structure of 2-bromo-4-(2-chlorophenyl)-9-methyl-6H-thieno-(3,2-f)(1,2,4)triazolo-(4,3-a)(1,4)diazepine (Brotizolam). *Acta Crystallogr.* **C41**, 265–266.

Butcher H. J., Hamor T. A., and Martin I. L. (1983) Structures of 7-bromo-1,3-dihydro-5-(2-pyridyl)-2H-1,4-benzodiazepine-2-one (Bromazepam) and 5-(2-fluorophenyl)-1,3-dihydro-1-methyl-7-nitro-2H-1,4-benzodiazepin-2-one (Flunitrazepam). *Acta Crystallogr.* **C39**, 1469–1472.

Cain M., Weber R. W., Guzman F., Cook J. M., Barker S. A., Rice K. C., Crawley J. N., Paul S. M., and Skolnick P. (1982) Beta-carbolines: Synthesis and neurochemical and pharmacological actions on brain benzodiazepine receptors. *J. Med. Chem.* **25**, 1081–1091.

Camerman A. and Camerman N. (1972) Stereochemical basis of anticonvulsant drug action. II. Molecular structure of diazepam. *J. Am. Chem. Soc.* **94**, 268–272.

Cazaux L., Vidal C., and Pasdeloup M. (1983) NMR of some benzodiazepine drugs: Structure elucidation of lanthanide-induced shifts of N1-substituted benzodiazepinones. *Org. Magn. Reson.* **21**, 190–195.

Chananont P., Hamor T. A., and Martin I. L. (1979) 5-(2-Chlorophenyl)-1,3-dihydro-7-nitro-2H-1,4-benzodiazepin-2-one (Clonazepam). *Cryst. Struct. Commun.* **8**, 393–400.

Chananont P., Hamor T. A., and Martin I. L. (1980a) The structure of 7-chloro-1,3-dihydro-1-(N-methylacetamido)-5-phenyl-2H-1,4-benzodiazepin-2-one. *Acta Crystallogr.* **B36**, 2115–2120.

Chananont P., Hamor T. A., and Martin I. L. (1980b) Structure of 7-chloro-2,3-dihydro-1-methyl-5-phenyl-1H-1,4-benzodiazepine hydrochloride (Medazepam hydrochloride). *Acta Crystallogr.* **B36**, 898–902.

Chang L. R-. and Barnard E. A. (1982) The benzodiazepine/GABA receptor complex: Molecular size in brain synaptic membranes and in solution. *J. Neurochem.* **39**, 1507–1518.

Chidichimo G., Longeri M., Menniti G., Romeo G., and Ferlazzo A. (1984) Proton NMR investigation of ring mobility of 1,5-benzodiazepine-2,4-diones. *Org. Magn. Reson.* **22**, 52–54.

Chiu T. H. and Rosenberg H. C. (1983) Multiple conformational states of benzodiazepine receptors. *Trends Pharmacol. Sci.* **4**, 348–350.

Codding P. W. and Muir A. K. S. (1985) Molecular structure of Ro15-1788 and a model for the binding of benzodiazepine receptor ligands. *Mol. Pharmacol.* **28**, 178–184.

Costa E., Guidotti A., and Mao C. C. (1975) Evidence for the involvement of GABA in the action of benzodiazepines: Studies in the cerebellum. *Adv. Biochem. Psychopharmacol.* **14**, 113–120.

Cowan P. J., Green A. R., Nutt D. J., and Martin I. L. (1981) Ethyl beta-carboline-3-carboxylate lowers seizure threshold and antagonises flurazepam induced sedation in rats. *Nature* (Lond.) **290**, 54–55.

Crippen G. (1982) Distance geometry analysis of the benzodiazepine binding site. *Mol. Pharmacol.* **22,** 11–19.

Danziger D. J. and Dean P. M. (1985) The search for functional correspondences in molecular structure between two dissimilar molecules. *J. Theor. Biol.* **116,** 215–224.

Decorte E., Toso R., Fajdiga T., Comisso G., Moimas F., Sega A., Sunjic V., and Lisini A. (1983) Chiral 1,4-benzodiazepines. XII. Conformation in a solution of 7-chloro-5-phenyl-3(S)-methyl-1,3-dihydro-2H-1,4-benzodiazepine. *J. Heterocycl. Chem.* **20,** 1321–1327.

Doble A., Martin I. L., and Richards D. A. (1982) GABA modulation predicts biological activity of ligands for the benzodiazepine receptor. *Br. J. Pharmacol.* **76,** 238P.

Dorow R., Horowski R., Paschelke G., Amin M., and Braestrup C. (1983) Severe anxiety produced by FG7142, a beta-carboline ligand for benzodiazepine receptors. *Lancet* 98–99.

Duax W. L., Weeks C. M., and Rohrer D. C. (1976) Crystal Structures of Steroids, in *Topics in Stereochemistry* vol. 9 (Allinger N. L. and Eliel E. L., ed.) Interscience, New York.

Dvorkin O. A., Simonov Y. O., Bogotskii O. V., Korotenko T. I., and Andronati S. A. (1982) Crystal and molecular structure of 7-chloro-5-phenyl-3-isopropyl-1,2-dihydro-3H-1,4-benzodiazepin-2-one. *Dopov. Akad. Nauk. Ukr. RSR, Ser. B: Geol. Khim. Biol. Nauki* 27–30.

Ferretti V., Bertolasi V., Gilli G., and Borea P. A. (1984) Structure of 2-(7-chloro-2(3H)-oxo-5-phenyl-1H-1,4-benzodiazepin-1-yl)ethyl hydrogen succinate N4-oxide. *Acta Crystallogr.* **C40,** 986–988.

Finner E., Zeugner H., and Milkowski W. (1984a) Conformational equilibria in flunitrazepam due to sp²-sp² carbon-carbon single bond rotational isomerism. *Arch. Pharm.* (Weinheim) **317,** 1050–1053.

Finner E., Zeugner H., and Milkowski W. (1984b) 1,4-Benzodiazepines and 1,5-benzodiazocines. IV. On the conformation of the 5-pyridyl substituent of bromazepam. Arch. Pharm. (Weinheim) **317,** 79–81.

Finner E., Zeugner H., and Milkowski W. (1985) On 1,4-benzodiazepines and 1,5-benzodiazocines. VIII. Conformation of medazepam in solution. *Arch. Pharm.* (Weinheim) **318,** 1135–1137.

Franklin T. J. (1980) Binding energy and the activation of hormone receptors. *Biochem. Pharmacol.* **29,** 853–856.

Freeman J. P., Duchamp D. J., Chidester C. G., Slomp G., Szmusczkovicz J., and Raban M. (1982) A new thermal rearrangement in the 4-isoxazoline system. Some chemical and stereochemical properties of a benzodiazepine oxide-ethyl propiolate adduct. *J. Amer. Chem. Soc.* **104,** 1380–1386.

Fryer R. I. (1983) Benzodiazepine Ligand Receptor Interactions, in *Benzodiazepines. From Molecular Biology to Clinical Practice* (Costa E., ed.) Raven, New York.

Fujimoto M., Hirai K., and Okabayashi T. (1982) Comparison of the effects of GABA and chloride ion on affinities of ligands for the benzodiazepine receptor. *Life Sci.* **30,** 51–57.

Galdecki Z. and Glowka M. L. (1980) The structure of a psychoactive agent: 7-chloro-3-hydroxy-1-methyl-5-phenyl-1,3-dihydro-2H-1,4-benzodiazepin-2-one (Temazepam). *Acta Crystallogr.* **B36,** 3044–3048.

Gee K. W. and Yamamura H. I. (1982) Differentiation of benzodiazepine receptor agonists and antagonists: Sparing of H-benzodiazepine antagonist binding following photolabelling of benzodiazepine receptors. *Eur. J. Pharmacol.* **82,** 239–241.

Gilli G., Bertolasi V., Sacerdoti M., and Borea P. A. (1977a) 7-Nitro-1,3-dihydro-5-phenyl-2H-1,4-benzodiazepin-2-one (Nitrazepam). *Acta Crystallogr.* **B33,** 2664–2667.

Gilli G., Borea P. A., Bertolasi V., and Sacerdoti M. (1977b) Stereochemical and electronic properties of benzodiazepines. Correlation with their anti-convulsant activity. *Eur. Cryst. Meeting, Oxford U.K.* abstract 38.

Gilli G., Bertolasi V., Sacerdoti M., and Borea P. A. (1978a) The crystal and molecular structure of 7-chloro-1,3-dihydro-3-hydroxy-5-phenyl-2H-1,4-benzodiazepin-2-one (Oxazepam). *Acta Crystallogr.* **B34,** 2826–2829.

Gilli G., Bertolasi V., Sacerdoti M., and Borea P. A. (1978b) 7-Chloro-2,3-dihydro-1-methyl-5-phenyl-1H-1,4-benzodiazepine (Medazepam). *Acta Crystallogr.* **B34,** 3793–3795.

Haefely W. (1984) Actions and Interactions of Benzodiazepine Agonists and Antagonists at GABAergic Synapses, in *Actions and Interactions of GABA and Benzodiazepines* (Bowery N. G., ed.) Raven, New York.

Haefely W., Kulcsar A., Mohler H., Polc P., and Schaffner R. (1975) Possible involvement of GABA in the central actions of the benzodiazepines. *Adv. Biochem. Psychopharmacol.* **14,** 131–151.

Haefely W., Kyburz Gerecke M., and Mohler H. (1985) Recent advances in the molecular pharmacology of benzodiazepine receptors and in the structure–activity relationships of their agonists and antagonists. *Adv. Drug Res.* **14,** 165–322.

Hamor T. A. and Martin I. L. (1983) The Benzodiazepines, in *Progress in Medicinal Chemistry* vol. 20 (Ellis G. P. and West G. B., eds.) Elsevier, Amsterdam.

Hamor T. A. and Martin I. L. (1984) Structure–Activity Relationships in a Series of 5-Phenyl-1,4-benzodiazepines, in *X-ray Crystallography and Drug Action* (Horn A. S. and De Ranter C. J., eds.) Clarendon, Oxford.

Haran R. and Tuchagues J. P. (1980) Carbon-13 and proton NMR studies of 1,4-benzodiazepines. *J. Heterocycl. Chem.* **17,** 1483–1488.

Herrnstadt C., Mootz D., Wunderlich H., and Mohrle H. (1979) Protonation sites of organic bases with several nitrogen functions: crystal structures of salts of chlordiazepoxide, dihydrazaline and phenformin. *J. Chem. Soc. Perkin Trans.* **2**, 735–740.

Hester J. B. Jr, Duchamp D. J., and Chidester C. G. (1971) A synthetic approach to new 1,4-benzodiazepine derivatives. *Tetrahedron Lett.* 1609–1612.

Hirsch J. D. (1982) Photolabelling of benzodiazepine receptors spares ^{3}H-propyl beta-carboline binding. *Biochem. Pharmacol. Behav.* **16**, 245–248.

Jensen L. H., Petersen E. N., and Braestrup C. (1983) Audiogenic seizures in DBA/2 mice discriminate sensitively between low efficacy benzodiazepine receptor agonists and inverse agonists. *Life Sci.* **33**, 393–399.

Jensen L. H., Petersen E. N., Braestrup C., Honore T., Kehr W., Stephens D, N., Schneider H., Seidelmann D., and Schmiechen R. (1984) Evaluation of the beta-carboline ZK 93426 as a benzodiazepine receptor antagonist. *Psychopharmacology* **83**, 249–256.

Kamiya K., Wada Y., and Nishikawa M. (1973) Molecular structures of 8-chloro-6-phenyl-4H-s-triazolo(4,3-a)(1,4)-benzodiazepine, 2-acetoxamino-4-acetyl-8-chloro-3,4-dihydro-6-phenyl-1,4,5-benzotriazocine and 8-chloro-4,11-diacetyl-4,11-dihydro-2-methyl-6-phenyloxazolo(4,5-b)(1,4,5)benzotriazocine. Formation of an eight-membered ring from a quinazoline derivative on treatment with hydrazine evidenced by X-ray analysis. *Chem. Pharm. Bull.* **21**, 1520–1529.

Karobath M. and Supavilai P. (1982) Distinction of benzodiazepine agonists and antagonists by photoaffinity labelling of benzodiazepine receptors in vitro. *Neurosci. Lett.* **31**, 65–69.

Kovar K. A., Linden D., and Breitmaier E. (1983) Benzodiazepines. II. NMR spectroscopic studies on conformations of benzodiazepines. *Arch. Pharm.* (Weinheim) **316**, 834–845.

Kuwayama T. and Yashiro T. (1985) The behaviour of 1,4-benzodiazepine drugs in acidic media. IV. Proton and carbon-13 nuclear magnetic resonance spectra of diazepam and fludiazepam in acidic aqueous solution. *Chem. Pharm. Bull.* **33**, 5503–5510.

Kuwayama T., Yashiro T., Kurono Y., and Ikeda K. (1986) The behaviour of 1,4-benzodiazepine drugs in acidic media. VI. Hydrogen exchange reaction and proton and carbon-13 nuclear magnetic resonance spectra of estazolam. *Chem. Pharm. Bull.* **34**, 2994–2998.

Linscheid P. and Lehn J. M. (1967) Nuclear magnetic resonance studies of rate processes and conformation. VII. Ring inversion in benzodiazepinones. *Bull. Soc. Chim. Fr.* 992–997.

Loew G. H., Nienow J. R., and Poulsen M. (1984) Theoretical structure activity studies of benzodiazepine analogues. Requirements for receptor affinity and activity. *Mol. Pharmacol.* **26,** 19–34

Loew G. H., Nienow J., Lawson J. A., Toll L., and Uyeno E. T. (1985) Theoretical structure activity studies of beta-carboline analogues. Requirements for benzodiazepine receptor affinity and antagonist activity. *Mol. Pharmacol.* **28,** 17–31.

Martin I. L. (1986) The Benzodiazepine Receptor, in *Neuromethods* vol. 4 *Receptor Binding* (Boulton A. A., Baker G. B., and Hrdina P. D., eds.) Humana, Clifton, New Jersey.

Martin I. L., Brown C. L., and Doble A. (1983) Multiple benzodiazepine receptors: Structures in the brain or structures in the mind. A critical review. *Life Sci.* **32,** 1925–1933.

MIMS (1987) *Monthly Index of Medical Specialities* (Medical Publications, London, February, 1987).

Mohler H. (1982) Benzodiazepine receptors: Differential interaction of benzodiazepine agonists and antagonists after photoaffinity labelling with flunitrazepam. *Eur. J. Pharmacol.* **80,** 435–436.

Mohler H. and Okada T. (1977) Benzodiazepine receptor: Demonstration in the central nervous system. *Science* **198,** 849–851.

Mohler H. and Okada T. (1978) The benzodiazepine receptor in normal and pathological human brain. *Br. J. Psychiatr.* **133,** 261–268.

Mohler H. and Richards J. G. (1981) Agonist and antagonist benzodiazepine receptor interactions in vitro. *Nature* (Lond.) **294,** 763–765.

Mohler H., Battersby M. K., and Richards J. G. (1980) Benzodiazepine receptor protein identified and visualised in brain tissue by a photoaffinity label. *Proc. Natl. Acad. Sci. USA* **77,** 1666–1670.

Motherwell W. D. S. and Clegg W. (1978) PLUTO78. Program for plotting molecular and crystal structures. Report 'Cambridge Crystallographic Files', University of Manchester Regional Computer Centre (Manchester, 1981).

Namasivayam S. and Dean P. M. (1986) Statistical method for surface pattern-matching between dissimilar molecules: Electrostatic potentials and accessible surfaces. *J. Mol. Graph.* **4,** 46–50.

Nielsen M. and Braestrup C. (1980) Ethyl beta-carboline-3-carboxylate shows differential benzodiazepine receptor interaction. *Nature* (Lond.) **286,** 606–607.

Nutt D. J., Cownan P. J., and Little H. J. (1982) Unusual interactions of benzodiazepine receptor antagonists. *Nature* (Lond.) **295,** 436–438.

Okada Y., Takebayashi T., Hashimoto M., Kasuga S., Sato S., and Tamura C. (1983) Formation of optically active compounds under achiral synthetic conditions. *J. Chem. Soc. Chem. Commun.* 784–785.

Olsen R. W. (1981) GABA-benzodiazepine-barbiturate interactions. *J. Neurochem.* **37**, 1–13.

Olsen R. W., Wong E. H. F., Stauber G. B., Murakami D., King R. G., and Fischer J. B. (1984) Biochemical properties of the GABA/Barbiturate/Benzodiazepine Receptor Chloride Ion Channel Complex, in *Neurotransmitter Receptors: Mechanisms of Action and Regulation* (Kito S., Segawa T., Kuriyami K., Yamamura H. I., and Olsen R. W., eds.) Plenum, New York.

Paul H. H., Sapper H., Lohmann W., and Kalinowski H. O. (1982) Analysis and application of carbon-13 NMR lanthanide induced shifts of 1,4-benzodiazepines. *Org. Magn. Reson.* **19**, 49–53.

Paul H. H., Sapper H., Lohmann W., and Kalinowski H. O. (1983) Carbon-13 NMR spectra of 1,4-benzodiazepines—influence of the 7-substituent. *Org. Magn. Reson.* **21**, 319–321.

Petcher T. J., Widmer A., Maetzel U., and Zeugner H. (1985) Structures of tifluadom (5-(2-fluorophenyl)-1-methyl-2-(3-thenoyl-aminomethyl)-2,3-dihydro-1H-1,4-benzodiazepine) hydrochloride, and of (+)-tifluadom p-toluenesulphonate, and the absolute configuration of the latter. *Acta Crystallogr.* **C41**, 909–912.

Petersen E. N., Jensen L. H., Honore T., Braestrup C., Kehr W., Stephens D. N., Wachtel H., Seidelmann D., and Schmiechen R. (1984) ZK 91296, a partial agonist at benzodiazepine receptors. *Psychopharmacology* **83**, 240–248.

Polc P., Bonetti E. P., Schaffner R., and Haefely W. (1982) A three state model of the benzodiazepine receptor explains the interactions between the benzodiazepine antagonist Ro 15-1788, benzodiazepine tranquillisers, beta-carbolines and phenobarbitone. *Naunyn Schmiedebergs Arch. Pharmacol.* **321**, 260–264.

Prado de Carvalho L., Grecksch G., Chapouthier G., and Rossier J. (1983) Anxiogenic and non-anxiogenic benzodiazepine antagonists. *Nature* (Lond.) **301**, 64–66.

Randall L. O. (1982) Discovery of Benzodiazepines, in *Pharmacology of Benzodiazepines* (Usdin E., Tallman J. F., Greenblatt D., and Paul S. M., eds.) MacMillan, London, U.K.

Randall L. O., Schallek W., Sternbach L. H., and Ning R. Y. (1974) Chemistry and pharmacology of 1,4-benzodiazepines. *Psychopharmacol. Agents* **3**, 175–281.

Romeo G., Aversa M. C., Giannetto P., Ficarra P., and Vigorita M. G. (1981) Nuclear magnetic resonance of psychotherapeutic agents. IV. Conformational analysis of 2,3-dihydro-1H-1,4-benzodiazepines. *Org. Magn. Reson.* **15**, 33–36.

Ruzic-Toros Z., Kojic-Prodic B., Bresciani-Pahor N., Nardin G., and Randaccio L. (1982) Structure of 7-chloro-5-phenyl-1-((S)-1-

phenylethyl)-1,3-dihydro-2H-1,4-benzodiazepin-2-one and its 3-methyl derivative. *Acta Crystallogr.* **B38**, 2977–2981.

Sadee W., Schwandt H. J., and Beyer K. H. (1973) On the conformation of unsymmetrically substituted 1,4-benzodiazepin-2-ones *Arch. Pharm.* **306**, 751–756.

Sarrazin M., Bourdeaux-Pontier M., and Briand C. (1976) Etude de quelques relations structure-activite de benzodiazepines. *Ann. Phys. Biol. Med.* **9**, 211–220.

Sarrazin M., Bourdeaux-Pontier M., Briand C., and Vincent E. J. (1975) NMR study of the ring inversion of some benzodiazepinones. *Org. Magn. Reson.* **7**, 89–93.

Sarrazin M., Faure R., Aubert C., and Vincent E. J. (1980) Carbon-13 NMR study of some benzodiazepinones. *J. Chim. Phys. Phys. Chim. Biol.* **77**, 91–93.

Scahill T. A. and Smith S. L. (1983a) Nitrogen-15 NMR studies of 1,4-benzodiazepines. 2. The triazolobenzodiazepines. *Org. Magn. Reson.* **21**, 662–665.

Scahill T. A. and Smith S. L. (1983b) Nitrogen-15 NMR studies of 1,4-benzodiazepines. 1. *Org. Magn. Reson.* **21**, 621–623.

Scahill T. A. and Smith S. L. (1985) Carbon-13 and hydrogen NMR data for a series of 1,4-benzodiazepines. *Magn. Reson. Chem.* **23**, 280–285.

Schauzu H. G. and Mager P. P. (1983) Free-Wilson approach applied to beta-carbolines, a series of potent benzodiazepine antagonists. *Pharmazie* **38**, 490.

Schmidt R. F., Vogel E., and Zimmerman M. (1967) Die Wirkung von Diazepam auf die prasynaptische Hemmung und andere Ruckmarks reflexe. *Naunyn Schmiedbergs Arch. Pharmacol.* **258**, 69–82.

Schoch P., Haring P., Takacs B., Stahli C., and Mohler H. (1984) As GABA/benzodiazepine receptor complex from bovine brain: purification, reconstitution and immunological characterisation. *J. Recept. Res.* **4**, 189–200.

Schofield P. R., Darlison M. G., Fujita N., Burt D. R., Stephenson F. A., Rodriguez H., Rhee L. M., Ramachandran J., Reale V., Glencorse T. A., Seeburg P. H., and Barnard E. A. (1987) Sequence and functional expression of the $GABA_A$ receptor shows a ligand-gated receptor super-family. *Nature* **328**, 221–227.

Sieghart W. and Karobath M. (1980) Molecular heterogeneity of benzodiazepine receptors. *Nature* (Lond.) **286**, 285–287.

Sigel E. and Barnard E. A. (1984) A gamma-aminobutyric acid/benzodiazepine receptor complex from bovine cerebral cortex: Improved purification with preservation of regulatory sites and their interactions. *J. Biol. Chem.* **259**, 7219–7223.

Sigel E., Stephenson E. A., Mamalaki C., and Barnard E. A. (1983) A gamma-aminobutyric acid/benzodiazepine receptor complex in bovine cerebral cortex. *J. Biol. Chem.* **258**, 6965–6971.

Sikirica M. and Vickovic I. (1982) 7-Chloro-5-phenyl-3(S)methyl-2H-1,4-benzodiazepin-2-one. *Cryst. Struct. Commun.* **11**, 1293–1298.

Skolnick P., Schweri M. M., Williams E. F., Moncada V. Y., and Paul S. M. (1982) An in vitro binding assay which differentiates between benzodiazepine agonists and antagonists. *Eur. J. Pharmacol.* **78**, 133–136.

Squires R. F., Benson D. I., Braestrup C., Coupet J., Klepner C. A., Myers V., and Beer B. (1979) Some properties of brain specific benzodiazepine receptors: New evidence for multiple receptors. *Pharmacol. Biochem. Behav.* **10**, 825–830.

Squires R. F. and Braestrup C. (1977) Benzodiazepine receptors in rat brain. *Nature* (Lond.) **266**, 732–734.

Stephenson F. A. and Barnard E. A. (1986) Purification and Characterisation of the Brain GABA/Benzodiazepine Receptor, in *Benzodiazepine/ GABA Receptors and Chloride Channels: Structure and Functional Properties* (Venter C. and Olsen R. W., eds.) Alan R. Liss, New York.

Sternbach L. H. (1973) Chemistry of 1,4-Benzodiazepines and Some Aspects of Their Structure Activity Relationships, in *The Benzodiazepines* (Garattini S., Mussini E., and Randall L. O., eds.) Raven, New York.

Sternbach L. H. (1982) The Discovery of CNS Active 1,4-Benzodiazepines (Chemistry), in *Pharmacology of Benzodiazepines* (Usdin E., Skolnick P., Tallman J. F., Greenblatt D., and Paul S. M., eds.) Macmillan, London.

Sternbach L. H., Randall L. O., and Gustavson S. R. (1964) 1,4-benzodiazepines (Chlordiazepoxide and related compounds). *Psychopharmacol. Agents* **1**, 137–224.

Study R. E. and Barker J. L. (1981) Diazepam and (–)-pentobarbital: Fluctuation analysis reveals different mechanisms for potentiation of gamma-aminobutyric acid responses in cultured central neurons. *Proc. Natl. Acad. Sci. USA* **78**, 7180–7184.

Sunjic V., Lisini A., Sega A., Kovac T., Kajfez F., and Ruscic B. (1979) Conformation of 7-chloro-5-phenyl-d5-3(S)-methyldihydro-1,4-benzodiazepin-2-one in solution. *J. Heterocycl. Chem.* **16**, 757–761.

Szmuszkovicz J., Chidester C. G., Duchamp D. J., MacKellar F. A., and Slomp G. (1971) Synthesis and proof of structure of a novel 1,4-benzodiazepine. *Tetrahedron Lett.* 3665–3668.

Tenen S. S and Hirsch J. D. (1980) Antagonism of diazepam activity by beta-carboline-3-carboxylic acid ethyl ester. *Nature* (Lond.) **288**, 609–610.

Unterhalt B. (1981) Nitrogen-15 NMR studies on selected 1,4-benzodiazepines. *Arch. Pharm.* (Weinheim) **314**, 733–735.

Wade P. C., Vogt B. R., Toeplitz B., Puar M. S., and Gougoutas J. Z. (1979) 1,2,4-Triazolo- and 1,2,5-triazino(4,3-d) (1,4)benzodiazepinone ring systems: Synthesis and barrier to ring inversion. *J. Org. Chem.* **44**, 88–99.

Weber K. H., Bauer A., Langbein A., and Daniel H. (1978) Heteroaromatic compounds with annelated seven-membered rings. III. Conversion of thienotriazolooxazepines into diazepines. *Liebigs Ann. Chem.* 1257–1265.

Index

59
1.389.400,-
129/90